Computational Modeling and Simulation of Advanced Wireless Communication Systems

The book covers the exploitation of computational models for effectively developing and managing large-scale wireless communication systems. The goal is to create and establish computational models for seamless human interaction and efficient decision-making in beyond 5G wireless systems.

Computational Modeling and Simulation of Advanced Wireless Communication Systems looks to create and establish computational models for seamless human interaction and efficient decision-making in the beyond 5G wireless systems. This book presents the design and development of several computational modeling techniques and their applications in wireless communication systems. It examines shortcomings and limitations of the existing computational models and offers solutions to revamp the traditional architecture toward addressing the vast network issues in wireless systems. The book addresses the need to design efficient computational and simulation models to address several issues in wireless communication systems, such as interference, pathloss, delay, traffic outage, and so forth. It discusses how theoretical, mathematical, and experimental results are integrated for optimal system performance to enhance the quality of service for mobile subscribers.

Further, the book is intended for industry and academic researchers, scientists, and engineers in the fields of wireless communications and ICTs. It is structured to present a practical guide to wireless communication engineers, IT practitioners, researchers, students, and other professionals.

Computational Modeling and Simulation of Advanced Wireless Communication Systems

Edited by

Agbotiname Lucky Imoize, Webert Montlouis,
Mohammad S. Obaidat, Segun I. Popoola, and
Mohammad Hammoudeh

CRC Press
Taylor & Francis Group
Boca Raton London New York

CRC Press is an imprint of the
Taylor & Francis Group, an **informa** business

First edition published 2025
by CRC Press
2385 NW Executive Center Drive, Suite 320, Boca Raton FL 33431

and by CRC Press
4 Park Square, Milton Park, Abingdon, Oxon, OX14 4RN

CRC Press is an imprint of Taylor & Francis Group, LLC

ISBN: 978-1-032-59799-7 (hbk)
ISBN: 978-1-032-60075-8 (pbk)
ISBN: 978-1-003-45742-8 (ebk)

DOI: 10.1201/9781003457428

Typeset in Times
by Newgen Publishing UK

Contents

Preface

Computational models are simulated mathematical models that investigate and simplify complex systems. The parameters of the mathematical models are tuned using computer simulations to observe various outcomes of systems, including wireless networks. Wireless communication systems find practical applications across different technologies and facilitate mobile connectivity to the Internet. Mobile computing impacts system design and enhances efficient wireless data and signal processing. However, a comprehensive understanding of how computational models and simulations aid the performance of a large-scale wireless communication system remains. Moreover, the interaction of billions of massive mobile devices poses a significant design complexity issue, and developing an appropriate computation-based model to ameliorate network design complexities is critical. In practice, the design should be robust to accommodate the characteristics of the wireless medium, the constraints from the mobile computing devices, the mobility of limited resources, complexity of the network, and the computational and overhead costs. In this light, practical systems can be studied dynamically and flexibly in a simulation environment, allowing changes in various aspects to be quickly evaluated without a physical experimental setting. This flexibility enables adjusting design parameters to improve system performance and minimize costs drastically. Additionally, efficient computational models are required to guarantee network coverage and extremely high throughput to enhance user mobility and quality of experience. However, ensuring high throughput and quality network coverage is quite challenging due to the size and the surging number of network nodes. In order to address this problem, this book proposes the application of computational modeling and simulations to design and develop cutting-edge wireless communication systems.

The book is an ideal reference to practitioners and researchers in computational modeling, wireless communications, simulations, wireless network design, artificial intelligence, machine learning, signal processing, security and privacy, as well as a good textbook for graduate and senior undergraduate courses in computational modeling, wireless communications, network simulations and informatics.

Special thanks to the reviewers of the original book proposal for their constructive suggestions. Thanks to all authors of the chapters for their insightful contributions. Many thanks to the editors and editorial assistants for their cooperation and support.

Agbotiname Lucky Imoize
Gelsenkirchen, North Rhine-Westphalia, Germany

Webert Montlouis
Baltimore, Maryland, United States

Mohammad S. Obaidat
Amman, Jordan

Segun I. Popoola
Cambridge, United Kingdom

Mohammad Hammoudeh
Dhahran, Saudi Arabia

About the Editors

Agbotiname Lucky Imoize received the B.Eng. degree (Hons.) in Electrical and Electronics Engineering from Ambrose Alli University, Nigeria, in 2008 and the M.Sc. degree in Electrical and Electronics Engineering from the University of Lagos, Nigeria, in 2012. He is a Lecturer in the Department of Electrical and Electronics Engineering at the University of Lagos, Nigeria. Before joining the University of Lagos, he was a Lecturer at Bells University of Technology, Nigeria. He was, until recently, a research scholar at the Ruhr University Bochum, Germany, under the sponsorship of the Nigerian Petroleum Technology Development Fund (PTDF) and the German Academic Exchange Service (DAAD) through the Nigerian-German Postgraduate Program.

He was awarded the Fulbright fellowship as a visiting research scholar at the Wireless@VT Laboratory, Bradley Department of Electrical and Computer Engineering, Virginia Tech, where he worked under the supervision of Prof. R. Michael Buehrer from 2017 to 2018. He worked as a core network products manager at ZTE, Nigeria, and as a Network Switching Subsystem Engineer at Globacom, Nigeria. His research interests cover the fields of 6G wireless communication, wireless security systems, and artificial intelligence. He has co-edited ten books and co-authored over 220 articles in peer review journals and conferences. He is an active reviewer and editor for over 50 international journals and conferences. He is the Vice Chair of the IEEE Communication Society, Nigeria chapter, a Registered Engineer with the Council for the Regulation of Engineering in Nigeria, and a member of the Nigerian Society of Engineers. He is a senior member of IEEE.

Webert Montlouis
Applied Physics Laboratory (APL) and Electrical and Computer Engineering Department, Johns Hopkins University, Baltimore. Dr. Montlouis received his BS, MS, and PhD in electrical and computer engineering from Northeastern University, Boston. At Johns Hopkins he has served as Chief Scientist at the Applied Physics Laboratory (APL) and faculty in the Electrical and Computer Engineering department. He has been the chair of the IEEE Massive MIMO standard development working group. He is also the co-Chair of the Massive MIMO working group. He has served as general co-Chair of the IEEE Massive MIMO workshop and served as session chair for many IEEE conferences. His research interests are in the areas Multi-Channel System Architecture, Sensing, Next Generation Radar Systems, Wireless Communications 5G and Beyond, Quantum Information Science, Digital signal Processing and Biomedical Signal Processing. He is a Fellow of the IEEE and a member of the IEEE Signal Processing Society and the IEEE Communications Society.

Distinguished Professor Mohammad S. Obaidat

He is recognized all over the World for his pioneering and lasting contributions to several areas, including Computing, Communications, AI, Cybersecurity, and IT domains. Professor Obaidat is a Life Fellow of IEEE, Fellow of AAIA, Fellow of FTRA, Fellow of AIIA, and Fellow of SCS. He is a distinguished professor at the King Abdullah II School of Information Technology, the University of Jordan, Amman. School of Computer and Communication Engineering, University of Science and Technology Beijing. Department of Computational Intelligence, School of Computing, SRM University, SRM Nagar, Kattankulathur, Tamil Nadu, India. School of Engineering, The Amity University, Noida, Uttar Pradesh, India.

Professor Obaidat is an internationally known academic, researcher, scientist, and scholar. He received his PhD degree in Computer Engineering with a minor in Computer Science from The Ohio State University, Columbus, Ohio, USA. He has received extensive research funding and published to date, over 1,200 refereed technical articles, more than half of the publications are journal articles, 110 books and about 70 book chapters. He is the editor-in-chief of three scholarly journals and an editor of many other international journals. He is the founding editor-in-chief of *Wiley Security and Privacy Journal* and is founder or co-founder of five prestigious international conferences.

Among his previous positions is the Advisor to the President of Philadelphia University for Research, Development and Information Technology, President and Chair of the Board of Directors of the Society for Modeling and Simulation International (SCS), Senior Vice President of SCS, Dean of the College of Engineering at Prince Sultan University, Chair, and tenured Professor at the Department of Computer and Information Science, and Director of the MS Graduate Program in Data Analytics at Fordham University, Chair and tenured Professor of the Department of Computer Science, and Director of the Graduate Program at Monmouth University, Tenured Full Professor at King Abdullah II School of Information Technology, University of Jordan, Founding Dean and Professor, College of Computing and Informatics at The University of Sharjah, UAE. He is also The PR of China Ministry of Education Distinguished Overseas Professor at the University of Science and Technology Beijing, an Honorary Distinguished Professor at Amity University – A Global University, and a Distinguished Professor at the SRMIST, TN, India and Indian Institute of Technology-Dhanbad, India.

He has chaired more than 185 international conferences and has given more than 185 keynote speeches worldwide. He has served as ABET/CSAB evaluator and IEEE CS Fellow Evaluation Committee. He has served as IEEE CS Distinguished Speaker/Lecturer and an ACM Distinguished Lecturer. Since 2004, he has been serving as an SCS Distinguished Lecturer. He received many best paper awards for his contributions, from IEEE ICC, IEEE GLOBECOM, AICSA, CITS, SPECTS, and DCNET International conferences. He also received best paper awards from *IEEE Systems Journal* in 2018 and in 2019. In 2020, he received four best paper awards from *IEEE Systems Journal*. In 2021, he received the IEEE Systems best paper award. In 2021, he was ranked by Guide2Research as the Number 1 Computer Scientist in UAE in terms of the number of publications. He has been ranked recently by research.com agency as number one Computer Scientist and number one Scholar in Electronics and Electrical Engineering in Jordan.

He has also received many other worldwide awards for his technical contributions, including The 2018 IEEE ComSoc-Technical Committee on Communications Software Technical Achievement Award for contribution to Cybersecurity, Wireless Networks, Computer Networks and Modeling, and Simulation, SCS prestigious McLeod Founder's Award, Presidential Service Award, SCS Hall of Fame –Lifetime Achievement Award for his technical contribution to modeling and simulation, and for his outstanding visionary leadership, and dedication to increasing the effectiveness and broadening the applications of modeling and simulation worldwide. He also received the SCS Outstanding

Service Award. He was awarded the IEEE CITS Hall of Fame Distinguished and Eminent Award. He is listed on the World's Top Career-long top 2% Scientists of the Stanford University/Elsevier recent list and has been ranked as the Top Scholar in Telecommunications in the world as per Scholar GPS recent ranking. In recognition of his significant scientific contribution, recently Springer published, a book honoring his contributions to computing, informatics, networking, and cybersecurity, entitled, *Advances in Computing, Informatics, Networking, and Cybersecurity – A Book Honoring Professor Mohammad S. Obaidat's Significant Scientific Contributions.*

Segun I. Popoola

School of Computing and Information Science, Anglia Ruskin University, Cambridge, UK. Dr. Segun I. Popoola is a senior lecturer in the school of computing and information science at Anglia Ruskin University, UK. He received the B.Tech. degree in electronic and electrical engineering from the Ladoke Akintola University of Technology in 2014, the M.Eng. degree in information and communication engineering from the Department of Electrical and Information Engineering, Covenant University, in 2018, and the PhD degree in cyber security and artificial intelligence from the Department of Engineering, Faculty of Science and Engineering, Manchester Metropolitan University, Manchester, UK, in 2022. His PhD thesis on federated deep learning for botnet attack detection in IoT networks was a product of an academic-industry partnership project jointly funded by the department of engineering at Manchester Metropolitan University and a cyber security company, Cyraatek Ltd UK. He is a Lecturer in the Department of Computing and Mathematics at Manchester Metropolitan University, UK. In June 2022, he was endorsed as a Global Exceptional Talent by The Royal Society. His research interests include wireless communications, deep learning, federated learning, cyber security, Internet of Things, and smart critical infrastructure. He is a registered engineer with the Council for the Regulation of Engineering in Nigeria (COREN) and a member of IEEE. He has published more than 100 research papers in reputable journals and conference proceedings, including IEEE Transactions on Consumer Electronics, IEEE *Internet of Things Journal*, IEEE Access, and IEEE Vehicular Technology Conference.

Mohammad Hammoudeh

Department of Information and Computer Science, King Fahd University of Petroleum and Minerals, Dhahran, Saudi Arabia. Professor Hammoudeh is the Saudi Aramco Cybersecurity Chair Professor at King Fahd University of Petroleum and Minerals. His research interest is in quantum communication technologies and quantum-safe protection methods and quantum risk assessment in both IT and OT systems. He has hands-on experience in working with quantum hardware and algorithms. He developed practical ways of thinking when evaluating the real applications of new technology such as digital twins and blockchains. Mohammad Hammoudeh received the B.Sc. degree in computer communications from Arts Sciences and Technology University, Beirut, Lebanon, in 2004, the M.Sc. degree in advanced distributed systems from the University of Leicester, Leicester, UK, in 2006, the Postgraduate Certificate in academic practice from Manchester Metropolitan University, Manchester, UK, in 2011, and the PhD degree in computer science from the University of Wolverhampton, Wolverhampton, UK, in 2008. As the current Saudi Aramco Chair Professor of Cyber Security with the Information and Computer Science Department, King Fahd University of Petroleum and Minerals, Dhahran, his research interests include the applications of zero trust security to Internet-connected critical national infrastructures, blockchains, the Internet of Things/cyber physical systems, and other complex, highly decentralized systems. Dr. Hammoudeh is the founder and co-editor-in-chief of ACM's journal *Distributed Ledger Technology: Research and Practice.*

Contributors

Ovye John Abari
Federal University Lokoja
Lokoja, Nigeria

Abdullateef Ola Adebayo
Kwara State University
Malete, Nigeria

Abayomi-Alli Adebayo
Federal University of Agriculture
Abeokuta, Nigeria

Abidemi Emmanuel Adeniyi
Bowen University
Iwo, Nigeria

Ibukun D. Adewale
University of Lagos
Akoka, Lagos, Nigeria

Adedotun Temitope Ajibare
Newbridge Graduate Insitute
Tyger Valley Campus
Cape Town, South Africa

Lateef Adesola Akinyemi
University of South Africa
Florida Campus, South Africa;
Lagos State University
Epe Campus, Lagos, Nigeria

G G Md Nawaz Ali
Bradley University
Peoria, Illinois, USA

Abubakar Aliyu
Federal University Lokoja
Lokoja, Nigeria

Halleluyah Oluwatobi Aworinde
Bowen University
Iwo, Nigeria

Joseph Bamidele Awotunde
University of Ilorin
Ilorin, Nigeria

Isaac Oniovosah Ayetuoma
Chrisland University
Abeokuta, Nigeria

Garba Alpha Baba
Kaduna State College of Education
Gidan Waya, Nigeria

Abdulrauf Olarenwaju Babatunde
University of Ilorin
Ilorin, Nigeria

Vladyslav Chevardin
Military Institute of Telecommunications and
    Information Technologies
Knyaziv Ostroz'kykh str., Kyiv, Ukraine

Jupeng Ding
Shandong Technology and Business University
Yantai, China
Xinjiang University
Urumqi, China

Stephen Obono Ekwe
Cape Peninsula University of Technology
Western Cape, South Africa

Emanuele Frontoni
University of Macerata
Via Crescimbeni, Macerata, Italy;
Marche Polytechnic University
Via Brecce Bianche
Ancona, Italy

Chih-Lin I
China Mobile Research Institute
Beijing, China

Agbotiname Lucky Imoize
University of Lagos
Akoka, Lagos, Nigeria

Rasheed Gbenga Jimoh
University of Ilorin
Ilorin, Nigeria

Taiwo Kolajo
Federal University Lokoja
Lokoja, Nigeria

Natalia Kryvinska
University of Comenius
Bratislava, Slovakia

Oleksandr Kuznetsov
University of Macerata
Via Crescimbeni, Macerata, Italy;
V. N. Karazin Kharkiv National University
Kharkiv, Ukraine; Comenius University
   Bratislava
Odbojárov, Bratislava, Slovakia

Maryam Sufiyanu Masari
Airforce Institute of Technology
Kaduna, Nigeria

Ernest Mnkandla
University of South Africa
Florida Campus, Johannesburg
South Africa.

Mykola Mormul
University of Customs and Finance
av. Volodymira Vernadsky
Dnipro, Ukraine

Ahmad Shehu Muhammad
Federal University Lokoja
Lokoja, Nigeria

Abdulwaheed Musa
Kwara State University
Malete, Nigeria

Deborah Olufemi Ninan
Obafemi Awolowo University
Ile-Ife, Nigeria

Solomon C. Nwaneri
University of Lagos
Akoka, Lagos, Nigeria

Emeka Ogbuju
Federal University Lokoja
Lokoja, Nigeria

Sunday Oladayo Oladejo
Stellenbosch University
South Africa

Francisca Oladipo
Thomas Adewumi University
Oko, Nigeria

Mukaila Olagunju
Federal University
Oye-Ekiti, Ekiti, Nigeria

Odunayo Dauda Olanloye
Bowen University
Iwo, Nigeria

Samuel Ibukun Olotu
Federal University of Technology
Akure, Nigeria

Folorunso Olusegun
Federal University of Agriculture
Abeokuta, Nigeria

Olamide Peter Oshinuga
Lagos State University
Epe Campus, Lagos, Nigeria

Oleg Poplavskiy
University of Customs and Finance
Dnipro, str. Volodymyr Vernadsky, Ukraine

Malik Adeiza Rufai
Federal University Lokoja
Lokoja, Nigeria

Mohammad Nazmus Sadat
Bradley University
Peoria, Illinois, USA

Wasswa Shafik
Dig Connectivity Research Laboratory
   (DCRLab)
Kampala, Uganda;
Universiti Brunei Darussalam
Jalan Tungku Link, Gadong
Bandar Seri Begawan, Brunei Darussalam

Oleksii Smirnov
Central Ukrainian National Technical
   University
University Ave.
Kropyvnytskyi, Ukraine

Jian Song
Tsinghua University
Beijing, China

Mbuyu Sumbwanyambe
University of South Africa
Florida Campus, South Africa

Hannah C. Ugo
Afe Babalola University
Ado-Ekiti, Nigeria

Olufunke Rebecca Vincent
Federal University of Agriculture
Abeokuta, Nigeria

Yun Wang
Bradley University
Peoria, Illinois, USA

Jintao Wang
Tsinghua University
Beijing, China

Daniel Dauda Wisdom
Chrisland University
Abeokuta, Nigeria

Yanping Zhang
Gonzaga University
Spokane, Washington, USA

Introduction

Agbotiname Lucky Imoize, Webert Montlouis, Mohammad S. Obaidat,
Segun I. Popoola and Mohammad Hammoudeh

The book discusses computational modeling and simulation techniques for effectively developing and managing large-scale wireless communication systems. First, the book critically looks at the limitations of the existing computational and simulation models and proffers solutions to revamp the traditional models toward addressing the prevailing network issues in wireless communication systems. Secondly, the book explores novel computational modeling and simulation methodologies, facilitating seamless human interaction and efficient decision-making in wireless communication systems. Thirdly, the book integrates theoretical, mathematical, and experimental simulations for optimal system performance to enhance the quality of wireless services for end users. Further, the book addresses the need to design efficient computational and simulation models to mitigate several wireless network issues such as interference, pathloss, congestion, delay, among others. Additionally, the book examines the legal requirements, regulatory policies, and the economics of computational modeling and simulations in advanced wireless communication systems. Last, the book provides researchers and academics with new insights into the real-world scenarios of the deployment, application, management, and the associated benefits of computational modeling and simulation of advanced wireless communication systems.

The book is well structured into sixteen chapters outlined as follows.

Chapter 1 presents an overview of computational modeling and simulations in wireless communication systems. The chapter elucidates the fundamental concepts and significance of computational modeling and simulations in this context and emphasizes how simulations can help to unravel the complexities inherent in wireless communication systems, providing insights into system behavior, facilitating efficient performance evaluation, and optimizing designs. Through effective network planning, protocol development, and performance analysis, computational modeling and simulations, the reliability and efficiency of wireless networks can be significantly enhanced. The critical challenges and limitations of computational modeling and simulations comprising difficulty in determining model accuracy and limited computational resources are elaborated. As wireless technology evolves, computational modeling and simulations are indispensable for advancing research, development, and deployment in wireless communications, promising a future shaped by innovation and efficiency.

Chapter 2 explores the historical perspectives on computational modeling and simulation in wireless communication systems. The chapter calls for the development of more sophisticated analytical approaches for designing and optimizing networks as systems complexity increases. The authors emphasized that tracing this intertwined history offers insights into the foundations of cutting-edge solutions to the proliferating issues in wireless communication systems. Additionally, the work explores pivotal development across fundamental science, transitional shifts, recent algorithms, and practical use cases spanning early quantitative inquiries in electromagnetism to increasingly versatile software abstractions stretched to their limits in 5G wireless networks and

DOI: 10.1201/9781003457428-1

beyond. Moreover, the study shows that by constructing a unified technical progression timeline linking wireless milestones with enabling modeling innovations, contrasting fledgling and contemporary capabilities, and elucidating the persistence of scientific visions fueling incremental improvement, enduring engineering themes are identified across generations. Key findings affirm a persistent symbiotic relationship between advancing modeling sophistication and wireless communication demands, hinting at exciting futures.

Chapter 3 focuses on computational modeling of communication systems and networks. In particular, the work explores the potential application of federated learning (FL) and machine learning (ML) in addressing computational modeling issues in emerging wireless networks. FL is currently attracting attention as an integral machine-learning technique to shift the computing load to edge devices. The chapter investigates FL in small-cell networks, where users and many base stations (BSs) are dispersed following a uniform Poisson point process (PPP) with different densities. The work examines how the placement of geographic nodes impacts the model aggregation in FL using a stochastic geometry-based approach. The study also considers how to build the closed-form expressions of coverage likelihood using feasible estimations and calculate the lowest required BS density for small-cell networks to achieve a good model aggregation rate. A suitable FL model is evolved in addition to the stochastic approach for modeling and simulating communication systems and networks. The moment-generating function (MGL) and probability-generating function (PGFL) were employed to relax the intractable problem and produce a closed-form expression for coverage probability. Numerical results for the outage probability against the signal-to-noise-plus-interference ratio (SINR), using the suggested scheme gives more appreciable performance results than the contenders.

Chapter 4 presents computational modeling and analysis of wireless sensor networks. The findings from the study reflect that Wireless Sensor Networks (WSNs) have experienced increasing complexity due to the wide range of civil and military applications they serve. The complexity is further intensified by the diverse types of sensors and the sheer network size in practice. Typically, hundreds to thousands, or even more sensors with limited capabilities are deployed in a Field of Interest (FoI) to locate application-specific events for intrusion detection. The interconnected sensors are also responsible for reporting the sensed events to a base station(s) in order to fulfil different application-specific requirements. Given the inherent complexity and diversity, it is essential to develop appropriate computational models and analytical techniques for facilitating network simulation, design, and optimization prior to practical deployment. Therefore, the chapter focuses on the design and development of computational modeling and analysis for WSN research from sensor deployments, sensing coverage, detection probability, and application requirements within the context of different circumstances and scenarios. The chapter explores the impact of different models on network performance and discusses their strengths, weaknesses, and interrelationships. Theoretical analysis, mathematical derivations, and visualized results are presented to enhance comprehension. Moreover, a literature review on related works is presented, and challenges and opportunities for future directions are discussed. The study aims to provide insights into selecting and developing the most appropriate model(s) to address challenges encountered in WSNs, and at the same time drive the advancement of research in the domain.

Chapter 5 reviews the security measures in computational modeling and simulations, exploring technical complexities and ethical concerns. The study explored the foundational concepts of computational modeling and simulations with a comprehensive review of related literature, security frameworks, and challenges as well as recent advancements. Also, the chapter deals with secure simulations and their growing significance in decision-making across fields, covering technological and non-technological measures such as multifactor identification, data safeguarding, and authentication techniques. In addition, the chapter examines cybersecurity threats, vulnerabilities, authentication methods such as multi-factor authentication (MFA) and role-based access control (RBAC), network security, intrusion detection, prevention systems

(IDS/IPS), and cloud-based simulation security along with cyber threat intelligence and secure software development practices. Last, the chapter suggested a cyber-attack model and highlights privacy regulations provided by GDPR and HIPAA in simulation software development, intrusion detection and prevention systems (IDPS), incident response, and ethical considerations for responsible data use.

Chapter 6 presents computational models for training, testing, and validating wireless networks and systems. Computational models do not function independently, but they rely on data and require training to address complex problems effectively. Training of computational models involves educating them to comprehend the distinct relationships within the data. The study utilizes computational models including, among others, Deep Neural Networks (DNN) and Linear Regression (LR), to predict the signal strength of wireless networks. To achieve this, a dataset containing attributes of wireless network metrics and location-based properties was employed to train the models. Accurate prediction of signal strength in wireless networks holds great significance as it enhances network performance and streamlines network troubleshooting. Following the predictions, the computational models undergo validation to gauge the accuracy of their predictions, ensuring their reliability. The dataset is further analyzed to reveal the connections between the attributes of the wireless network metrics. Evaluation metrics, such as Mean Absolute Error (MAE) and Mean Square Error (MSE), are employed to gauge the effectiveness of the models in predicting wireless network signal strength. Among the computational models tested, the Linear Regression emerged as the best contender, achieving the lowest MAE score of 1.22 and the k-Nearest Neighbors proved to be the least effective in predicting the signal strength of wireless networks.

Chapter 7 investigates modeling and simulation of non-Lambertian beams based vehicular visible light communications in 6G and beyond. Due to the serious spectrum shortage and complex interference issues of the well-known vehicular radio frequency (RF) technology paradigm, vehicular visible light communication (VLC) is becoming one powerful enabling technology for 6G and beyond wireless networks, thanks to the abundant and unregulated visible light spectrum resource and the ubiquitous deployment of light emitting diodes (LED)-based lighting sources. Typically, vehicular-to-vehicular VLC transmission techniques have been introduced and explored to enhance the connection performance of vehicular wireless networks. Nevertheless, the current works are usually limited to vehicular connection scenarios, employing well-discussed LED transmitter with Lambertian beam patterns, which fail to characterize the vehicular VLC scenarios with distinctive non-Lambertian transmitters. For filling this fundamental research gap, vehicular-to-vehicular VLC links employing typical non-Lambertian optical beams is proposed. Unlike the conventional Lambertian vehicular VLC research paradigm, the proposed vehicular-to-vehicular connection scheme employs the commercially available non-Lambertian transmitter to configure the vehicular VLC links. Numerical results illustrate that, for typical outdoor lane scenario, compared with about 17.10 dB average signal-to-noise (SNR) of the benchmark Lambertian vehicular VLC links configuration, up to about 4.0 dB average SNR gain could be achieved by the proposed typical non-Lambertian vehicular VLC links configuration.

Chapter 8 dwells on the application of computational modeling in electronics devices. Computational modeling has remained pivotal to the understanding of the behavior of various scientific and engineering systems. A comprehensive understanding of these systems is necessary for optimizing the design of electronic circuits and devices. The chapter reviews related studies on computational modeling of electronic devices. The modeling of Silicon Carbide MOSFETs was extensively investigated in the LTSPICE electronic design automation (EDA) application software, making use of STMicroelectronics library to simulate the Non-linear I-V characteristics of the MOSFET. The drain-to-source voltage Vds was made to vary between 0 and 80 Volts in steps of 0.01 Volts, and a swept voltage of 10–20 Volts in incremental steps of 2 Volts was supplied to the gate. The results showed significant differences for the current values at the different temperatures. Further, the chapter briefly discussed the applications of electronics sensing and object tracking system in visual surveillance systems, and wearable electronics systems modeling and simulation were discussed.

Chapter 9 considers high-speed stream ciphers for wireless communication systems, focusing on the design and simulation aspects. The chapter explores the intricate world of high-speed stream ciphers, pivotal in securing wireless communication systems. The focus is on the development and computational modeling of these ciphers, which are essential for ensuring data integrity and confidentiality in high-throughput environments. The chapter begins by outlining the fundamental principles of stream ciphers, emphasizing their relevance in the context of massive mobile device interactions and the consequent complexity in network design. It then transitions into a detailed exploration of the design methodologies for high-speed stream ciphers, highlighting the challenges and considerations unique to wireless communication systems. A significant contribution of the work lies in the comprehensive computational modeling and simulation of these ciphers. This includes a thorough analysis of various modeling techniques and their effectiveness in replicating real-world scenarios, thereby providing a deeper understanding of the ciphers' performance under various network conditions. The research fills a critical gap in the existing literature by providing empirical evidence of the efficiency and security of various stream ciphers in wireless communication systems. The experimental results, combined with a rigorous discussion on the findings, offer valuable insights into the selection and implementation of appropriate cryptographic solutions in wireless networks. The chapter not only contributes to the theoretical understanding of stream ciphers, it also serves as a practical guide for researchers and practitioners in the field of wireless communication security.

Chapter 10 explores the legal frameworks regulating computational models in wireless communication systems. The landscape of contemporary connectivity has changed because of the rapid development of computational models in wireless communication networks. Hence, as these technologies grow to become increasingly integrated into our daily lives, robust legal frameworks should be established in order to assure their responsible development, implementation, and use. The chapter examines the complex legal environment that surrounds computational models used in wireless communication systems, offering a thorough analysis of the legal issues and ethical considerations. The chapter explores the essential components of the legal frameworks controlling wireless communication systems, looking at the functions of the government organizations in charge of licensing, technical standards, and spectrum allocation. It emphasizes data protection regulations, informed consent, and strong security measures to protect user information and ensure fairness in wireless communication systems. Additionally, the chapter addresses moral issues like justice, bias, and inclusivity while giving a full discussion of compliance with regulations for international cooperation, global norms, and cross-border data transfers in wireless communication systems. Conclusively, comprehensive legal frameworks are crucial for the responsible advancement of computational models in wireless communication systems for the promotion of transparency, user rights, and data privacy.

Chapter 11 takes a critical look at the government policies and economics of computational modeling in wireless networks. Government policies and economics play a pivotal role in shaping the landscape of computational modeling and simulations in wireless networks. As wireless communications become increasingly vital in modern society, efficient management and optimization of wireless networks are of paramount importance. Computational modeling holds significant promise, with its capacity to optimize network resources, improve user experience, and support emerging technologies like the Internet of Things (IoT), 5G wireless networks and beyond. The chapter examines three main facets of the implications of government policies and economics in computational modeling in advanced wireless networks. First, the adoption, implementation, and commercialization of computational models in wireless networks are examined, focusing on government policies and economic variables. Further, the work explores investment and commercialization prospects in computational models for wireless communications, exploring healthcare, smart cities, agriculture, and environmental monitoring. Secondly, regulatory structures that control spectrum allocation, network security, net neutrality, data protection, and competition are discussed elaborately, focusing on how they affect computational modeling and simulation techniques. Thirdly, the financial effects of computational models in dense wireless networks are analyzed considering the cost–benefit analysis for stakeholders, such as edge computing/IoT device manufacturers, and network management solution providers. The chapter concludes with the significance of government regulations, economics, and computational modeling interactions in wireless network design, development, deployment, and how stakeholders are enabled by them to make informed decisions for a more efficient ecosystem.

Chapter 12 presents an enhanced lightweight cryptographic algorithm towards securing wireless networks and big data. Wireless networks face challenges such as security, efficiency, and energy consumption. Despite the development of energy-efficient cryptographic methods, wireless networks operating in open channels still suffer from authentication, adaptability, key administration, and resource management issues. The chapter presents a study on potential methods for securing wireless networks using computational models. It proposes an enhanced lightweight cryptographic algorithm with robust security features for wireless networks and big data. The algorithm is designed as an extension of existing lightweight encryption algorithms, addressing the need for increased security in the digital environment. It enables secure communication with high throughput, low latency, and strong resistance against several attacks. The algorithm is suitable for applications like streaming services, network monitoring, and medical data storage. The comparative analysis is based on encryption time and delay, power consumption, and network lifetime delay and encryption time in addition to power consumption and network lifetime. The results show that the proposed scheme can effectively extend the network lifetime by 90% and 95%, when compared with AES, and TEA ciphers, respectively. The results show that the algorithm achieves high level of security and reliability, making it suitable for real-time applications.

Chapter 13 focuses on the computational models enabling smart teaching and learning in wireless communication systems. The integration of wireless communication systems into education has transformed the way we teach and learn. The advent of wireless technologies has ushered in a new era of smart teaching and learning, offering unprecedented opportunities for personalized and interactive educational experiences. However, harnessing the full potential of these wireless systems in education requires careful planning, optimization, and innovative approaches. Computational models have emerged as a critical tool in achieving these objectives. Wireless communication systems, including Wi-Fi networks, Bluetooth, and emerging technologies such as 5G and beyond, have become ubiquitous in educational settings. They facilitate seamless connectivity, enabling students and educators to access digital resources, collaborate in real-time, and engage with content from anywhere within the coverage area. Additionally, the proliferation of smart devices such as tablets, laptops, and smartphones has further accelerated the integration of wireless technologies into teaching and learning processes. The concept of "smart teaching and learning" in wireless communication systems goes beyond the mere use of wireless technologies in classrooms. It encompasses a holistic approach that leverages wireless infrastructure to create dynamic, adaptive, and data-driven educational environments. Smart teaching and learning involve tailoring educational content and experiences to individual learners, promoting active engagement, and enhancing the overall quality of education. The chapter is important as it addresses a critical component of modern communication technology. By exploring the role of computational techniques in wireless systems, the chapter contributes to the development of more efficient, reliable, and high-capacity networks. The study offers an interactive and immersive learning environment, showing that computational modeling and simulations have dramatically altered education. These technologies enable students and instructors to visualize complicated ideas, experiment with scenarios, and comprehend system dynamics without being constrained by the limits of real-world experimentation. By allowing learners to alter factors and witness consequences in a virtual context, simulation promotes deeper comprehension, encourages critical thinking, and improves problem-solving abilities. As a result, it bridges the theoretical and practical divide, making learning more interesting and successful.

Chapter 14 reviews the application of stochastic geometry and federated learning in computational modeling of communication systems. One important system that affects our daily lives is the communication system. It is an important system that enables communication and sharing of information and resources from any location around the world over different sophisticated communication networks. The description of the architecture and the functionality of these communication networks can assist network administrators and designers to get better understanding on their functionalities and operations. One modeling approach that has been applied to communication systems and networks stochastic geometry. It helps to describe their operations and the distribution of users properly. An advantage of the stochastic geometry technique is its ability to accurately model the random patterns of the nodes in the communication systems. A recent type of tool applicable to wireless communication systems is the federated learning (FL) that combines distributed computing and machine learning techniques to be trained locally and avoids the upload of data directly to the cloud server. The learning technique gives users an opportunity to benefit from the shared model that is trained from rich data without centrally storing it. The chapter presents a review of existing research works in federated learning schemes and stochastic geometry modeling of wireless communication systems. The review poses significant benefits to researchers in communication network and system modeling, leveraging stochastic geometry and federated learning.

Chapter 15 delves into the intricate world of cryptographic security in advanced wireless communication systems. It focuses on the pivotal role of high-speed stream ciphers, essential for ensuring data integrity and confidentiality in high-throughput environments. The chapter begins by reviewing related work in the field, setting the stage for a comprehensive exploration of cryptographic challenges and solutions. A significant portion of the chapter is dedicated to discussing the essence of cryptographic security in wireless networks. It highlights the importance of stream ciphers and the challenges they face, such as the need for randomness and resistance to various forms of cryptanalysis. The chapter emphasizes the role of computational modeling as a crucial tool in cryptographic security, demonstrating how it aids in the evaluation and development of cryptographic algorithms. The methodologies of NIST STS and DIEHARD for assessing the statistical security of stream ciphers are explored in detail. These testing suites are crucial for evaluating the randomness and robustness of cipher outputs, ensuring their effectiveness in wireless communication systems. The chapter presents the results of computational experiments using both NIST STS and DIEHARD, offering empirical insights into the performance of various cryptographic ciphers. In conclusion, the chapter provides a holistic view of cryptographic security in wireless networks, blending theoretical perspectives with practical insights, and discusses the current state and future directions of cryptographic security, emphasizing the continuous need for innovation in the face of evolving cyber threats.

Finally, Chapter 16 offers an in-depth examination of computational modeling and simulation techniques for enhancing spread-spectrum multiple-access asynchronous methods in wireless communication systems. The primary objective is the development of sophisticated codes that substantially increase network capacity and efficiency, thereby meeting the growing demands of contemporary wireless networks. This advancement is especially vital given the rapid increase in mobile device usage and the resulting complexities in network design and management. Central to this study is the creation of innovative spread-spectrum codes, strategically designed to be statistically uncorrelated at random starting points. This characteristic is crucial for reducing interference and maximizing bandwidth efficiency, ensuring robust and effective wireless communication. The chapter explores the mathematical and theoretical foundations of these codes, utilizing the Welch bound to clarify the interplay between code cardinality, length, and cross-correlation. Enhanced by computational methods for code generation, the study includes a comprehensive analysis of code parameters. A key aspect of the chapter is the application of computational modeling and simulation. These approaches allow for the dynamic and flexible examination of practical systems in a simulated setting, facilitating quick assessment and optimization of various network parameters without physical experimental setups. The chapter showcases experimental results and simulations that illustrate the improved performance of these codes in real-world scenarios, underscoring their role in improving mobile subscriber service quality. Additionally, the chapter tackles the challenges of ensuring extensive network coverage and high throughput in densely populated wireless networks. Comparative analysis with existing methods reveals that the proposed codes can significantly expand the capacity of spread-spectrum multiple-access systems with only a slight increase in correlation modulus. This feature is particularly advantageous for future radio control systems and smart grids, supporting the concept of 'soft capacity,' where base stations can adjust subscriber capacity with minimal service quality degradation. In summary, the chapter contributes significantly to both the theoretical and practical realms of wireless communication and paves the way for future research, highlighting the necessity for ongoing innovation in computational models to keep pace with the evolving wireless technology landscape.

1 An Overview of Computational Modeling and Simulations in Wireless Communication Systems

Wasswa Shafik

1.1 INTRODUCTION

Wireless communication systems (WCSs) have emerged as essential components of contemporary society, serving as the fundamental infrastructure for worldwide connection and facilitating diverse applications [1]. These applications encompass various domains, including mobile telecommunications, the Internet of Things (IoT), and critical infrastructure. These systems' optimization and efficient operation are paramount to address society-increasing needs effectively. The utilization of computational modeling and simulations has become essential in this pursuit, providing the necessary resources to comprehend, evaluate, and enhance the efficacy of wireless communication systems [2]. This research aims to examine the underlying principles and elucidate the crucial role of computational modeling and simulations in the progression of wireless communication technologies. This study analyzes the uses, advantages, and obstacles associated with these technologies, explaining their impact on wireless communication current and future landscape [3].

Wireless communication networks have emerged as a vital component of contemporary society, serving as the foundation for our globally networked environment. The proliferation of wireless networks has revolutionized various aspects of our lives, including communication, work, and daily living. This transformation is evident in the widespread use of smartphones, laptops, smart homes, and autonomous cars, which facilitate the smooth transmission of information [4]. The importance of addressing the growing need for quicker, more dependable, and efficient wireless services has made it crucial to prioritize the design, optimization, and comprehension of these intricate systems [5]. By examining the underlying principles, practical implementations, and associated difficulties, this analysis aims to elucidate the pivotal role of these tools in driving the progress of wireless technologies, hence influencing the trajectory of our digital landscape.

The emergence of wireless technology has initiated a period of unparalleled interconnectedness, surpassing limitations imposed by physical distance and facilitating immediate and seamless communication. Cellular networks, Wi-Fi hotspots, satellite communication, and upcoming technologies, for example, fifth generation (5G) and beyond, serve as the conduits for data transmission, facilitating a wide range of applications, including global commerce, healthcare, entertainment, and transportation [6]. In the present environment, computational modeling and simulations offer a methodical and organized strategy for addressing the complexities inherent in wireless systems. Computational models play a significant role as effective abstractions in an environment characterized by intricate relationships, various variables, and practical obstacles. These components enable us to condense the fundamental aspects of wireless communication networks, providing a framework that facilitates

DOI: 10.1201/9781003457428-2

a more profound comprehension of their functionality [7]. These models are beneficial tools for understanding the complexities of signal propagation, interference, and network protocols. They enable researchers to gain insights that are difficult to obtain only from traditional experimentation.

Comprehensive system performance evaluation is fundamental to computational modeling and simulations in wireless communication. Simulations offer a controlled and economically efficient setting to evaluate diverse scenarios, allowing researchers, engineers, and network planners to analyze data throughput, signal quality, and network capacity [8]. By conducting simulated experiments, valuable insights can be obtained into the performance of wireless systems under various settings, hence informing decision-making processes. In addition, computer modeling and simulations have a broader scope beyond retrospective analysis, encompassing innovation and optimization. Engineers and researchers employ these tools to create and optimize wireless networks. Various configurations, antenna placements, communication protocols, and network topologies are systematically tested to identify the most efficient and reliable solutions. Implementing this proactive method guarantees that wireless networks are responsive and capable of adapting to the dynamic requirements of users and applications [9].

As society progresses and adjusts to the challenges of an increasingly interconnected global environment, a pressing necessity arises for sophisticated tools and procedures to effectively design, enhance, and resolve issues within these systems [10]. Computational modeling encompasses the development of mathematical or computational representations of real-world systems, whereas simulations utilize these models to mimic the system behavior under different circumstances. These tools provide a methodical framework for investigating and examining intricate networks within wireless communication systems [11]. This empowers engineers, researchers, and decision-makers to acquire valuable perspectives, assess efficiency, and enhance designs.

In addition, computational modeling and simulations present a distinct advantage within wireless communication systems owing to their inherent versatility and adaptability. Customization options are available to satisfy unique research inquiries, enabling the examination of different situations and variables within a regulated setting [12]. Adaptability is essential in an industry characterized by swift technological advancements and the potential for network modifications to have wide-ranging implications. One of the primary uses of computational modeling and simulations in wireless communication systems is network planning and optimization. The strategic positioning of base stations, access points, and other infrastructure components becomes crucial as wireless networks expand to accommodate the needs of an expanding user population. Simulations facilitate network planners' evaluation of various deployment techniques, thereby ensuring optimal coverage, capacity, and resource allocation, as expounded in the subsequent sections.

1.1.1 Key Contribution of the Chapter

This chapter presents the following contributions as listed below:

- The study offers an overview of computational modeling and simulations in wireless communication systems, including key features and components, transmission medium, transmitter and receivers, frequency bands, antennas, protocols and standards, network infrastructure, security measures, and applications within Section 1.2.1.
- Enabling technologies of computational modeling and simulations are presented with an illustration of application in the WSCs in Section 1.2.2, for example, high-performance computing, parallel and distributed computing.
- The importance of computational modeling and simulations in wireless communication, including understanding complex systems, performance evaluation, optimization, and design, are also presented.

- This section examines the pivotal significance of tools and equipment in WCSs, including simulation software tools for system-level analysis, simulation devices and hardware for realistic testing, device-to-device simulations for emerging technologies, and integration with testing and measurement equipment in Section 1.4.
- The chapter illustrates the applications of computational modeling and simulations in several fields, including the possibility of determining the optimal placement of base stations, access points, and other infrastructure elements to provide maximum coverage and capacity, assisting in developing and testing communication protocols in Section 1.5.
- The chapter demonstrates the challenges and limitations to the accuracy of model computational resources including model validity and accuracy, computational resources, model validation, complex system, uncertainty and sensitivity, and ethical and privacy concerns.
- The chapter discusses the role of simulations in innovation and optimization, highlighting how these tools drive advancements in wireless technology. This includes improving system performance, enhancing energy efficiency, and addressing emerging field challenges, as detailed in Section 1.6.
- Lastly, it underscores the continued relevance of computational modeling and simulations in the evolving landscape of wireless communication. It recognizes that these tools remain critical in addressing the demands of 5G and beyond, the IoT, and emerging wireless technologies shown in lessons learned from the chapter in Section 1.7.2.

1.1.2 Organization of the Chapter

The remainder of this chapter is structured as follows. Section 1.2 presents wireless Communication Systems on how to enable the exchange of information over a wireless medium for physical connections. These systems encompass various technologies, including cellular networks, Wi-Fi, Bluetooth, and satellite communication. It further illustrates an overview of computational modeling and simulations that involve the creation of mathematical or algorithmic representations of real-world systems. Simulations, on the other hand, utilize these models to mimic the behavior of the actual system under different conditions. Section 1.3 details the Importance of Computational Modeling and Simulations in wireless communication, understanding complex Systems, performance evaluation, optimization, and design, among others. Section 1.4 examines the pivotal significance of tools and equipment in wireless communication systems, including simulation software tools for system-level analysis, simulation devices and hardware for realistic testing, device-to-device simulations for emerging technologies, and integration with testing and measurement equipment. Section 1.5. illustrates the applications of computational modeling and simulations in several fields, including possibilities of determining the optimal placement of base stations, access points, and other infrastructure elements to provide maximum coverage and capacity, assisting in developing and testing communication protocols. Researchers can simulate the behavior of new protocols in various scenarios to assess their performance and robustness and analyze the performance of wireless communication systems in the presence of interference, fading, and other real-world challenges; all are detailed. Section 1.6 demonstrates existing challenges and limitations like the accuracy of models' computational resources. Section 1.7 presents the future trends and lessons learned from these developments are presented. Finally, Section 1.8 entails the conclusion of the chapter.

1.2 WIRELESS COMMUNICATION SYSTEMS

This section explains WCSs to provide a proper understanding of computational modeling and simulations, since it enables the exchange of information over a wireless medium without the need for physical connections. WCSs cover various technologies, such as cellular networks,

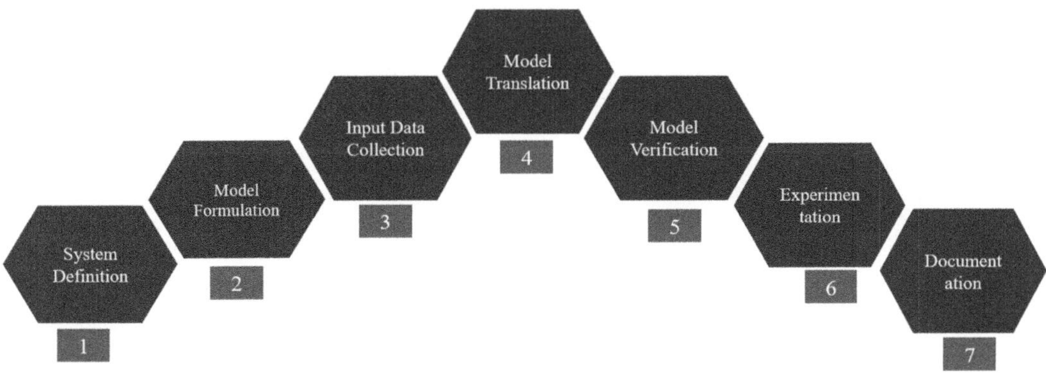

FIGURE 1.1 The Seven System Simulation Processes.

Wi-Fi, Bluetooth, satellite communication, and upcoming paradigms like 5G and beyond; types of these WCSs are presented in Figure 1.1. These systems are vital in various industries, including telecommunications, healthcare, transportation, and other sectors.

1.2.1 Key Features and Components of Wireless Communication Systems

The system characteristics discussed herein comprise various technologies, such as cellular networks, Wi-Fi, Bluetooth, satellite communication, and other related technologies. These technologies are the fundamental basis of our contemporary society, providing energy to many applications and sectors. The research provides notable characteristics and elements supported by illustrative instances to enhance comprehension. They are employed in diverse industries, like telecommunications, IoTs, healthcare, transportation, and entertainment.

1.2.1.1 Transmission Medium

The transmission medium in wireless communication systems refers to the mechanism by which data is transmitted between devices. The primary focus of this field of study revolves around the utilization of electromagnetic waves, specifically radio waves and microwaves. These waves propagate via air or vacuum, enabling the establishment of wireless communication links. For example, Wi-Fi networks employ radio waves to transmit data between a router and a laptop, facilitating wireless connectivity to the Internet [2]. Similarly, establishing satellite communication is contingent upon utilizing microwave signals to establish a connection between terrestrial stations and satellites in orbit. This facilitates the provision of worldwide coverage for various services, such as satellite television.

1.2.1.2 Transmitters and Receivers

Wireless gadgets are outfitted with transmitters and receivers to speed up communication. Transmitters are tasked with modulating data onto electromagnetic waves to facilitate their transmission, while receivers are responsible for capturing and subsequently demodulating these waves to extract the original data. In the context of cellular communication, it is observed that a smartphone transmitter functions by transmitting voice and data messages in the form of radio waves to a cell tower located nearby [4]. Subsequently, the receiver within the tower effectively catches and demodulates the signals, facilitating the transmission of the call or data to its intended recipient. In contrast, within the realm of radio broadcasting, the transmission of audio signals is facilitated by a radio station transmitter. These signals are subsequently captured and transformed into audible content by various radio receivers, including vehicle and home receivers.

TABLE 1.1

Key Aspects in Wireless Communication Systems: Modeling, Simulation, and Future Trends (5G, Fifth Generation)

Reference	Computational Modeling	Simulation	Wireless Communication	Wireless Systems	Current Development	Future Trends
[1]	Mathematical algorithms	Network simulations	Cellular communication	Base stations	5G deployment	6G+ deployment
[2]	Signal propagation	Wireless channel models	Wi-Fi technology	Routers	IoT connectivity	Spectrum exploration
[3]	Path loss modeling	Monte Carlo simulations	Satellite communication	Mobile devices	Network optimization	AI in network management
[4]	Channel capacity	Radio wave simulations	Antenna technology	Access points	Low latency solutions	Green wireless tech
[5]	MIMO systems	Spectrum allocation	Beamforming	Wireless protocols	Network slicing	Security enhancements
[6]	Propagation models	Network traffic models	IoT devices	Network architecture	Edge computing	Quantum communication
[7]	Channel estimation	Data packet modeling	Wireless standards	Antenna arrays	Beamforming advancements	Terahertz communication
[8]	Interference modeling	Radio resource allocation	5G and 6G networks	Mobile network infrastructure	Small cell deployments	Holographic networks
[32]	Spectrum sharing models	Mobility models	Satellite constellations	Wireless access points	mmWave technology	Bio-inspired networks
[9]	Error correction coding	Cross-layer simulations	Wireless security	Network management	Software-defined networks	Resilient wireless tech
[10]	Channel capacity bounds	Fading channel models	Vehicle-to-Everything	Wireless sensor networks	Cloud RAN	Connectivity in space

1.2.1.3 Frequency Bands

Various wireless technologies function within designated frequency bands assigned by regulatory bodies to minimize interference and optimize spectrum utilization. The Wi-Fi standard's 2.4 Gigahertz (GHz) band is frequently utilized for wireless networks in residential settings, while the 5 GHz band is employed to achieve faster connection speeds. In satellite communication, the "Ku band" is commonly employed for satellite broadcasting purposes, but the X band finds utility in military and scientific contexts, such as weather radar systems [6]. Table 1.1 demonstrates the current and future trends of computational simulations and modeling in wireless communication systems.

1.2.1.4 Antennas

Antennas play a vital role in wireless communication systems by facilitating effective transmission and reception of electromagnetic waves. These system designs exhibit variations contingent upon the specific application and frequency range. In the context of Global Positioning System (GPS) navigation systems, compact antennas integrated within smartphones or specialized GPS devices receive signals from several satellites in orbital paths, enabling the device to determine its precise geographical coordinates accurately. At a broader scope, broadcasting antennas, such as those

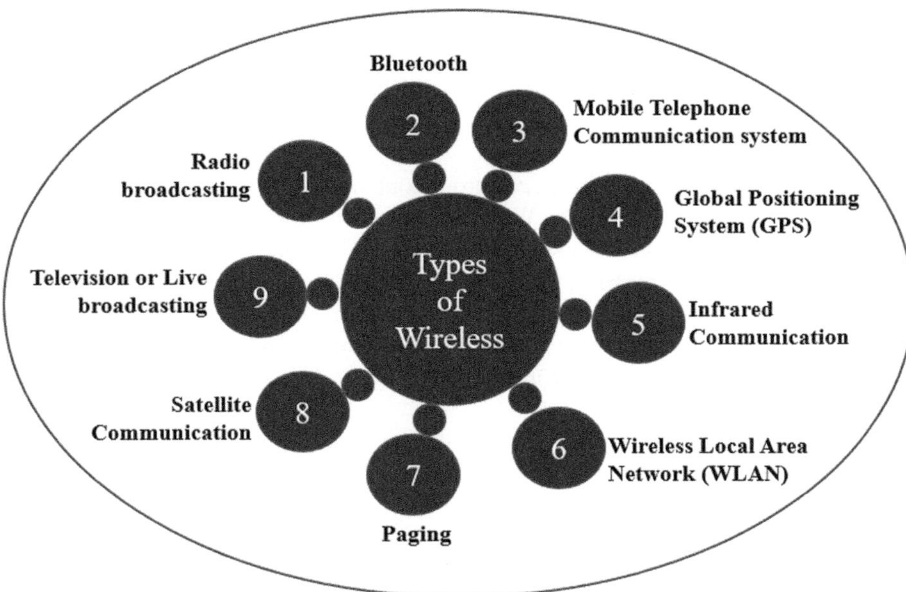

FIGURE 1.2 Types of Wireless Communication System.

employed by FM (Frequency Modulation) radio stations, provide the transmission of signals across extensive geographical regions, thereby assuring that listeners receive a coherent and uninterrupted broadcast; types of the WCSs are illustrated in Figure 1.2.

1.2.1.5 Protocols and Standards

Wireless communication systems depend heavily on precisely specified protocols and standards to ensure smooth interoperability and optimize data sharing. The IEEE 802.11 standard regulates the implementation of Wi-Fi technology, delineating the protocols and procedures for devices to establish connections with wireless networks, handle data transmission, and facilitate the negotiation of security configurations. The standardization process facilitates the utilization of Wi-Fi-enabled devices manufactured by different companies, ensuring interoperability and encouraging customer widespread acceptance and usage of such devices.

1.2.1.6 Network Infrastructure

The network infrastructure components are the fundamental framework for wireless communication systems, facilitating the essential infrastructure required for data transfer. Base stations, also called cell towers, are significant in cellular networks. In the context of 4G (fourth generation) LTE networks, base stations are deliberately positioned to provide coverage for specific geographical regions. As customers relocate, their devices smoothly transition between neighboring base stations, guaranteeing continuous connectivity. The network infrastructure facilitates users in voice communication, transmitting text-based messages, and gaining access to online resources from a wide range of locations.

1.2.1.7 Security Measures

The prioritization of securing and preserving the privacy of wireless data transmission is of utmost importance. Wireless communication systems utilize a variety of security mechanisms, such as encryption, authentication, and secure communication protocols. Sophisticated encryption measures safeguard sensitive financial transactions in mobile banking applications [13]. When a user initiates a transaction to transfer funds or inquire about their account balance via a mobile application, the transmitted data undergoes encryption to safeguard against illegal access or modification [14].

1.2.1.8 Applications

Wireless communication systems are utilized throughout various industries to improve operational efficiency and provide added convenience. Within the automotive industry, implementing V2X communication systems facilitates the exchange of information between vehicles and between vehicles and the surrounding roadside infrastructure [15]. This technology facilitates road safety improvement by establishing a communication network among cars, enabling the interchange of crucial information regarding their respective positions, velocities, and intended actions [16]. Consequently, this system is pivotal in accident prevention and mitigating traffic congestion.

1.2.1.9 Emerging Technologies

The continuous emergence of wireless technology is reshaping the landscape. The advent of 5G networks has brought about a range of transformative capabilities, including notably enhanced data rates and remarkably low latency. These advancements have paved the way for novel applications, such as remote surgery and immersive virtual reality experiences [17]. Furthermore, as demonstrated by enterprises such as SpaceX Starlink, satellite constellations endeavor to provide broadband Internet connectivity to regions lacking adequate access, thereby mitigating the digital divide.

1.2.1.10 Regulatory Framework

Regulatory entities, such as the Federal Communications Commission (FCC) in the United States and the Uganda Communications Commission (UCC), hold significant importance in the oversight of the radio frequency spectrum and the facilitation of equitable and effective distribution of spectrum resources. Rules and guidelines are established to mitigate interference and assign spectrum licenses to suppliers of wireless services [18]. As an illustration, the FCC conducts auctions to award spectrum bands to mobile network operators such as UCC in Uganda. This process serves to facilitate the implementation of cutting-edge wireless technology. As mentioned earlier, the elaborations shed light on the complex characteristics of wireless communication systems and their significant influence on diverse facets of human existence [19]. These systems persistently develop through establishing protocols, utilizing reliable network infrastructure, emphasizing security measures, adopting developing technology, offering novel functionalities, and stimulating innovation in various sectors.

1.2.2 Enabling Technologies of Computational Modeling and Simulations

The foundational technologies that provide computational modeling and simulations in wireless communication systems serve as a crucial catalysts for innovation and advancement within the discipline. These technologies offer the required tools and capacities for creating precise and efficient simulations, influencing the advancement and enhancement of wireless networks.

1.2.2.1 High-Performance Computing

Wireless communication simulations heavily rely on high-performance computing (HPC) clusters, supercomputers, and cloud-based resources as fundamental components. These systems provide significant computational capabilities, typically quantified in teraflops or petaflops, facilitating the execution of intricate simulations. HPC enhances the efficiency of handling extensive datasets, enables the simulation of complex wireless environments, and speeds up the attainment of simulation outcomes [20]. Furthermore, this technology allows researchers to investigate various network situations, encompassing extensive Multiple-Input Multiple-Output (MIMO) configurations and quantum communication models, with unparalleled velocity and precision.

1.2.2.2 Parallel and Distributed Computing

Using parallel and distributed computing technologies is crucial in effectively managing the computational requirements associated with wireless communication simulations. Parallelism enables the partitioning of simulations into smaller units of work that can be executed concurrently, resulting in a substantial reduction in simulation duration. Contemporary computer designs, such as multicore CPUs (Central Processing Units), Field-Programmable Gate Arrays (FPGAs), and Graphics Processing Units (GPUs), are integral in attaining parallelism [21]. Distributed computing utilizes a network of networked devices to allocate simulation workloads, improving scalability and performance. These technologies guarantee the ability of simulations to scale and accurately mimic wireless networks of significant magnitude successfully.

1.2.2.3 Software-Defined Networking and Network Function Virtualization

Network Function Virtualization (NFV) and Software-Defined Networking (SDN) are influential technologies that facilitate simulation by providing software-centric network infrastructure management. SDN is a networking paradigm that facilitates the separation of the control plane and the data plane inside a network. This decoupling allows for the dynamic configuration and programmability of the network [15]. Researchers can replicate diverse network designs and evaluate alternative routing and switching algorithms through simulations. In contrast, NFV facilitates the virtualization of network functions, enabling researchers to implement and evaluate network services in software components [22]. Utilizing virtualization techniques facilitates the optimization of network emulation within simulations, hence enhancing the assessment of the effects of network functions and services on wireless communication systems.

1.2.2.4 Advanced Simulation Frameworks and tools

Efficient and accurate modeling of wireless communication systems necessitates the utilization of specialized simulation frameworks and tools specifically designed for this purpose. Various simulation tools such as OMNeT++, Network Simulator version 3 (NS-3), OPNET, and MATLAB/ Simulink provide users with pre-existing models and libraries, streamlining the simulation procedure. These frameworks can serve as a fundamental basis for researchers to develop customized simulations, include real-world data, and evaluate the efficacy of novel algorithms and technologies [23]. Moreover, these technologies frequently support scripting and programming interfaces, facilitating substantial customization and experimentation.

1.2.2.5 Massive Dataset Handling

Simulations of wireless communication frequently incorporate comprehensive datasets, encompassing channel models, mobility traces, and propagation data. Efficient handling, processing, and management of these extensive datasets is necessary. Technological advancements include big data analytics platforms, distributed databases, and data streaming solutions [24]. These technologies enable researchers to effectively store, retrieve, and analyze extensive quantities of data within simulations, hence ensuring the fidelity of simulations in properly representing real-world circumstances.

1.2.2.6 Machine Learning and Artificial Intelligence

The integration of ML and AI is becoming more prevalent in simulations, augmenting their realism and cognitive capabilities. In simulations, AI-assisted agents replicate dynamic behaviors, like navigating autonomous cars via wireless networks or adapting smart gadgets to varying network conditions. ML methods are utilized in various applications, for example, forecasting network behavior, improving simulation settings, and producing synthetic data to augment datasets with few samples [25]. The utilization of AI-driven upgrades facilitates the replication of real-world difficulties with more precision in simulations.

1.2.2.7   Visualization and Analytics

Sophisticated visualization and analytics technologies offer academics the capability to assess simulation outcomes efficiently. Using 3D visualization technologies facilitates the generation of intricate network topologies and immersive settings, enabling the thorough examination of network behavior. Heatmaps and graphical representations are valuable tools for academics to visually depict performance indicators, signal propagation patterns, and sources of interference [26]. Analytics solutions provide functionalities for deriving significant insights from simulation data, facilitating data-driven decision-making and optimization techniques.

1.2.2.8   Quantum Computing Simulation

The advancement of quantum communication technologies necessitates the inclusion of quantum computing simulation to comprehensively comprehend their effects on wireless communication networks. Quantum simulators and quantum computing platforms enable the representation and analysis of quantum phenomena through computational simulations. These technologies facilitate investigating the characteristics and dynamics of quantum-entangled particles, quantum channels, and quantum key distribution (QKD) procedures [27]. Simulations about quantum computing facilitate the advancement and evaluation of safe and efficient quantum communication protocols, hence ensuring the preparedness of wireless networks for quantum advancements.

1.2.2.9   Augmented Reality and Virtual Reality

Integrating augmented reality (AR) and virtual reality (VR) technology enhances immersion and interactivity in wireless communication simulations. These technologies enable researchers to visually represent simulations in a three-dimensional spatial context, hence facilitating the creation of virtual environments that facilitate in-depth exploration of wireless networks and devices. Academic researchers can engage with simulated scenarios, facilitating a more profound comprehension of network behaviors and configurations [28]. The degree of immersion AR and VR apps provide holds significant value in assessing user experience and enhancing network performance. These enabling technologies are illustrated in Figure 1.3.

1.2.3   AN OVERVIEW OF COMPUTATIONAL MODELING AND SIMULATIONS

In this section, we introduce computational modeling, which entails the development of mathematical or algorithmic representations that simulate real-world systems. Simulations, however, employ these models to replicate the behavior of the actual system under varying situations.

1.2.3.1   An Overview of Computational Modeling

Using computational modeling is a critical methodology in several scientific and technical fields to acquire knowledge, formulate predictions, and enhance comprehension of intricate systems and events. Computational modeling fundamentally encompasses developing mathematical or algorithmic representations that mirror real-world phenomena. This allows researchers to mimic and examine these phenomena inside a controlled virtual setting. The process of computational modeling commences by formulating mathematical equations, algorithms, or logical representations that effectively encapsulate the fundamental characteristics of a system existing in the actual world, The construction of these models is predicated upon the principles, laws, or facts that control the

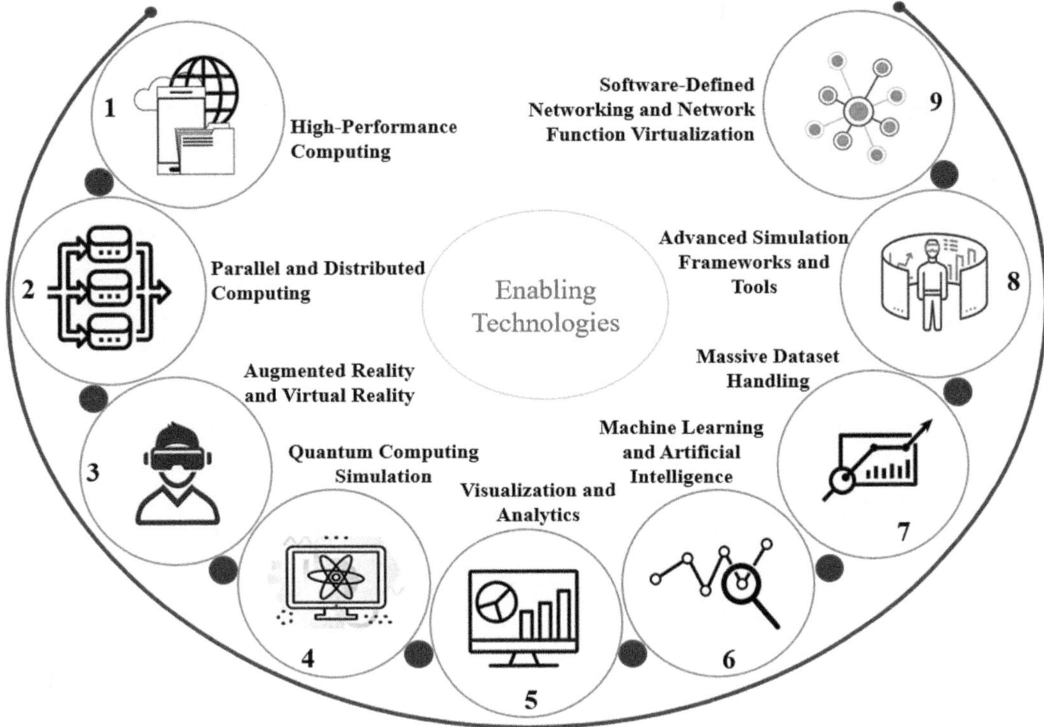

FIGURE 1.3 Enabling Technologies of Computational Modeling and Simulations.

system being studied. The system complexity under investigation determines how much models can vary, ranging from rudimentary representations to intricate simulations.

Computational modeling is an influential and adaptable methodology employed in various disciplines to replicate and examine intricate real-world systems, occurrences, and processes. Computational modeling, in its essence, entails the development of mathematical, logical, or algorithmic representations of systems to capture their fundamental properties and behaviors. These models function as virtual laboratories, facilitating the examination, experimentation, and comprehension of intricate systems behavior for researchers and professionals, eliminating the necessity for expensive or unfeasible real-world trials [29].

The fundamental elements of computational modeling encompass two primary constituents: the mathematical or logical model and the simulation. The model establishes the principles, connections, and factors that govern the system under investigation, whereas the simulation implements these models to reproduce the system behavior across different contexts or scenarios. Models can exhibit a spectrum of complexity, ranging from basic equations to elaborate algorithms. Conversely, simulations can be conducted using various computational resources such as PCs, supercomputers, or dedicated simulation hardware [30]. The utilization of computational modeling is prevalent across various domains in research and engineering. In disciplines such as physics and engineering, system analysis facilitates comprehension and enhancement of intricate systems, such as using fluid dynamics simulations in aircraft design or examining structural integrity in the context of building analysis. Similarly, computer models are utilized in medical research to replicate biological processes, thereby supporting medication discovery and treatment planning [31]. Likewise, environmental scientists use models to forecast climate change, analyze ecosystem dynamics, and anticipate occurrences of natural disasters.

There are many advantages associated with the utilization of computational modeling. Utilizing virtual simulations promotes cost efficiency by minimizing reliance on costly physical experiments, mitigating risks by examining several scenarios inside a controlled environment, and facilitating the expeditious generation of results. Furthermore, this approach facilitates the attainment of accuracy in capturing nuanced nuances and behaviors of systems that could pose difficulties in direct observation [32]. Although computational modeling offers numerous benefits, it encounters several problems, including the requirement for precise model validation, significant computer resources for intricate simulations, and dependence on accurate input data. However, with the continuous advancement of computer power and modeling techniques, the outlook for computational modeling in the future seems optimistic [33]. This ongoing technology progress and utilization are anticipated to catalyze novel ideas, progressions, and a more profound comprehension of intricate systems within various sectors and fields of study.

1.2.3.2   An Overview of Computational Simulations

Computational simulations comprise a crucial scientific study and engineering methodology, employing computer-based approaches to recreate and analyze intricate real-world phenomena. Simulations offer a methodological approach to investigate these phenomena inside a controlled virtual setting, facilitating researchers to acquire valuable insights, forecast potential outcomes, and enhance systems without requiring resource-intensive or time-consuming physical trials [34]. Computational simulations function as virtual laboratories, providing a pragmatic and potent means of comprehending and engaging with complex systems.

The fundamental elements of computer simulations consist of two crucial components: the mathematical or algorithmic model and the simulation itself. The model functions as a computer representation of the real-world system, encompassing its laws, relationships, and behavior. Simulations facilitated by high-performance computing execute these models, producing outcomes that closely resemble the behavior of the natural system. The models in this context exhibit a broad spectrum, varying from relatively straightforward equations to intricate numerical algorithms. The choice of model complexity is contingent upon the specific characteristics and intricacies of the phenomena under investigation.

Computational simulations possess a wide range of applications across various sectors and industries. Within the domains of physics and engineering, these disciplines enable the examination of fluid dynamics, structural mechanics, and electromagnetic interactions, hence providing essential assistance in developing various structures ranging from aircraft to bridges. Moreover, simulations play a crucial part in the healthcare field as they simulate intricate biological processes, facilitate the advancement of pharmaceuticals, and empower the implementation of individualized therapy. Simulations play a crucial role in environmental science by facilitating climate change prediction, the simulation of ecosystems, and the assessment of the damages caused by natural disasters. These applications exemplify only a tiny portion of the vast range of practical uses for computational simulations.

The benefits of computer simulations are many and varied. They provide cost-effective alternatives as substitutes for conventional experimental procedures, reducing equipment, materials, and facilities expenditures. Additionally, they offer secure settings for exploring events that may be excessively dangerous or expensive to recreate in the tangible realm. Moreover, these rapid generation technologies of extensive datasets facilitate the acceleration of research processes. This enables researchers to iterate, optimize, and refine their models efficiently. Simulations provide scientists and engineers with enhanced precision and control, enabling them to understand intricate systems profoundly. Consequently, this capability facilitates more informed decision-making and fosters innovation.

Although computational simulations provide significant advantages, they are not devoid of obstacles. Persistent challenges in the field include model validation, the accessibility of high-performance computing resources, and the requirement for precise input data. Nevertheless, the current issues are being tackled using enhanced computer capacity and advanced modeling tools. The potential of computational simulations in various domains, such as artificial intelligence, materials research, and urban planning, is highly promising. The ongoing advancement of computing capabilities will ensure that simulations retain their pivotal role in scientific investigation and engineering advancement, facilitating progress across various fields of study.

1.3 COMPUTATIONAL MODELING AND SIMULATIONS MERITS IN WIRELESS COMMUNICATION SYSTEMS

This section presents identified advantages of computational modeling and simulations within the WCS perspective.

1.3.1 COMPLEXITY MANAGEMENT

Wireless communication systems are complex, comprising several elements such as transmitters, receivers, antennas, and protocols. Using computational modeling offers a systematic methodology for comprehending the intricacies associated with this phenomenon. In the context of a vast cellular network, it is possible to employ a mathematical model to depict the intricate dynamics between many mobile devices and a substantial number of base stations. Simulations investigate the influence of many parameters on network performance, including user mobility patterns, signal propagation characteristics, and interference levels. Including these intricacies in the model makes enhancing network design and operation possible for optimal performance.

1.3.2 PERFORMANCE EVALUATION

Simulations play a crucial role in conducting a complete evaluation of the performance of wireless systems. Consider, for instance, the advancement of a novel wireless communication protocol. Simulations enable researchers to replicate real-world circumstances accurately. Inside the framework of a vehicular communication system, simulations can accurately reproduce the intricate dynamics of vehicles inside traffic, their intricate interconnections, and the nuanced behavior of the communication protocol across a range of traffic densities and diverse road conditions. Using simulations offers significant contributions in enhancing the understanding of protocol efficiency, latency, and reliability, hence facilitating design optimization before the implementation phase.

1.3.3 PROTOTYPING AND DESIGN

Computational modeling and simulations play a crucial role during the design phase. The establishment of a wireless sensor network for environmental monitoring warrants consideration. Engineers can construct a comprehensive model encompassing sensor nodes, their communication protocols, and the operational environment. Subsequently, simulations imitate several sensors' deployment over a vast expanse and forecast how data traverses the network. This facilitates optimizing the location of sensors, techniques for managing power, and routing data protocols, guaranteeing that the system achieves its intended objectives.

1.3.4 SCALABILITY AND FUTUREPROOFING

The concept of scalability holds significant importance in wireless communication networks, especially when anticipating future demands. As the development of 5G networks progresses, simulations can be utilized to forecast the ability of the system to handle the rapid increase in the number of connected devices and the volume of data traffic. By using simulations involving a substantial quantity of IoT gadgets concurrently accessing the network, scholars can evaluate the occurrence of network congestion, discern possible congestion points, and formulate approaches to uphold superior service quality as the network expands.

1.3.5 Enhancing Reliability and Robustness

Simulations play a crucial role in identifying and mitigating vulnerabilities within wireless networks. Simulations within the context of an intelligent grid can effectively replicate many failure scenarios, including cyberattacks or equipment breakdowns. By implementing simulated disruptions, the grid can undergo rigorous testing and enhancement of security measures while also facilitating the development of contingency plans. This approach aims to bolster the reliability and resilience of the grid in response to actual threats.

1.3.6 Resource Optimization

The utilization of computational modeling facilitates optimizing the allocation of resources, specifically in the case of spectrum resources inside wireless networks. Let us contemplate a multi-user MIMO system within the context of a Wi-Fi network. Simulations can evaluate various beamforming algorithms and ascertain the most favorable transmit power distribution to optimize data throughput and minimize interference. The employment of simulations aids in the optimization of spectrum allocation, resulting in increased network capacity and higher customer satisfaction.

1.3.7 Risk Mitigation and Cost Reduction

Simulations are crucial in cost reduction and risk mitigation while implementing wireless systems. Before implementing an extensive IoT network for agricultural monitoring, it is advisable to employ simulations to forecast the network performance across diverse weather conditions and crop densities. The early identification of possible issues enables the implementation of cost-effective improvements, such as optimizing sensor location or power management measures [35]. This proactive approach also helps to limit the risks associated with unforeseen operational obstacles. In conclusion, computational modeling and simulations play a crucial role in wireless communication systems by facilitating a more profound comprehension of intricate networks, enhancing performance optimization, guaranteeing reliability, and preparing for forthcoming improvements.

1.4 AN OVERVIEW OF COMPUTATIONAL MODELING AND SIMULATION TOOLS AND DEVICES

The rapid evolution of wireless communication technology has resulted in their integration into our daily lives and vital infrastructure. Engineers and researchers primarily depend on computational modeling, simulation methods, and devices to build, optimize, and evaluate these systems efficiently. This section examines the pivotal significance of tools and equipment in wireless communication systems, including simulation software tools for system-level analysis, simulation devices and hardware for realistic testing, device-to-device simulations for emerging technologies, and integration with testing and measurement equipment.

1.4.1 Simulation Software Tools for System-Level Analysis

This section presents some simulation software tools crucial to wireless communication research. These tools play a crucial role in developing system-level models for analyzing and evaluating many elements of wireless systems.

1.4.1.1 MATLAB and Simulink

The NS-3 network simulator is an open-source tool widely used to evaluate wireless network protocols and applications. This tool enables the examination of various wireless communication scenarios, such as wireless LANs, cellular networks, and ad-hoc networks. Use of NS-3 by researchers allows for assessing network performance parameters like throughput, latency, and packet loss [36]. This facilitates the optimization of protocols and network topologies to address wireless communication system requirements effectively.

1.4.1.2 OPNET (Riverbed Modeler)

OPNET, recently integrated into Riverbed Modeler, is a robust simulation platform that provides extensive features for simulating network performance. Within wireless communication systems, it facilitates the examination of wireless network behavior, the influence of network topologies on performance, and the enhancement of network parameters [37]. This tool is valuable in evaluating the behavior of wireless communication protocols and applications inside intricate and ever-changing network settings. These tools and devices are illustrated in Figure 1.4.

1.4.1.3 Wireshark

Wireshark, an open-source packet analyzer, is an essential tool for comprehending the dynamics of wireless communication protocols and traffic dynamics. Wireshark is a commonly employed tool by researchers to capture and analyze packets transferred within wireless networks. This enables researchers to gain valuable insights into various aspects, such as protocol interactions, packet flow, and network security [38]. This tool holds significant importance in identifying problems and enhancing the performance of wireless systems.

1.4.1.4 QualNet

Another tool is a specialized network simulation tool that has been developed to facilitate the modeling and analysis of wireless communication systems. The platform provides diverse wireless models and modules, enabling researchers to explore various wireless technologies such as Wi-Fi, WiMAX, and mobile cellular networks [39]. Researchers can evaluate the efficacy of these

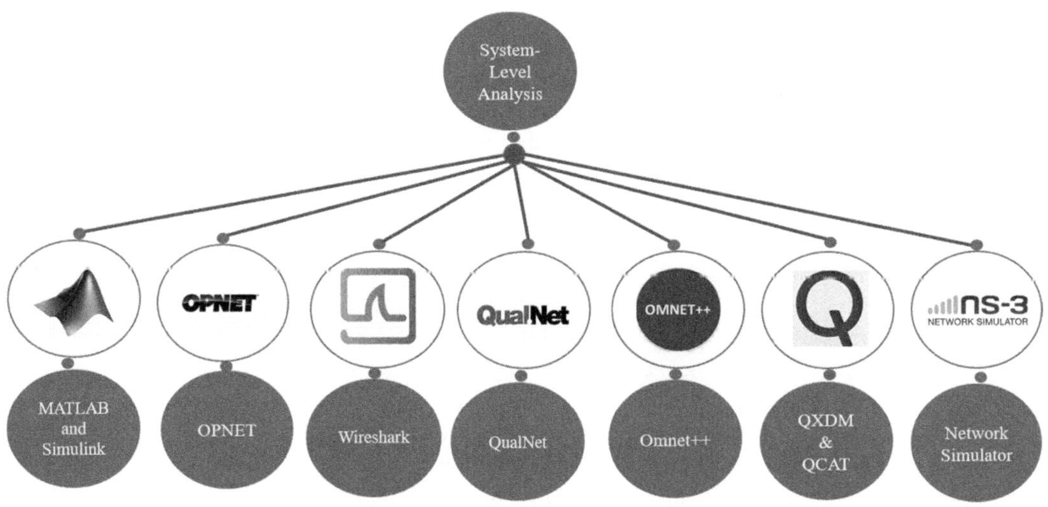

FIGURE 1.4 Computational Modeling and Simulation Tools and Devices.

technologies in various situations, empowering them to make well-informed choices about network design and implementation tactics.

1.4.1.5 Omnet++

OMNeT++ is a modular discrete event simulation framework available as open-source software. Wireless communication systems are commonly employed to model wireless network protocols and assess the efficacy of mobile ad hoc networks, wireless sensor networks, and vehicular communication systems [40]. OMNeT++ offers a versatile framework for exploring intricate wireless communication scenarios.

1.4.1.6 Qualcomm's Qualcomm Advanced Diagnostic Tool (QXDM) and Qualcomm eXtensible Diagnostic Monitor (QCAT)

Qualcomm's tools, namely QXDM and QCAT, are imperative in examining and resolving issues on wireless communication systems that rely on Qualcomm chipsets. Engineers are granted the capability to gather and evaluate performance data derived from mobile devices, hence facilitating the identification and resolution of issues about the wireless stack [41]. This process also ensures that the devices adhere to established industry standards.

1.4.1.7 Network Simulator Tools (Ns-2 and beyond)

The network simulator is a prevalent discrete event simulator utilized in networking research. The software has a substantial user base and offers comprehensive support for simulating wireless communication systems [42]. Researchers utilize network simulators to examine the effectiveness of routing algorithms, analyze network congestion, and measure the influence of mobility on network performance inside wireless contexts, as summarized in Table 1.2. Besides these tools, the following section demonstrates devices and hardware for realistic testing.

1.4.2 SIMULATION DEVICES AND HARDWARE FOR REALISTIC TESTING

These devices provide a link between simulations that are solely based on software and actual hardware deployments, providing a method to imitate and evaluate the effectiveness of wireless technology.

1.4.2.1 Software-defined Radios

Software-defined radios (SDRs) refer to a class of radios that can transmit and receive a wide range of wireless signals owing to their programmable nature. These tools provide opportunities for flexibility and versatility in conducting experiments. SDRs are utilized in developing, testing, and validating wireless communication protocols, waveforms, and signal-processing methodologies. The implementation and evaluation of sophisticated technologies such as cognitive radio and dynamic spectrum access heavily rely on their presence [43]. SDRs allow researchers to generate personalized waveforms, replicate diverse propagation scenarios, and assess the efficacy of wireless devices and networks.

1.4.2.2 Channel Emulators

Channel emulators are devices designed to simulate wireless channel conditions in real-world scenarios accurately. These emulators can replicate fading, interference, and multipath effects. Channel emulators are employed to verify the efficacy of wireless devices in real-world channel scenarios, guaranteeing their ability to function optimally in demanding surroundings [44]. These devices provide comprehensive testing of wireless networks and aid in identifying and mitigating signal quality and reliability issues.

TABLE 1.2
Simulation Software Tools for System-Level Analysis

Reference	Simulation Software Tool	Description	Key Applications	Role in Wireless Communication	Benefits
[36]	MATLAB and Simulink	A versatile numerical computing platform	Signal processing, communication protocols, MIMO	System modeling, simulation, and analysis of wireless technologies	A comprehensive environment for wireless analysis
[37]	Network Simulation (NS-3)	An open-source network simulator	Wireless LANs, cellular networks, ad-hoc networks	Evaluation of wireless network protocols, network performance analysis	Extensive support for wireless network research
[38]	OPNET (Riverbed Modeler)	A network performance modeling tool	Performance evaluation, protocol analysis	Analysis of wireless network behavior, network optimization	Comprehensive network modeling capabilities
[39]	Wireshark	An open-source packet analyzer	Protocol debugging, network troubleshooting	Packet capture and analysis of wireless communication traffic	Deep insights into wireless protocol behavior
[40]	QualNet	A network simulation tool	Wireless technology assessment, network optimization	Modeling and analysis of wireless communication systems	Wide range of wireless models and modules
[41]	OMNeT++	An open-source discrete event simulator	Mobile ad-hoc networks, wireless sensor networks	Simulation of wireless network protocols and scenarios	Flexible platform for complex wireless simulations
[41]	Qualcomm's QXDM and QCAT	Diagnostic and analysis tools	Testing and validation of Qualcomm-based devices	Analysis and debugging of wireless systems using Qualcomm chipsets	Specialized tools for Qualcomm chipset analysis
[42]	Network Simulator 2	A discrete event simulator	Routing protocol evaluation, network congestion	Network simulation for wireless communication research	A popular and extensive network simulator

1.4.2.3 Multiple-Input, Multiple-Output Testbeds

MIMO testbeds enable the conduct of experiments using sophisticated antenna configurations and signal processing techniques, such as beamforming and spatial multiplexing. MIMO testbeds are employed by researchers to evaluate the tangible advantages of MIMO technology and to enhance the efficiency of wireless systems that depend on the utilization of multiple antennas [45]. MIMO testbeds facilitate the empirical verification of MIMO-enabled wireless communication systems, encompassing wireless local area networks (LANs), cellular networks, and upcoming technologies such as 5G and beyond.

1.4.2.4   Fading Testers

Fading testers are utilized to create controlled fading channels to assess the performance of wireless devices under fading conditions. Fading tests are employed to evaluate wireless receivers across a range of channel models, hence facilitating the assessment of communication systems' resilience and dependability in real-world contexts [46]. Using these tools, researchers and investors can gain insights into the effects of fading on signal quality, packet error rates, and link dependability. This enables the optimization of wireless systems to handle propagation conditions that pose challenges effectively.

1.4.2.5   Noise Generators and Impairment Simulators

These devices are designed to introduce deliberate noise and impairments into the signal stream to replicate the effects of interference, noise, and distortion. Noise generators and impairment simulators play a crucial role in evaluating the efficacy of wireless communication systems in adverse conditions, including but not limited to urban and industrial environments [47]. These tools assist researchers in identifying vulnerabilities, assessing error-correcting techniques, and optimizing wireless systems to function efficiently amid diverse impairments.

1.4.2.6   Vehicular Emulators

Vehicular emulators replicate the wireless communication conditions in vehicular ad-hoc networks (VANETs) and linked automobiles. Researchers employ vehicular emulators to conduct testing and enhance communication protocols on autonomous driving, V2V communication, and V2I communication [48]. These emulators facilitate the evaluation of safety-critical applications in real-world automotive settings and the advancement of dependable communication networks for intelligent transportation. And Figure 1.5 illustrates the simulation devices and hardware for realistic testing.

1.4.3   Device-to-Device Simulations for Emerging Technologies

Simulations involving Device-to-Device (D2D) communication play a crucial role in wireless communication research, particularly within new technologies and applications. D2D communication is establishing wireless communication links directly between user devices, including smartphones, IoT devices, and cars, without the need to route the data through a centralized network infrastructure.

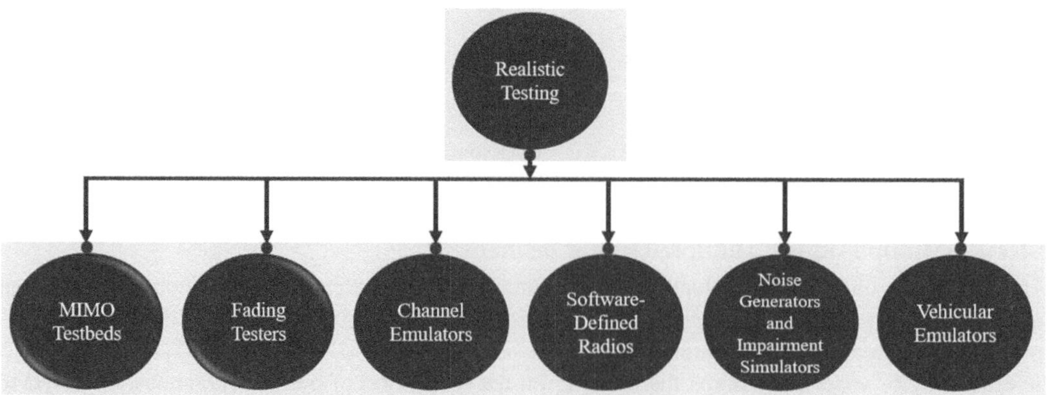

FIGURE 1.5   Simulation Devices and Hardware for Realistic Testing.

The simulation of D2D situations is crucial to understanding and optimizing several facets of wireless communication networks.

1.4.3.1   Internet of Things Simulators

The IoT is a burgeoning domain that is seeing tremendous expansion. Within this realm, D2D communication is pivotal in facilitating the connection and exchange of data among IoT devices. IoT simulators facilitate the simulation of D2D interactions within a diverse range of IoT devices. Simulators are employed for testing the connection of IoT devices, evaluating network coverage, assessing power consumption, and optimizing IoT protocols to enhance the efficiency of device-to-device communication. IoT simulators offer valuable insights into the scalability and dependability of D2D connections across extensive IoT deployments. These simulators enable researchers to design and optimize IoT solutions for many applications, ranging from smart homes to industrial IoT.

1.4.3.2   Vehicular-to-Everything (V2X) Simulations

V2X communication, which falls under the category of D2D communication, plays a vital role in facilitating connectivity and autonomy in vehicles. Simulations about V2I scenarios aim to accurately reproduce the dynamic interactions among vehicles, infrastructure, and other entities on the road. Researchers and engineers employ V2X simulators to assess communication protocols for improving road safety, traffic management, and vehicle autonomy [38]. This encompasses evaluating cooperative adaptive cruise control, collision avoidance systems, and traffic signal communication. V2X simulations evaluate the dependability and time delay of D2D communication among automobiles, offering valuable insights for enhancing V2X systems and guaranteeing the safety and effectiveness of emerging automotive technology [25].

1.4.3.3   Mobile Edge Computing Simulations

The Mobile Edge Computing (MEC) framework incorporates cloud computing functionalities at the periphery of the mobile network, enabling efficient and rapid D2D communication between devices and neighboring edge servers. MEC simulations are employed to evaluate the effects of MEC on D2D communication, optimize the placement of edge servers, and assess the decrease of latency in real-time applications like VR and AR [38]. The utilization of MEC simulations contributes to the advancement of low-latency D2D services, improves the user experience in new applications, and facilitates the exploration of novel D2D use cases, like augmented reality, AR-assisted navigation and gaming.

1.4.3.4   Drone-to-Drone (D2D) Simulations

Drones, also known as unmanned aerial vehicles (UAVs), utilize D2D communication to effectively synchronize their operations and exchange data in various applications such as aerial surveillance, search and rescue missions, and package transportation [49]. The utilization of D2D simulations in the context of drones is of utmost importance as it allows for evaluating communication protocols and collision avoidance systems, optimizing drone swarming tactics, and assessing interference effects [50]. Using simulations facilitates the secure and effective implementation of D2D-enabled devices in diverse scenarios, promoting the drone sector's expansion and guaranteeing dependable wireless communication within the airspace.

1.4.3.5   5G D2D Simulations

D2D communication plays a crucial role within the framework of 5G networks. Simulations conducted within the realm of 5G technology primarily examine the intricate dynamics between cellular communication and D2D communication, thereby facilitating direct connections between

TABLE 1.3
Simulation Devices and Hardware for Realistic Testing

Reference	Simulation Device/ Hardware	Description	Key Applications	Role in Wireless Communication	Benefits
[43]	Software-Defined Radios	Programmable radios for signal generation and reception	Protocol testing, waveform development, dynamic signal generation	Testing wireless devices and protocols, emulating real-world signals	Versatility and adaptability for various wireless tests
[44]	Channel Emulators	Replicates real-world channel conditions	Testing devices in fading, interference, and multipath scenarios	Simulating realistic propagation environments for device testing	Ensures devices can operate in challenging conditions
[45]	Multiple-Input, Multiple-Output (MIMO) Testbeds	Experimental setups for MIMO technology	Assessing MIMO performance, optimizing wireless networks	Investigating advanced antenna configurations and signal processing	Real-world validation of MIMO technology
[46]	Fading Testers	Generate controlled fading channels	Testing receiver performance, validating error correction mechanisms	Evaluating devices under various channel conditions and fading models	Insights into signal quality and robustness
[47]	Noise Generators and Impairment Simulators	Introduce noise and impairments	Testing device robustness in noisy environments, error correction	Simulating interference, noise, and signal degradation for testing	Identifying vulnerabilities and improving reliability
[48]	Vehicular Emulators	Simulate vehicular communication environments	Validating communication protocols for autonomous vehicles, safety	Testing V2X and vehicular communication systems, including V2V and V2I	Ensuring the reliability of vehicular communication

devices within the framework of 5G networks [51]. Simulations of D2D communication in the context of 5G cellular networks are crucial in evaluating many aspects, such as interference control, resource allocation, and network architecture of D2D services. D2D communication is crucial for enhancing communication efficiency in urban, rural, and industrial settings. Using simulations plays a crucial role in comprehending the capabilities of 5G D2D communication [50]. This enables efficient communication with conventional cellular networks while facilitating a wide range of applications, such as public safety, mobile gaming, and content sharing, as concise in Table 1.3.

1.4.4 Integration with Testing and Measurement Equipment

The integration facilitates validating, analyzing the performance, and conducting tests on different components included in the wireless system. The following is a comprehensive examination of the incorporation of testing and measuring equipment within the framework of WCS systems.

1.4.4.1 Signal Generators and Analyzers

Signal generators generate radio frequency (RF) signals, while signal analyzers are applied to measure and evaluate these signals. Fundamental tools play a crucial role in creating and testing wireless devices. Wireless communication systems employ these instruments to evaluate the operation of transmitters, check the quality of signals, and quantify parameters such as frequency, amplitude, and modulation characteristics. The incorporation of signal generators and analyzers facilitates the adherence of wireless devices to specifications and standards, hence enabling a meticulous assessment of their performance.

1.4.4.2 Protocol Testers

The role of protocol testers is to assess and verify the adherence and effectiveness of communication protocols. Protocol messages are exchanged to evaluate the functionality and adherence to established standards. In wireless communication systems, using protocol testers is imperative to validate the adherence of devices and networks to established industry standards. This guarantees interoperability and the facilitation of dependable communication [52]. Protocol testers play a crucial role in identifying and addressing implementation difficulties. They are vital in diagnosing problems specific to protocols and aiding the creation of wireless devices that are both interoperable and consistent with established standards.

1.4.4.3 Spectrum Analyzers

Spectrum analyzers are utilized to examine the frequency domain of radio frequency (RF) transmissions, enabling the identification of signal interference, spectrum attributes, and potential origins of signal degradation. Spectrum analyzers possess significant utility in the identification of signal interference and the quantification of occupied bandwidth. Within the realm of wireless communication, these devices play a crucial role in the optimization of frequency allocation and the mitigation of interference-related challenges [53]. The integration of spectrum analyzers enables engineers to effectively detect and address potential sources of interference, guaranteeing wireless communication systems' integrity and dependability.

1.4.4.4 Network Analyzers

Network analyzers evaluate network and device performance by examining impedance, network parameters, and signal transmission characteristics. Network analyzers play a crucial role in evaluating the efficacy of antennas, transmission lines, and network elements in wireless communication, facilitating optimal signal propagation [54]. Incorporating network analyzers facilitates developing and optimizing wireless networks and checks that network components function within specified parameters.

1.4.4.5 Oscilloscopes

Oscilloscopes serve as instruments for the visual representation and examination of electrical waveforms in a temporal dimension, hence facilitating the quantification of signal attributes such as voltage, amplitude, and frequency. Oscilloscopes are crucial in wireless communication since they are indispensable tools for analyzing signal quality, detecting signal distortion, and identifying signal fidelity-related problems [55]. Oscilloscopes serve as valuable instruments in identifying signal anomalies, aiding engineers in the precise localization and resolution of issues that can potentially impact wireless communication quality.

1.4.4.6 Channel Emulators

Channel emulators are devices designed to accurately reproduce the wireless channel characteristics encountered in real-world scenarios, encompassing fading, multipath propagation, and interference. Their purpose is to facilitate the realistic evaluation and testing of wireless devices. Channel emulators are commonly employed in wireless communication systems to assess the efficacy of devices under adverse channel circumstances and to ascertain their ability to function dependably in

TABLE 1.4
Simulation Devices and Hardware for Realistic Testing

Reference	Simulation Device/Hardware	Description	Key Applications	Role in Wireless Communication Systems	Benefits
[49]	Signal Generators and Analyzers	Programmable radios for signal generation and reception	Transmitter performance evaluation, frequency analysis	Performance testing, signal quality verification	Ensures adherence to signal specifications
[52]	Protocol Testers	Validate protocol compliance and performance	Testing wireless device and network adherence to standards	Protocol conformance testing, interoperability assessment	Ensures reliable communication and interoperability
[53]	Spectrum Analyzers	Analyze frequency spectrum and signal characteristics	Detecting signal interference, optimizing frequency allocation	Interference detection, frequency analysis, spectral measurement	Identifies sources of signal degradation and interference
[54]	Network Analyzers	Analyze network parameters and transmission properties	Network component evaluation, network parameter optimization	Network and device performance analysis, impedance measurement	Ensures efficient signal transmission and network health
[55]	Oscilloscopes	Visualize and analyze electrical waveforms	Signal quality assessment, signal distortion detection	Signal analysis, waveform visualization	Helps diagnose and rectify signal anomalies
[44]	Channel Emulators	Replicate real-world wireless channel conditions	Validation of device performance under various conditions	Realistic testing of wireless devices in challenging channels	Assesses device resilience in different environments

practical environments, as established in Table 1.4. The utilization of channel emulators facilitates the evaluation of the resilience and dependability of wireless communication systems, specifically in challenging propagation conditions.

1.5 APPLICATIONS OF COMPUTATIONAL MODELING AND SIMULATIONS

These examples underscore the vast utility of computational modeling and simulations across various disciplines. As technology advances, their applications expand, driving innovation, problem-solving, and a deeper understanding of complex systems and phenomena.

1.5.1 Network Planning and Optimization

Simulations are crucial in aiding wireless service providers in planning and optimizing their network coverage and capacity. Through signal propagation modeling, analysis of the terrain, and assessment of interference patterns, operators can ascertain optimal placements for base stations and antennas, ensuring consistent and dependable network performance. In deploying 5G networks, simulations play a crucial role in enhancing the effectiveness of small cell placement and beamforming techniques, hence enhancing coverage and capacity in urban environments. Spectrum management involves employing computational models to simulate the allocation and utilization of spectrum resources. These simulations are employed by wireless regulators and telecommunications businesses to guarantee the optimal utilization of the radio frequency spectrum. Through spectrum-sharing situations, interference can be mitigated, and spectrum allocation can be optimized for various wireless technologies, for example, Wi-Fi and cellular networks.

1.5.2 WIRELESS PROTOCOL DEVELOPMENT

Simulations are vital in developing and testing wireless communication protocols, particularly protocol testing and optimization. Engineers employ simulations to assess the efficacy of protocols such as Wi-Fi, Bluetooth, and 4G/5G in diverse scenarios, including those characterized by high user density or interference from other wireless devices. This facilitates the optimization of protocols to ensure smooth data transfer. IoT networks frequently exhibit distinctive communication requisites for their associated devices. Simulations facilitate the development and evaluation communication protocols specifically designed for IoT applications. As an illustration, simulations can replicate the functioning of low-power, wide-area network (LPWAN) protocols employed to establish connections between sensors and devices within smart cities and industrial Internet of Things (IIoT) implementations.

1.5.3 ANTENNA DESIGN AND OPTIMIZATION

The antenna performance analysis is significant in designing and optimizing wireless communication systems, with simulations as a valuable tool for engineers. The models above represent antennas' radiation patterns, gain, and directivity. In the context of 5G antenna array advancement, simulations optimize beamforming methodologies, resulting in enhanced data transmission speeds and improved network coverage [56]. MIMO systems: Using multiple antennas at both the transmitter and receiver in MIMO technology improves wireless communication. Simulations play a vital role in examining and enhancing MIMO system performance across many scenarios, including but not limited to massive MIMO for 5G networks and beamforming in Wi-Fi systems.

1.5.4 INTERFERENCE AND COEXISTENCE ANALYSIS

Coexistence testing aims to evaluate the simultaneous operation of multiple wireless technologies within shared frequency bands through simulations. Simulations are employed to ascertain the operational feasibility of Wi-Fi and Bluetooth devices in a smart home setting, ensuring their compatibility and absence of interference with each other and with other wireless technologies such as Zigbee. The efficient utilization of available spectrum is contingent upon implementing dynamic spectrum sharing. Simulations mimic various scenarios in which diverse wireless services coexist and share spectrum resources, guaranteeing each service function without inducing detrimental interference. The significance of this matter is particularly pronounced within the context of shared spectrum bands, such as the broadband radio service (CBRS) in the United States.

1.5.5 SECURITY ANALYSIS

Using simulations in cybersecurity testing enables evaluating the security levels of wireless communication systems in the face of potential cyber threats and attacks. Security researchers employ simulations to simulate and analyze diverse attack scenarios, encompassing jamming, eavesdropping, and data breaches. These simulations serve to devise and evaluate security systems, including encryption and intrusion detection. Using simulations is employed in vulnerability assessments to detect weaknesses within wireless networks and assess their capacity to withstand attacks. As an illustration, one potential use involves the simulation of distributed denial-of-service (DDoS) assaults on mobile networks to evaluate the network resilience against such security threats.

1.5.6 IoT AND SENSOR NETWORKS

Simulations are crucial in researching large-scale sensor deployments within the IoT and sensor networks. These tools aid in evaluating network performance, energy usage, and data routing schemes. Simulations play a crucial role in enhancing the efficiency of deploying wireless sensor nodes in precision agriculture for soil monitoring and in industrial settings for predictive maintenance. The applications underscore the essentiality of computational modeling and simulations as important tools in creating, optimizing, and operating wireless communication systems [57]. These tools empower researchers, engineers, and wireless service providers to address intricate difficulties, improve systems performance, and guarantee wireless networks' dependability and security within a swiftly advancing technological environment.

1.6 CHALLENGES AND LIMITATIONS

This section presents detailed challenges and limitations that underscore the complexity and nuances associated with computational modeling and simulations, emphasizing the need for careful consideration, validation, and interpretation of results in various scientific and engineering applications.

1.6.1 MODEL VALIDITY AND ACCURACY

Simplifications and assumptions are often used in studying real-world systems due to their inherent complexity, encompassing many factors and interactions. To render simulations computationally feasible, models frequently depend on simplifications and assumptions. In the context of fluid dynamics simulations, it is common practice to introduce simplifying assumptions regarding the compressibility and viscosity of the fluid. The simplifications mentioned above may not comprehensively encompass real-world phenomena complexities, resulting in potential disparities between simulated outcomes and observed data [58]. The accuracy of simulations is contingent upon the correctness of the input data. In disciplines like climate modeling, acquiring precise initial circumstances and procuring correct data for boundary conditions can provide significant challenges [59]. The presence of inaccurate data within a simulation can propagate errors, compromising the dependability of the obtained conclusions.

1.6.2 COMPUTATIONAL RESOURCES

The computational demands of simulating intricate systems or conducting high-resolution simulations frequently require significant processing power. Access to supercomputers or high-performance computing clusters is essential for conducting large-scale simulations, as commonly observed in astrophysics or molecular dynamics. Smaller entities and researchers may have constraints concerning computational resources and accessibility [51]. Similarly, some simulations exhibit a significant temporal investment, necessitating several hours, days, or weeks to complete. The elongated duration can hinder the expeditious process of experimentation and iteration required for specific applications, such as instantaneous decision-making in autonomous cars.

1.6.3 Model Validation

The process of validating a computer model encompasses comparing the model outputs with empirical data from real-world observations. However, locating suitable benchmarks or experimental data for validation might be a considerable obstacle, especially in new or specialized domains where complete data may be limited [60]. Overfitting is a phenomenon that can occur in models characterized by their ability to achieve high performance on the training data, but then failing to generalize effectively to novel, unknown instances. This phenomenon may result in the development of models that exhibit excessive complexity and excessive customization to the training dataset.

1.6.4 Complex Systems

Emergent behavior refers to the phenomenon observed in complex systems, such as ecosystems or social networks, where patterns and outcomes develop due to the interactions among numerous separate components. Predicting or elucidating such behavior inside a computational model presents challenges, as it may not be readily discernible only from the behavior of individual components. Nonlinear dynamics, for example, are systems characterized by nonlinear dynamics that can potentially display chaotic behavior, rendering long-term forecasts untrustworthy. Weather systems and turbulent fluid flows are instances in which little alterations in initial conditions can result in significantly divergent outcomes as time progresses.

1.6.5 Lack of Transparency and Validation Against Real-World Complexity

Black-box models refer to specific simulations incorporating machine learning or intricate algorithms, which can be opaque or inscrutable. This implies that although the model may yield precise outcomes, its internal mechanisms are not readily interpretable or explainable, constraining the comprehension of the fundamental processes involved. The complexity of real-world phenomena poses significant challenges when attempting to represent it accurately within a model. The task at hand is finding a harmonious equilibrium between the simplification of the model and the inclusion of sufficient complexity to ensure the simulation holds significance.

1.6.6 Uncertainty and Sensitivity

The propagation of uncertainty occurs when uncertainty is present in the model parameters or initial conditions, which propagates through the simulation process, resulting in uncertainty in the obtained findings. The management and quantification of uncertainty play a crucial role in enabling informed decision-making that relies on the outcomes of simulations. Conducting sensitivity analysis to ascertain the parameters that exert the most substantial influence on the outcomes can be a computationally demanding and intricate task. Identifying pivotal aspects that influence the model behavior is paramount in comprehending its behavior and enhancing its precision.

1.6.7 Ethical and Privacy Concerns

Using real-world data in simulations can potentially give rise to privacy problems, particularly within healthcare and social sciences. Effectively anonymizing and safeguarding sensitive data while preserving its usefulness for simulations might present significant challenges. Using simulations for decision-making purposes, particularly within domains such as autonomous vehicles or healthcare, can potentially give rise to ethical quandaries [61]. Achieving a harmonious equilibrium between safety, efficiency, and ethical issues is multifaceted and intricate.

1.7 FUTURE TRENDS AND LESSONS LEARNED

This section details the explanations that underscore each future trend's technical intricacies and significance in computational modeling and simulations within WCSs. They emphasize the role of simulations in addressing the complex challenges and opportunities presented by evolving wireless technologies below.

1.7.1 FUTURE TRENDS

These detailed technical futures highlight the intricacies of each future trend in computational modeling and simulations within wireless communication systems, emphasizing their importance in addressing the complexities and challenges of advanced wireless technologies.

1.7.1.1 Sixth Generation and Beyond

The advancement of 6G (sixth generation) networks signifies the forthcoming frontier in wireless communication systems. The utilization of computational modeling and simulations is expected to impact the development and implementation of 6G technology significantly. The simulations will primarily concentrate on frequencies in the millimeter-wave range, specifically investigating intricate propagation properties and optimizing antenna designs. The efficacy of advanced beamforming algorithms will be subjected to thorough evaluation through simulations, focusing on enhancing beam-steering procedures, particularly in the context of massive MIMO configurations [20]. Furthermore, simulations will investigate incorporating various communication technologies, such as terahertz and optical wireless communication, into 6G networks. This will guarantee uninterrupted and high-quality connectivity in forthcoming wireless systems.

1.7.1.2 Massive MIMO and Beamforming

MIMO systems and beamforming are poised to revolutionize wireless communication. The simulations will explore the complexities of these technologies, explicitly emphasizing adaptive beamforming techniques. The simulations will utilize real-time channel state information to optimize beamforming weights. This optimization will enable interference suppression, support for user mobility, and appropriate adaptability to the spatial characteristics of the wireless environment [21]. It is imperative to employ comprehensive modeling techniques to fully use the capabilities of massive MIMO and beamforming in improving network performance.

1.7.1.3 IoT and Edge Computing

The exponential rise of the IoT necessitates the development of simulations that can optimize the deployment of IoT networks. Simulations are a valuable tool for tackling intricate difficulties related to edge computing, encompassing the allocation of workloads, adherence to latency limitations, and optimization of energy use. Using simulations by researchers will enable the identification of the most effective positioning of edge servers and the implementation of dynamic resource allocation [22]. This will ensure IoT devices can process data efficiently at the network edge. The comprehensive analysis and modeling of these elements will be of utmost importance to achieve a smooth integration of the IoT into wireless communication systems.

1.7.1.4 Spectrum Sharing and Dynamic Spectrum Access

The optimization of radio frequency allocation is a paramount objective in the field of wireless communication. The simulations will prioritize examining and enhancing dynamic spectrum access (DSA) techniques. The simulations will assess the performance of DSA in practical situations, considering variables such as spectrum availability, the management of interference, and the coexistence of different wireless technologies within shared frequency bands [23]. Simulations have the potential to enable the adaptive and efficient utilization of existing spectrum resources by effectively addressing the intricacies associated with this matter.

1.7.1.5 Artificial Intelligence-enhanced Simulations

Integrating AI into simulation processes is expected to play a crucial role in boosting the adaptability and intelligence of those processes.. The utilization of AI-driven simulations has the potential to automate the process of parameter tweaking and model selection, resulting in enhanced convergence speed and decreased computing burden. In addition, generative AI models can produce synthetic data that may be used to enhance existing real-world datasets [24]. This augmentation process serves to enrich simulations and enhance their resilience. The integration of AI will facilitate the dynamic evolution of simulations, allowing them to adapt to the increasing needs of wireless communication systems.

1.7.1.6 Quantum Communications

Quantum communications signifies a state-of-the-art progression in the realm of wireless security. Simulations will play a crucial role in advancing and evaluating quantum communication protocols. These simulations aim to explore the complex dynamics of quantum entanglement by creating models that depict its transmission and interactions within quantum networks. The security and effectiveness of QKD methods will be subjected to thorough evaluation in diverse network scenarios [25]. In addition, the simulations will investigate the viability of quantum repeaters, which play a crucial role in the realization of quantum communication over great distances, and assess their influence on the network performance.

1.7.1.7 Network Slicing and Virtualization

Network slicing, a fundamental aspect of 5G and future generations of wireless communication, will be the subject of thorough investigation and analysis in simulation studies. Simulations might be utilized to optimize network slices to fulfill distinct application demands, including enhanced Mobile Broadband (eMBB), Ultra-Reliable Low Latency Communication (URLLC), and massive Machine Type Communication (mMTC) [26]. Thorough testing of advanced network slicing algorithms will be conducted to verify the ability of each slice to meet the required requirements in terms of bandwidth, latency, and reliability. This will facilitate the development of wireless networks that are incredibly flexible and efficient.

1.7.1.8 Resilience and Security

As the landscape of cybersecurity threats constantly evolves, simulations will be employed to replicate and analyze complex security scenarios. These simulations evaluate the effectiveness of quantum-safe encryption systems, threat detection mechanisms powered by artificial intelligence, and autonomous security actions. The assessment of wireless network resilience against quantum attacks and complex cyber threats will be conducted by researchers through the utilization of simulations [28]. The primary objective of these simulations will be to quantify the impact of the dangers and to create effective solutions.

1.7.1.9 Autonomous and Connected Vehicles

Detailed simulations will examine vehicular communication systems, which are important for connected and autonomous vehicles. The simulations will examine the relationship between Vehicles-to-Vehicle (V2V) and Vehicle-to-Everything (V2I) in intricate urban settings. Integrating real-time vehicle dynamics, sensor data, and network interactions within simulations aims to enhance safety, minimize latency, and improve dependability for autonomous vehicles. This integration is crucial for facilitating the seamless incorporation of autonomous vehicles into future transportation systems.

1.7.2 Lessons Learned

Some lessons learned from the chapter are presented as recommendations for researchers and stakeholders.

1.7.2.1 Role of Simulation as a Vital Tool

Recently, there has been a growing recognition of the pivotal role played by computational modeling and simulations in driving advancements in wireless communication systems. These tools provide a regulated and cost-efficient setting for investigating intricate scenarios, assessing novel technologies, and enhancing network performance. It has been observed by researchers and engineers that simulations are not merely supplementary to real-world testing but rather are essential elements inside the wireless technology development cycle. Sandbox environments are established to foster innovation by facilitating the swift iteration of ideas and the practical evaluation of their feasibility prior to resource-intensive real-world deployment. Simulation has emerged as a fundamental component in the repertoire of techniques to design and optimize wireless communication systems.

1.7.2.2 Importance of Realistic Models

The significance of realistic models in wireless communication simulations has been emphasized by the lessons that have been learned. The utilization of tractability in simulations can enhance computational tractability; nonetheless, it is essential to note that straightforward models have the potential to yield deceptive outcomes. Accurate representations of real-world conditions, encompassing channel characteristics, device behavior, and traffic patterns, are essential for simulations to yield practical and applicable insights. Scholars have acknowledged the necessity of achieving a harmonious equilibrium between the intricacy of a model and its computing efficiency. This is crucial to ensure that the simulation accurately encompasses the fundamental attributes of the wireless environment while being feasible and easily comprehensible.

1.7.2.3 Validation is Critical

The significance of validation in wireless communication simulations has been firmly ingrained in the field. Regardless of their level of sophistication, simulations are not devoid of fallibility. To foster confidence in the accuracy of simulation outcomes, it is imperative to conduct thorough validation procedures that involve rigorous comparison with empirical data. This procedure entails comparing simulation results with empirical measurements verifying the WSC model to replicate observed occurrences accurately. The validation process is not a singular occurrence but rather a continuous endeavor to enhance the precision and fidelity of simulations. Researchers and practitioners can enhance their trust in the dependability of simulation-based judgments and designs by ensuring that the outputs of simulations are aligned with real-world observations.

1.7.2.4 Adaptability to Emerging Technologies

The continuous development of wireless communication technology has demonstrated the importance of being able to adjust simulations. With the increasing prevalence of 5G networks and the anticipation of 6G, the IoT, and quantum communication, there is a need to advance simulations to accommodate these coming technologies. Researchers have recognized that the design of simulations should prioritize flexibility and extensibility, enabling them to adapt to the intricate and complicated nature of forthcoming wireless systems. The capacity to adapt ensures that simulations continue to be pertinent and efficacious in influencing future wireless networks.

1.7.2.5 Optimization and Efficiency

Incorporating lessons in optimization and efficiency has significantly transformed the landscape of wireless communication simulators. Simulations have evolved beyond their traditional role as exploratory tools and have become potent optimizers. Academics have widely adopted iterative simulation-based optimization procedures, which involve refining network parameters, protocols, and algorithms. The utilization of these simulations leads to increased efficiency, resulting in enhanced network capacity, decreased latency, and improved reliability. The observation mentioned earlier is evident: simulations possess analytical value and play a crucial role in attaining optimal wireless network performance.

1.7.2.6 Spectrum Management and Interference

Spectrum management and interference mitigation principles have become indispensable in wireless communication simulators. In a future where the spectrum is limited and holds significant value, simulations are critical in evaluating and enhancing methods designed for sharing spectrum resources. DSA algorithms are evaluated to achieve optimal spectrum usage, minimize interference, and facilitate coexistence among various wireless technologies. The simulations have provided valuable insights into the potential for enhanced adaptability and efficiency in utilizing the wireless spectrum. This lesson holds excellent value in a society that is becoming increasingly interconnected.

1.7.2.7 Security and Resilience Testing

As cybersecurity threats continue to evolve, lessons in security and resilience have prompted a renewed focus on simulations. Simulations are the battleground for testing security mechanisms and assessing a resilient network against cyber threats, including quantum attacks. Lessons learned here underscore the need to identify vulnerabilities, validate security protocols, and develop robust countermeasures in the simulated realm before they are deployed in real-world networks. Simulation-based testing of security and resilience is the foundation upon which trustworthy wireless communication systems are built.

1.7.2.8 Ethical Considerations

The wireless communication community has learned the importance of ethical considerations in using simulations. Privacy concerns, data anonymization, and ethical dilemmas related to decision-making in autonomous systems have become paramount. Lessons have highlighted the need to balance technological advancement and ethical responsibility. Simulations must be technically sound and adhere to ethical principles, ensuring that advancements in wireless communication are made responsibly and with respect for the rights and values of individuals and societies.

1.7.2.9 Collaboration and Interdisciplinary Approaches

Lessons have demonstrated that effective simulation-driven research often requires collaboration between experts from diverse fields. Mathematics, computer science, telecommunications, and physics converge in developing simulations for wireless communication. These interdisciplinary approaches harness the collective expertise needed to address complex challenges. Researchers have come to appreciate that the fusion of knowledge from different domains is beneficial and often necessary to tackle the multifaceted issues presented by wireless communication systems.

1.8 CONCLUSION AND FUTURE SCOPE

Simulations and computational modeling in wireless communication systems have become essential engines of innovation and optimization. They reflect real-world conditions, balancing complexity and efficiency. Simulations have evolved from analytical instruments to optimizers, driving efficiency gains, network capacity enhancement, and latency reduction. They are crucial in spectrum management, interference mitigation, security, and resilience. Ethical considerations are also important, with interdisciplinary collaboration addressing complex wireless challenges. AR and VR applications prioritize low-latency communication, ensuring seamless user experiences. These lessons guide stakeholders in delivering efficient, secure, ethical, and user-centric wireless technologies for a dynamic future. Security and resilience lessons underscore the need for simulations to identify vulnerabilities, validate security protocols, and develop robust countermeasures in an ever-evolving threat landscape. Ethical considerations remind us to uphold ethical simulation usage principles, safeguard privacy, and promote responsible technological advancement. Interdisciplinary collaboration has highlighted the value of converging expertise from various fields to address complex wireless challenges.

REFERENCES

[1] T. Shiina, and K. Yamazaki, "Numerical modeling and assessment of human exposure to radiofrequency electromagnetic fields used for wireless communication system installed on a smart meter," *Electronics and Communications in Japan,* vol. 103, no. 9, 2020. https://doi.org/10.1002/ecj.12257

[2] T. Nagao, and T. Hayashi, "Geographical clustering of path loss modeling for wireless emulation in various environments," 2022 16th European Conference on Antennas and Propagation, EuCAP 2022. https://doi.org/10.23919/eucap53622.2022.9769198

[3] W. Shafik, S. M. Matinkhah, and F. Shokoor, "Cybersecurity in unmanned aerial vehicles: A review," *International Journal on Smart Sensing and Intelligent Systems*, vol. 16, no. 1, 2023. https://doi.org/ 10.2478/ijssis-2023-0012

[4] D. Sun, J. Zhao, B. Chen, H. Wu, and J. Wu, "Cross-level dependability assessment with a distributed split mechanism for wireless communication systems," *IEEE Transactions on Network Science and Engineering,* 2022, https://doi.org/10.1109/TNSE.2022.3194535

[5] S. Liu, G. Yu, R. Yin, J. Yuan, and F. Qu, "Communication and computation efficient federated learning for internet of vehicles with a constrained latency," *IEEE Transaction on Vehicle Technology,* 2023. https://doi.org/10.1109/TVT.2023.3309088.

[6] D. W. Matolak, "Channel modeling for vehicle-to-vehicle communications," *IEEE Communications Magazine*, vol. 46, no. 5, 2008. https://doi.org/10.1109/MCOM.2008.4511653

[7] S. A. Khan, R. Kumar, O. Kaiwartya, R. A. Khan, and M. Faisal, Computational intelligent security in wireless communications. 2022. https://doi.org/10.1201/9781003323426

[8] D. K. Verma, C. Verma, and J. Aslam, Computational modelling and simulations for designing of corrosion inhibitors: Fundamentals and realistic applications. 2023. https://doi.org/10.1016/ C2021-0-02414-7

[9] F. P. Morais, and J. M. R. Curto, "Challenges in computational materials modelling and simulation: A case-study to predict tissue paper properties," *Heliyon*, vol. 8, no. 5, 2022. https://doi.org/10.1016/ j.heliyon.2022.e09356

[10] J. G. Kim, J. H. Lim, and P. Persson, Computational Modeling and Simulation of Solids and Structures: Recent Advances and Practical Applications. (p. 196). MDPI-Multidisciplinary Digital Publishing Institute. 2023. https://doi.org/10.3390/books978-3-0365-7749-4

[11] M. Goudarzi, D. Grazioli, and A. Simone, "An efficient computational approach for three-dimensional modeling and simulation of fibrous battery electrodes," *International Journal for Numerical Methods in Engineering,* vol. 123, no. 7, 2022. https://doi.org/10.1002/nme.6881

[12] N. D. Filipovic, Computational modeling and simulation examples in bioengineering. 2021. https:// doi.org/10.1002/9781119563983

[13] Y. Jun et al., "Artificial intelligence application in cybersecurity and cyberdefense," *Wireless Communications and Mobile Computing,* 2021. https://doi.org/10.1155/2021/3329581

[14] L. Zhao et al., "Artificial intelligence analysis in cyber domain: A review," *International Journal of Distributed Sensor Networks*, vol. 18, no. 4. 2022. https://doi.org/10.1177/15501329221084882

[15] W. Shafik, "Making cities smarter: IoT and SDN applications, challenges, and future trends," in Opportunities and Challenges of Industrial IoT in 5G and 6G Networks, 2023. https://doi.org/10.4018/ 978-1-7998-9266-3.ch004

[16] W. Shafik, "Cyber security perspectives in public spaces: Drone case study," *in Handbook of Research on Cybersecurity Risk in Contemporary Business Systems,* 2023. https://doi.org/10.4018/978-1-6684-7207-1.ch004

[17] W. Shafik, "A Comprehensive Cybersecurity Framework for Present and Future Global Information Technology Organizations," 2023. https://doi.org/10.4018/978-1-6684-9018-1.ch002

[18] W. Shafik, S. M. Matinkhah, and F. Shokoor, "Recommendation system comparative analysis: Internet of Things aided networks," *EAI Endorsed Transactions on Internet of Things,* vol. 8, no. 29, 2022. https://doi.org/10.4108/eetiot.v8i29.1108

[19] W. Shafik, S. M. Matinkhah, F. Shokoor, and M. Nur Sanda, "Internet of things-based energy efficiency optimization model in fog smart cities," *International Journal on Informatics Visualization,* vol. 5, no. 2, 2021. https://doi.org/10.30630/joiv.5.2.373

[20] D. He, B. Ai, K. Guan, L. Wang, Z. Zhong, and T. Kürner, "The design and applications of high-performance ray-tracing simulation platform for 5G and beyond wireless communications: A tutorial," *IEEE Communications Surveys and Tutorials,* vol. 21, no. 1. 2019. https://doi.org/10.1109/COMST.2018.2865724

[21] J. Weng, X. Tu, Z. Lai, S. Salous, and J. Zhang, "Indoor Massive MIMO channel modelling using ray-launching simulation," *International Journal of Antennas Propagation,* vol. 2014. 2014. https://doi.org/10.1155/2014/279380

[22] S. Hao, Y. Hong, and Y. He, "An energy-efficient routing algorithm based on greedy strategy for energy harvesting wireless sensor networks," *Sensors,* vol. 22, no. 4, 2022. https://doi.org/10.3390/s22041645

[23] Y. Guo, F. R. Yu, J. An, K. Yang, C. Yu, and V. C. M. Leung, "Adaptive Bitrate streaming in wireless networks with transcoding at network edge using deep reinforcement learning," *IEEE Transaction on Vehicle Technology,* vol. 69, no. 4, 2020. https://doi.org/10.1109/TVT.2020.2968498

[24] M. Nekovee, and R. S. Saksena, "Simulations of large-scale WiFi-based wireless networks: Interdisciplinary challenges and applications," *Future Generation Computer Systems,* vol. 26, no. 3, 2010. https://doi.org/10.1016/j.future.2008.05.007

[25] Food and Drug Administration (FDA), "Assessing the credibility of computational modeling and simulation in medical device submissions," *Draft Guidance,* no. 301, 2021, www.regulations.gov/docket/FDA-2021-D-0980

[26] R. Lesage, M. Van Oudheusden, S. Schievano, I. Van Hoyweghen, L. Geris, and C. Capelli, "Mapping the use of computational modelling and simulation in clinics: A survey," *Frontiers in Medical Technology,* vol. 5, 2023, https://doi.org/10.3389/fmedt.2023.1125524

[27] R. A. Mangkuto, T. P. Simamora, D. P. Pratiwi, and M. D. Koerniawan, "Computational modelling and simulation to mitigate the risk of daylight exposure in tropical museum buildings," *Energy and Built Environment,* 2022, https://doi.org/10.1016/j.enbenv.2022.09.001

[28] M. C. Davis, H. P. N. Hughes, A. McKay, M. A. Robinson, and C. N. van der Wal, "Ergonomists as designers: computational modelling and simulation of complex socio-technical systems," *Ergonomics,* vol. 63, no. 8, 2020. https://doi.org/10.1080/00140139.2019.1682186

[29] J. Fine, M. J. McShane, G. L. Coté, and C. G. Scully, "A computational modeling and simulation workflow to investigate the impact of patient-specific and device factors on hemodynamic measurements from non-invasive photoplethysmography," *Biosensors,* vol. 12, no. 8, 2022. https://doi.org/10.3390/bios12080598

[30] R. Baggio, "Computational modelling and simulations in tourism: A primer," *Annals of Tourism Research Empirical Insights,* vol. 1, no. 1, 2020. https://doi.org/10.1016/j.annale.2020.100005

[31] A. Mourato et al., "Computational modelling and simulation of fluid structure interaction in Aortic Aneurysms: A systematic review and discussion of the clinical potential," *Applied Sciences,* vol. 12, no. 16. 2022. https://doi.org/10.3390/app12168049

[32] M. Maghsoudi-Ganjeh, L. Lin, X. Yang, and X. Zeng, "Computational modeling and simulation of bioinspired nacre-like composites," *Journal of Material Research,* vol. 36, no. 13, 2021. https://doi.org/10.1557/s43578-021-00124-6

[33] K. N. Aronis, R. L. Ali, J. A. Liang, S. Zhou, and N. A. Trayanova, "Understanding AF mechanisms through computational modelling and simulations," *Arrhythm Electrophysiol Rev,* vol. 8, no. 3, 2019. https://doi.org/10.15420/aer.2019.28.2

[34] A. J. Magana, H. W. Fennell, C. Vieira, and M. L. Falk, "Characterizing the interplay of cognitive and metacognitive knowledge in computational modeling and simulation practices," *Journal of Engineering Education,* vol. 108, no. 2, 2019. https://doi.org/10.1002/jee.20264

[35] E. Vargas et al., "An emerging consensus on voltage-dependent gating from computational modeling and molecular dynamics simulations," *Journal of General Physiology*, vol. 140, no. 6, 2012. https://doi.org/10.1085/jgp.201210873

[36] K. Kossenas, S. K. Podilchak, and M. Beveridge, "Microwave system development for wireless communications inside oil and gas well pipelines," *IEEE Journal of Microwaves,* vol. 3, no. 2, pp. 553–69, 2023. https://doi.org/10.1109/JMW.2022.3232032

[37] U. Panahi, and C. Bayılmış, "Enabling secure data transmission for wireless sensor networks based IoT applications," *Ain Shams Engineering Journal,* vol. 14, no. 2, pp. 101866, 2023. https://doi.org/10.1016/j.asej.2022.101866

[38] M. A. Găină, M. Popescu, R. A. Dobre, and D. N. Vizireanu, "Traffic sniffing in wireless communication, using Kali Linux and Wireshark," In *Advanced Topics in Optoelectronics, Microelectronics, and Nanotechnologies,* vol. 12493, pp. 678–683, 2023, SPIE. https://doi.org/10.1117/12.2643055

[39] K. Sathish, M. Hamdi, R. Chinthaginjala, G. Pau, and A. Ksibi et al., "Data transmission in underwater wireless sensor networks using a cluster-based routing protocol endorsed by member nodes," *Electronics,* vol. 12, no. 6, pp. 1287, 2023. https://doi.org/10.3390/electronics12061287

[40] M. K. Baazaoui, I. Ketata, A. Fakhfakh, and F. Derbel, "Modeling of packet error rate distribution based on received signal strength indications in OMNeT++ for Wake-Up Receivers," *Sensors,* vol. 23, no. 5, pp. 2394, 2023. https://doi.org/10.3390/s23052394

[41] E. Gambi, L. Senigagliesi, A. Barbaresi, M. Mellini, and A. De Santis, "A WKNN-based approach for NB-IoT sensors localization," *Digital Communications and Networks,* vol. 9, no. 1, pp. 175–82, 2023. https://doi.org/10.1016/j.dcan.2022.04.033

[42] V. Muttillo, L. Pomante, M. Santic, and G. Valente, "System C-based co-simulation/analysis for system-level hardware/software co-design," *Computers and Electrical Engineering,* vol. 110, pp. 108803, 2023. https://doi.org/10.1016/j.compeleceng.2023.108803

[43] M. P. Angert et al., "Frontier radio – multi-lingual – a next generation space software defined radio," *IEEE Aerospace Conference, Big Sky, MT, USA,* 2023, pp. 1–12, 2023. https://doi.org/10.1109/AERO55745.2023.10115782

[44] Ó. Seijo, I. Val, and J. A. López-Fernández, "Portable full channel sounder for industrial wireless applications with mobility by using sub-nanosecond wireless time synchronization," *IEEE Access,* vol. 8, pp. 175576–175588, 2020. https://doi.org/10.1109/ACCESS.2020.3025896

[45] T. Fang, Y. Gao, C. Suo, G. Sun, and P. Chen et al., "A multi-beam XL-MIMO testbed based on hybrid CPU-FPGA architecture," *Electronics,* vol. 12, no. 2, pp. 380, 2023. https://doi.org/10.3390/electronics12020380

[46] Y. Liu, X. Li, H. Gong, Z. Guo, and C. Zhang, "Analysis of the potential fading mechanism of sweet cherry after freezing and thawing using untargeted metabolomics," *LWT,* vol. 178, pp. 114633, 2023. https://doi.org/10.1016/j.lwt.2023.114633

[47] R. Michev, Y. Shu, D. Werbunat, J. Hasch, and C. Waldschmidt, "Adaptive compensation of hardware impairments in digitally modulated radars using ML-based behavioral models," *IEEE Transactions on Microwave Theory and Techniques,* pp. 1–14, 2023. https://doi.org/10.1109/TMTT.2023.3285438

[48] S. Masood et al., "Detecting and preventing false nodes and messages in vehicular ad-hoc networking (VANET)," *IEEE Access,* vol. 11, pp. 93920–93934, 2023. https://doi.org/10.1109/ACCESS.2023.3308035

[49] J. Shen, W. Zhang, C. Wu, X. Wu, E. Fang, and X. Liu, "A compact duplexing rectenna with wireless power transfer and harmonic feedback capabilities," *IEEE Antennas and Wireless Propagation Letters,* vol. 22, no. 7, pp. 1751–1755, 2023. https://doi.org/10.1109/LAWP.2023.3262706

[50] D. M. Mielke, D. Becker, and M. Walter, "Path-based statistical modeling of multipath components in propagation channels for wireless communications in unmanned aviation," *2023 Integrated Communication, Navigation and Surveillance Conference (ICNS), Herndon, VA, USA,* pp. 1–8, 2023. https://doi.org/10.1109/ICNS58246.2023.10124256

[51] O. Yasar, and J. Maliekal, "Computational pedagogy: A modeling and simulation approach," *Computing in Science & Engineering,* vol. 16, no. 3, 2014. https://doi.org/10.1109/MCSE.2014.60

[52] S. Buzura, A. Peculea, B. Iancu, E. Cebuc, V. Dadarlat, and R. Kovacs, "A hybrid software and hardware sdn simulation testbed. *Sensors,* vol. 23, no. 1, pp. 490, 2023. https://doi.org/10.3390/s23010490

[53] A. Vijayan et al., "5G Wireless channel characterization in indoor factory environments: simulation and validation," *2023 IEEE Wireless Communications and Networking Conference (WCNC), Glasgow, United Kingdom,* pp. 1–6, 2023. https://doi.org/10.1109/WCNC55385.2023.10118661

[54] V Rajavel, and D. Ghoshal "A compact triband antenna using artificial magnetic conductor for wireless body area network communications Front," *Wireless Networks,* vol. 28, pp. 1–23, 2023. https://doi.org/10.1007/s11276-023-03354-0

[55] O. Laudatu, M. Iordache, M. Stanculescu, L. Bobaru, D. Niculae, and O. Drosu, "Wireless power transmission system design. A practical approach," *2023 13th International Symposium on Advanced Topics in Electrical Engineering (ATEE), Bucharest, Romania,* pp. 1–4, 2023, https://doi.org/10.1109/ATEE58038.2023.10108155

[56] G. Ceder, M. Doyle, P. Arora, and Y. Fuentes, "Computational modeling and simulation for rechargeable batteries," *MRS Bull,* vol. 27, no. 8, 2002. https://doi.org/10.1557/mrs2002.198

[57] M. Green, Y. N. G. Hong, J. Roh, and B. J. Fregly, "Computational modeling and simulation of closed chain arm-robot multibody dynamic systems in OpenSim," *Multibody System Dynamics,* vol. 56, no. 4, 2022. https://doi.org/10.1007/s11044-022-09847-8

[58] A. K. Kandy, K. Rossi, A. Raulin-Foissac, G. Laurens, and J. Lam, "Comparing transferability in neural network approaches and linear models for machine-learning interaction potentials," *Physical Review B*, vol. 107, no. 17, pp. 174106, 2023. https://doi.org/10.1103/PhysRevB.107.174106

[59] X. S. Yang, S. Koziel, and L. Leifsson, "Computational optimization, modelling and simulation: Recent trends and challenges," *Procedia Computer Science,* 2013. https://doi.org/10.1016/j.procs.2013.05.250

[60] M. B. Kivi, Y. Hong, and M. A. Zaeem, "A review of multi-scale computational modeling tools for predicting structures and properties of multi-principal element alloys," *Metals,* vol. 9, no. 2. 2019. https://doi.org/10.3390/met9020254

[61] F. Shokoor, W. Shafik, and S. M. Matinkhah, "Overview of 5G & beyond security," *EAI Endorsed Transactions on Internet of Things,* vol. 8, no. 30, 2022. https://doi.org/10.4108/eetiot.v8i30.1624

2 Historical Perspectives on Computational Modeling and Simulation in Wireless Communication Systems

Abdulwaheed Musa and Abdullateef Ola Adebayo

2.1 INTRODUCTION

The rapid evolution of wireless communication technologies over the past century has been enabled by significant advances in computational modeling and simulation techniques. As wireless systems have grown more complex, sophisticated computer-based modeling approaches have become indispensable for designing, analyzing, and optimizing networks [1].

Simulation provides a virtual prototyping environment to evaluate system performance under different conditions without costly real-world deployment [2]. Computational modeling offers mathematical abstractions to study the underlying theoretical principles governing radio wave propagation, network protocols, traffic patterns, and other phenomena. Together, computational modeling and simulation have become deeply intertwined with discoveries in the wireless domain [3].

This chapter explores the historical development of computational modeling methods applied to wireless communications. By studying the key milestones and pioneering innovations that shaped this interdisciplinary field, we can gain insight into the technological trajectories that led to present-day solutions.

The origins of wireless technologies can be traced back to 19th-century experiments in electromagnetism, the development of the theoretical foundations in signals and systems, and early analog computing [4]. The advent of digital computers and programming languages catalyzed more sophisticated models of radio wave propagation and cellular network standards. Recent exponential advances in software, hardware, and algorithms have enabled high-fidelity emulation of intricate real-world environments.

2.1.1 CONTRIBUTION

The key contributions of this chapter are as follows:

i. *Unified technical evolution timeline linking wireless milestones with enabling modeling innovations.* Covered in Sections 2.2 through 2.5, which present a comprehensive timeline tracing the parallel advancement of key wireless communication milestones and innovations in modeling/simulation techniques. For example, Section 2.2 links early work in electromagnetism to the origins of wireless modeling philosophies, while Section 2.5 ties together multi-generational case studies focused on radio propagation and cellular system modeling, contrasting techniques across generations.

DOI: 10.1201/9781003457428-3

ii. *Highlight of analog-digital shift as a seminal disruptive paradigm modernization.* Section 2.3 explicitly explores the progression from mechanical to electronic analog computation, discussing how the disruptive onset of digital computers sparked a revolution in scientific computing capabilities that displaced established analog approaches and reshaped the role of simulations.

iii. *Synthesis of integrated historical trajectories for holistic appreciation.* This chapter on structuring and content flow presents a unified timeline synthesizing an integrated historical trajectory tying together the salient innovations in wireless communication systems and modeling/simulation tools for a holistic perspective.

iv. *Showcasing of sophistication progression by contrasting recent and early techniques.* Section 2.5 provides illustrative case studies focused specifically on showcasing the progression in radio propagation models and cellular network modeling sophistication by qualitatively and quantitatively contrasting recent state-of-the-art vs earlier primitive methods.

v. *Elucidating of scientific foundations underlying modern infrastructure powering connectivity.* Sections 2.2 and 2.4 focus on elucidating the pioneering experiments and theories that laid the bedrock for contemporary wireless infrastructure, while Section 2.5 contrasts multi-generational techniques to showcase how modern approaches build firmly upon these scientific foundations.

vi. *Analyses interplay between theories, technologies, and analyses driving progress.* This key objective driving this chapter is realized across the entire narrative tracing the parallel intertwined progression of theoretical breakthroughs, concrete technological innovations, and advances in analytical capabilities that collectively contributed to driving wireless communication progress over generations. For example, Section 2.2 highlights the vital linkage between Maxwell's theoretical unification of electromagnetism to Hertz's experimental rig validating predictions that would spur Marconi's early radio prototypes demonstrating practical functionality. Similarly, Section 2.5 explores multi-generational case studies focused specifically on contextualizing the advancement of radio modeling techniques and cellular network architectures as the interplay between evolving simulation platforms, wireless communication technologies, deployment needs, and research insights.

2.1.2 CHAPTER ORGANIZATION

The rest of this chapter is structured as follows. Section 2.2 covers fundamental theories and early experimentations establishing the imperative for analytical modeling. Section 2.3 discusses the progression of computational tools from mechanical to electronic eras enabling simulations. Section 2.4 explores recent wireless-focused innovations in computational capabilities. Section 2.5 ties concepts together through multi-generational case studies focused on radio modeling and cellular systems analysis. Section 2.6 captures the lessons learned, while Section 2.7 concludes with a summary, recommendations, and directions building on the continuum of engineering ingenuity across generations.

2.2 EARLY DISCOVERIES IN WIRELESS TECHNOLOGY AND COMPUTATION

This section provides an overview of the pioneering work on electromagnetism, radio wave transmission, electronic amplifiers, and foundational mathematics that laid the basis for wireless communications and computational modeling techniques that are still in use today. Understanding these early discoveries is key to appreciating the origins of modern wireless systems.

2.2.1 Overview of Pioneering Work in Electromagnetism

The nineteenth-century origins of wireless technology established key imperatives that would spawn the practices of computational modeling and simulation that evolved later. Faraday's empirical methodology demonstrated that wireless forces obey inherent mathematical relationships governable by deterministic laws waiting to be uncovered through experimentation [5]. This finding gave birth to modeling philosophies centered on quantifying dynamics through reproducible equations.

Maxwell's breakthrough theoretical unification then directly fulfilled Faraday's vision by elegantly encapsulating electromagnetic interactions into computable expressions. His differential equations enabled numeric simulation of wave propagation by completely specifying constituent interdependencies [6]. Hertz's rig further cemented the relevance of Maxwell's abstractions by manifesting them in physical hardware realizations that could replicate predicted phenomena. This affirmed simulation viability by exhibiting aligning conceptual models with tangible set-ups.

Collectively, their interlinked contributions revealed wireless systems that intrinsically lend themselves to deterministically constructed simulations mirroring real-world interactions. The seeds for mathematically emulating physical dynamics through software models have been planted [7]. Thus, the origins of wireless modeling and simulation as engineering practices trace back to these pioneering works that established key techniques of quantifying dynamics through scientific modeling and physical measurement. By elucidating governing laws expressed numerically, they set the stage for computer-based analytical approaches.

2.2.2 Marconi's Experiments on Radio Waves for Wireless Telegraphy

Building directly upon the wireless transmission breakthroughs of Hertz, the ingenious Italian inventor, Guglielmo Marconi, successfully transmitted and received the first radio telegraphic messages across distances up to 1.5 miles in 1895, using a primitive spark-gap rig. Rather than focusing on the theoretical annotations governing wireless propagation, Marconi's research methodology centered on building end-to-end functionality by empirically configuring black-box components like oscillators and wave detectors. Yet, his prototypes inherently manifested the same system-level modeling notions that would later become commonplace [8].

By realizing wireless transmission through carefully cascaded building blocks of spark signal generators, antennae for emission and reception, and coherer detectors, Marconi's work echoed modeling philosophies relying on modular abstractions of functional elements. His configurations exhibited key concepts like amplitude modulation, amplification, demodulation, and detection by methodically combining components [9].

This empirically validated modularity of reliably combining stages for an overall wirelessly connected system set the stage for modular modeling architectures built by isolating the mathematical interactions between constituent blocks. By demonstrating successful end-to-end signal transfer through orchestrated components, Marconi's prototypes affirmed the potential for computational models relying on functional black boxes linked via standardized interfaces [9].

The modeling visions seeded by Marconi's experimental rigs would blossom into full-fledged wireless system simulations over the coming decades as computing techniques matured in tandem. But the early visions took root in his pioneering radio wave transmission apparatus.

2.2.3 Vacuum Tube and Early Electronic Amplifiers and Oscillators

The early twentieth century engendered critical innovations in electronic amplifier and oscillator contraptions, which would serve as indispensable building blocks for modeling wireless communication systems. John Ambrose Fleming's two-electrode vacuum tube diode, called the Fleming valve, invented in 1904, allowed current to only pass in one direction, enabling reliable envelope detection functionality. Reginald Fessenden then significantly advanced radio communications by conceiving the three-electrode Audion triode amplifier tube in 1906, which allowed reliable signal reception through voltage amplification by harnessing cathode-plate electron flows [10].

Such electronic amplifier modules using vacuum tubes offered, for the first time, reliable and reproducible off-the-shelf components that performed essential functions such as modulation, detection, or amplification, predictably. Their parameters such as voltage gains, input-output power ranges, and so forth, were reliably quantified [11].

This meant that system designers could now incorporate trusted representations of these elements within higher-level system models to simulate wireless channels by modeling the individual blocks using parameterized mathematical relationships. The amplifier modules served as tangible realisations of conceptual functional blocks, enabling modular abstractions.

By providing versatile, reproducible electronic hardware artifacts that offer standardized functionality, vacuum tube circuits firmly seeded perspectives and frameworks for mathematically modeling wireless systems as interconnections of quantifiable blocks, ushering in circuit-level modeling and simulation eras.

2.2.4 FOUNDATIONAL MATHEMATICS BRIDGING THEORY WITH COMPUTATION ON SIGNAL ANALYSIS AND TRANSMISSION

Alongside the pivotal advancements in physical experimentation, the mathematical theories underpinning signal analysis and processing were developed by pioneers like Fourier, Nyquist and Shannon in the nineteenth and twentieth centuries [12]. These established vital scientific bridges that would link conceptual theories to software-based computational simulation and modeling architectures.

The work of French mathematician, Baron Jean Baptiste Fourier, demonstrated that any periodic waveform could be expressed as a sum of sine and cosine waves, using the elegant Fourier Series formulations. This enabled complex time-domain signals to be decomposed into constituent frequencies, forming the foundations for spectral analysis critical in wireless communications.

Harry Nyquist's subsequent findings on sampling theorems cemented minimum sampling rates required for distortion-free signal reconstruction, redundancies, and capacity. Claude Shannon later essentially launched the discipline of information theory through his 1948 papers quantifying the maximum error-free channel capacity limits in noisy environments.

Together, these mathematical theories formally developed pivotal software machinery including spectral decomposition, sampling, source coding, and channel capacity encoding that would subsequently become integral components of any wireless network simulations [13]. By introducing algorithmic building blocks synthesizing real-world communications functionality, these breakthroughs paved the way for flexible software application development and simulation-driven wireless system designs.

2.3 EVOLUTION OF ANALOG COMPUTATION

This section traces the development of analog computation from early mechanical calculating devices to electronic analog computers, showing their importance as precursors to digital simulation methods. Examining this progression sheds light on foundational ideas that persist even in today's digital paradigms.

2.3.1 MECHANICAL CALCULATORS AS EARLIEST MODELING TOOLS

The earliest known computational modeling tools were mechanical arithmetic calculators like the Antikythera Mechanism (100 BCE) used for astrological predictions, and Blaise Pascal's

Pascaline (1642 CE) for tax tabulations. These automated tedious numeric calculations using precision gear trains and lever mechanisms governed by the mathematical relationships they encoded.

Handcranks enabled iterative counting while gears ensured carries across orders. But rigid metal constructions only offered fixed functions with minimal versatility [14]. These rudimentary machines however inaugurated notions of emulating real-world dynamics computationally using physical components obeying mathematical models. But wireless communications involve continuous signal propagation, necessitating more advanced analog modeling techniques. Electromechanical Analyzers for Equation Solving were introduced in the early 1900s. General-purpose differential analyzers were developed that could solve complex differential equations and mathematically model dynamical systems electromechanically [6].

These consisted of integration wheels, discs, differentiators, and mechanical linkages arranged to emulate target equation terms. Accuracy was limited by manufacturing precision and equation conversion heuristics. The typical solution time for a second-order ordinary differential equations was 15 mins. Nonetheless, such analyzers enabled acceptable modeling.

While limited for wireless systems involving high-frequency carriers and signals, electromechanical analyzers demonstrated the applicability of custom analog computers built to mimic target dynamical phenomena using appropriate mathematical transformations. This philosophy persisted in subsequent electronic analog eras.

2.3.2 Electronic Analog Computers for Wireless System Modeling

By the 1930s, general purpose electronic analog computers had superseded earlier mechanical calculators and differential analyzers for dynamically modeling complex systems by adopting vacuum tube based electronic amplifiers, integrators, multipliers, and comparators to emulate required mathematical operations.

These allowed the real-time simulation of Ordinary Differential Equations (ODEs) and differential-integral equations governing system behaviours using relatively simpler reconfigurable modular hardware units. Interconnected electronic circuits represented mathematical variables and operations [15]. Electrical signals and voltages encoded continuous magnitudes. Stability and accuracy exceeded mechanical artifacts.

For modeling wireless communications, the inherent continuity and parallelism of analog electronics was well-suited to simulating the dynamics of linear time-invariant systems typically used for radio frequency components like oscillators and filters. Sinusoidal voltage-controlled oscillators readily modeled high-frequency carriers. Also, summing amplifiers enabled the superposition of signals. Modulation/demodulation operations were realizable using transistor-based mixers and product detectors.

However, limitations included inflexibility in testing different channel configurations due to fixed wiring, limited precision from component tolerances, signal distortion and drift over time. The maximum complexity was also restricted by the electronic hardware combinatorial limitations. Verifying and validating models was constrained by one-to-one correspondence between model and real-world set-up. The prowess of electronic analog computers in emulating deterministic systems by constructing electronic circuits matching conceptual block diagrams was harnessed in multiple domains until digital computers ultimately disrupted modeling and simulation paradigms.

Table 2.1 shows the key highlights in improvements in computational speeds, accuracy, flexibility, and complexity handling from the shift to electronic analog computers while retaining the continuous-time signal processing capabilities. The constraints moved from physical rigidity to issues around interconnect scaling.

TABLE 2.1
Comparison of Capabilities and Performance between Mechanical and Electronic Analog Computers

Parameter	Mechanical Analog Computers	Electronic Analog Computers
Capabilities	• Could solve complex differential equations and mathematically model dynamical systems using shafts, gears, integrators, and mechanical linkages • Limited to modeling continuous deterministic systems with a few state variables • Equation conversion to mechanical system required manual mapping heuristics	• Can simulate a wider range of system behaviours and dynamics using electronic circuits with amplifiers, integrators etc. • More flexible – circuits reconfigured by changing connections • Can include stochastic elements by incorporating noise sources
Accuracy	• Precision limited by manufacturing tolerances in gears and disc mechanisms • Typical accuracy of 1-0.1% full scale reading • Suffers from mechanical friction and backlash	• Accuracy limited by electronic component tolerances • Typical accuracy of 0.1-0.02% for high precision models • Vulnerable to thermal drifts and nonlinearities
Solution Times	• Around 15 mins required to solve a second-order ODE • Configuration changes needed manual rebuilding	• Computational speeds up to 1000 times faster than mechanical systems • Reconfiguration requires minor circuit changes
Complexity Limits	• Equation conversion complexity limited to 4th-order linear ODEs	• Can handle complex nonlinear systems • Computational load depends on amplifier gains
Key Constraints	• Inflexibility due to specialized wired mesh architecture • Restricted extensibility due to mechanical rigidity	• Reconfigurability enabled incremental extensibility • Scalability limited by electronic modular interconnections

2.3.3 DIGITAL COMPUTERS ENABLE SOFTWARE-BASED MODELING

The advent of fully electronic digital computers in the 1940s sparked a seminal revolution in scientific computing capabilities that disrupted long-standing analog simulation techniques. Leveraging discrete transistors, logic gates and microsecond cycle-time binary arithmetic-logic units, they ushered unprecedented versatility and analytical power.

Unlike hardwired analog topologies constrained by modular hardware units, digital computers offered inexhaustible flexibility by functionally emulating mathematical models in software [16]. Logical bit-patterns represented abstract informational entities that were manipulated by stored instruction sets to mimic model dynamics. Any functional block could exchange messages with others.

For modeling wireless networks, this allowed the simulated interaction of software-defined communication system components executed sequentially – like binary data sources, encoders, modulators, channels with noise and interferers, demodulators, decoders, and sinks. The runtime program states precisely mirrored target model behaviour. Modeling pragmas like SDL, SPEC, UML provided high-level abstractions [17].

The capacities for unlimited precision, quick design iterations through code tweaks, deriving mathematical insights into model dynamics analytically, boundless problem complexity and computational parallelism fostered an absolute historical shift from analog electro-mechanical tools to omnipotent software-based wireless modeling and simulation ubiquitous today.

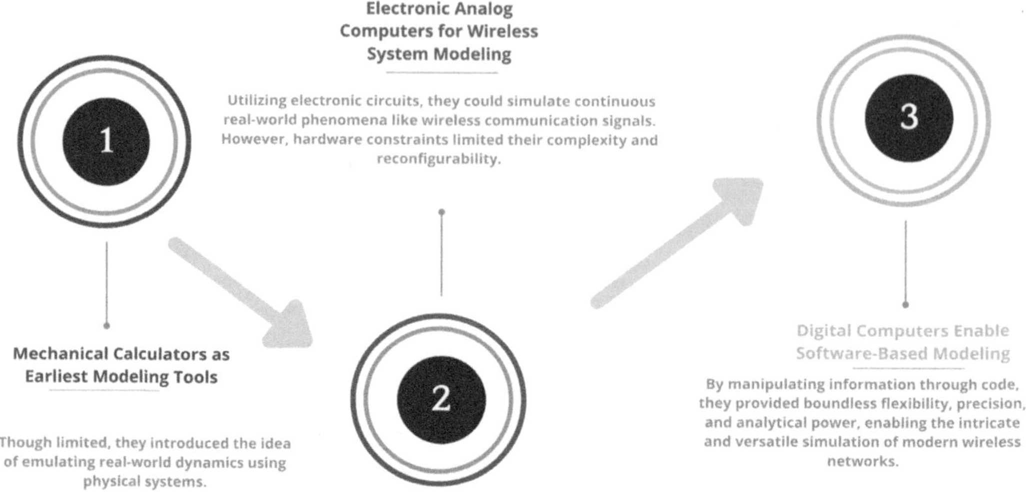

FIGURE 2.1 Evolution of Analog Computation.

Figure 2.1 depicts the progression of analog computation from early mechanical calculators to electronic analog computers, highlighting their importance as precursors to digital simulation techniques. Mechanical calculators automated numeric calculations but had limited versatility. Electromechanical analyzers could solve complex differential equations. Electronic analog computers offered higher accuracy and speed for dynamical system modeling using electronic circuits, though faced interconnect scaling constraints. While limited for wireless systems, analog computers demonstrated emulating real-world dynamics computationally. This established approaches later realized through digital computers, sparking a seminal revolution in modeling and simulation capabilities.

2.4 RECENT INNOVATIONS IN COMPUTATIONAL MODELING

This section explores major recent advances in computational power, algorithms, and modeling capabilities that have vastly expanded the scale and complexity of wireless network simulations. Appreciating these innovations provides context about the state-of-the-art tools available now for modeling sophisticated communication scenarios.

2.4.1 EMERGENCE OF POWERFUL DIGITAL COMPUTING TECHNOLOGIES POST-1970s

Following the significant advancements in computational power and algorithmic development witnessed in the late twentieth century, the field of computational modeling and simulation (CM&S) has continued to evolve rapidly in the twenty-first century. This section explores some of the most notable recent innovations in this domain.

A. Moore's Law and its Impact

Moore's Law, formulated by Gordon Moore in 1965, states that the number of transistors on integrated circuits doubles approximately every two years. This exponential growth in computing power has been a major driver of innovation in CM&S. Early computational models for wireless networks were often limited by the available computational resources, making it difficult to accurately simulate complex network scenarios [18]. However, with the dramatic increase in processing power over the past few decades, researchers have been able to develop more sophisticated and detailed models that can capture the intricate dynamics of wireless networks [19].

B. High-Performance Computing (HPC)

The development of high-performance computing (HPC) clusters has further expanded the capabilities of CM&S. HPC systems are powerful computers that can be used to perform large-scale simulations that would be intractable on traditional desktop computers [20, 21]. This has allowed researchers to study the behaviour of wireless networks under realistic conditions, such as with thousands of mobile devices and complex network topologies.

C. Cloud Computing and Virtualization

Cloud computing and virtualization technologies have also emerged as valuable tools for CM&S. These technologies provide researchers and engineers with access to on-demand computational resources, eliminating the need for costly investments in hardware infrastructure [22]. This has made it easier and more affordable to conduct large-scale simulations and to share and collaborate on research projects.

2.4.2 APPLICATIONS OF COMPUTATIONAL MODELING AND SIMULATION TO MOBILE WIRELESS NETWORKS

A. Network Planning and Optimization

Mobile wireless networks are indeed one of the most significant applications of computational modeling and simulation (CM&S). This process encompasses the design and optimization of network topology, resource allocation, and network parameter tuning to achieve desired performance metrics such as coverage, capacity, and quality of service (QoS) [23]. The application of CM&S in network planning and optimization involves the use of advanced algorithms and simulation techniques to model and analyse complex wireless network scenarios, enabling the evaluation of various network configurations and optimization strategies.

The design of network topology using CM&S involves the creation of simulation models that represent the spatial layout of base stations, access points, and other network elements. These models consider factors such as terrain, building structures, and user distribution to simulate realistic wireless propagation environments. By incorporating real-world environmental data and wireless channel models, CM&S enables the accurate prediction of signal coverage, interference patterns, and path loss characteristics, which are essential for effective network planning. Resource allocation, including frequency assignment and base station placement, is another critical aspect of network planning and optimization. CM&S facilitates the dynamic allocation of frequencies and the placement of base stations to minimise interference, maximise spectrum utilisation, and ensure efficient coverage and capacity. Through simulation, different resource allocation strategies can be evaluated and optimised to meet the specific requirements of the wireless network under consideration [24].

Network parameter optimization, such as power control and scheduling algorithms, plays a vital role in achieving optimal network performance. CM&S allows for the modeling and simulation of various power control schemes and scheduling policies to assess their impact on network capacity, QoS provisioning, and energy efficiency [25]. By iteratively adjusting and evaluating these parameters within a simulated environment, CM&S enables the identification of optimal configurations that balance trade-offs between coverage, capacity, and QoS metrics.

B. Coverage Analysis and Prediction

The application of computational modeling and simulation (CM&S) in coverage analysis and prediction for mobile wireless networks is a critical aspect of network planning and optimization. By utilising CM&S, network operators can predict the radio coverage of a network and identify areas with insufficient signal strength. This information is crucial for planning base station placement and ensuring adequate coverage for subscribers. Computational models enable the simulation of wireless propagation characteristics, taking into account factors such as terrain, building structures, and environmental conditions to accurately predict signal coverage and identify potential coverage gaps [26]. This allows network operators to make informed decisions regarding the placement of base stations and the optimization of antenna configurations to enhance overall network coverage and quality of service.

CM&S plays a pivotal role in capacity analysis and optimization for mobile wireless networks. Network capacity, defined as the maximum amount of data that can be transmitted over a network within a specific period, is a critical performance metric for wireless communication systems. By leveraging CM&S, network operators can analyse network capacity under different traffic conditions and identify potential bottlenecks that may limit network performance. Through simulation, various network parameters, such as power control, resource allocation, and scheduling algorithms, can be optimised to improve network capacity and accommodate increasing traffic demands. This enables network operators to proactively address capacity constraints and ensure the efficient and reliable delivery of data services to subscribers.

C. Capacity Analysis and Optimization

Mobile wireless networks are a critical aspect of network performance management. Network capacity, defined as the maximum amount of data that can be transmitted over a network within a specific period, is a key determinant of the network's ability to handle increasing traffic demands and deliver services efficiently. CM&S play a pivotal role in analysing network capacity under varying traffic conditions and identifying potential bottlenecks that may limit network performance. By leveraging CM&S, network operators can optimise network parameters and deploy appropriate resource allocation strategies to enhance network capacity and effectively manage escalating traffic demands. One notable study by [27] provides a comprehensive analysis of network capacity under different traffic scenarios using CM&S, highlighting the impact of varying user densities, data rates, and network configurations on network performance. This study underscores the importance of CM&S in evaluating network capacity and identifying potential performance limitations, providing valuable insights for the optimization of network parameters to accommodate increasing traffic demands and improve overall network capacity.

Furthermore, the work by [28] focuses on the development of resource allocation strategies using CM&S to optimise network capacity and mitigate potential bottlenecks in wireless communication systems. By simulating different resource allocation schemes and traffic patterns, the study evaluates the effectiveness of dynamic resource management strategies in enhancing network capacity and ensuring efficient data transmission. This research emphasises the significance of CM&S in optimising resource allocation to address capacity constraints and improve the network's ability to handle increasing traffic demands.

Moreover, the research by [29] addresses the dynamic nature of network capacity in the context of emerging 5G networks, leveraging CM&S to model and analyse capacity limitations under diverse usage scenarios. The study demonstrates the potential of CM&S in developing adaptive capacity optimization techniques, enabling network operators to proactively manage network capacity and ensure optimal performance in the face of evolving traffic demands and network dynamics. This research highlights the application of CM&S in addressing capacity challenges in next-generation wireless networks, contributing to the development of efficient capacity optimization strategies.

D. Mobility Management

Mobility management for mobile networks is a critical aspect of network performance optimization. User mobility in mobile networks introduces dynamic changes in user location and network connections, necessitating the analysis of its impact on network performance and the development of efficient mobility management strategies. CM&S provides a powerful framework for simulating and analysing user mobility patterns, enabling the evaluation of handover techniques between base stations, routing of data packets, and power management for mobile devices. In addition, [30] presents a comprehensive analysis of user mobility patterns in mobile networks using CM&S, highlighting the impact of mobility on network performance and the effectiveness of different handover strategies in maintaining seamless connectivity for mobile users. This study underscores the importance of CM&S in understanding and optimising mobility management techniques to ensure uninterrupted communication for mobile users as they move across different network areas.

Furthermore, the work by [31] focuses on the development of efficient routing algorithms for data packets in mobile networks, considering user mobility patterns and network dynamics. By leveraging CM&S, the study demonstrates the effectiveness of dynamic routing strategies in adapting to user mobility and network changes, thereby improving the efficiency of data packet delivery in mobile environments. This research emphasises the significance of CM&S in optimising routing protocols to accommodate user mobility and enhance network performance. The research by [32] addresses the power management challenges for mobile devices in dynamic network environments, utilising CM&S to model and analyse power consumption patterns based on user mobility and network conditions. The study highlights the potential for CM&S to optimise power management strategies, prolonging the battery life of mobile devices while maintaining seamless connectivity during user mobility. This research showcases the application of CM&S in addressing power management concerns in mobile networks, contributing to the development of energy-efficient mobility management solutions.

E. Interference and Resource Management

The application of computational modeling and simulation (CM&S) in interference and resource management for wireless communication systems is crucial for optimising network performance in the presence of shared communication channels and potential interference. Wireless communication channels are often shared by multiple users, leading to interference and competition for resources. CM&S provides a powerful framework for analysing and predicting interference patterns and developing resource allocation strategies to mitigate interference and enhance network performance. In turn, [33] demonstrates the use of CM&S to analyse interference patterns in shared wireless communication channels, highlighting the impact of user density, transmission power, and channel conditions on interference levels. The work focuses on the development of resource allocation strategies using CM&S to mitigate interference and optimise spectrum utilisation in wireless communication systems. By simulating various resource allocation schemes and interference scenarios, the study evaluates the effectiveness of dynamic resource allocation strategies in minimising interference and maximising network capacity. Next, [34] addresses the dynamic nature of interference in cognitive radio networks, leveraging CM&S to model and analyse interference patterns in dynamic spectrum access scenarios. The study demonstrates the potential of CM&S in developing cognitive interference management techniques, enabling adaptive resource allocation and interference mitigation strategies to improve spectrum utilisation and network performance in dynamic communication environments. This research highlights the application of CM&S in addressing interference challenges in cognitive radio networks, contributing to the development of intelligent interference management solutions.

Figure 2.2 highlights key applications of computational modeling and simulation (CM&S) in mobile wireless networks. By creating virtual representations of wireless environments and traffic, CM&S enables in-depth evaluation of network configurations before deployment. As shown, CM&S aids mobile network analysis and optimization in areas like planning, capacity,

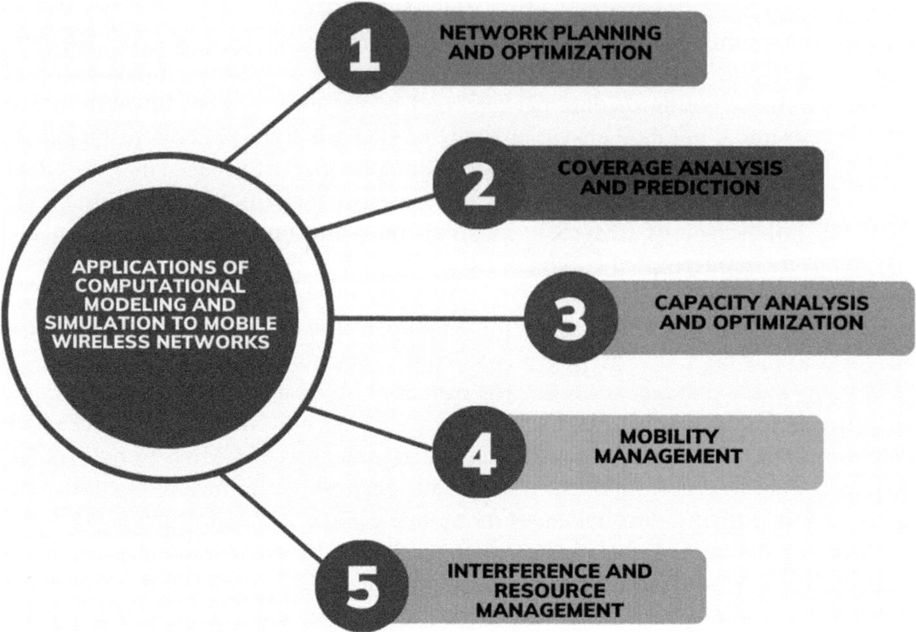

FIGURE 2.2 Applications of Computational Modeling and Simulation to Mobile Wireless Networks.

coverage, mobility management, interference mitigation, and resource allocation. The insights gained help meet quality of service targets in a cost-effective manner while addressing technical complexities.

2.4.3 ADVANCED TECHNIQUES: MACHINE LEARNING, ARTIFICIAL INTELLIGENCE, AND EDGE COMPUTING

2.4.3.1 Machine Learning for Network Modeling and Optimization

Machine learning (ML) algorithms have emerged as powerful tools for developing more accurate and predictive models of wireless networks. These algorithms can learn from large datasets of network data and identify hidden patterns and relationships that can be exploited to improve network performance. Some key applications of ML in CM&S include:

- *Channel modeling:* ML algorithms can be used to develop data-driven models of radio propagation channels, which can be more accurate than traditional physics-based models, especially in complex environments.
- *Traffic prediction:* ML models can analyse historical traffic data and predict future traffic patterns, enabling network operators to proactively allocate resources and optimise network settings.
- *Anomaly detection and network troubleshooting:* ML algorithms can be trained to detect unusual network behaviour that may indicate potential problems, helping network operators to identify and troubleshoot issues quickly.
- *Resource allocation and optimization:* ML-based algorithms can be used to optimise resource allocation in real-time, ensuring that network resources are used efficiently and effectively to meet changing traffic demands.

2.4.3.2 Artificial Intelligence for Network Management and Automation

Artificial intelligence (AI) techniques, such as deep learning and reinforcement learning, are being increasingly used to automate various tasks in network management and operation. AI can do the following:

- *Automate network configuration and optimization:* AI algorithms can learn from experience and optimise network parameters, such as power control, scheduling algorithms, and routing protocols to achieve desired performance goals.
- *Proactively manage network resources*: AI can predict future resource needs and proactively allocate resources to prevent congestion and service disruptions.
- *Automate network security*: AI-powered security systems can detect and respond to cyberattacks in real-time, protecting networks from intrusion and data breaches.

2.4.3.3 Edge Computing for Real-time Network Analysis and Control

Edge computing brings computational resources closer to the network edge, enabling real-time processing and analysis of data collected from mobile devices and network infrastructure. This has several benefits for CM&S, and they include:

- *Reduced latency*: By processing data locally at the edge, edge computing reduces the latency associated with transmitting data to the cloud for analysis, enabling faster and more responsive network management.
- *Improved scalability*: Edge computing allows for distributed processing of data across multiple edge nodes, making it possible to handle large volumes of data generated by dense networks with numerous devices.
- *Enhanced privacy and security*: By processing data locally, edge computing reduces the need to transmit sensitive information to the cloud, improving privacy and security.

2.5 CASE STUDIES DEMONSTRATING EVOLUTION OF MODELING IN WIRELESS SYSTEMS

This section provides illustrative case studies highlighting the progression of modeling techniques applied to coverage analysis, cellular network protocols, and simulation platforms over generations of wireless systems. These examples showcase in practice the continual advancement of modeling approaches and impact on design.

2.5.1 Early Radio Propagation Models for Coverage Analysis

Before the advent of powerful computational models, engineers relied on early radio propagation models to predict signal strength and coverage areas in wireless networks. These models, though often simple by today's standards, played a crucial role in the early days of wireless communication and paved the way for the sophisticated CM&S tools we use today.

The Free Space Path Loss (FSPL) model represents a foundational concept in wireless communication, providing a fundamental understanding of signal attenuation based solely on distance [35]. This model serves as a cornerstone in the early development of radio propagation models and has significantly contributed to the understanding of signal behaviour in wireless communication systems. When considering the propagation of signals between two antennas, the FSPL model calculates the attenuation of the signal as it travels through free space [36]. In this idealised scenario, the model assumes a clear line of sight between the transmitting and receiving antennas, disregarding any obstacles or environmental factors that may affect signal propagation [37].

While simplistic in nature, the FSPL model offers essential insights into the basic principles of signal degradation over distance in unobstructed environments. The FSPL model is fundamental in providing a theoretical basis for understanding the impact of distance on signal strength, forming the basis for more advanced radio propagation models. By establishing a clear relationship between signal attenuation and distance, the FSPL model has laid the groundwork for subsequent developments in radio propagation modeling, serving as a starting point for more sophisticated models that consider additional environmental factors and obstacles [38].

The Okumura-Hata model has significantly advanced the understanding of signal propagation in wireless communication systems, particularly in urban environments [39]. Developed in the 1960s, this empirical model introduced the concept of adjusting signal strength based on factors such as urban, suburban, and rural environments, thereby providing a more realistic representation of signal behaviour in real-world networks. The model goes beyond the idealised conditions of free space and incorporates the effects of terrain and vegetation, addressing the impact of environmental factors on signal propagation. By considering the influence of different environments, the Okumura-Hata model provides a more accurate representation of signal behaviour, especially in urban areas where buildings and other structures significantly affect signal propagation [40].

The Okumura-Hata model has been widely utilised in various studies to analyse and predict signal propagation in diverse scenarios. For example, [41] demonstrated the use of the Okumura-Hata model in a theoretical coverage study, providing expected power-level maps obtained from an accurate radio-network planning tool. This application showcases the model's relevance in predicting signal coverage in real-world scenarios. The model's superior fit to measured data compared to other propagation models, emphasising its effectiveness in predicting signal behaviour in practical settings. Additionally, [42] presented a case study in Dar-es-Salaam, Tanzania, where the Okumura-Hata model was optimised based on measurement data using the least square method. This approach resulted in modified models for rural, suburban, and urban environments, enhancing the model's applicability in diverse settings. The Okumura-Hata model has significantly advanced the understanding of signal propagation in wireless communication systems, particularly in urban environments. Its incorporation of environmental factors and its widespread application in practical scenarios have solidified its importance in the field of radio propagation modeling.

The Walfisch-Ikegami model has significantly advanced the understanding of signal propagation in wireless communication systems, particularly in urban environments. Developed in the 1980s, this model refined radio propagation modeling by considering the impact of buildings, building penetration, and diffraction effects, allowing for more accurate predictions in dense urban environments, where buildings often obstruct the path of radio waves [43].

The Walfisch-Ikegami model has been widely utilised in various studies to analyse and predict signal propagation in diverse scenarios. For example, [44] aimed to optimise the capacity of radio resources in cellular coverage based on a COST-231 Walfisch-Ikegami semi-empirical propagation model for power control in the upward direction. This application showcases the model's relevance in predicting signal coverage in real-world scenarios, particularly in urban areas, where signal propagation is influenced by complex environmental factors. Furthermore, the model has been applied in studies to evaluate the impact of obstacles on the capacity of LTE networks in urban areas. [44] demonstrated the model's effectiveness in optimising the capacity of radio resources in cellular coverage, highlighting its relevance in practical settings.

In conclusion, the Walfisch-Ikegami model has significantly advanced the understanding of signal propagation in wireless communication systems, particularly in urban environments. Its incorporation of building penetration and diffraction effects, and its widespread application in practical scenarios, have solidified its importance in the field of radio propagation modeling.

2.5.2 Impact of Computational Modeling and Simulation Evolution in Early Wireless Communication Systems

The early impact on computational modeling and simulation (CM&S) evolution in wireless communication systems has been instrumental in shaping the trajectory of wireless network analysis and design. The introduction of mathematical formulas for quantifying radio wave propagation and signal attenuation marked a significant advancement, enabling engineers to move beyond qualitative estimations and conduct quantitative analysis of network coverage. These early models laid the foundation for the development of more complex and comprehensive models that could address the growing diversity of environments and technologies in wireless networks. The success of these early models in predicting network coverage demonstrated the value of CM&S in wireless communication systems, paving the way for further research and development in this field.

This early impact has been pivotal in the advancement of modeling tools and techniques, leading to the sophisticated and accurate CM&S tools available today. One notable study by Rappaport delved into the early development of radio wave propagation models, emphasising the significance of these models in quantifying signal attenuation and enabling engineers to conduct quantitative analysis of network coverage [45]. This study highlighted the foundational role of early models in advancing the understanding of radio wave propagation and signal behaviour in wireless communication systems.

The work by [46] provided insights into the value of CM&S in wireless communication systems, showcasing the success of early models in predicting network coverage and demonstrating the importance of further research and development in this field. This study underscored the pivotal role of early CM&S in shaping the trajectory of wireless network analysis and design. In conclusion, the early impact on CM&S evolution has been pivotal in advancing the understanding and analysis of wireless communication systems. The introduction of early models for quantifying radio wave propagation and signal attenuation has laid the foundation for the development of more advanced and comprehensive models, demonstrating the value of CM&S in predicting network coverage and shaping the trajectory of wireless network analysis and design.

2.5.3 Limitations of Computational Modeling and Simulation Evolution in Early Wireless Communication System

The early models in wireless communication systems, while pioneering, were not without limitations. These limitations have been instrumental in shaping the trajectory of computational modeling and simulation in wireless communication systems, leading to advancements in addressing these challenges and developing more accurate and comprehensive models. One notable limitation of early models was their restricted scope, which hindered their ability to accurately represent the diverse environments and technologies present in wireless networks [47]. These models were often limited in their capacity to address the complexities of real-world scenarios, such as urban, suburban, and rural environments, where environmental factors significantly impact signal propagation.

The limited scope of early models necessitated the development of more advanced and comprehensive models that could effectively capture the diverse and dynamic nature of wireless communication environments. Furthermore, early models often lacked the capability to accurately mimic real physiological conditions and monitor the developmental course of certain phenomena, such as the early development of atherosclerosis. This limitation hindered the models' ability to provide accurate predictions and insights into the progression of complex physiological processes, highlighting the need for more sophisticated modeling approaches that could address the intricacies of biological systems.

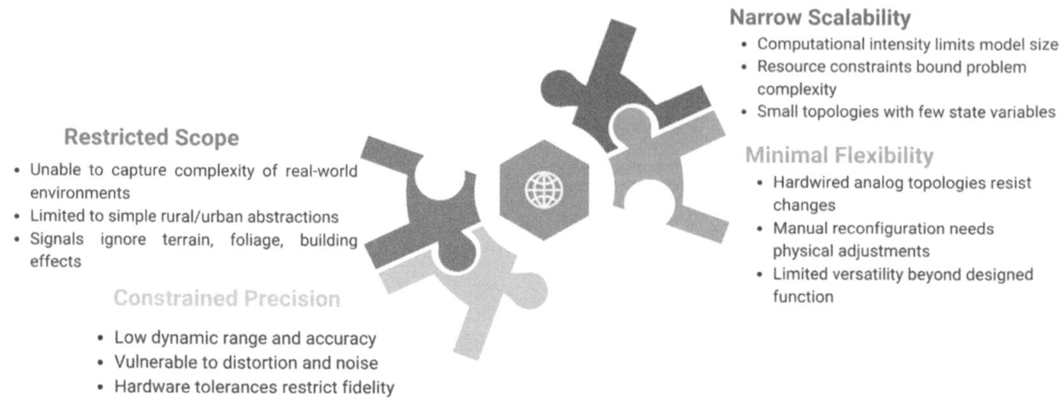

FIGURE 2.3 Limitations of Early Wireless Communication Models.

In addition, the computational intensity and resource requirements of early models posed significant challenges, particularly in scenarios where very large, computationally intensive models were necessary to correctly classify complex examples [48]. The limitations in computational efficiency and scalability hindered the widespread applicability of these models, emphasising the need for more efficient and scalable modeling approaches to address the computational demands of wireless communication systems. The limitations of early models have underscored the need for more advanced and accurate modeling tools and techniques in wireless communication systems. These challenges have paved the way for further research and development in the field of CM&S, leading to the development of more sophisticated and comprehensive models that can effectively address the complexities and dynamics of wireless communication environments.

The limitations of early models in wireless communication systems have been instrumental in shaping the trajectory of CM&S, leading to advancements in addressing these challenges and developing more accurate and comprehensive models that can effectively capture the diverse and dynamic nature of wireless communication environments [49].

While pioneering modeling and simulation developments in the early stages contributed significantly to the analysis of wireless communication systems, they were also significantly limited in many respects as described above. Figure 2.3 illustrates a high-level perspective of the key restrictions along four salient dimensions that constrained their applicability and fidelity for effectively capturing intricate real-world wireless propagation environments.

2.5.4 Modeling of Cellular Network Architectures and Protocols

Modeling the evolution of cellular network architectures and protocols from 1G to 5G has been a critical area of research and development in the field of wireless communication. The advancements in computational modeling and simulation have played a pivotal role in adapting to the changing landscape of cellular networks, enabling researchers and engineers to optimise network performance and design future generations of cellular networks. The following elaboration provides a comprehensive overview of the evolution of CM&S models in the context of cellular network architectures and protocols.

i. *1G and 2G Modeling*: During the early stages of cellular networks, CM&S models focused on simulating handovers, channel fading, and interference effects in analog and digital cellular systems [50]. These models provided valuable insights into the performance of early cellular networks, laying the foundation for subsequent advancements in wireless communication technologies.

ii. *3G and 4G Modeling*: As cellular networks transitioned to 3G and 4G, more complex CM&S models were developed to simulate packet-based traffic, radio resource management algorithms, and advanced features such as multimedia streaming. These models enabled researchers to analyse and optimise the performance of evolving cellular networks, addressing the increasing demand for data services and more efficient resource management. [51] which provided insights into the value of CM&S in wireless communication systems, showcasing the success of early models in predicting network coverage and demonstrating the importance of further research and development in this field.

iii. *5G and Beyond Modeling*: The advent of 5G networks has brought about a new set of challenges and requirements, prompting the development of current CM&S models that address complex scenarios, including massive multiple-input multiple-output (MIMO), millimetre wave propagation, and network slicing. These advanced models have enabled researchers and engineers to analyse and optimise the performance of 5G networks, supporting the diverse requirements of ultra-reliable low-latency communication (URLLC), enhanced mobile broadband (eMBB), and massive machine-type communication (mMTC) [52].

The evolution of CM&S models in the context of cellular network architectures and protocols has been instrumental in advancing the understanding and optimization of wireless communication systems. The adaptation and advancement of CM&S models have played a crucial role in addressing the challenges and requirements of evolving cellular networks, paving the way for the design and optimization of future generations of wireless communication technologies.

2.5.5 Simulation Platforms for Wireless Research

The evolution of simulation platforms such as OPNET, NS-2, and OMNeT++ has significantly impacted wireless research, enabling the analysis and optimization of wireless communication systems. The following elaboration provides a comprehensive overview of the impact of these simulation platforms on wireless research.

i. *OPNET Simulation Platform:* The OPNET simulation platform has been instrumental in enabling researchers to analyse and optimise wireless communication systems. [53] highlighted the significance of the OPNET simulation platform in developing a 6LoWPAN model for machine-to-machine communications, emphasising the challenges in interfacing the built-in IP module with custom modules in OPNET. This underscores the platform's importance in addressing complex wireless communication scenarios.

ii. *NS-2 Simulation Platform*: The NS-2 simulation platform has played a crucial role in simulating wireless network protocols and optimising network performance. [54] demonstrated the integration of the OPNET network protocols simulator with the MATLAB development environment, showcasing the platform's capability to execute simulations of network traffic with predetermined parameter values and optimise parameters using mathematical algorithms. This highlights the platform's versatility in facilitating network protocols optimization.

TABLE 2.2

Contrasting Popular Wireless Network Simulation Platforms on Scalability and Interoperability Metrics

Feature	NS-2	OPNET	OMNeT++
Scalability	• Simulations limited to hundreds of nodes • Model complexity constraints from memory limits	• Supports large-scale models (thousands of nodes) • Distributed simulation for very large topologies	• Excellent scalability using parallel distributed execution • Tested with million node simulations
Interfaces	• Supports C++ and OTcl • Need glue code between layers	• Uses proprietary interfaces • Some standard imports supported	• Flexible architecture integrates with other tools • Open interfaces for model coupling
Platform Compatibility	• Linux and Unix platforms	• Windows and Linux	• Platform independent (Windows, Linux, Mac)
Extensibility	• Adding new protocols requires coding • Constraints around layered architecture	• User definable modules enable extensions • Good customization capabilities	• Highly modular framework simplifies component addition • Open source enables custom plug-ins
Learning Curve	• Steep learning curve for new users	• High complexity with long training times	• Clean and simple model development process • Faster proficiency

iii. *OMNeT++ Simulation Platform*: OMNeT++ has been widely used for simulating traditional wired and wireless networks, providing a versatile platform for wireless research. [55] emphasised the extensive capabilities of the OMNeT++ simulation environment, which has been utilised in numerous domains, including wireless and ad-hoc network simulations, business process simulations, and peer-to-peer network simulations. This demonstrates the platform's adaptability to diverse wireless research applications.

OMNeT++ offers the best scalability through distributed processing as well as a flexible simulation environment enabling interfacing with external tools and simpler extension of models. NS-2 is more constrained in scaling while OPNET enables larger topologies through its commercial simulation engine. OMNeT++ strikes a good balance between scalability and interface openness. Table 2.2 shows the feature comparison of network simulation platforms and their capabilities related to scalability and interface support.

2.6 LESSONS LEARNED

The historical vantage exposing the co-evolution of wireless communication systems with sophisticated analytical modeling tools that enabled unlocking their utmost potential crystallises overarching lessons for the future. Lessons that could be learned include:

i. *Interdependence*: Material progress manifested symbiotically across scientific theories, concrete systems, and computational models uplifting each other iteratively;

ii. *Fluidity*: Disruptive modernization sprang from inflection points when tools reoriented to overcome limitations by embracing new paradigms;

iii. *Ingenuity:* Prevailing trends of scientific inquiry and engineering creativity persisted determinedly as the driving ethos over generations;

iv. *Adaptability:* Continuous incorporation of emerging techniques, contexts and requirements persisted as a hallmark enabling perpetual self-reinvention;

v. *Upliftment:* Each quantum leap potentiated next-generation capabilities while retaining collective accumulation of knowledge, methods, and visions.

2.7 TECHNICAL OUTLOOK AND EMERGING TRENDS

The rapid advancement of wireless communications and modeling techniques shows no signs of abating. Key areas of expected progress driving the next frontiers of innovation include:

i. *Ultra-fast and energy-efficient 6G networks powered by AI and quantum computing:* The next frontier of wireless innovation will leverage emerging techniques like AI and quantum computing to realize new heights in speed and efficiency. Key developments include harnessing AI and machine learning for real-time intelligent optimization across all network layers, applying quantum techniques to solve extremely complex challenges in massive MIMO beamforming, dynamic routing, and scheduling. This will enable pushing throughput speeds into the multi-terabit range through new mmWave and THz spectrum bands combined with advanced beamforming techniques and 3D heterogeneous networking. Radical gains in energy efficiency are also targeted from femto sleep modes to novel backscatter communications harvesting ambient signals.

ii. *High-fidelity digital twin simulations fusing multi-domain data streams:* High-fidelity emulation of real-world environments will be enabled by converging digital clones merging telemetry data flows from associated physical systems with virtual and simulated models. This facilitates explorations into future configurations using predictive analytics to forecast user mobility, traffic patterns, radio propagation, interference variability, and so forth. Immersive visualizations will enhance interactions while optimization loops between digital twins and testbeds enable accelerated innovation cycles.

iii. *Real-time edge analytics and closed-loop automation:* The proliferation of edge computing down to end devices coupled with advances in tiny machine learning chips will unleash a wave of distributed real-time analytics, creating locally aware network nodes. This lays the foundations for autonomous closed-control loops enabling self-optimization and instantiation of services dynamically adapted to user context. The coordination of such hyper-flexibility across edge, fog, and central cloud resources will become pivotal.

iv. *Lifecycle verifiability, trust, and sustainability:* Holistic concerns around transparency, inclusiveness, and universal accessibility will drive research into formal verification techniques establishing system correctness, trust mechanisms ensuring accountability and sustainability considerations assessing social, economic, and environmental footprints. Networks will need to be neutral, non-discriminating platforms uplifting collective potential.

v. *Democratization expanding access and participatory engagement:* Connectivity expansion remains imperative with broadband access as a fundamental right in unlocking individual and community potential. Grassroots community networking buildouts enhancing last-mile availability will accelerate alongside platforms promoting user co-creation. Continued abstractions hiding complexity behind simple interfaces holds the key to unlimited creativity.

2.8 CONCLUSION

The intertwined history of wireless communications and computational modeling seen through early quantitative inquiries into electromagnetism by nineteenth-century pioneers who established the imperative of measurements, modularity, and validation transitioned rapidly into functional electronic component abstractions and then software-based platforms. Their unprecedented versatility, precision, and complexity-scaling abilities disrupted established analog approaches and essentially revolutionised the role of simulations in system design and analysis. The technology co-evolution persisting unrelentingly as evidenced by radio propagation and cellular network models being stretched repeatedly to their limits by demands from real-world wireless problem complexities thus requiring researchers to expand modeling frontiers through orders-of-magnitude gains in algorithmic efficiency, hardware capabilities, interface expressiveness, workflow automation and methodological maturity as exemplified by emerging capabilities to handle vast scales, uncertainty, and sustainability considerations during model construction. This reveals a continuity of vision, spanning generations unified by the enduring spirit of science manifested over a century of innovation that promises to breach yet unforeseen horizons as the intellectual marathon progressing from fundamental discoveries in electromagnetism now powering modern civilization continues persevering, fueled intrinsically by insatiable curiosity.

REFERENCES

[1] D. Kotz, C. Newport, R. S. Gray, J. Liu, Y. Yuan, and C. Elliott, "Experimental Evaluation of Wireless Simulation Assumptions," *Proceedings of the 7th ACM International Symposium on Modeling, Analysis and Simulation of Wireless and Mobile Systems*, 2004. doi: 10.1145/1023663.1023679

[2] K. Srinivas, and D. K. Rao, "Modeling and Computer Simulation of Nanostructured Devices," *American Journal of Nanoscience and Nanotechnology*, vol. 2, no. 3, 2014. doi: 10.11648/j.nano.20140203.12

[3] S. A. Akinboro, O. Olaniyan, G. A. Aderounmu, E. A. Olajubu, A. O. Ajayi, and I. K. Ogundoyin, "An Improved Computational Model for Adaptive Communication Channel Estimation," *International Journal of Computers Communications &Amp; Control*, vol. 6, no. 2, 2011. doi: 10.15837/ijccc.2011.2.2167

[4] W. Guo, C.-C. Hung, R. K. Shepherd, Z. Ma, and S. Kulkarni, "Intelligent Techniques for Simulation and Modeling," *Mathematical Problems in Engineering*, vol. 2014, 2014. doi: 10.1155/2014/179457

[5] A. Sihvola, "Johan Jacob Nervander and the Quantification of Electric Current [Historically Speaking]," *IEEE Antennas and Propagation Magazine*, vol. 63, no. 1, 2021. doi: 10.1109/map.2020.3039803

[6] W. Vegt, "The Origin of Gravity, A Second Order Lorentz Transformation for 'Accelerated Electromagnetic Fields', Generating a Gravitational Field and the Property of Mass," *International Research Journal of Pure and Applied Physics*, vol. 9, no. 1, 2022. doi: 10.37745/irjpap.13/vol9n11252

[7] S. Iyer *et al.*, "A Survey on Semantic Communications for Intelligent Wireless Networks," *Wireless Personal Communications*, vol. 129, no. 1, 2022. doi: 10.1007/s11277-022-10111-7

[8] B. S. Olanrewaju, and O. Osunade, "Design of a Mathematical Model for Spectrum Utilisation in Cognitive Radio," *International Journal of Computer Applications*, vol. 180, no. 19, 2018. doi: 10.5120/ijca2018916436

[9] M. Leone, and N. Robotti, "Guglielmo Marconi, Augusto Righi and the Invention of Wireless Telegraphy," *The European Physical Journal H*, vol. 46, no. 1, 2021. doi: 10.1140/epjh/s13129-021-00021-w

[10] A. Hajimiri, and T. H. Lee, "A General Theory of Phase Noise in Electrical Oscillators," *IEEE Journal of Solid-State Circuits*, vol. 33, no. 2, 1998. doi: 10.1109/4.658619

[11] C. A. Duarte, "The Classical Geometrization of the Electromagnetism," *International Journal of Geometric Methods in Modern Physics*, vol. 12, no. 09, 2015. doi: 10.1142/s0219887815600221

[12] K. Fujikawa, M.-L. Ge, Y. Liu, and Q. Zhao, "Uncertainty Principle, Shannon–Nyquist Sampling and beyond," *Journal of the Physical Society of Japan*, vol. 84, no. 6, 2015. doi: 10.7566/jpsj.84.064801

[13] T. Blumensath, and M. E. Davies, "Sampling Theorems for Signals From the Union of Finite-Dimensional Linear Subspaces," *IEEE Transactions on Information Theory*, vol. 55, no. 4, 2009. doi: 10.1109/tit.2009.2013003

[14] Y. Gueye, "Modeling and Simulation Tools in Mechanical Design Education: What Didactical Determinants?," *Turkish Journal of Computer and Mathematics Education (TURCOMAT)*, vol. 12, no. 4, 2021. doi: 10.17762/turcomat.v12i4.605

[15] M. Goldenbaum, H. Boche, and S. Stańczak, "Harnessing Interference for Analog Function Computation in Wireless Sensor Networks," *IEEE Transactions on Signal Processing*, vol. 61, no. 20, 2013. doi: 10.1109/tsp.2013.2272921

[16] A. Siebold, and M. Valleriani, "Digital Perspectives in History," *Histories*, vol. 2, no. 2, 2022. doi: 10.3390/histories2020013

[17] A. Dahlgren, and A. Wasielewski, "Cultures of Digitization: A Historiographic Perspective on Digital Art History," *Visual Resources*, vol. 36, no. 4, 2020. doi: 10.1080/01973762.2021.1928864

[18] L. Chang, Y. K. Choi, J. Kedzierski, N. Lindert, P. Xuan, J. Bokor, C. Hu, and T. J. King. Moore's law lives on [CMOS transistors]. *IEEE Circuits and Devices Magazine*, vol. 19, no. 1, 35–42, 2003. doi: 10.1109/MCD.2003.1175106

[19] J. Shalf, "The Future of Computing beyond Moore's Law," *Philosophical Transactions of the Royal Society A: Mathematical, Physical and Engineering Sciences*, vol. 378, no. 2166, 2020. doi: 10.1098/rsta.2019.0061

[20] R. Blaheta *et al.*, "High Performance Computing Applications," *Cybernetics and Information Technologies*, vol. 17, no. 5, 2017. doi: 10.1515/cait-2017-0050

[21] H. Cornelius, "The Future of High-Performance Computing (HPC)," *Encyclopedia of Information Science and Technology, Fourth Edition*, 2018. doi: 10.4018/978-1-5225-2255-3.ch347

[22] H. Cheng, Z. Chen, N. Sun, F. Chen, and M. Wang, "Evaluation Framework of Virtualization Systems for Cloud Computing," *2012 IEEE Asia Pacific Cloud Computing Congress (APCloudCC)*, 2012. doi: 10.1109/apcloudcc.2012.6486510

[23] S. Talwar, S.P. Yeh, N. Himayat, K. Johnsson, G. Wu, and R.Q. Hu, "Capacity and Coverage Enhancement in Heterogeneous Networks," *Heterogeneous Cellular Networks*, 2013. 51–65.

[24] D Benyamina, A. Hafid, M. Gendreau, and N. Hallam, "Optimization Models For Planning Wireless Mesh Networks: A Comparative Study," *2009 IEEE Wireless Communications and Networking Conference*, 2009. doi: 10.1109/wcnc.2009.4917871

[25] C. Prommak, and C. Wechtaison, "Network Planning and Optimization for Multi-Hop Relay Placement in WiMAX Networks," *Journal of Computer Science*, vol. 8, no. 9, 2012. doi: 10.3844/jcssp.2012.1414.1421

[26] Q. Ai, W. Feng, W. Wu, Y. Liu, G. Liu, and T. Bao, "Design of Power Wireless Private Network Coverage Prediction System," *Proceedings of the 3rd International Conference on Mechatronics Engineering and Information Technology (ICMEIT 2019)*, 2019. doi: 10.2991/icmeit-19.2019.56

[27] X. Xiong, Q. Li, X. Ge, and H. Lou, "Capacity Modeling and Performance Analysis of OAM-OFDM Wireless Communication Systems," *IEEE Access*, vol. 8, 2020. doi: 10.1109/access.2020.3021822

[28] B. Zheng, and D. Yun, "A Wireless Network Communication Capacity Control Technology Based on Fuzzy Wavelet Neural Network," *Wireless Communications and Mobile Computing*, vol. 2021, 2021. doi: 10.1155/2021/9994200

[29] O. Salih, C.-X. Wang, B. Ai, and R. Mesleh, "Adaptive Generative Models for Digital Wireless Channels," *IEEE Transactions on Wireless Communications*, vol. 13, no. 9, 2014. doi: 10.1109/twc.2014.2325028

[30] H. Shimada, S. Tagashira, T. Nakanishi, and A. Fukuda, "Evaluation of a Location Management System for Wireless Communicating Mobile Computers," *Information Networking: Wireless Communications Technologies and Network Applications*, 2002. doi: 10.1007/3-540-45801-8_16

[31] N. Uncu, and R. Erol, "Capacity Optimization in Dynamically Routing Computer Network Systems," *Tehnički Glasnik*, vol. 16, no. 1, 2022. doi: 10.31803//tg-20201215233216

[32] H.-S. Jang, and J.-H. Baek, "Mobility Management Scheme with Mobility Prediction in Wireless Communication Networks," *Applied Sciences*, vol. 12, no. 3, 2022. doi: 10.3390/app12031252

[33] A. Omri, and M. O. Hasna, "Interference Management Schemes for Multi-User Cooperative Wireless Networks," *Wireless Communications and Mobile Computing*, vol. 16, no. 9, 2015. doi: 10.1002/wcm.2593

[34] Y. Mao, J. Zhang, S. Song, and K. B. Letaief, "Stochastic Joint Radio and Computational Resource Management for Multi-User Mobile-Edge Computing Systems," *IEEE Transactions on Wireless Communications*, vol. 16, no. 9, 2017. doi: 10.1109/twc.2017.2717986

[35] B. Myagmardulam *et al.*, "Path Loss Prediction Model Development in a Mountainous Forest Environment," *IEEE Open Journal of the Communications Society*, vol. 2, 2021. doi: 10.1109/ojcoms.2021.3122286

[36] A. D. M. Africa, "Radio Wave Propagation: Simulation of Free Space Propagation Path Loss," *International Journal of Emerging Trends in Engineering Research*, vol. 8, no. 2, 2020. doi: 10.30534/ijeter/2020/07822020

[37] D. O. Oluseun, S. Thomas, O. Idowu-Bismark, P. Nzerem, and I. Muhammad, "Absorption, Diffraction and Free Space Path Losses Modeling for the Terahertz Band," *International Journal of Engineering and Manufacturing*, vol. 10, no. 1, 2020. doi: 10.5815/ijem.2020.01.05

[38] A. S. Yaro, and A. Z. Sha'ameri, "Effect of Path Loss Propagation Model on the Position Estimation Accuracy of a 3-Dimensional Minimum Configuration Multilateration System," *International Journal of Integrated Engineering*, vol. 10, no. 4, 2018. doi: 10.30880/ijie.2018.10.04.006

[39] Z. Nadir, M. Bait-Suwailam, and M. Idrees, "Pathloss Measurements and Prediction using Statistical Models," *MATEC Web of Conferences*, vol. 54, 2016. doi: 10.1051/matecconf/20165405006

[40] I. O. Akhideno, and O. Eguasa, "Investigating the Path Loss of Cellular Mobile Network in Suburban Areas of Benin Metropolis of Nigeria at 910mhz," *Fudma Journal of Sciences*, vol. 7, no. 2, 2023. doi: 10.33003/fjs-2023-0702-1080

[41] M. I. Nashiruddin, M. T. B. Sihotang, and M. A. Murti, "Comparative Study of Low Power Wide Area Network Based on Internet of Things for Smart City Deployment in Bandung City," *Indonesian Journal of Electrical Engineering and Computer Science*, vol. 25, no. 1, 2022. doi: 10.11591/ijeecs.v25.i1.pp425-439

[42] M. S. Mollel, and M. Kisangiri, "An Overview of Various Propagation Model for Mobile Communication," *Proceedings of the 2nd Pan African International Conference on Science, Computing and Telecommunications (PACT 2014)*, 2014. doi: 10.1109/scat.2014.7055150

[43] L. F. P. Martínez, C. Hernández, and E. Rodríguez-Colina, "A Spectral Opportunities Forecasting Method in a Mobile Network Based on the Integration of COST 231 Walfisch-Ikegami and Wavelet Neural Models," *Contemporary Engineering Sciences*, vol. 10, 2017. doi: 10.12988/ces.2017.68149

[44] W. Yamada, M. Sasaki, and N. Kita, "Extended Walfisch-Bertoni Propagation Model to Cover Short Range and Millimeter-Wave Bands," *Radio Science*, vol. 56, no. 3, 2021, doi: 10.1029/2020rs007161

[45] Ł. Januszkiewicz, "Model for Ray-Based UTD Simulations of the Human Body Shadowing Effect in 5G Wireless Systems," *International Journal of Antennas and Propagation*, vol. 2018, 2018. doi: 10.1155/2018/9084830

[46] O. Salih, C.-X. Wang, B. Ai, and R. Mesleh, "Adaptive Generative Models for Digital Wireless Channels," *IEEE Transactions on Wireless Communications*, vol. 13, no. 9, 2014. doi: 10.1109/twc.2014.2325028

[47] M. Celaya-Echarri *et al.*, "Wireless Channel Characterization and System Analysis of Complex Utility Tunnel Environments," *The 6th International Electronic Conference on Sensors and Applications*, 2019. doi: 10.3390/ecsa-6-06559

[48] B. Aminian, J. Araújo, M. Johansson, and K. H. Johansson, "GISOO: A Virtual Testbed for Wireless Cyber-Physical Systems," *IECON 2013 – 39th Annual Conference of the IEEE Industrial Electronics Society*, 2013. doi: 10.1109/iecon.2013.6700049

[49] M. Szczodrak, Y. Yang, and L. P. Carloni, "AN Open Framework to Deploy Heterogeneous Wireless Testbeds for Cyber-Physical Systems," *2013 8th IEEE International Symposium on Industrial Embedded Systems (SIES)*, 2013. doi: 10.1109/sies.2013.6601494

[50] N. M. Hassan, and X. Fernando, "Massive MIMO Wireless Networks: An Overview," *scite.ai*. Accessed: Aug. 05, 2023. [Online]. Available: https://doi.org/10.32920/14639130.v1

[51] R. Borgaonkar, L. Hirschi, S. Park, and A. Shaik, "New Privacy Threat on 3G, 4G, and Upcoming 5G AKA Protocols," *Proceedings on Privacy Enhancing Technologies*, vol. 2019, no. 3, pp. 108–127. Jul. 2019, doi: 10.2478/popets-2019-0039

[52] K. A. Bakare, L. I. Bagiwa, M. M. Nafisa, and A. Abdulsalam, "A Review on 5G Wireless Network Implementation Strategies in Nigeria," *Fudma Journal of Sciences*, vol. 5, no. 2, Art. no. 2, Jul. 2021. doi: 10.33003/fjs-2021-0502-633

[53] D. Chen, J. Y. Khan, J. Brown, M. A. Javed, and Y. Zhuang, "A 6LoWPAN OPNET Simulation Model for Machine-to-Machine Communications," *Transactions on Emerging Telecommunications Technologies*, vol. 31, no. 11, 2020, doi: 10.1002/ett.4120

[54] A. Ceco, and S. Mrdović, "Test Bed for Network Protocols Optimization," *Telfor Journal*, vol. 11, no. 1, 2019. doi: 10.5937/telfor1901014c

[55] A. Varga, and R. Hornig, "An Overview of the OMNeT++ Simulation Environment," *Proceedings of the First International ICST Conference on Simulation Tools and Techniques for Communications Networks and Syste*, 2008. doi: 10.4108/icst.simutools2008.3027

3 Computational Modelling of Communication Systems and Networks

Lateef Adesola Akinyemi, Mbuyu Sumbwanyambe and Ernest Mnkandla

3.1 INTRODUCTION

The goal of stochastic geometry (SG), a branch of practical probability, is to give comprehensible mathematical frameworks and suitable statistical techniques to investigate and evaluate unpredictability in two-dimensional space R^2 [1] or in bigger dimensions. Applications in a variety of scientific fields, including forestry, image analysis, geophysics, neurophysiology, cardiology, finance, and economics, served as the primary impetus for its advancement and development.

In the context of communication networks, the locations of user equipment (UE) and base stations (BSs) are dispersed at random over a huge number of possibilities, making it difficult and time-consuming to design the system for any network realisation [2–3]. Instead, the position of nodes, that is, user equipment (UEs) and/or base stations (BSs) [2–9] is evaluated statistically to examine their interaction, which inherently takes into account all feasible network realisations and encapsulates the primary network connectivity performance dependencies (capacity/throughput and reliability).

The terms "stochastic geometry models of wireless networks" are used in both the telecommunications and mathematical fields to describe mathematical models that are based on stochastic geometry and are designed to represent particular characteristics of wireless communication networks. The associated study investigates these models with the aim of better understanding wireless communication networks to forecast and control different performance metrics [5–8]. Furthermore, assuming that the idea of employing a statistical distribution to abstract the range of alternative topologies for networks is substantially analogous to the method of taking into account a statistical distribution to describe the infinite opportunities of multiple-path fading and shadowing, then this is generally understood.

Moreover, their use in research on numerous wireless network technologies, including hybrid cellular networks, ad hoc networks for mobile devices, networks of sensors, automobile ad hoc networks, intelligent radio networks, and wireless ad hoc networks significantly increased as a result [6–9]. Key performance and quality of service criteria are commonly defined using the theory of information principles, such as the signal-to-interference-plus-noise ratio (SINR), which provides the mathematical basis for building network connection and coverage.

Furthermore, a collection of (data-theoretic) links that are physically connected and share a frequency spectrum constitute a wireless network. Each channel is a collection of transmitting and receiving elements cooperating to transfer data. The most straightforward sort of channel is the point-to-point channel, which transmits data from a single transmitter to a single receiver. The broadcast channel is described in the theory of information as a complex situation where a single transmitter seeks to send many types of data to several recipients. For instance, the downlink of a cellular

DOI: 10.1201/9781003457428-4

network may experience this [10–11]. On the other hand, a multiple-access channel employs a large number of transmitters to deliver various types of data to a single receiver [10]. Using a many-to-one, for instance, this situation happens during a cellular network for the uplink mode scenario.

3.1.1 MOTIVATION

The motivation for the study in this chapter emanates from the fact that the subject of machine learning, specifically federated learning (FL), is one of the powerful tools that can be utilised to solve the problem of multiple user-friendly relationships, which have been covered in several works [12–14]. They have concentrated on putting forth fresh transmission designs to lessen the hierarchical nature of FL communication overhead in small-scale cellular networks. It is challenging to immediately implement their innovative communication architectures to existing cellular networks with distance-based user association and inter-cell interference, which is why they have been presented for telecommunication-efficient FL.

Additionally, when taking into account the geographic deployment of nodes, their performance study results cannot readily provide insights into massive amounts of mobile network performance. This study looks at multi-user association-based model aggregation for FL in large-scale cellular networks as a result. Instead of putting forth a replacement for the current cellular networks, our effort in this study focuses mainly on analysing the model aggregation performance and optimising a few system parameters such as spectral efficiency, transmission efficiency and so on for communication-efficient FL.

3.1.2 CONTRIBUTION OF THE CHAPTER

The improvements of this research are summarised as follows, taking into account the previously mentioned reasons in this book chapter:

1. The in-depth review offered here provides current works on stochastic geometry applications and machine learning in the computational modelling of wireless networks, providing a hands-on for researchers who desire to conduct further studies in this area.
2. This book chapter describes a few potential applications of diverse stochastic geometry in wireless communication and machine learning methods.
3. To the greatest degree of our understanding, this serves as the primary study that analyses the model aggregation performance, optimises certain system parameters for communication-efficient federated learning (FL) within existing cellular networks, and takes the geometry of wireless communication networks into account when analysing machine learning techniques, for instance, FL performance;
4. Using an algorithm centred on stochastic geometry, we construct the closed-form mathematical expressions for the estimated coverage probability for a few distinct circumstances in small-cell networks where each base station is capable of receiving updates from numerous connected devices with orthogonal bandwidth allotment;
5. Using the closed-form expressions, we provide a method that iterates through the application of federated learning and other schemes to optimise communication variables to achieve a target simulation aggregation rate. This can assist in minimising the amount of time and maximising the spectral efficiency and bandwidth needed for FL in substantial cellular networks.
6. Assessment, evaluation, and simulation outcomes provide valuable insights into the behaviours of FL across large-scale small-cell networks and act as a roadmap for creating cellular networks that support wireless FL. In the model aggregation of FL under the current cellular network with distance-based user connection, the objective is to maximise the total aggregate rate, transmission capacity/efficiency, communication throughput and outage probability.

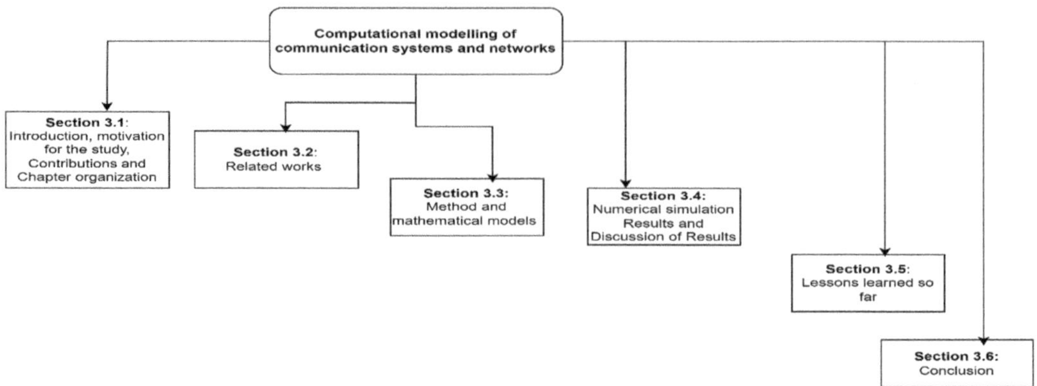

FIGURE 3.1 The organisational layout of the chapter.

7. The novel aspect of this study is further enhanced by the application of the FL algorithm and the inclusion of stochastic geometry. As far as the authors are aware, this is the first time that an investigation of this nature has been conducted.

3.1.3 Organisation of the Chapter

Section 3.1 discusses the introductory section and contributions of the book chapter on computational modelling of wireless systems and networks. Furthermore, Section 3.2 discusses the related work regarding the modelling of wireless networks and systems using the stochastic geometry approach and a summary of the reviewed works is equally presented in a tabular form stating the strengths and weaknesses. The mathematical modelling of wireless networks using stochastic geometry is formulated and presented in Section 3.3. Numerical simulation results are presented and discussed in Section 3.4. More importantly, lessons that have been learned and subsequently discussed in the computational modelling of wireless networks and systems are presented in Section 3.5. Section 3.6 completes the book chapter by making conclusions regarding the modelling of wireless networks and systems computationally. The organisational structure for this chapter is depicted in Figure 3.1.

3.2 RELATED WORK ON STOCHASTIC GEOMETRY-BASED MODELS IN WIRELESS NETWORKS

This section highlights the up-to-date review in line with the title of the chapter of this study. Furthermore, the literature review explores state-of-the-art research on system and mathematical modelling, application of stochastic geometry, optimisation in wireless communication and finally the use of artificial intelligence such as federated learning in wireless networks.

Furthermore, numerous intriguing questionnaires, surveys, books, technical conferences and journal papers on the application of stochastic geometry (SG) in wireless communications have been developed due to its mathematical flexibility and rich theoretical background, on the one hand, and the rapid development of novel communication ideas and methods, on the contrary [15–20, 21–26]. In this regard, the research conducted in [15] is the initial research effort to provide a summary of the primary SG models and techniques employed during the assessment concerning communication networks.

The previous references discussed in this piece, where SG techniques were very useful, are given special attention. (1) Static link systems to determine the key numerical characteristics of wires linking mobile users and concentrating locations; these characteristics are then applied to determine the monetary value of the equipment as an indicator of the terminal population; (2) wireless networks to examine how network structure affects critical performance indicators centred on their signal-to-interference-plus-noise ratio (SINR) phases, such as service coverage, handover, and paging; and (3) ad hoc networks, which are used to investigate connection characteristics of the stochastic networks predicated on the SINR.

The study in [16] is a guide that looks at the use of analytical techniques from SG, percolation theory, and random geometric graphs to assess disruption in huge-scale networks of ad hoc devices and subsequently extract the associated performance metrics. A thorough review of the research on the modelling and evaluation of wireless networks concerning SG is provided in [17]. Given its growing significance in the wireless communications of the future, the work focuses on multitier and cognitive networks. Although quite a few other radio access network (RAN) technologies, such as unmanned aerial vehicle(UAV)-aided communication networks [19], centralized radio access networks (C-RAN) [27–28], and Fog radio access networks (F-RAN) [20], have evolved since the study of [17].

Additionally, various fifth-generation and beyond (5G/B5G) technological enablers including the non-orthogonal multiple access (NOMA) schemes [29], In-band full-duplex (IBFD) communications [30], and physical layer security [31] have proliferated. With a focus on the fading of Rayleigh and Poisson point process (PPP) dispersed nodes, the instructional paper in [18] introduced generative analytical tools widely utilised in the literature to generate the SINR distribution. A new modelling paradigm from SG will be crucial to capture the implications of such heterogeneity, such as cell association, uplink-downlink relationship, and nodes' mobility. The works in [21] and [22] emphasised the increasing pattern in contemporary wireless networks regarding variability and level of complexity.

Additionally, the study published in [20] talked about important methods that are employed to construct the geographical distribution if considering a pair of generative cases because communication in wireless networks is closely related to the distribution separation within the receiving end and the transmitting node/s along with the dispersion of the infringement power. Nodes deployed freely and evenly within a restricted region of the 2-D surface R^2 and nodes spread in the two-dimensional plane R^2 following a PPP. The study in [26] is a pedagogical piece that demonstrates how SG has been carefully employed to characterise interruption in wireless networks.

At the same time, papers in [24] and [25] evaluated existing findings on how SG approaches have been investigated to represent accurately the interference impact in ad hoc networks. Therefore, Table 3.1 gives a summary of the reviewed works in this study with their respective strengths and weaknesses where applicable to reveal and identify the research gap in the literature
Thus, by reviewing the most recent publications in books and papers with the corresponding fields of application of this study, the first two contributions of this research have been satisfied.

From Figure 3.2, it can be observed that the related works carried out in this chapter centre on computational modelling of communication wireless networks and systems based on the types of the environment, large-scale fluctuations, small-scale fluctuations, stochastic geometry approach in different space ranging from one-dimensional to multi-dimensional form, transceiver characteristics, the channel of the transmission, and the various methods employed in the literature. More importantly, this illustration in Figure 3.2 is provided to ease the understanding of the basic and underlying concept as regards the reviewed works on computational modelling of communication systems and networks which is the main subject of this chapter.

TABLE 3.1
Summary of the Literature Review

Reference	Contributions/Description	Techniques	Weaknesses	Strengths
S. Zuyev (2009) [15]	How telecommunications problems can be solved using stochastic geometry as an analytical and computational tool for contemporary communication networks	Emphasis on stochastic geometry (SG) concerning wireless networks	Did not consider addressing the application of machine learning for the actual implementation of wireless networks	Comprehensive and rigorous discussion on the benefits and challenges of applying stochastic geometry scheme to wireless communications
Haenggi et. al. (2009) [16]	The use of random graphs and stochastic geometry to design and proffer solutions to wireless networks challenges	For the examination and development of wireless networks, stochastic geometry and random graphs are employed in the study	Failed to explore the use of either metaheuristic algorithms or machine learning approaches to solve wireless-based network systems	The demonstration of the use of random graph and stochastic geometry techniques formed the greatest strengths of the study with a thorough review
El Sawy et. al. (2013) [17]	An Overview of Stochastic Geometry for Multi-Tier and Intelligent Cellular Wireless Networks Modelling, Evaluation, and Implementation was the main contribution of the study	The use of probabilistic topology or stochastic geometry for the architecture, evaluation and modelling of wireless networks	The issue of probability outage and coverage using any machine learning algorithms was not discussed nor addressed.	The strength of the survey lies in the area of mobile communication networks that are multi-tiered and intelligent
Andrews et al. (2016) [18]	Employing Stochastic Geometry, the analysis of the primer Cellular Networks was investigated.	In computing the SINR using a stochastic geometry scheme, three cases were considered, namely: uplink, downlink and multi-tier-based downlink	The issues such as latency, delay and spectral efficiency in addition to the application of machine learning schemes were not addressed.	Employing stochastic geometry to simulate and study cellular network functionality. Calculating the signal-to-interference-plus-noise ratio (SINR) distribution is our main concern.
Mozaffari et. al. (2019) [19]	A tutorial on unmanned aerial vehicles (UAVs) for Wireless Networks: Uses, Difficulties, and Unresolved Challenges was presented in this study	Different conceptual structures and mathematical techniques were discussed, including the theory of games, stochastic geometry, optimisation theory, algorithms for learning, and stochastic geometry. It was discussed how to use these instruments to solve particular UAV issues.	The detailed mathematical simulation and implementation of this problem using AI or ML were not considered.	A comprehensive review of the use of stochastic geometry to solve UAV-related problems was presented.

Reference	Description	Limitation	Remarks
Li et al.(2018) [20]	A thorough analysis of the most advanced caching technologies more recently created for cellular networks, including macro-cellular networks, heterogeneous networks, device-to-device networks, cloud-radio access networks, and fog-radio access networks was presented	The shortcoming of the study was that no numerical results nor implementation of this excellent work regarding the application of machine learning algorithms presented	The detailed study of the use of SG and other schemes was discussed comprehensively with future directions presented
Andrews et al. (2013) [21]	Heterogeneous networks (HetNets): Seven Cellular Paradigmatic Shifts in one technology were presented using SG and other novel schemes	The use of any ML/AL or its associated federated learning was not employed in the work.	The spectral performance and the outage/coverage probability are two commonly employed measurements presented in the study with striking results
Andrews et al. (2010) [22]	A preamble on the Spatial modelling and Analysis in Wireless Networks was presented using SG	The work failed to incorporate the use of federated learning or ML/AI	A thorough analysis of the obtained numerical results was presented with the future direction of spatial models
Moltchanov Dimitri (2012) [23]	Distributions of distance in random networks were investigated.	The study did not consider the application of ML/AI to solve wireless-based network problems. No numerical results were presented despite good classical and rigorous ideas put forward	It explored how to combine the homogeneous Poisson process with independently dispersed points in a specific area of two-dimensional space. More importantly, a well-detailed mathematically-inspired was presented.
Haenggi et al. (2008) [24]	The study focuses on large-scale wireless networks with interference using a Poisson point process scheme. The network topology (spatial geographic distribution of simultaneously broadcasting nodes) and the path degradation law (signal degradation with length) are the two fundamental factors that affect interference in which SG is employed to solve the problem in the study	The study failed to apply any metaheuristic algorithm or ML/AI technique.	Characterising the interference statistics is essential. The network topology (spatial geographic distribution of simultaneously broadcasting nodes) and the path degradation law (signal degradation with length) are the two fundamental factors that affect interference.

(continued)

TABLE 3.1 (Continued)
Summary of the Literature Review

Reference	Contributions/Description	Techniques	Weaknesses	Strengths
Wu et al. (2018) [31]	This study provides the most up-to-date overview of the physical layer security research on several promising 5G technologies, including physical layer security coding, massive multiple-input multiple-output, millimetre wave	It focuses on secrecy in physical layers using different schemes of codes including SG	The work did not consider the application of federated learning to solving wireless network issues	A thorough review was presented for different cases in the physical layer most especially secrecy in mobile networks
Shoewu et al. (2022) [32]	5G New Radio Wireless Standards for UAV Cellular Communication was the main focus of the work	The use of metaheuristic algorithms were employed in radio resource management for 5G networks	The work did not consider the application of ML/AI or federated learning to the formulated problem	The strength of the work lies in the application of metaheuristic algorithms to UAV-based cellular networks with striking numerical results obtained.
Akinyemi et al.(2023) [33]	Utilising artificial intelligence to address 6G privacy and security concerns was the focus of the work	The use of supervised learning algorithms such as linear regression (LR), Random Forest (RF), K-nearest neighbors (KNN) and Support Vector Machine (SVM) was utilised in the study.	The study failed to demonstrate the use of SG and by extension federated learning in the work	A robust multi-objective function was formulated and solved using supervised learning algorithms with striking numerical simulation results obtained and presented.

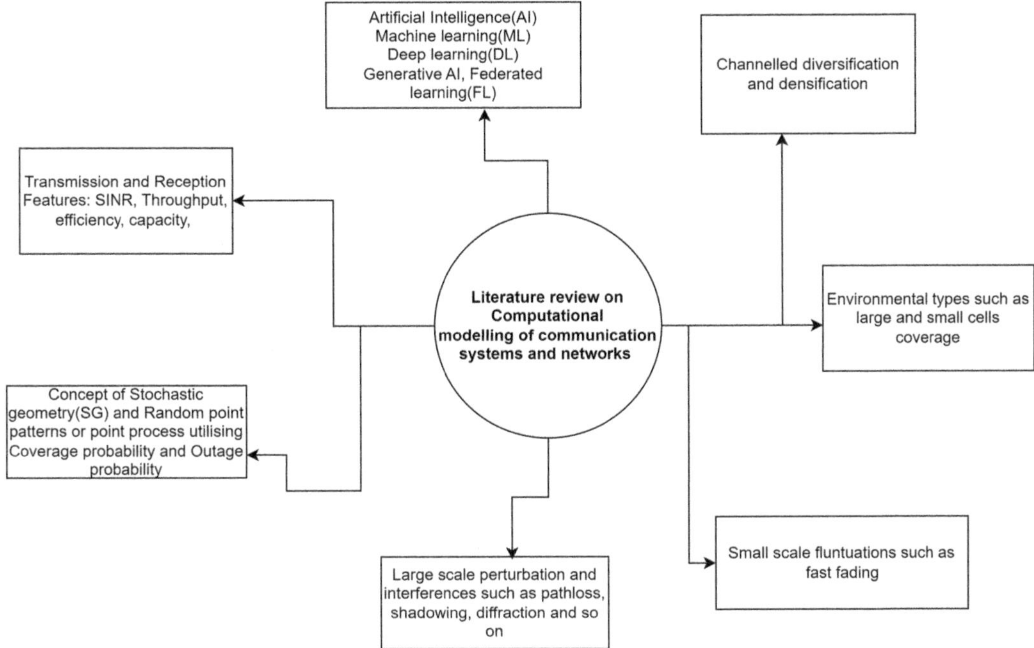

FIGURE 3.2 The organogram nature of the literature review carried out in this book chapter.

3.3 METHOD AND MATHEMATICAL MODELS: SG AND FEDERATED LEARNING IN WIRELESS NETWORKS

This section provides the mathematical modelling, methods and algorithm alongside the flowchart employed in obtaining the numerical results presented in the subsequent section in this chapter. Subsections 3.3.1 through 3.3.5 cover model foundations, propagation model, node positioning model, Poisson process modelling, and mathematical models of Stochastic geometry and federated learning in wireless networks from a computational standpoint, respectively.

3.3.1 MODEL FUNDAMENTALS FOR WIRELESS NETWORKS

Some underlying wireless systems and networks have similar fundamental models, but others have different ones depending on the channel and environment that are being considered. Nodes, which can be either transmitters or receivers or both depending on the system, are the building blocks of a wireless network and are responsible for producing, relaying, or consuming data throughout the network. As an illustration, base stations and users in a cellular phone network or sensor nodes in a sensor network. The mathematical representation of the signal propagation and node positioning is necessary before stochastic geometry wireless models can be developed. How signals travel from transmitters to receivers is depicted by the propagation model [34].

The positions of the nodes are idealised and shown as point processes in the node location or positioning model. The cellular network's context and choice of models are determined by these factors. The specific network design (for example, cellular) and the channel or medium access control (MAC) protocol, which regulates the channels and, in turn, the network's communicative structures, determine the network type. The MAC protocol specifies, based on specific rules, when transmitter-receiver pairs can access the network both in time and space, which also influences the active node placement model. This is done, in particular, to prevent the collision of transmissions in the network.

3.3.2 Propagation Model

The propagation model plays a pivotal role in the computational modelling of wireless communication systems and networks. The propagation of electromagnetic signals (or waves) through different media, such as air, requires appropriate and manageable models that account for multipath propagation, which is produced by signals clashing with obstacles like buildings and results in refraction, reflection, diffraction, and dispersion. A component of the stochastic geometry wireless network model is the propagation model. Consideration of propagation models with two distinct portions that include the random and deterministic (or non-random) components of signal propagation is a popular strategy [34].

When modelling the power decay of electromagnetic signals, the deterministic component is typically represented by a path-loss or attenuation function that takes into account the signal's propagation distance (from its source). A fast-decaying exponential function, a straightforward power-law function (such as the Hata model), a combination of both, or any decreasing function could make up the distance-dependent path-loss function. Because it is tractable, models frequently include the power-law function, $l(\|x - y\|) = |x - y|\beta$ where the path-loss exponent $\beta > 2$, and $|x - y|$ represents the distance between the point y and the signal source at x.

3.3.3 Node Positioning Model

Node placement in wireless networks and systems is crucial for better coverage and performance, which improves the quality of service that mobile users receive. The selection of a mathematical model for the position of the network nodes is a crucial step in stochastic geometry network models. A (spatial) point process is a stochastic or random structure. It is typically assumed that the nodes are represented by (idealised) points in some space (commonly Geometric R^n, and even regularly in the plane R^2) [35].

3.3.4 Poisson Process Modelling

In this subsection, the location of wireless network nodes can be modelled using a variety of point procedures. The Poisson process, which results in a Poisson network model, is the most widely utilised of these [34]. Due to its highly tractable and well-studied nature, the Poisson process, in general, is frequently employed as a mathematical model across a wide range of disciplines [35–36]. The Poisson process is frequently thought to be homogenous with a fixed node density, indicating that it is stationary. This implies that the probability of n points or nodes in a bounded region $\mathbf{Q}$ is given by a process called Poisson on the plane in Eq. (3.1):

$$P(m) = \frac{(\lambda|Q|)^m}{m!} e^{-\lambda|Q|},\tag{3.1}$$

where $|Q|$ denotes the area of Q, m! represents m factorial. Eq. (3.1) can be extended to the R^3 scenario by replacing the term area with the volume form.

Furthermore, it should be known that Poisson models' "complete independence," which essentially states that two (or more) disjoint (or non-overlapping) bounded regions each contain two (or more) a Poisson number of points that are independent of each other is the primary factor that makes them computationally easy or easy to work with. This key trait, which is often used to characterise the Poisson process, sets it apart [37]. Poisson processes' complete autonomy or 'randomness' [38] quality results in several beneficial characteristics and outputs of point process activities, such as the superposition property: The superposition of m Poisson processes with densities varying from λ_1 to λ_m is another Poisson process with density compactly stated as in Eq. (3.2):

$$\lambda = \sum_{k}^{m} \lambda_k \tag{3.2}$$

3.3.5 Mathematical Models for SG and FL in Wireless Networks: A System Model

This subsection essentially discusses and presents the system model regarding the algorithm of federated learning, the mathematical formulation involving the stochastic approach, the coverage probability and the transmission capacity of the considered case in this chapter.

In particular, the mathematical formulation of the network regarding probability coverage and the system model for the examined wireless networks with a federated learning scheme is covered in Subsections 3.3.5.1 and 3.3.5.2, respectively.

When FL is employed in wireless networks, specifically small-cell networks, the convergence centre (CC) or focus centre generates the global framework by integrating the periodically modified models that end-users send to the CC via BSs. In its wired broadband links, the connections between the BSs and CC are thought to have limitless bandwidth. The uniform PPPs BS with density BS (ω_{BS}) and UE (μ_u,) with density μ_u, presuppose that BSs and users are positioned in the Geometric plane. The received signal-to-noise-plus-interference (θ_u =SINR) of a user can be described by considering interference as additive noise, which is then added to the received signal. The Dirichlet tessellation diagram for the deployment of mobile users is depicted in Figure 3.3. Figure 3.3 essentially shows the multi-based users with single base stations and multi-driven base stations with multi-users for the area of coverage under consideration in this study. Therefore, the application of FL in wireless communication is highlighted in this study. The following succinct statement sums up the concept of a federated learning scheme (FLS) in wireless (small-cell) networks:

Assuming a macro base station is positioned at the origin, that is, 0 ∈ MBS. The received signal of such BS can be compactly represented in wireless communications as:

1. At the start of update cycle m, CC uses BSs to communicate the entire model's variables to the dispersed subscribers/users. Because BSs transmit an identical signal, it is expected that each user can effectively pick up the intended signal with no hindrance;
2. Every subscriber/user communicates their local update model to the connected BS after updating the model settings based on their local information.
3. Each connected subscriber broadcasts its local update model with capacity B, where B represents overall bandwidth if there are U users associated with BS;
4. Every BS calculates the mean of the local modifications it obtained from subscribers it connects with and communicates it to CC;
5. Using the variables combined with the BSs, CC modifies the global model. If the point convergence requirement fails to be fulfilled, the algorithm then updates round (m+1) by starting at stage 1.

Assuming a macro base station is positioned at the origin, that is., 0 ∈ MBS. The received signal of such BS can be compactly represented in wireless communications as in Eq. (3.3):

$$\aleph = r_u s_u^{-w/2} a_u + \sum_{u' \in M_{BS}\, 0,\, \infty} r'_u s_{u'}^{-w/2} a_{u'} + t \tag{3.3}$$

where u denotes users associated with macro base station (MBS), $r_u \sim CN(0,1)$ represents the Rayleigh fading channel gain of user u, s_u represents the connection distance between user u and MBS. The w stands for the path loss factor, a_u represents the transmitting signal of user u,

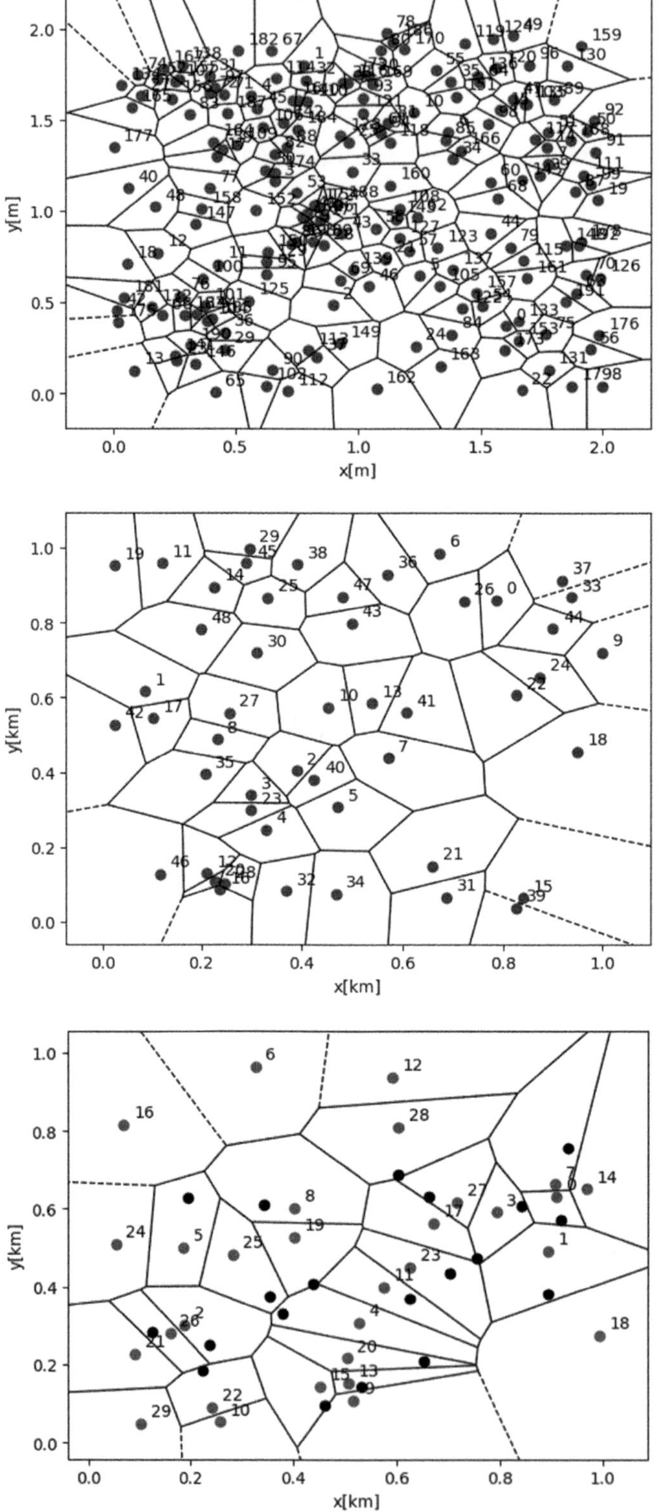

FIGURE 3.3   Deployment of mobile users in a cellular network for single- and multi-tiered systems.

and $t \sim CN(0,\sigma^2)$ denotes the additive white Gaussian noise. In this study, a constant power κ is assumed for all users. The $M_{BS}\,0,\infty$ contains the other cell users disrupting the signal reception of the BS. More importantly, the state information channel is considered to be available at the MBS.

By viewing interference as a set of additive disturbances that are then added to the receiving signal, interference can be used to characterise the received signal-to-noise-plus-interference (SINR) of a user and expressed as in Eq. (3.4):

$$\theta_u = \frac{|r_u|^2\, s_u^{-w}\, \kappa}{\sum_{u' \in M_{BS}\,0,\,\infty}|r_{u'}|^2\, s_{u'}^{-w}\, \kappa + t^2},\tag{3.4}$$

For the rate of transmission denoted as χ, the coverage probability conditionally assuming $x>0$ users within the Dirichlet patterns or tessellations (Voronoi diagram) as in Eq. (3.5):

$$\Lambda(u) = P_r\left[\frac{Q}{X}\log_2(1+\psi) \ge u | X = x\right] = \left[\psi \ge 2^{\frac{u\chi}{Q}} \mid X = x\right],\tag{3.5}$$

where $T = 2^{\frac{u\chi}{Q}} - 1$ represents the SINR of the network, ψ is the total SINR, and Q represents the bandwidth. Therefore, the expected number of total bits of locally trained models in transmission time can be cast compactly as in Eq. (3.6):

$$\mathfrak{I} = \chi\omega_u \sum_{x=1}^{\infty} P_r[X = x]\Lambda(x) = \chi\omega_u \Lambda(x).\tag{3.6}$$

Hence, assuming the densities of the base station (BS) and users are denoted as ω_{BS} and μ_u, respectively. The rate of transmission can be enhanced to improve the model total rate. It should be noted that the bandwidth and time used for a complete model aggregation are inversely related to $\mathfrak{I}$. The optimisation problem can be formulated to maximise the total aggregate rate of the network as Eq. (3.7):

$$maximise\ \mathfrak{I}\tag{3.7}$$
$$(\chi)$$

Furthermore, the density of the macro base station as well as χ can be enhanced to minimise the needed BS distribution to obtain a threshold aggregate rate $\mathfrak{I}$. The formulation is cast as in Eq. (3.8):

$$miniimise\ \varpi_{BS}\tag{3.8}$$
$$(\omega_{BS},\chi)$$
$$s.t.: \mathfrak{I} \ge \mathfrak{I}\_threshold$$

According to a limit on the allowed probability of outages (where an outage happens when the SINR at the receiving end falls within a threshold), capacity for transmission measures the spatial intensity of successful transmissions per unit area in the context of wireless networks. The transmission capacity (i.e., the maximum spatial throughput in terms of densification of networks such that the outage probability ρ_o must be less than or equal to a given condition say ξ that is very small in number) of a network is closely related to the area spectral efficiency of the network. The spectral efficiency can be cast compactly as in Eq. (3.9):

$$\vartheta_{SE} = \rho\omega\log_2\left(1+\psi\right)e^{-\rho\omega\pi r^2\psi^{2/\alpha}\zeta} \tag{3.9}$$

And the transmission capacity $[bit\,s^{-1}Hz^{-1}m^{-2}]$ is expressed as in Eq. (3.10):

$$\phi_{TC} = \max SE = \vartheta_{SE} = \rho\omega\log_2\left(1+\psi\right)e^{-\rho\omega\pi r^2\psi^{2/\alpha}\zeta}, \tag{3.10}$$

$$\text{s.t.}:\rho_o = 1-\rho_s = e^{-\rho\omega\pi r^2\psi^{\frac{2}{\alpha}}\zeta} \leq \xi$$

where ρ_s denotes the probability of successful transmission and r is the distance in kilometres.

Additionally, Eq. (3.7) is equivalent to the optimisation problem of maximisation of the network throughput concerning transmission rate. Using some of the techniques in literature, with further simplification, Eq. (3.8) can be reformulated as in Eq. (3.11):

$$maximise\,f\left(\chi,\omega_{BS}\right) \tag{3.11}$$

$$\left(\chi\right)$$

Eq. (3.11) can be solved computationally and numerically using a technique by Bisection as presented in **Algorithm 1** and a two-step incremental scheme is employed to solve Eq. (3.8) with **Algorithm 2.**

3.3.5.2 Mathematical Formulation of Probability Coverage of Network

In this subsection, authors in [41–46] employed various schemes ranging from artificial intelligence to metaheuristic-based algorithms and the new human-based metaheuristic called deep sleep optimiser [46] to model communication wireless networks and systems in a variety of environments, from single to multi-inspired form. Furthermore, considering its increasing importance in the future of wireless communications, the works in [47–51] concentrate on multi-tier, multi-domain, radiofrequency electromagnetic radiation in wireless networks and socially inspired energy efficiency in wireless communication networks by employing a variety of strategies to solve computational modelling problems, such as genetic algorithms and game theory. The study in this book chapter merits further investigation because those works did not use the FL to solve this problem. The diverse outcomes derived from those investigations indicate a noteworthy enhancement concerning the performance metrics utilised to evaluate their models. Nevertheless, from the standpoint of stochastic geometry-inspired federated learning, the works do not tackle fundamental issues, like transmission capacity, outage probability and so on.

The coverage probability [39–40] is denoted as in Eq. (3.12) and Eq. (3.13):

$$P_{cov} = P_r\{\theta_u > H\} \tag{3.12}$$

where H is the set SINR threshold.

$$\theta_u = \frac{\left|r_u\right|^2 s_u^{-w}\,\kappa}{t^2 + I_{agg(r_o)}} = \frac{\left|r_u\right|^2 s_u^{-w}\,\kappa}{\sum_{u'\in M_{BS}\,0,\,\infty|}\left|r_{u'}\right|^2 s_{u'}^{-w}\,\kappa + t^2}$$

$$P_{cov} = P_r\{ \frac{|r_u|^2 \, s_u^{-w} \, \kappa}{t^2 + I_{aggs(s_u)}} > H \}$$

$$P_{cov} = \{ |r_u|^2 > \left(t^2 + I_{agg(s_u)} \right) \kappa^{-1} H s_u^w \}$$ (3.13)

Note that the channel gain in Eq. (3.13) can be expressed as in Eq. (3.14):

$$\left(|r_u|^2 \sim \exp \gg \right) = E\_Iagg \left\{ e^{\left(-\left(t^2 + I_{agg(s_u)} \right) \kappa^{-1} H s_u^w \right)} \right\}$$ (3.14)

The moment generating function (MGF) and expectation value (E) of the function can be cast as in Eq. (3.15):

$$MGF_X(y) + E_X\left\{ e^{-yX} \right\} = E_{su}\left\{ e^{\left(-\kappa^{-1} H s_u^w t^2 \right)} * MGF_{Iagg}(s_u)\left(\kappa^{-1} H s_u^w \right) \right\}$$ (3.15)

Therefore, the coverage probability can be compactly represented as in Eq. (3.16):

$$P_{cov} = \int_0^{+\infty} E_{su}\left\{ e^{\left(-\kappa^{-1} H \gamma^w t^2 \right)} * MGF_{Iagg}(s_u)\left(\kappa^{-1} H \gamma^w \right) PDF_{su}(\gamma) d\gamma, \right.$$ (3.16)

where PDF is the probability density function.

It should be noted that it seems trivial so far, but where is the primary magic in this case? The answer to this question lies in the concept of stochastic geometry. Stochastic geometry provides us with the mathematical tools for computing, in closed form, the MGF and the probability density function (PDF) of the equation above in Eqs. (3.17 and 3.18):

$$e^{\vartheta x} = 1 + \frac{\vartheta x}{1!} + \frac{x^2 \vartheta^2}{2!} + \frac{\vartheta^3 x^3}{3!} + \dots + \frac{\vartheta^n x^n}{n!}$$ (3.17)

Hence,

$$M_X(\vartheta) = E\left(e^{\vartheta x} \right) = 1 + \frac{\vartheta E(x)}{1!} + \frac{E(x^2)\vartheta^2}{2!} + \frac{\vartheta^3 E(x^3)}{3!} + \dots + \frac{\vartheta^n E(x^n)}{n!}$$

$$M_X(\vartheta) = 1 + \vartheta m_1 + \frac{\vartheta^2 m_2}{2!} + \frac{\vartheta^3 m_3}{3!} + \dots + \frac{\vartheta^n m_n}{n!} + \dots,$$ (3.18)

where m_n is the nth moment. The moment function can be expressed as in Eq. (3.19):

$$M_X(\vartheta) = E\left[e^{\vartheta x} \right] = \int_{-\infty}^{+\infty} dx \, e^{\vartheta x} f_X(x)$$ (3.19)

The aggregate of other cell interference constitutes a marked PPP, where the marks are the channel power gains denoted as $I_{agg}(s_u)$, the PDF of the closest distance follows from the null probability of spatial PPPs represented as $PDF_{su}(\gamma)$, the MGF of the aggregate other-cell interference follows from the probability geometry functional (PGFL) of marked PPPs denoted as $MGF_{Iagg}(s_u)(\vartheta)$ and probability generating functional of marked Poison point process (PPP) *PGFL* are compactly expressed as follows in Eqs. (3.20, 3.21, 3.22, 3.23):

$$I_{agg}(s_u) = \sum_{u \in U \setminus MBS} \kappa |r_u|^2 s_u^{-w},$$
(3.20)

$$PDF_{su}(\gamma) = 2\pi\omega e^{-\pi\omega\varsigma^2}$$
(3.21)

$$MGF_{Iagg}(y) = E_U\left\{|r_u|^2\right\}\left\{e^{-\left(y\sum_{u \in U \setminus MBS} \kappa |r_u|^2 s_u^{-w}\right)}\right\}$$
(3.22)

$$PGFL = e^{\left(-2\pi\omega\int_{-\infty}^{+\infty}\left(1-E_{|r_u|^2}\left\{e^{\left(-y\kappa|r_u|^2\gamma_u^{-w}\right)}\right\}\right)\gamma_u d\gamma_u\right)} = e^{\left(\pi\omega r_u^2\left(1-2F_1\left(1,-\frac{2}{w},1-\frac{2}{w},\frac{-y\kappa}{r_u^w}\right)\right)\right)}$$

$$PGFL = e^{\left(\pi\omega r_u^2\left(1-2F_1\left(1,-\frac{2}{w},1-\frac{2}{w},\frac{-y\kappa}{r_u^w}\right)\right)\right)}$$
(3.23)

From Eq. (3.13), the coverage probability can be compactly expressed after some mathematical manipulation and simplification as in Eq. (3.24):

$$P_{cov} = \int_0^\infty e^{-\left(\kappa^{-1}H\gamma^w t^2\right)} * e^{\left(\pi\omega\gamma^2\left(1-F_1\left(1,-\frac{2}{w},1-\frac{2}{w},-H\right)\right)\right)} * 2\pi\omega\gamma e^{-\pi\gamma^2} d\gamma$$
(3.24)

By ignoring the first term in the integral or equating it to unity, the expression becomes in Eq. (3.25):

$$P_{cov} = \int_0^\infty e^{\left(\pi\omega\gamma^2\left(1-F_1\left(1,-\frac{2}{w},1-\frac{2}{w},H\right)\right)\right)} * 2\pi\omega\gamma e^{-\pi\omega\gamma^2} d\gamma$$

$$P_{cov} = P_{cov}(H,w) = \frac{1}{1-F_1\left(1,-\frac{2}{w},1-\frac{2}{w},-H\right)} = \int_0^\infty MGF_{Iagg}\left(\kappa^{-1}H\gamma^w\right)PDF_{ru}(\gamma)d\gamma \quad (3.25)$$

The rate can be computed as in Eq. (3.26):

$$\Gamma = \frac{Q_w}{\ln 2}\int_0^\infty \frac{P_{cov}(x)}{1+x}dx \left[\frac{bit}{s}\right]$$
(3.26)

The potential spectral efficiency (PSE) can be computed as in Eq. (3.27):

$$PSE = \omega Q_w \log_2(1+H)P_{cov}(H)\left[\frac{\frac{bit}{sec}}{m^2}\right]$$
(3.27)

Additionally, the flowchart for the proposed federated learning scheme for this study is depicted in Figure 3.4 describing the major steps taken to obtain numerical results.

Algorithm 1: Technique by Bisection

---

1. $function\,TB\left(f, v_{low}, v_{high}, c,\ n\right)$
2. $while\, v_{high} - v_{low} > n,\ do$
3. $v_{thres} \leftarrow \left(v_{low} + v_{high}\right) * 0.5$
4. $if\left(f\left(v_{low}\right) - c\right)\left(f\left(v_{thres}\right) - c\right) < 0\ then$
5. $v_{high} \leftarrow v_{thres}$
6. $else$
7. $v_{low} \leftarrow v_{thres}$
8. $end\,if$
9. $end\,while$
10. $return\,v_{thres}$
11. $end\,function$

Algorithm 2: Two-step-inspired incremental scheme

---

1. $Initialiase\,\chi_{min}, \chi_{max}, \omega_{BS}, min, \omega_{BS}, max$
2. $Initialise\,\xi_1,\ \xi_2,\ \chi_{new} \leftarrow \chi_{max}$
3. $Initialise\,\omega_{BS}, new \leftarrow \omega_{BS}, max$
4. $Equalise\,\chi_{thres} = \chi_{new}$
5. $Set\,\omega_{BS,thres} \leftarrow \omega_{BS}, thres$
6. $Set\,g_1\left(\chi\right) = \dfrac{\partial f}{\partial \chi}\left(\chi, \omega_{BS},\ thres\right)$
7. $Set\,\chi_{new} \leftarrow TB\left(g_1\left(\chi\right), \chi_{min}, \chi_{max}, \xi_1, 0\right)$
8. $Equate\,g_2\left(\omega_{BS}\right) = g\left(\chi_{new},\ \omega_{BS}\right)$
9. $Also,\,set\,\omega_{new} \leftarrow TB(g_2\left(\omega_{BS}\right), \omega_{BS\,min}, \omega_{BS}, max, \Im_{thres}\,/\,\mu_u, \xi_2)$
10. $Until\,\left|\chi_{new} - \chi_{thres}\right| < \xi_1\,and\,\left|\varpi_{BS,new} - \varpi_{BS}, thres\right| < \xi_2$
11. $\textbf{Return}\,\omega_{BS}, new, \chi_{new}$

The numerical simulation results for these mathematical derivations are presented in Section 3.4 and discussed afterwards

3.4   NUMERICAL SIMULATION RESULTS AND DISCUSSION OF RESULTS

This section of the chapter presents and discusses the numerical simulation results obtained with its physical interpretation. To support the suggested FL approach for the optimisation-resource

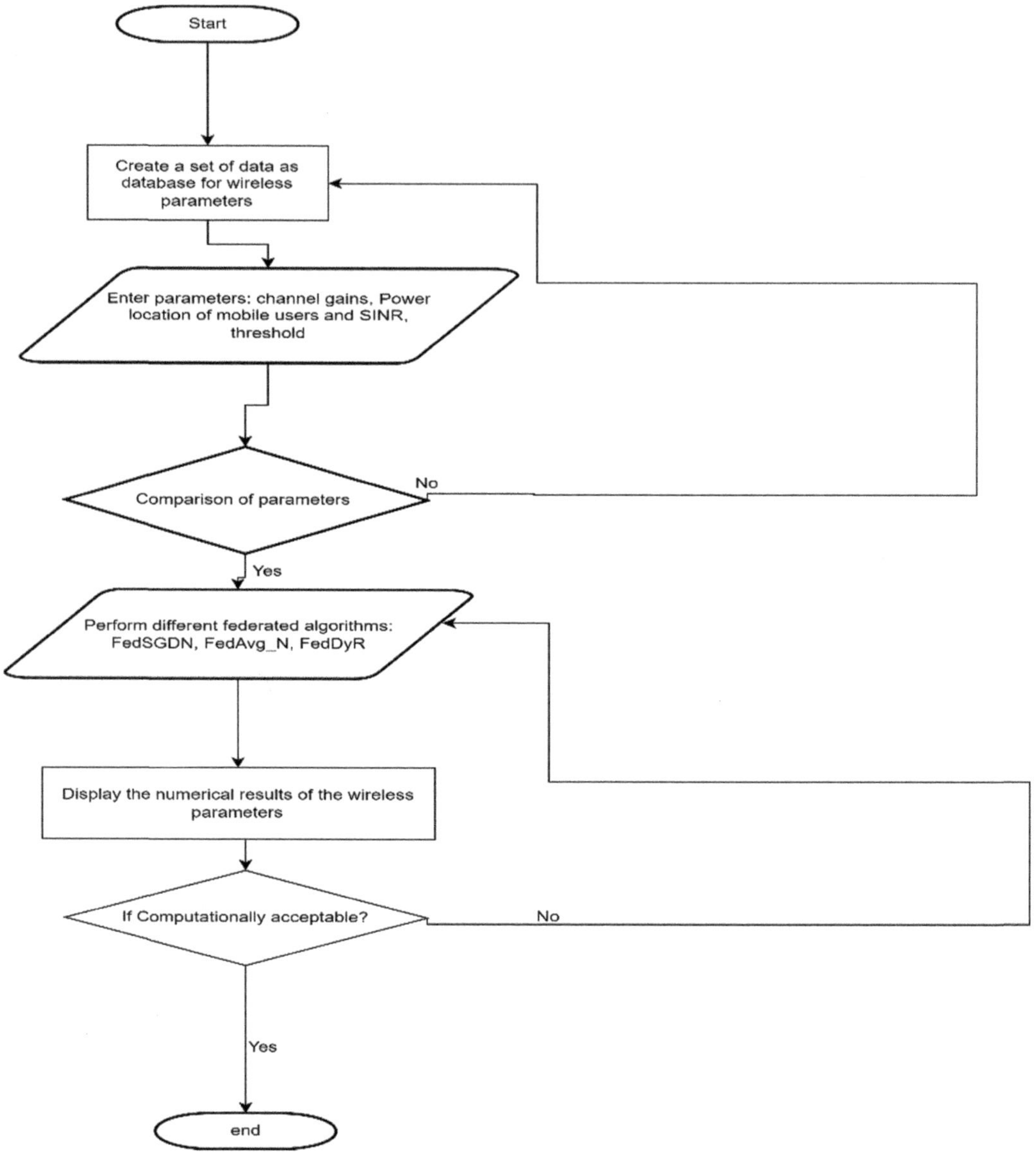

FIGURE 3.4 The flowchart for the proposed scheme: Federated learning.

allocation issue under consideration, the numerical simulation results obtained are presented and discussed accordingly. This study takes into account a single-tier and multi-tiered macrocell with a variable radius that can accommodate a variable number of mobile-user types of equipment and is deployed uniformly and arbitrarily using PPPs and SG.

Python and MATLAB are used to implement our suggested approach. The deployment of users with a single- and multi-tiered network is depicted in Figure 3.3. In wireless communications, the signal-to-interference-plus-noise ratio, or SINR, is regarded as a crucial performance parameter based on information-theoretic assumptions. According to the Poisson network models described, the SINR-based k-coverage probability of an average user (or UE) in a multi-tier network with constant base station density (for each tier) and a common path-loss exponent. The findings apply to

TABLE 3.2
Numerical Simulation Parameters

Symbolic Notation	Description	Value(Unit
α	Path loss factor	2-6
Q	Bandwidth	25(MHz)
μ_u	User density	100(users/)
χ	Transmission rate	20(Mbps)
κ	Transmit power	25(dBm)
t^2	Additive Gaussian noise	-103.5(dBm)
ω_{BS}	Mobile base station density	4(MB/100 users)

any fading, shadowing, or propagation effects as well as the entire SINR domain, that is, not simply when the SINR threshold is H > 1.

For low H values, such as less than 0.1 (or -10 dB), the calculations can take noticeably longer. The integral is multiplied to form the equation for the k-coverage coverage probability as denoted in (Eq. 3.24). More importantly, in the first part of this section, we present the simulation parameters as shown in Table 3.2. The concept of outage probability and coverage probability of the typical user is then explored as a function of the number of mobile density and user's density, SINR threshold, fading severity, and distance between the user equipment (UE) and the mobile base station (MBS). Finally, we validate our numerical findings using FLS, GA, and analytical models.

Except when otherwise noted, we employ the following simulation parameters to all of our numerical results as tabulated in Table 3.2. This integral can be reduced to a closed-form solution for an alleged "interference-limited network" where the noise is set to zero (i.e., $t^2 =0$) as expressed in (Eq. 3.25). If there is network noise, the numerical integration approach fails for large values corresponding to very tiny values of the SINR threshold -13 dB. Because of a characteristic sometimes referred to as propagation invariance based on a Poisson process for the base station configuration that states that propagation impacts exclusively revolve around one crucial moment, these conclusions hold for any fading and/or shadowing probability distribution as long as they comply with a moment condition. The multi-tiered work makes use of closely related factorial moment measures of the SINR process. Of course, in multi-tier, results for a single tier are produced when the number of tiers is set to 1. Single-tier networks are scale-invariant (i.e., unaffected by base station density, omega) when there is no noise (i.e., $t^2 = 0$).

When there is no noise, the ratios between the base station densities (and not their absolute values) matter for multi-tier networks. Hence, it demonstrates that spectral efficiency rises in tandem with the number of users. This has the implication that, within the limits of the predetermined threshold Signal-to-Noise ratio, a channel is used more effectively the more users it carries. To further clarify the idea and make the underlying principle of this study easier to understand, a flowchart is shown in Figure 3.5 that illustrates the systematic approach of the concept and methods involved in obtaining the numerical results from the first stage of identifying the problem to the numerical results.

3.4.1 Discussion of Numerical Results

This subsection discusses the obtained numerical results with their interpretation. In this subsection, the numerical results of spectral efficiency, transmission efficiency, probability outage and so on are examined and discussed.

Figure 3.6 shows the relationship between spectral efficiency (SE) and the base station's SINR (ψ) and density. It is noted that all SINR and density values are equal and growing arithmetically

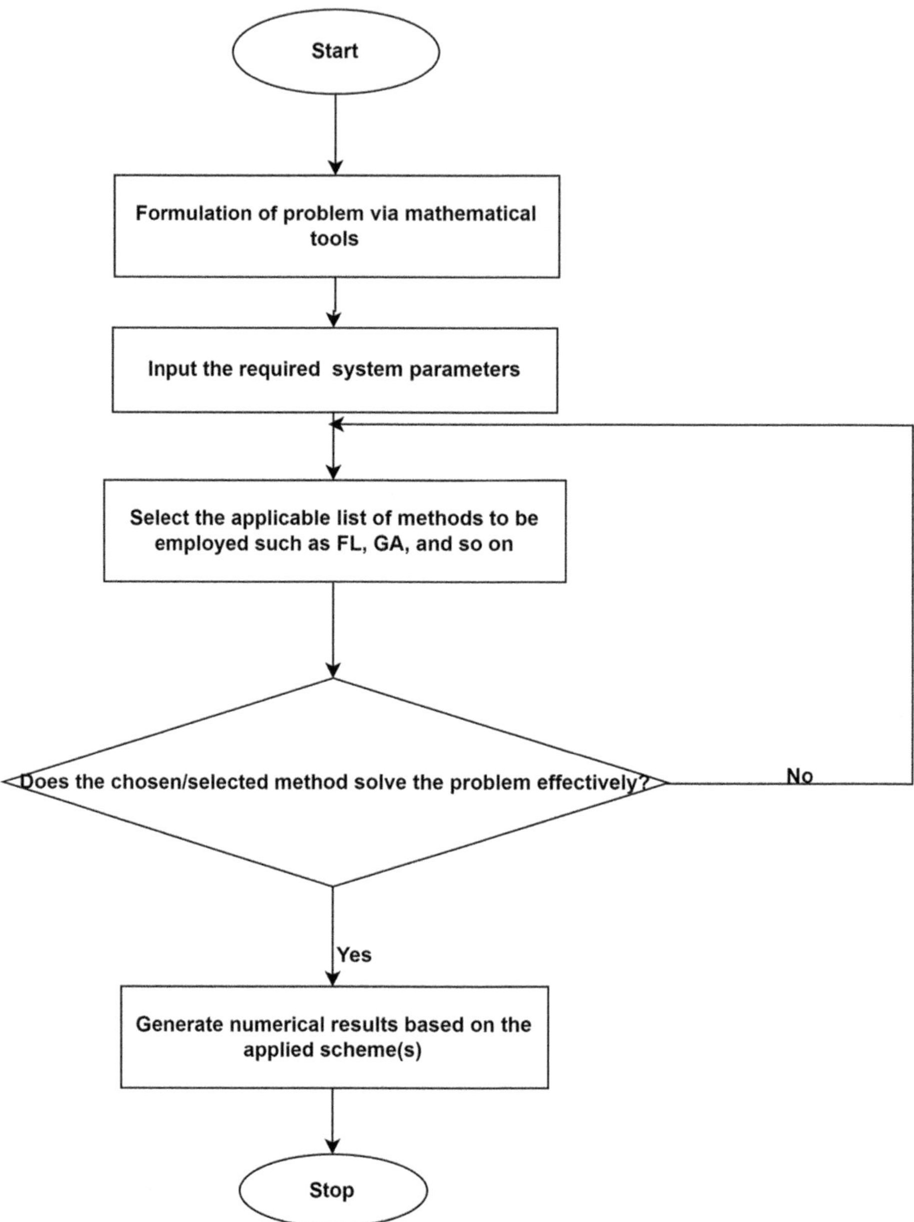

FIGURE 3.5 The flowchart of the systematic approach of the steps involved in obtaining results.

at the zero point. The apex of the curve in relation to the SE is roughly 0.025 for a density ω=0.15. Additionally, the peak yields about 0.027 and 0.028, respectively, for ω=0.10 and 0.05. The SE curve begins to progressively decrease as the SINR values rise, with the possibility of convergence at some time. On the other hand, the same justification may be provided for different values of SINR when SE is plotted against the density of the base station.

In Figure 3.7, the relationship between transmission efficiency (TE) and base station density and SINR is shown. At the zero point, it is demonstrated that all SINR and density values are equal and growing arithmetically. The peak of the curve for the TE is approximately 0.0025 for the density

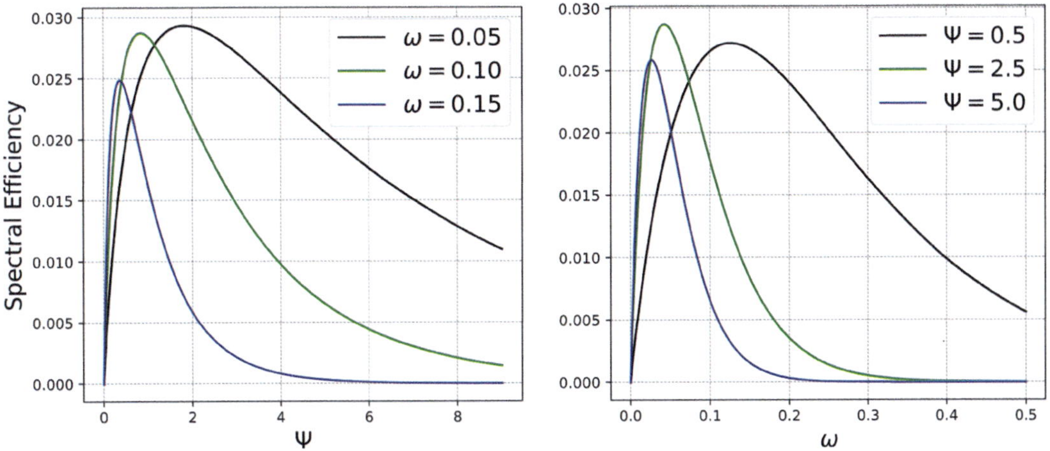

FIGURE 3.6 Spectral efficiency versus SINR and density of networks.

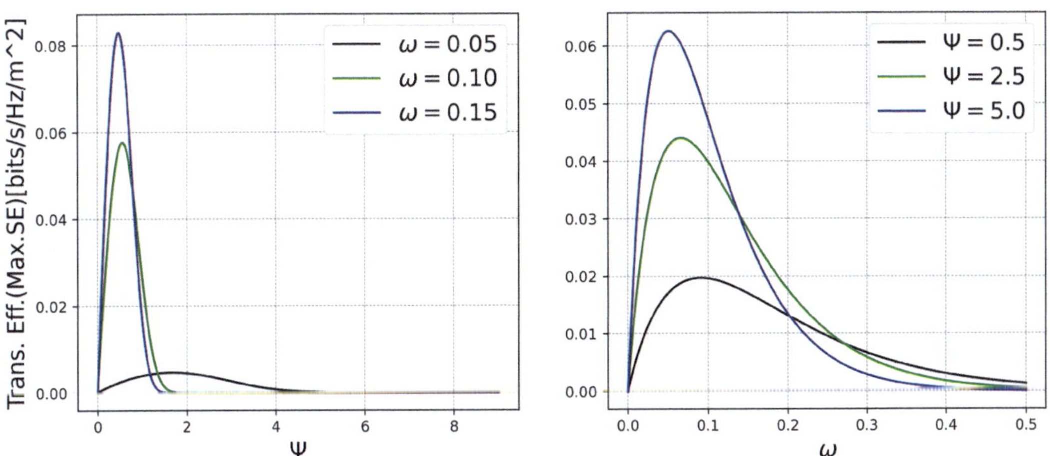

FIGURE 3.7 Transmission Efficiency versus SINR and density of networks.

ω =0.15. The peak yields approximately 0.0597 and 0.082, respectively, for ω =0.10 and 0.05. As the SINR value rises, the TE curve begins to steadily erode, with the intersection at 0.04 and 0.002 having the potential to eventually converge. If TE is plotted versus base station density for a range of SINR values, on the other hand, the same justification can be made.

Figure 3.8 shows the relationship between the data efficiency (informational efficiency) and the densification of mobile users. It has been noticed that each graph's peak increases when more varied distances are used in the experiment. Thus, the distance affects the effectiveness of data transmission. Additionally, the curve begins to decline at a certain value as the density of mobile users rises before peaking at another value as the distance is increased, and eventually, it starts to converge.

As seen in Figure 3.9, the throughput versus distance is also similarly displayed. The genetic algorithm, the federated learning scheme, and the analytical approach are all employed in this case. As seen in Figure 3.8, it is apparent that the proposed FL performs better than alternative methods. Additionally, the outage probability is similarly shown in Figures. 3.10 and 3.11 against the SINR and distance. Notably, as both the SINR and distance grow, it is clear that the FS performs

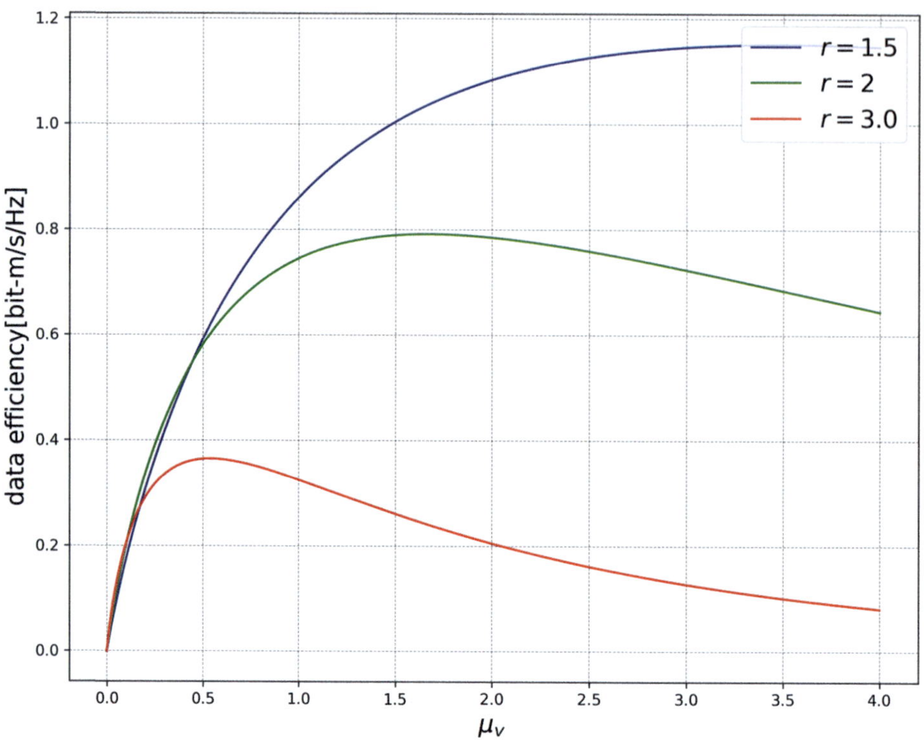

FIGURE 3.8   Data efficiency versus mobile-user density of networks for varying distance.

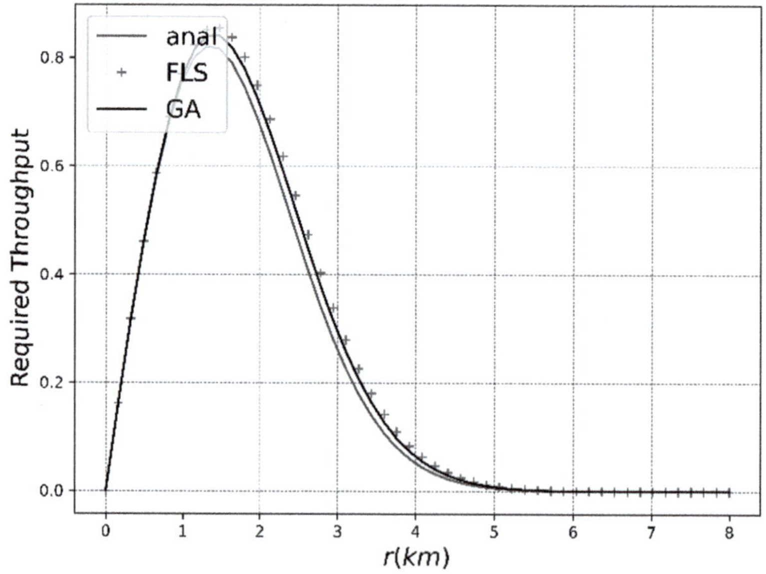

FIGURE 3.9   Required throughput versus distance.

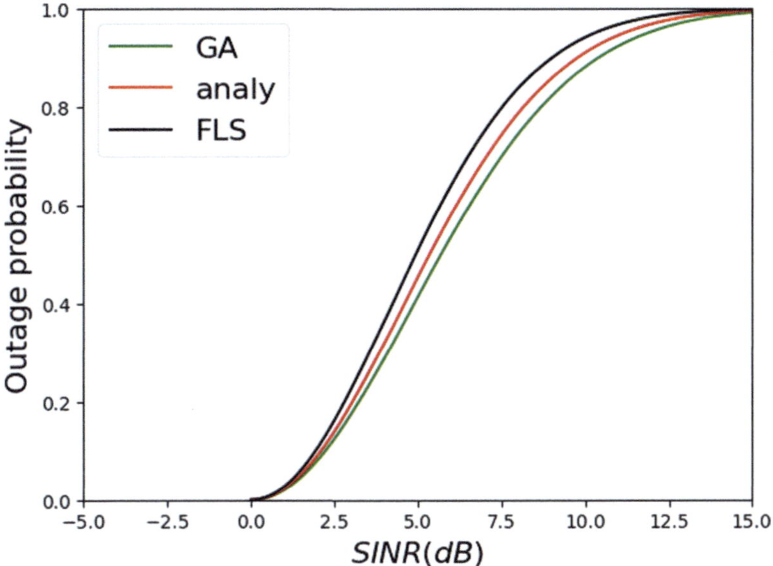

FIGURE 3.10   Outage probability versus SINR.

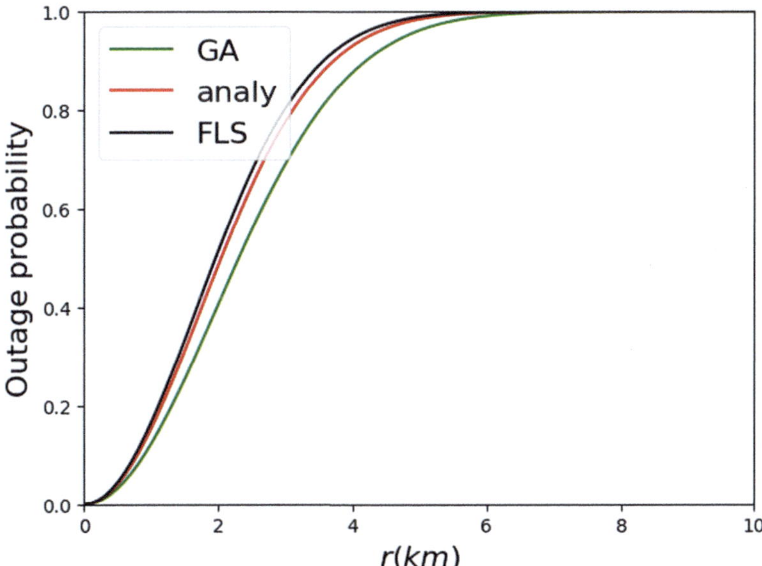

FIGURE 3.11   Outage probability versus distance.

noticeably better than the comparable schemes for both outage likelihood versus distance and outage versus distance. The solution of computational communication networks and systems can thus be accomplished via a federated learning scheme (FLS).

The coverage rate is finally displayed against various SINR values in Figure 3.12. As shown in Figure 3.12, it is noted that the introduction of noise has a significant impact on the coverage rate. The coverage rate is significantly impacted by network noise. In contrast to the existence of noise

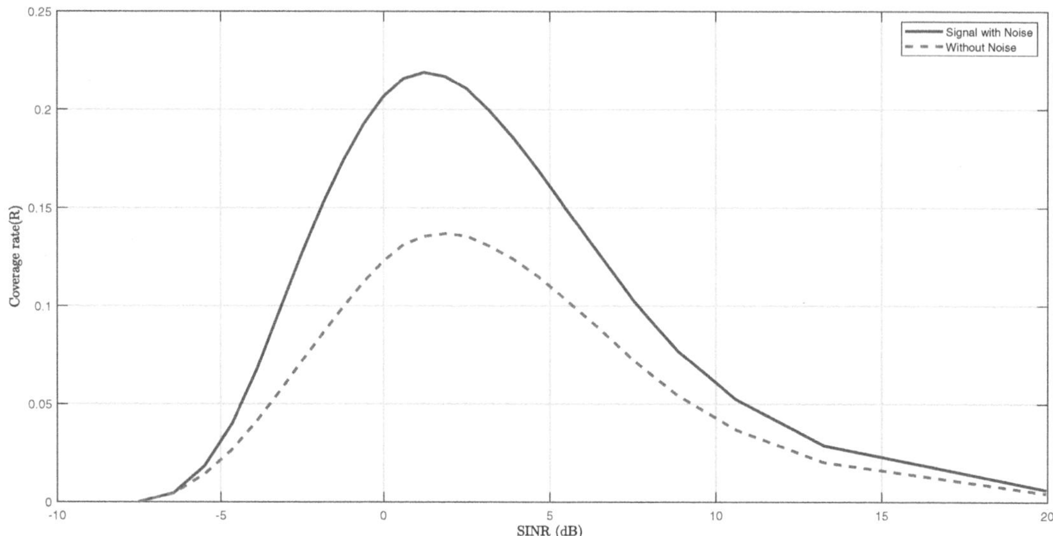

FIGURE 3.12 Coverage rate versus SINR.

in the network channel, it can be seen that the outcome is better and the convergence is faster when the noise is removed.

In light of this, the final five contributions mentioned in the introductory section have been addressed and accomplished through the formulation and numerical outcomes of the suggested scheme.

3.5 LESSON LEARNED

Following the related review on stochastic geometry for FL-based computational problem solving, some lessons and takeaways are learned in this section. The key lessons learned during this study are highlighted below:

- Machine learning, in particular federated learning, has been investigated and is now widely used in a variety of fields, including financial services, complex systems (engineered systems), healthcare systems, and computational wireless communications. Recently, during the pandemic, AI and its subsets, including FL, were crucial in the fight against the illness. But more crucially, they have assisted in offering answers to problems based on physical layers. As a result, the following are the most important lessons or highlights from using stochastic geometry schemes and FL as a subset of AI and ML to model and simulate communication systems and networks computationally: The primary issue of outage likelihood and coverage probability with the communication systems and networks that mobile users and subscribers currently face might be resolved, which would greatly enhance the network's performance and increase the value of the services offered by the communications providers to consumers. Several lessons are highlighted in this section, including the following:
- The mathematical model incorporating federated learning (FL) approaches and other schemes need additional in-depth investigation given how swiftly the subject is expanding. Communications networks and systems can benefit from federated learning. The mobile user's or subscriber's information can be gathered for the accuracy and performance of the entire network when paired with other methods, such as the stochastic geometry approach. In advance of proposing the concept of the proposed framework, it is essential to do in-depth

research, especially the use of federated learning techniques in the computational modelling of wireless systems and networks, which may also be one of our future research objectives. The ability of these models to address other physical layer difficulties that are not covered in this study gives an outline of the potential future direction.

- There is still more opportunity for development in the other areas of communication systems and networks, such as the security and privacy considerations and resource allocation management motivated by FL. Implementing the newest algorithms and using a rigorous validation method could be the main topics of future investigations. It is impracticable to determine the advantages and disadvantages of utilising the Bisection approach and two-step procedure for every user's density and the density of mobile base stations for the capacity for transmission and overall throughput for this study area. In contrast to the conventional methodology, the AI-driven methods, in particular the FL, might be utilised to implement and solve previous problems more efficiently, enhancing the system's throughput and efficiency.

- The process of computing various performance metrics required to evaluate the effectiveness and accuracy of good communication systems and networks is thought to be accelerated by federated learning powered by SG when compared to alternative computing techniques and tools. Future research should focus on how to interface these algorithms with existing systems and how to boost the operating speed of the algorithms powered by AI-inspired schemes like FL. Enhancing the models and computational properties of wireless systems is also of the utmost importance if communication systems and networks powered by artificial intelligence-powered solutions are to be commercially deployed in the telecommunications industry and other industries such as healthcare.

3.6 CONCLUSION

To represent communication systems and networks using computational inspiration, this chapter focuses on the application of federated learning schemes, a subset of artificial intelligence (AI), and stochastic geometry. Additionally, a thorough review of current advancements in the modelling of communication systems and networks is provided in this chapter. There is a lot of opportunity for improving the way that mobile customers receive telecommunications services by simulating communication networks and systems with FL and SG. The chapter provides an overview of the related research on the usage of PPPs and SG to address wireless network difficulties by using bisection and two-step-inspired incremental techniques to address a well-posed optimisation problem. This serves to show the utility of the stochastic geometry method inspired by FL in wireless communication systems and networks. However, careful deployment, continuing assessment, and addressing ethical issues are crucial in this regard for the successful adoption and realisation of FL, ML, AI, and blockchain technology's full potential in communication systems and networks. Importantly, FL, ML, and AI algorithms have been used in the telecommunications industry, making it simpler to get reliable results. The FL findings are computationally compared with other metaheuristic algorithms, such as the genetic algorithm (GA), and it is discovered that the proposed FL-inspired approach outperforms other schemes significantly concerning the defined optimisation problem. The future scope of this study will be to apply the newly developed human-based metaheuristic algorithm termed deep sleep optimiser (DSO) with other evolutionary schemes with their computational analysis in communication systems and networks. Additionally, future studies should concentrate on how to integrate these algorithms with current systems and accelerate the algorithms driven by AI-inspired schemes such as FL, DSO, and so on.

REFERENCES

[1] D. Stoyan, W. Kendall, and J. Mecke, "Stochastic geometry and its applications," 2nd ed. Hoboken, NJ, USA: Wiley, 1996.

[2] F. Baccelli, M. Klein, M. Lebourges, and S. Zuyev, "Stochastic geometry and architecture of communication networks," *J. Telecommun. Syst.,* vol. 7, no. 1–3, pp. 209–227, Jun. 1997.

[3] F. Baccelli, and S. Zuyev, "Stochastic geometry models of mobile communication networks," in *Frontiers Queueing.* Boca Raton, FL, USA: CRC Press, 1997, pp. 227–243.

[4] F. Baccelli, and B. Blaszczyszyn, *Stochastic Geometry and Wireless Networks in Foundations and Trends in Networking,* vol. 1. Boston, MA, USA: Now, 2009.

[5] F. Baccelli, and B. Blaszczyszyn, *Stochastic Geometry and Wireless Networks in Foundations and Trends in Networking,* vol. 2. Boston, MA, USA: Now, 2009.

[6] M. Haenggi, *Stochastic Geometry for Wireless Networks.* Cambridge, U.K.: Cambridge Univ. Press, 2012.

[7] Ravi, Sreenivasan. "New Perspectives in Stochastic Geometry." pp. 845–846., 2011.

[8] D. J. Daley, and D. Vere-Jones, "An introduction to the theory of point processes, elementary theory and methods," vol. 1. New York, NY, USA: Springer-Verlag, 2003.

[9] J.-F. Coeurjolly, J. Møller, and R. Waagepetersen, "A tutorial on palm distributions for spatial point processes," *Int. Stat. Rev.,* vol. 85, no. 3, pp. 404–420, Dec. 2017.

[10] T. M. Cover, "Elements of information theory" John Wiley & Sons, 1999.

[11] D. Tse, and P. Viswanath, "Fundamentals of wireless communication," Cambridge University Press., 2005.

[12] S.Liu, G.Yu, X. Chen, M. Bennis, "Joint user association and resource allocation for wireless hierarchical federated learning with IID and non-IID data," *IEEE Wirel Commun.,* vol. 21, no. 10, pp. 7852–66, Apr 4, 2022.

[13] W. Y. B. Lim, J. S. Ng, Z. Xiong, D. Niyato, C. Miao, and D. I. Kim, " Dynamic edge association and resource allocation in self-organizing hierarchical federated learning networks," *IEEE J Sel Areas Commun.,* vol. 39, no. 12, pp. 3640–3653, 2021.

[14] Q. Wu, X. Chen, T. Ouyang, Z. Zhou, X. Zhang, S. Yang, and J. Zhang, "Hiflash: Communication-efficient hierarchical federated learning with adaptive staleness control and heterogeneity-aware client-edge association, *IEEE Trans. Parallel Distrib. Syst.,* vol. 34, no. 5, pp. 1560–1579, 2023.

[15] S. Zuyev, "Stochastic geometry and telecommunications networks," in *Stochastic Geometry: Highlights, Interactions and New Perspectives,* W. S. Kendall and I. Molchanov, Eds. London, U.K.: Oxford Univ. Press, 2009.

[16] M. Haenggi, J. G. Andrews, F. Baccelli, O. Dousse, and M. Franceschetti, "Stochastic geometry and random graphs for the analysis and design of wireless networks," *IEEE J. Sel. Areas Commun.,* vol. 27, no. 7, pp. 1029–1046, Sep. 2009.

[17] H. ElSawy, E. Hossain, and M. Haenggi, "Stochastic geometry for modeling, analysis, and design of multi-tier and cognitive cellular wireless networks: A survey," *IEEE Commun. Surveys Tuts.,* vol. 15, no. 3, pp. 996–1019, 3rd Quart., 2013.

[18] J. G. Andrews, A. K. Gupta, and H. S. Dhillon, "A primer on cellular network analysis using stochastic geometry," Oct. 2016, arXiv:1604.03183. [Online]. Available: http://arxiv.org/abs/1604.03183

[19] M. Mozaffari, W. Saad, M. Bennis, Y.-H. Nam, and M. Debbah, "A tutorial on UAVs for wireless networks: Applications, challenges, and open problems," *IEEE Commun. Surveys Tuts.,* vol. 21, no. 3, pp. 2334–2360, 3rd Quart., 2019.

[20] L. Li, G. Zhao, and R. S. Blum, "A survey of caching techniques in cellular networks: Research issues and challenges in content placement and delivery strategies," *IEEE Commun. Surveys Tuts.,* vol. 20, no. 3, pp. 1710–1732, 3rd Quart., 2018.

[21] J. G. Andrews, "Seven ways that HetNets are a cellular paradigm shift," *IEEE Commun. Mag.,* vol. 51, no. 3, pp. 136–144, Mar. 2013.

[22] J. Andrews, R. Ganti, M. Haenggi, N. Jindal, and S. Weber, "A primer on spatial modeling and analysis in wireless networks," *IEEE Commun. Mag.,* vol. 48, no. 11, pp. 156–163, Nov. 2010.

[23] D. Moltchanov, "Distance distributions in random networks," *Ad Hoc Netw.,* vol. 10, no. 6, pp. 1146–1166, Aug. 2012.

[24] M. Haenggi, and R. Ganti, "Interference in large wireless networks," in *Foundations and Trends in Networking*, vol. 3, no. 2. Boston, MA, USA: Now, 2008, pp. 127–248.

[25] P. Cardieri, "Modeling interference in wireless ad hoc networks," *IEEE Commun. Surveys Tuts*, vol. 12, no. 4, pp. 551–572, 4th Quart, 2010.

[26] H. ElSawy, A. Sultan-Salem, M.-S. Alouini, and M. Z. Win, "Modeling and analysis of cellular networks using stochastic geometry: A tutorial," *IEEE Commun. Surveys Tuts*, vol. 19, no. 1, pp. 167–203, 1st Quart, 2017

[27] China Mobile, "C-RAN: The road towards green RAN," *Version 2.5, China Mobile Res. Inst.*, Beijing, China, White Paper, Oct. 2011.

[28] A. Checko, H. L. Christiansen, Y. Yan, L. Scolari., G. Kardaras, M. S. Berger, and L. Dittmann, "Cloud RAN for mobile networks—A technology overview," *IEEE Commun. Surveys Tuts.*, vol. 17, no. 1, pp. 405–426, 2014.

[29] S. M. R. Islam, N. Avazov, O. A. Dobre, and K.-S. Kwak, "Power-domain non-orthogonal multiple access (NOMA) in 5G systems: Potentials and challenges," *IEEE Commun. Surveys Tuts*, vol. 19, no. 2, pp. 721–742, 2nd Quart. 2017.

[30] A. Sabharwal, P. Schniter, D. Guo, D. W. Bliss, S. Rangarajan, and R. Wichman, "In-band full-duplex wireless: Challenges and opportunities," *IEEE J. Sel. Areas Commun.*, vol. 32, no. 9, pp. 1637–1652, Sep. 2014.

[31] Y.Wu, A. Khisti, C. Xiao, G. Caire, K.-K.Wong, and X. Gao, "A survey of physical layer security techniques for 5G wireless networks and challenges ahead," *IEEE J. Sel. Areas Commun.*, vol. 36, no. 4, pp. 679–695, Apr. 2018.

[32] O. O. Shoewu, L. A. Akinyemi, and R. Edozie. "UAV cellular communication in 5G new radio wireless standards," In *Unmanned Aerial Vehicle Cellular Communications*, pp. 25–45. Cham: Springer International Publishing, 2022.

[33] L. A. Akinyemi, O. O. Shoewu, C. O. Folorunso, O. Ipinnimo, A. A. Ajasa, Q. A. Mumuni, and J. F. Orimolade, "Alleviating 6G security and privacy issues using artificial intelligence," *Book on Security and Privacy Schemes for Dense 6G Wireless Communication Network*, 2023. doi.org/10.1049/PBSE021E_ch14

[34] J. G. Andrews, R. K. Ganti, M. Haenggi, N. Jindal, and S. Weber, "A primer on spatial modelling and analysis in wireless networks," *Communications Magazine, IEEE*, vol. 48, no. 11, pp. 156–163, 2010.

[35] D. Stoyan, W. S. Kendall, S. N. Chiu, and J.Mecke, "Stochastic geometry and its applications". John Wiley & Sons, 2013.

[36] E. Gilbert, and H. Pollak. Amplitude distribution of shot noise. *Bell Syst. Tech. J*, vol. 39, no. 2, pp. 333–350, 1960.

[37] J. F. C. Kingman, "Poisson processes," vol. 3. Oxford University Press, 1992.

[38] D. J. Daley, and D.Vere-Jones, "An introduction to the theory of point processes: Volume I: elementary theory and methods", Springer New York; 2003.

[39] S. Khadanga, M. Kashyap, K. Verma, and P. Sudharsan, "Coverage Probability Analysis of Amplify and Forward Based Multi-user Hybrid Satellite Terrestrial Relay Network", In *Journal of Physics: Conference Series* (vol. 2466, no. 1, pp. 012002). IOP Publishing, 2023.

[40] V. B.Kumaravelu, A. L.Imoize, F. R. C.Soria, P. G. S.Velmurugan, S. J.Thiruvengadam, D.T.Do, and A. Murugadass, A., "RIS-assisted fixed NOMA: Outage probability analysis and transmit power optimization," *Future Internet*, vol. 15, no. 8, pp. 249, 2023.

[41] V. B. Kumaravelu, A. L. Imoize, F. R. C. Soria, P. G. S.Velmurugan, S. J. Thiruvengadam, D.T.Do, and A. Murugadass," RIS-Assisted Fixed NOMA: outage probability analysis and transmit power optimization", *Future Internet*, vol. 15, no. 8, pp. 249, 2023.

[42] J. Isabona, A. L. Imoize, S. Ojo, O. Karuwi, Y. Kim, C. C.Lee, and C. T. Li, "Development of a multi-layer perception neural network for optimal predictive Modeling in urban microcellular radio environments", *Applied Sciences*, vol. 2, no. 11, pp. 5713, 2022.

[43] T. S. Ajani, A. L. Imoize, and A. A. Atayero," An overview of machine learning within embedded and mobile devices-optimization and applications", *Sensors*, vol. 21, no. 13, pp. 4412, 2021.

[44] A. L. Imoize, A. E. Ibhaze, A. A. Atayero, and K. V. N. Kavitha, "Standard propagation channel models for MIMO communication systems", *Wireless Communications and Mobile Computing*, pp. 1–36, 2021.

[45] A. Abdulkarim, N. Faruk, E. Alozie, O. A. Sowande, I. F. Y. Olayinka, A. D. Usman, and L. S. Taura, "Application of machine learning algorithms to path loss modeling: A review," 5th Information Technology for Educational and Development(ITED), pp. 1–6, 2022.

[46] S. O. Oladejo, S. O. Ekwe, L. A. Akinyemi, and S. A. Mirjalili, "The deep sleep optimiser: A human-based metaheuristic approach," *IEEE Access*, vol. 11, pp. 83639–83665, 2023.

[47] S. O. Ekwe, S. O. Oladejo, L. A. Akinyemi, and N. Ventura, "A socially-inspired energy-efficient resource allocation algorithm for future wireless network," In *2020 16th International Computer Engineering Conference (ICENCO)*, pp. 168–173. IEEE, 2020.

[48] S. O. Oladejo, S. O. Ekwe, and L. A. Akinyemi, "Multi-tier multi-domain network slicing: A resource allocation perspective." In *2021 IEEE AFRICON*, pp. 1–6. IEEE, 2021.

[49] S. O. Oladejo, S. O. Ekwe, and L. A. Akinyemi, " Multi-tier multi-tenant network slicing: A multi-domain games approach," *ITU Journal on Future and Evolving Technologies, vol. 2, Issue 6 – Wireless communication systems in beyond 5G era*, pp. 57–82, 2021.

[50] S. O.Ekwe, L. A. Akinyemi, S. O. Oladejo, and Ventura," Social-Aware Joint uplink and downlink resource allocation scheme using genetic algorithm," In *2021 IEEE AFRICON*, pp. 1–6, 2021.

[51] A. T. Ajibare, S. O. Oladejo, S. O. Ekwe, L. A. Akinyemi, and D. Ramotsoela, "Radiofrequency electromagnetic radiation exposure assessment, analysis, computation, and minimization technique in 5G networks: A perspective on QoS Trade-Offs," in *SAIEE Africa Research Journal*, vol. 114, no. 4, pp. 114–127. doi: 10.23919/SAIEE.2023.10319380, 2023

4 Computational Modeling and Analysis of Wireless Sensor Networks

Yun Wang, Mohammad Nazmus Sadat,
G. G. Md. Nawaz Ali and Yanping Zhang

4.1 INTRODUCTION

Wireless Sensor Networks (WSNs) have gained significant adoption [1] from civil applications such as healthcare monitoring [2–3], industrial automation [4–6], smart cities [7– 9], to the military fields, including battlefield monitoring and surveillance [10–11], intrusion detection [12–13] and so forth. Given the inherent complexity and diversity arising from various applications and circumstances, the development of appropriate computational models and analytical techniques is crucial to facilitate network simulation, design, and optimization before practical deployment.

A wireless sensor network (WSN) is typically modeled as a graph, with vertices representing sensors and edges representing communication links. Figure 4.1 depicts a typical random WSN, where each sensor is represented by a grey dot and the communication range is depicted by green dotted lines. In addition, each sensor is equipped with a sensing range to detect application-specific events such as a fire, water leakage, moving intruder(s), and so forth. The sensing range of each sensor is represented by a red circle with a grey shade in the figure, following the Boolean disk model to be defined in Section 4.3.

It should be noted that, in Figure 4.1, $N = 100$ sensors are randomly and independently deployed in a field of interest (FoI) with side length $L = 1000$ and each sensor is assumed to have a sensing range of $R_s = 50$ and communication range $R_c = 125$, and $R_c = 2.5R_s$. Different from most research work focusing on full sensing coverage and complete communication connectivity, Figure 4.1 illustrates a WSN with partial sensing coverage and partial connectivity. Such a network configuration is able to prevent an intruder from invading and crossing the field from left to right.

Different applications may have distinctly different requirements, environmental factors, and variations in sensor types, and counts. This results in the increasing complexity of WSN design and makes it imperative to explore and develop computational models for WSN simulation, performance evaluation, and optimization. These efforts are essential before the practical sensor deployment to ensure performance, minimize cost, and mitigate risks. In view of this, the present chapter focuses on the design and development of computational modeling and analysis for WSN research from sensor deployments, sensing coverage, and application requirements within the context of different circumstances and scenarios. It also explores the impact of various models on network performance and discusses their strengths, weaknesses, and interrelationships. Theoretical analysis, mathematical derivations, and visualized results are presented to enhance the understanding. The study aims to provide insights into selecting and developing the most appropriate model(s) to address challenges encountered in WSNs, and at the same time drive the advancement of research in the field. To facilitate ease of cross-referencing and ensure consistency, Table 4.1 illustrates the notations, abbreviations, and descriptions that we have utilized in this chapter.

DOI: 10.1201/9781003457428-5

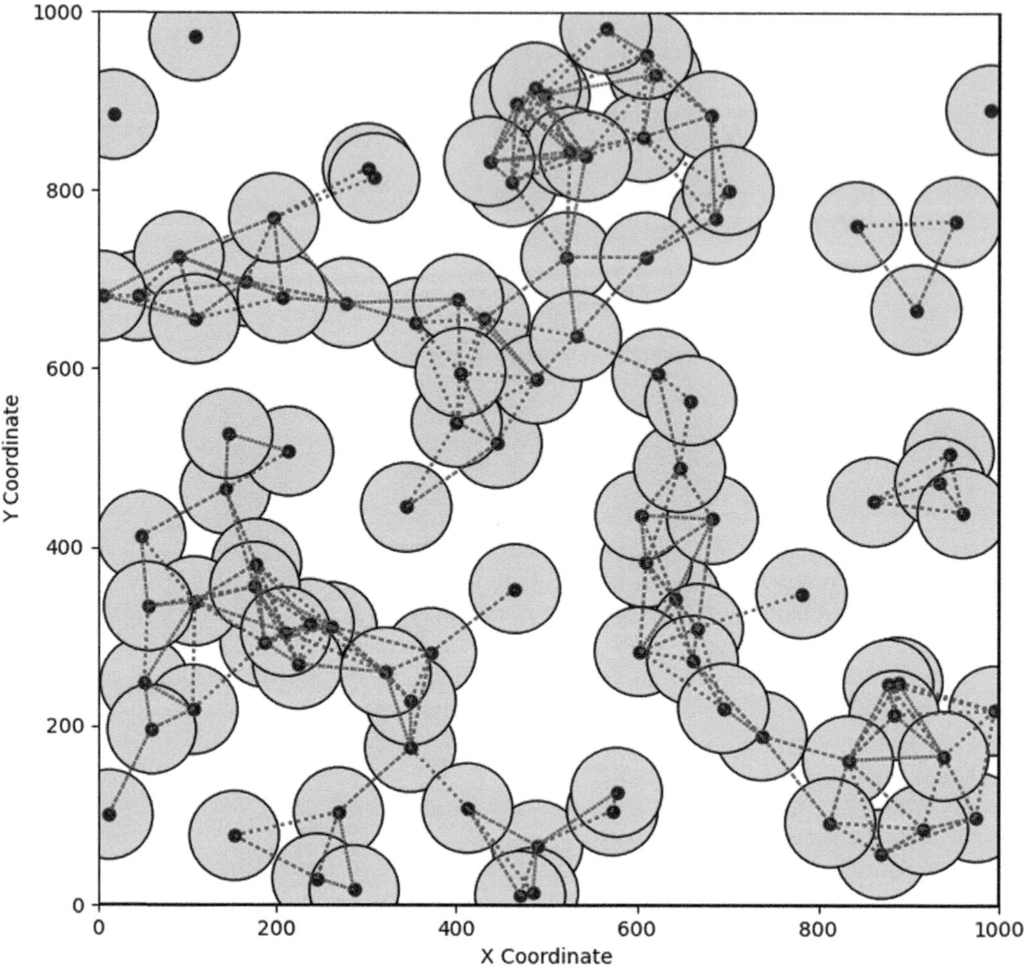

FIGURE 4.1   A 2D random WSN where $N = 100$, $R_s = 50$, and $R_c = 125$.

TABLE 4.1
Notations, Abbreviations, and Their Descriptions in this Chapter

Notation	Description
N	Number of deployed sensors
R_s	Maximum sensing range of a sensor
R_c	Maximum communication range of a sensor
r_s	Minimum sensing range of a sensor
r_s	Minimum communication range of a sensor
L	Side-length of the field of interest
FoI	Field of interest
d	Euclidian distance between two points/sensors
(x, y)	Coordinates of a point in a 2D FoI
(x, y, z)	Coordinates of a point in a 3D FoI
$\grave{\imath}_r$	Mean of a random variable r
$\acute{o}_r$	Standard deviation of a random variable r

The remainder of the chapter is organized as follows. Section 4.2 presents the typical sensor deployment models. Section 4.3 defines both deterministic and probabilistic models in simulating a sensor's capability in terms of sensing and communications, along with a discussion on their merits and shortcomings. Section 4.4 discusses the impact of different models on the performance evaluation under various scenarios. Section 4.5 exhibits recent trends in the field of WSN research, and the chapter is concluded in Section 4.6.

4.2 MODELING OF SENSOR DEPLOYMENTS

Sensors are usually deployed in a deterministic or random manner according to the application requirements and environments.

4.2.1 DETERMINISTIC SENSOR DEPLOYMENTS

In deterministic deployments, sensors are positioned in repeatable patterns such as squares, triangles, hexagons, and so forth, in a 2D field or cubes, tetrahedrons, hexahedrons, and so forth, in a 3D domain. This approach is often employed in small-scale and human-friendly circumstances, where sensors can be precisely placed to achieve optimal performance. Figure 4.2 showcases a deterministic 3D WSN, where 125 sensors are precisely deployed following a cubic pattern and are evenly spaced in X, Y, and Z dimensions. Recent surveys on deployment techniques and their characteristics can be found in [14–16].

Deterministic sensor deployments facilitate a structured and organized placement of sensors, ensuring systematic coverage, connectivity, and application-specific requirements. They are widely utilized in civil applications, including environmental monitoring, surveillance systems, industrial automation, structural health monitoring, underwater sensing, the smart city, and so forth [17–18].

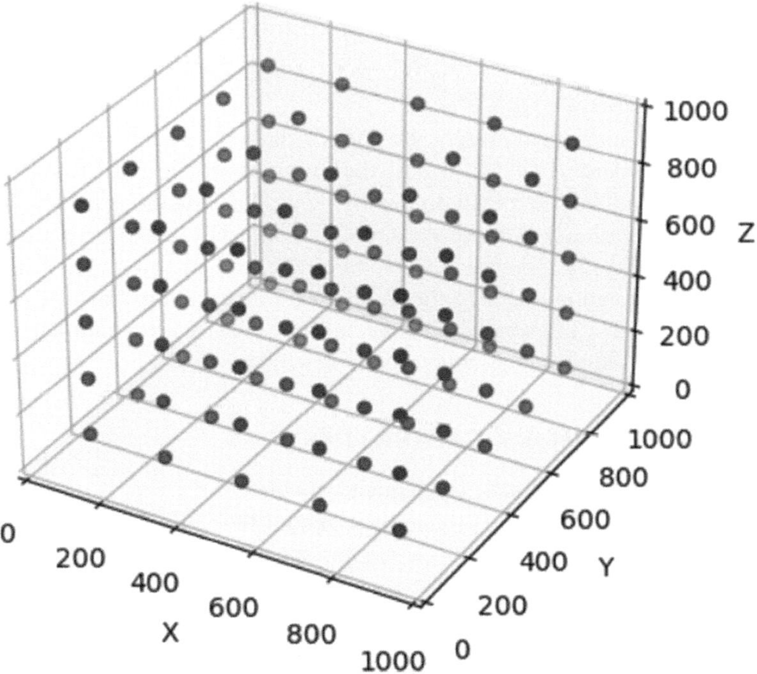

FIGURE 4.2 Example 3D WSNs where sensors are precisely deployed following a cubic pattern.

4.2.2 Random Sensor Deployments

On the other hand, random strategies are utilized in large-scale networks or hostile scenarios to enable fast deployments. In general, two widely adopted distributions in modeling random WSN deployments are the Uniform distribution and the Gaussian distribution [14, 19–24]. In both random WSN models, a number of sensors are assumed to be randomly and independently deployed in a given field of interest (FoI).

In a random WSN following a uniform distribution, the PMF (Probability Mass Function) that a sensor is deployed at an arbitrary point (x, y) is constant over the entire field of interest F as:

$$f(x, y) = 1 / F. \tag{1}$$

The field F is commonly represented as a square with an area of L^2 or as a disk with an area of πL^2. For the 3D case, it is often defined as a cubic space with a volume of L^3 or as a sphere with a volume of $\dfrac{4\pi L^3}{3}$.

In a random WSN following a Gaussian distribution, the joint Probability Density Function (PDF) of a sensor being deployed at an arbitrary point (x, y) in a 2D field is:

$$f(x, y) = \frac{1}{2\pi\sigma_x\sigma_y} e^{-\left(\frac{(x-\mu_x)^2}{2\sigma_x^2} + \frac{(y-\mu_y)^2}{2\sigma_y^2}\right)}, \tag{2}$$

and that of a sensor being deployed in a 3D field with at the location of (x, y, z) can be represented as:

$$f(x, y) = \frac{1}{2\pi\sqrt{2\pi}\sigma_x\sigma_y\sigma_z} e^{-\left(\frac{(x-\mu_x)^2}{2\sigma_x^2} + \frac{(y-\mu_y)^2}{2\sigma_y^2} + \frac{(x-\mu_y)^2}{2\sigma_z^2}\right)}. \tag{3}$$

In both equations, x, y, and/or z represent the horizontal, vertical, and/or altitude position in the 2D/3D coordinate system, and μ_r and σ_r denote their corresponding mean and standard deviation in each dimension resulting from the random deployments. The equations assume that X, Y, and Z are independent and uncorrelated. A more general Gaussian PDF definition considering non-zero corrections can be found in [25]. Figure 4.3 compares the two common WSN deployments: one following a Uniform distribution and the other a Gaussian distribution. Each deployment consists of 100 sensors deployed in a 2D square Field of Interest (FoI) with a side-length of $L = 1000$. Additionally, Figure 4.4 illustrates their counterparts in a 3D cubic FoI.

4.3 MODELING OF SENSORS' CAPABILITIES

For theoretical analysis, the sensing and communication range of a sensor is commonly represented by a deterministic Boolean model or a probabilistic model [26–29]. Probabilistic models include the Elfes model, multi-layer model, shadow-fading model, sigmoid model, and others [30].

4.3.1 Boolean Model

The Boolean model is a simplified representation of a sensor's capability for sensing and communication operations. In this model, each sensor has a pre-determined sensing range R_s and

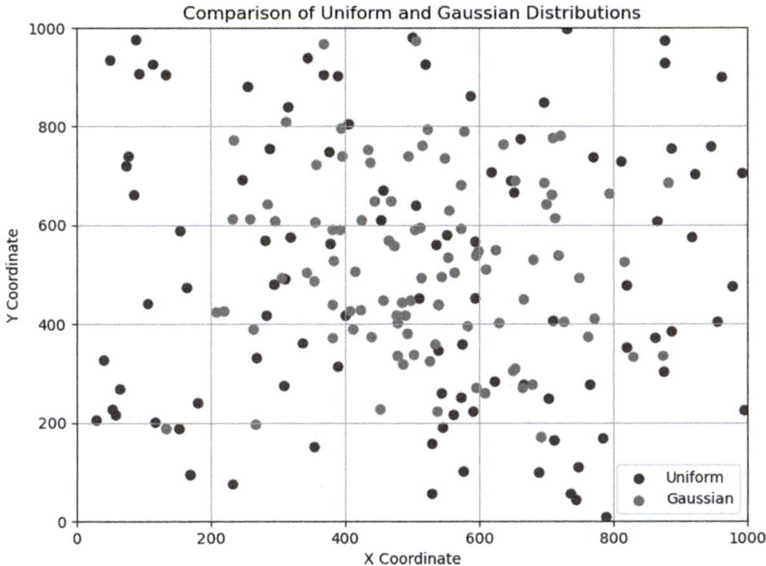

FIGURE 4.3 Uniform versus Gaussian distribution of a random 2D WSN.

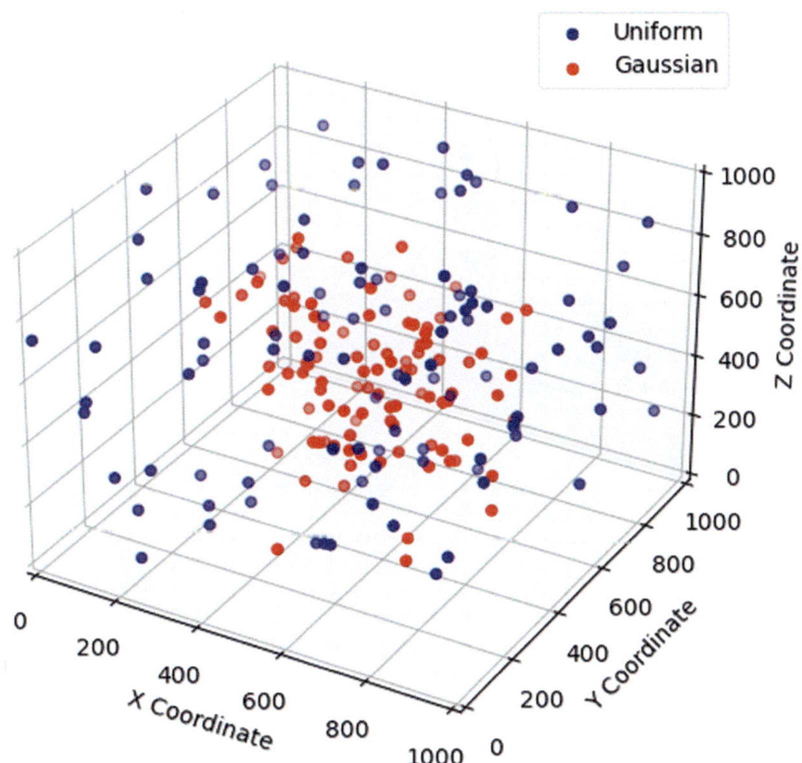

FIGURE 4.4 Uniform versus Gaussian distribution of a random 3D WSN.

communication range R_c. The sensor is able to sense or detect an event occurring within the distance R_s to it and communicate with a sensor if their Euclidean distance is less than or equal to R_c.

Assume d represents the Euclidean distance between the sensor and the location of the event being monitored for sensing or a neighboring sensor for communication. Mathematically, the model can be represented as follows, where R_x can be the sensing range R_s or communication range R_c for a given sensor.

$$p(d) = \begin{cases} 1, & \text{if } d \leq R_x \\ 0, & \text{if } d > R_x \end{cases} \tag{4}$$

The Boolean model is usually used in abstracting the sensing and communication range of sensors in WSNs to enable theoretical deviation and analysis. Figure 4.1 showcases a WSN where the Boolean model is adopted to simulate the sensing and communication range of deployed sensors. Figure 4.5 illustrates the connectivity of a 3D WSN where $N = 100$, $L = 1000$, and $R_c = 0.20 * L$ following the Boolean communication model.

The Boolean model has been widely adopted in WSN research and education due to its numerous benefits, including conceptual clarity, analytical tractability, and computational efficiency. It helps establish a baseline understanding and enables the derivation of analytical expressions and theoretical results to gain insights into the behavior and performance of a WSN under various circumstances. However, the simplified Boolean model cannot capture the characteristics of wireless links, which are subject to various signal attenuation factors and are not isotropic or symmetric in reality. This has been revealed by empirical measurements [31, 32].

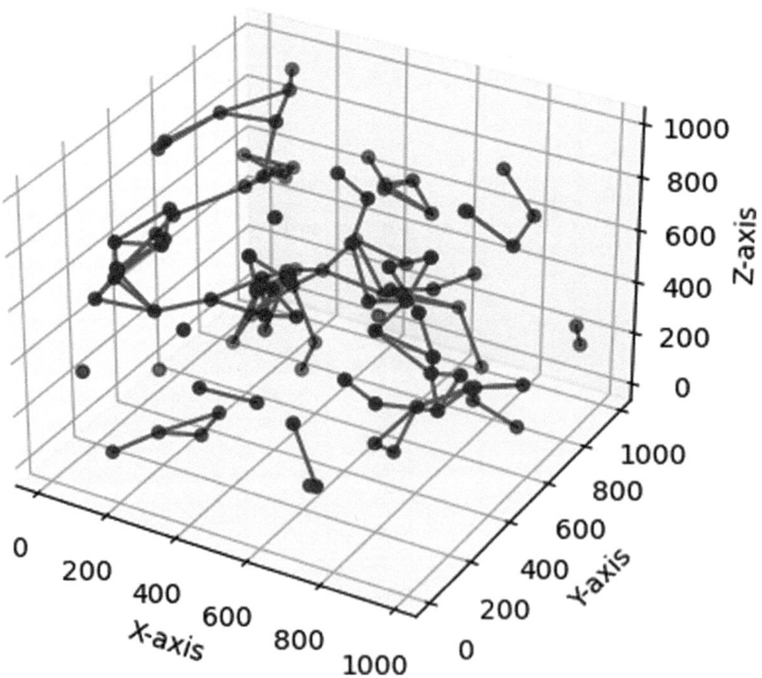

FIGURE 4.5 3D WSN connectivity following a Boolean communication model.

Thus, it is necessary to develop more sophisticated models that incorporate realistic factors and dynamics without introducing significant complexity. Probabilistic Models including Elfes and Multi-level are the way forward in this endeavor.

4.3.2 ELFES MODELING

Following the Elfes model [33, 28], a sensor is able to detect or sense or communicate events occurring in its surroundings with distance-specific probabilities as follows:

$$p(d) = \begin{cases} 1, & \text{if } d \leq d_{min} \\ e^{-\alpha(d-d_{min})^{\beta}}, & \text{if } d_{min} < d < d_{max} \\ 0, & \text{if } d \geq d_{max} \end{cases} \tag{5}$$

Figure 4.6 illustrates the probability for varying distances following the probabilistic Elfes Model, where $d_{min} = 20$ and $d_{max} = 40$ with different settings of α and β. Note that we use the same value for α and β in this example, that is, $\alpha = \beta$, but they can be different to meet the needs of simulating a real-life sensor's capability.

This Elfes model offers a more realistic representation of real-life sensors and incorporates uncertainties present in their sensing and communication processes, such as noise, interference, fading, and more. Research results based on this model are likely to provide deeper insights into the underlying processes and performance of a WSN. However, it is important to acknowledge that this realism comes at the cost of increased complexity and the challenges associated with theoretical deviations and analysis.

4.3.3 MULTI-LEVEL MODEL

The multi-level model serves as a trade-off between the simplicity of the deterministic Boolean model and the complexity of the probabilistic Elfes model. It provides a bridge between them by

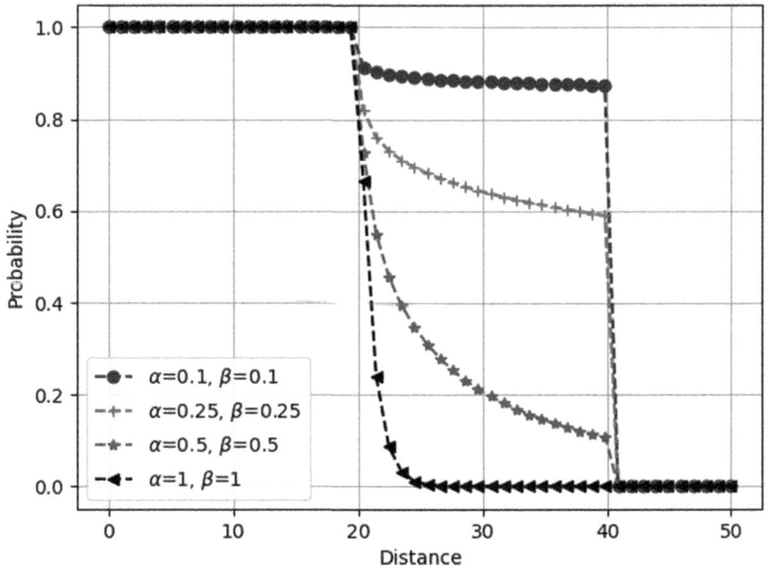

FIGURE 4.6   Probability for varying distances following the Probabilistic Elfes Model.

incorporating multiple levels of abstraction and captures more realism than the deterministic model while offering a simpler representation than the probabilistic Elfes model. It allows for a gradual increase in complexity and sophistication.

In this model, three fixed probabilities (1, η, and 0) are assigned based on the Euclidean distance between sensors or points or sensors to points. When the distance is less than d_{min}, a probability of 1 is assigned. For distances between d_{min} and d_{max}, a probability of $\eta(0 < \eta < 1)$ is assigned. And when the distance exceeds d_{max}, a probability of 0 is assigned. It can be represented as:

$$p(d) = \begin{cases} 1, & \text{if } d \leq d_{min} \\ \eta \ (0 < \eta < 1), & \text{if } d_{min} < d < d_{max} \\ 0, & \text{if } d \geq d_{max} \end{cases} \tag{6}$$

It is worth noting that, mathematically, the probabilistic Elfes Model can be simplified to the multi-level model, which in turn can be further reduced to the deterministic Boolean model. Researchers can start with the deterministic Boolean model, then progress to the multi-level model, and eventually transition to the probabilistic Elfes model to gradually advance the level of sophistication in the study.

4.4 IMPACT OF MODELS ON THE WSN DETECTION PROBABILITY

Sensor deployments, sensing, and communication models are all of crucial importance for simulating and evaluating the properties of WSNs. Understanding the impact of the chosen models on network performance help in providing insights into the performance evaluation of WSNs for diverse real-world situations.

4.4.1 Partial Sensing Coverage in Deterministic WSNs

In this analysis, we examine and compare the partial sensing coverage of a square-based WSN, as depicted in Figure 4.7, adopting the three common sensing models as defined in Section 4.3. We aim to evaluate how the selection of each model affects the assessment of the WSN's coverage under the same circumstances and to what extent.

Rather than focusing on the full sensing coverage, which has been extensively investigated in the literature, however, full sensing coverage may not be necessary for many applications such as intrusion detection. Figure 4.7 showcases a scenario with partial sensing coverage, where $L = 1000$, $r_{max} = 50, N = 1000$. However, the WSN is able to detect any intruder attempting to enter the field of interest.

More specifically, we assume a number of N sensors precisely deployed following a Square pattern in a 2D field of interest with side-length L. The sensing ranges of neighboring sensors are adjacent, following their maximum range, while ensuring no overlapping in their sensing coverage. In order words, each sensor's sensing range is carefully positioned to avoid overlap with its neighboring sensors to maximize coverage while minimizing node redundancy. In other words, partial sensing coverage is adequate to provide complete intrusion detection in this case, assuming full connectivity.

Based on the network settings depicted in Figure 4.7, we can derive the partial sensing coverage, under the three sensing models as follows:

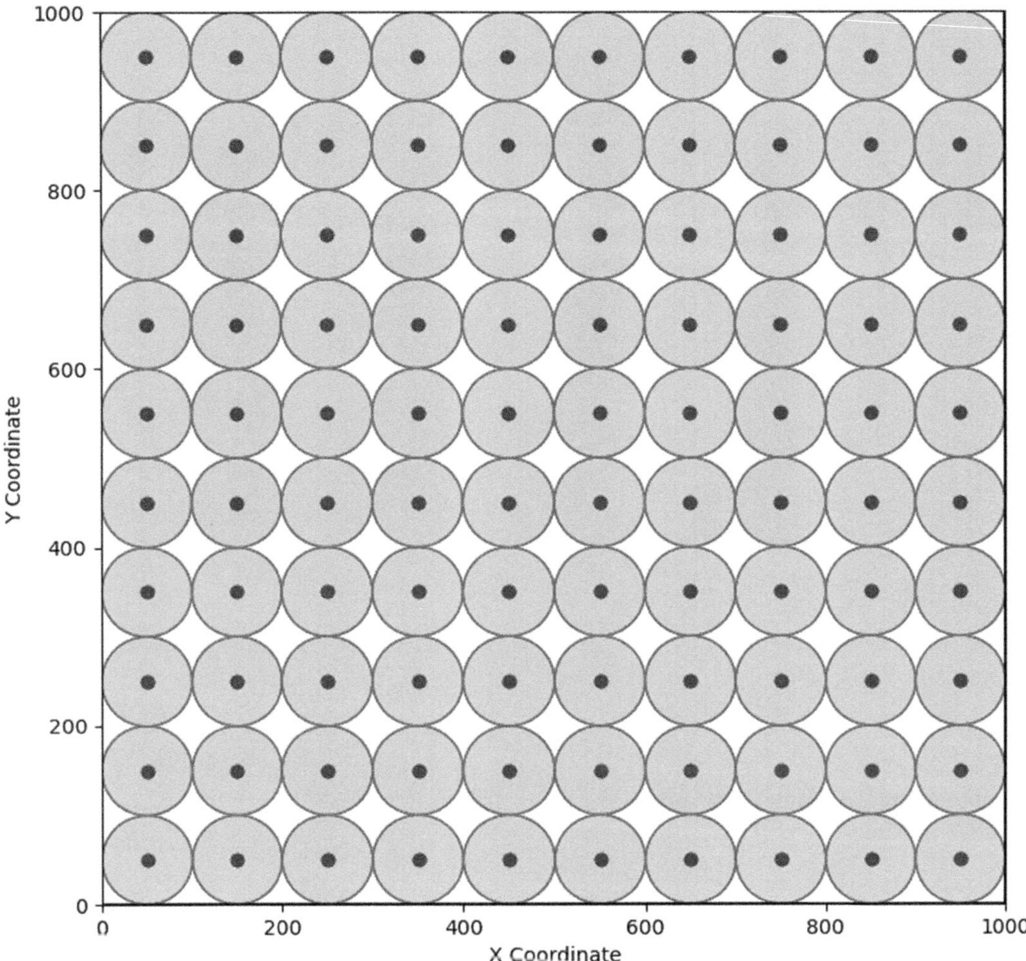

FIGURE 4.7   Partial coverage of a deterministic square-based WSN.

$$
P_{cov} = \begin{cases} \dfrac{N * \pi r_{max}^2}{L * L} & \text{if Boolean Model,} \\[2ex] \dfrac{N \pi r_{min}^2 + N \beta \pi \left(r_{max}^2 - r_{min}^2 \right)}{L * L} & \text{if Multi - level Model,} \\[2ex] \dfrac{N \pi r_{min}^2 + N \int_0^{2\pi} \int_{d_{min}}^{d_{max}} e^{-\alpha(r - d_{min})\gamma} dr\, d\theta}{L * L} & \text{if Elfes Model.} \end{cases} \tag{7}
$$

Figure 4.8 compares and contrasts the partial sensing coverage probability using the three models, where $L = 1000$, $r_{max} = 50$, $N = 100$. Table 4.2 presents detailed results for clear comparison and easy reference and allows for easy identification of trends or differences. We observe significant differences in the outcomes when employing different sensing models. For instance, the sensing coverage employing the Elfes model is only 14.9% of that of adopting the Boolean model under the same network settings when $d_{min} = 20$, and $d_{max} = 50$ as shown in the Table.

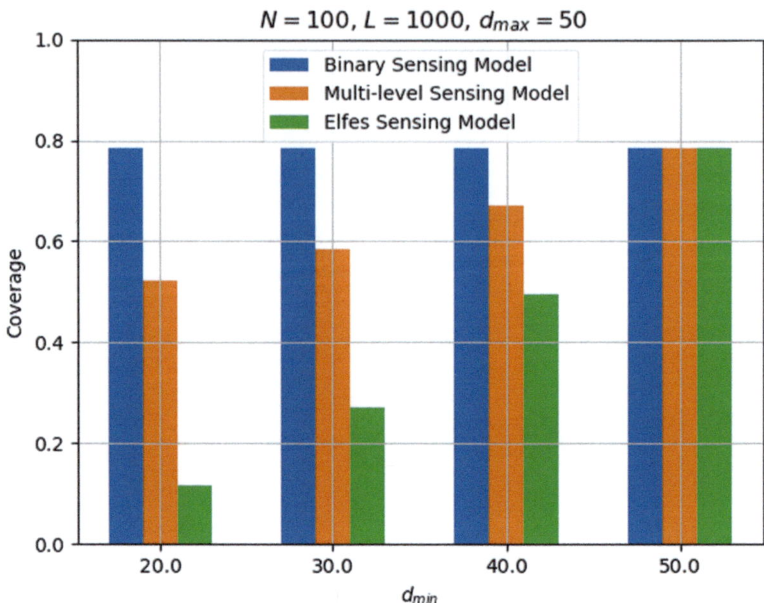

FIGURE 4.8 Partial sensing coverage of a square-based deterministic WSN using binary, multi-level, and Elfes models.

TABLE 4.2
Partial Sensing Coverage of a Square-Based Deterministic WSN Using Binary, Multi-level, and Elfes Models

Case #	Condition Settings	Binary $d_0 = max$	Multi-level $\eta = 0.6$	Elfes $\alpha = 0.5, \beta = 0.5$
1	$d_{min} = 20, d_{max} = 50$	0.785	0.522	0.117
2	$d_{min} = 30, d_{max} = 50$	0.785	0.584	0.272
3	$d_{min} = 40, d_{max} = 50$	0.785	0.672	0.496
4	$d_{min} = 50, d_{max} = 50$	0.785	0.785	0.785

The results substantiate our assertion that the sensing model plays a vital role in the performance evaluation, and the three models exhibit a convergent relationship under certain settings. To be specific, the probabilistic Elfes model can be transformed into a probabilistic multi-level model. Both probabilistic models can be converted into the deterministic Boolean model under certain conditions or configurations.

4.4.2 Partial Sensing Coverage in Random WSNs

Figure 4.9 showcases the partial sensing coverage of a randomly and uniformly distributed WSN with $N = 100$ sensors, a field side-length of $L = 1000$, and a maximum sensing range of $R_s = 50$. These network parameters correspond to those used in Figure 4.7, which represents a deterministic WSN, enabling a meaningful comparison between the two scenarios as deterministic and random counterparts. Intuitively, the WSN with partial sensing coverage, as shown in Figure 4.9, is capable of detecting an intruder before it traverses from the left side to the right side of the FoI, assuming it follows a straight-line invading path.

Using the Boolean sensing model, the sensing coverage or the intrusion detection probability is derived as follows:

$$P_{cov}^{B} = 1 - P\left(0, \pi R_s^2\right) = 1 - \frac{e^{-\lambda \pi R_s^2} \cdot \left(\lambda \pi R_s^2\right)}{0!} = 1 - e^{-\lambda \pi R_s^2} \tag{8}$$

where λ represents the network node density, and $\lambda = \dfrac{N}{L*L}$ in a 2D domain.

Proof: Consider a random point (x, y) within the field of interest. The probability that a sensor can sense an intrusion occurring at this point is equal to the probability of having at least one sensor deployed within the circular area centered at (x, y) with a radius of R_s^2. This circular area represents the sensing and detection coverage around the point (x, y) and has an area of πR_s^2. The probability of having no sensor deployed within the circular area of πR_s^2 can be calculated using the Poisson distribution [34]:

$$P\left(0, \pi R_s^2\right) = \frac{e^{-\lambda} \pi R_s^2 \cdot \left(\lambda \pi R_s^2\right)}{0!} = e^{\lambda \pi R_s^2}.$$

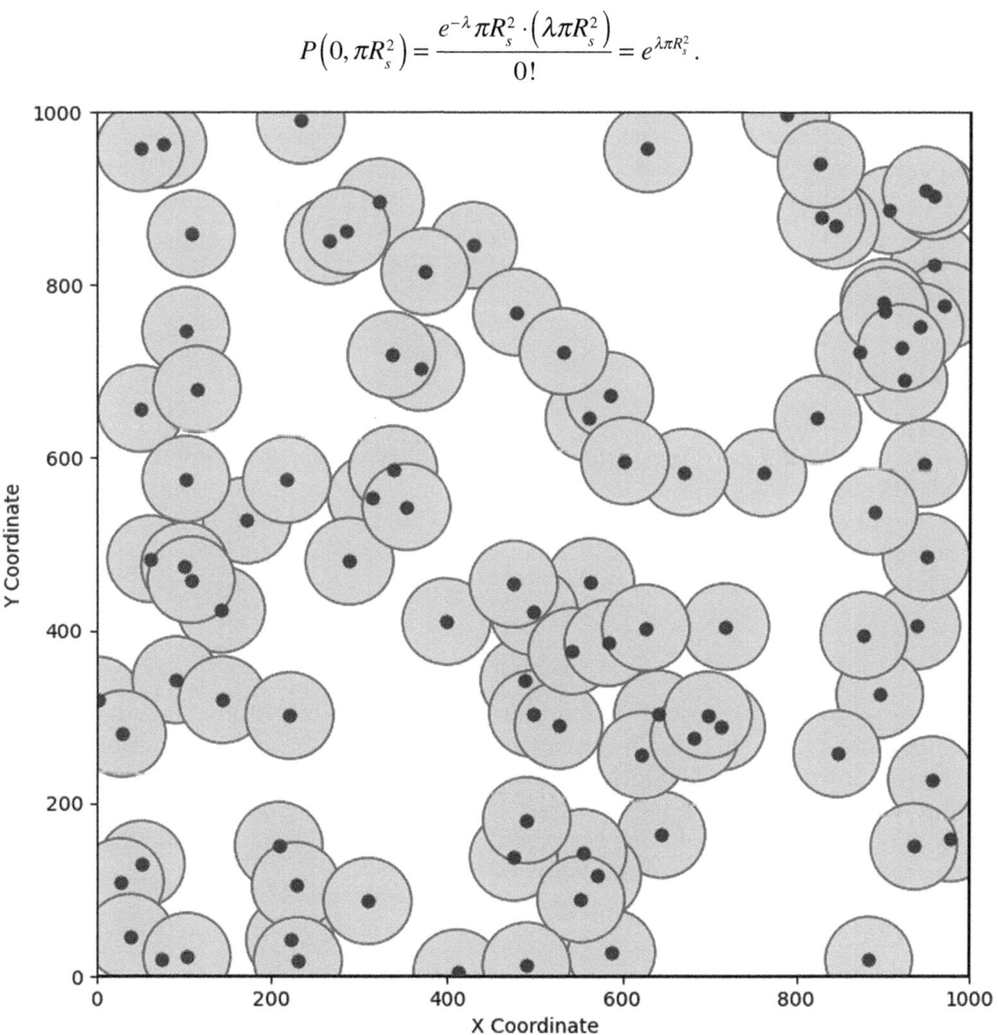

FIGURE 4.9 Partial sensing coverage of a randomly and uniformly distributed WSN.

Thus, the probability that this point (x, y) can be sensed by at least one sensor is $\left(1 - e^{-\lambda \pi R_s^2}\right)$.

Elfes Sensing Model for a randomly and uniformly distributed WSN with node density $\lambda = N/A$, the coverage using the Elfes sensing model, is quite complex.

Again, consider a random point t in the network and assume P_t is the probability that a random event occurring at t is sensed by at least one sensor with its sensing range. Adopting the Elfes sensing model, the probability P_t that any event at t is detected by a deployed sensor is derived as [35]:

$$
P_t^E = \frac{\pi \cdot r_s^2}{A} + \frac{2\pi}{A} \cdot \alpha^{-1/\beta} \cdot 1/\beta.
$$
$$
\left\{ r_s \cdot \gamma\left(\frac{1}{\beta}, \alpha \cdot (R_s - r_s)^\beta\right) + \alpha^{-\frac{1}{\beta}} \cdot \gamma\left(\frac{2}{\beta}, \alpha \cdot (R_s - r_s)^\beta\right) \right\},
$$
(9)

where $\gamma(s, w)$ is the incomplete gamma function as follows:

$$
\gamma(s, w) = \int_0^w e^{-t} \cdot t^{s-1} dt.
$$

Thus, $(1 - P_t)$ is the probability that no sensor is deployed around position t within their sensing range to detect an event(s) happening at position t. In the randomly and uniformly distributed WSN, the probability that all N sensors are deployed outside the coverage area with respect to position t is $(1 - P_t)^N$. Finally, the sensing coverage of the WSN adopting Elfes sensing model is therefore:

$$
P_{cov}^E = 1 - \left(1 - P_t^E\right)^N,
$$
(10)

where $P(t)$ is given in Equation (9).

Figure 4.10 depicts the sensing coverage of a WSN with a random and uniform distribution, utilizing both Elfes and Boolean sensing models under different configurations. The parameters used in all the examined scenarios are as follows: $L = 1000$, $R_s = 50$, and r_s set to 20, 35, and 50 respectively. It is important to note that when r_s equals R_s, the probabilistic Elfes model simplifies to the Boolean disk model, and the obtained results confirm this observation.

4.4.3 Intrusion Detection in Random WSNs

In our earlier work [36], we derived the intrusion detection probability for an intruder following a straight-line intrusion path in partially covered random WSNs. This detection probability refers to the likelihood of detecting the intruder within a maximal allowable intrusion distance, denoted as ξ. The intruder is assumed to start at a distance R from its target, which is located at the center of the WSN. More specifically, let $P_1[D \leq \xi]$ represent the probability that the intruder can be detected within the maximum allowable intrusion distance ξ by a single sensor. This $P_1[D \leq \xi]$ can be derived as follows:

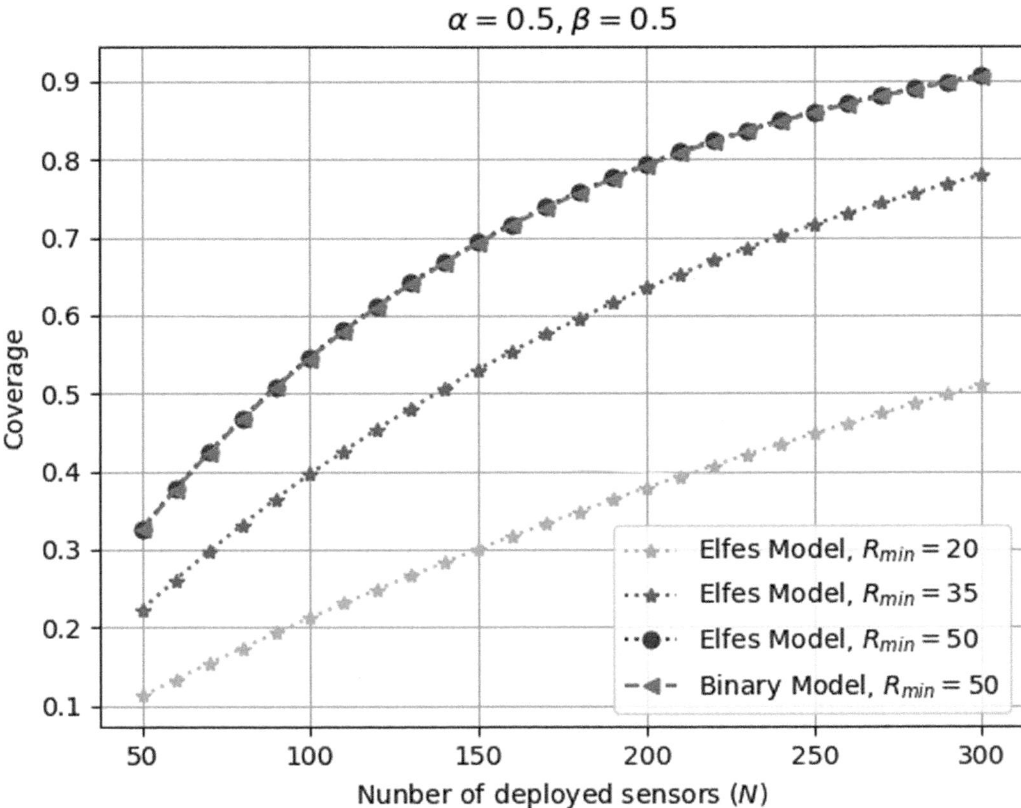

FIGURE 4.10 Partial coverage of a randomly and uniformly distributed WSN.

$$P_1[D \le \xi] = 1 - \left\{ 1 - \int_{R-\xi}^{R} \int_{-r_s}^{r_s} f(x,y)\,dydx \right.$$
$$\left. - \int_{R-\xi-r_s}^{R-\xi} \int_{-\sqrt{r_s^2-(x-R+\xi)^2}}^{\sqrt{r_s^2-(x-R+\xi)^2}} f(x,y)\,dydx \right.$$
$$\left. - \int_{R}^{R+r_s} \int_{-\sqrt{r_s^2-(x-R)^2}}^{\sqrt{r_s^2-(x-R)^2}} f(x,y)\,dydx \right\}^N,$$

where $f(x, y)$ is defined as in Equation 1 for a uniformly distributed WSN, and for a Gaussian distributed WSN, it is defined in Equation 2.

Figure 4.11 compares and contrasts the intrusion detection probability of a random WSN following Gaussian and Uniform distributions under the same network settings, adopting the Binary sensing model where $R_s = r_s$. Specifically, $N = 100$ and $R = 100$. For Gaussian-distributed WSNs, $\sigma_x = \sigma_y = 50$, and $\mu_x = \mu_y = 0$. The results demonstrate that network deployment and its modeling significantly impact network performance. Moreover, a WSN with partial sensing coverage can fulfill application requirements, such as detecting intruders within the maximum intrusion detection distance ξ, while conserving network resources and increasing energy efficiency. For instance, as shown in Figure 4.11, with $R_s = 15$, the intrusion detection probability is close to 1 for $\xi=30$, even though the sensing coverage is only about 0.9, which is equivalent to the detection probability of $\xi=0$ in a Uniformly distributed WSN.

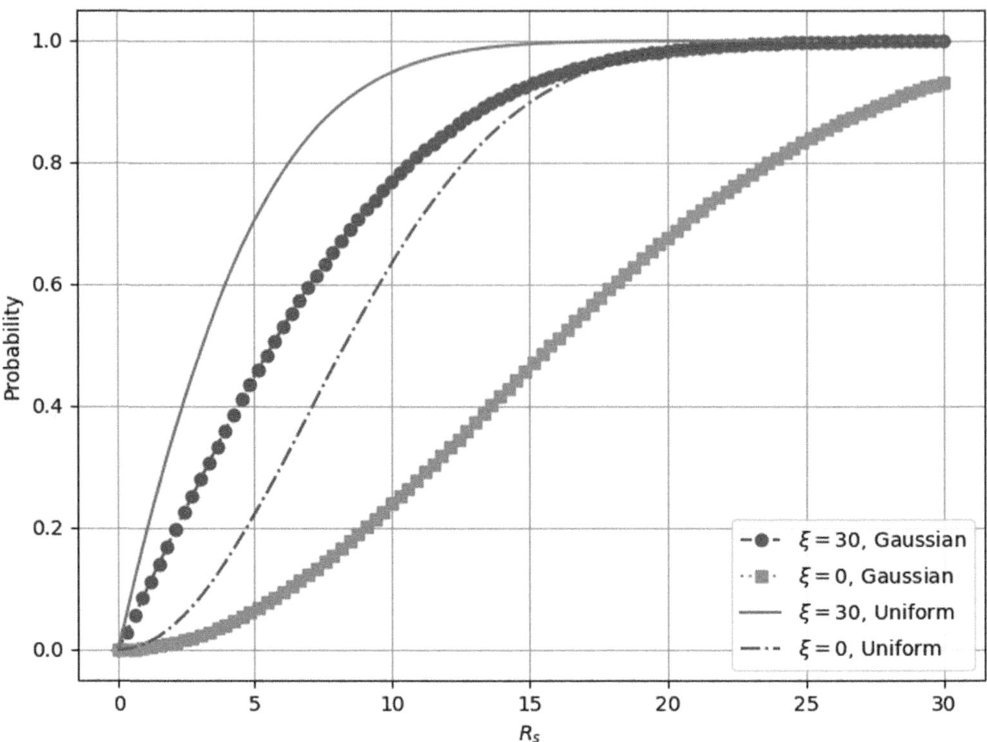

FIGURE 4.11   Intrusion detection probability of a random WSN following Gaussian and Uniform distributions.

4.5 RELATED WORKS

Numerous researchers have proposed solutions to address the issue of intrusions in wireless sensor networks. Recently, a notable trend has been in utilizing machine learning (ML) and artificial neural networks (ANN), particularly deep learning, for intrusion detection. Although both traditional machine learning and deep learning models have been employed in this domain, studies demonstrated that deep learning tends to achieve higher accuracy. To enhance the effectiveness of machine learning models, several studies have investigated more efficient feature selection techniques, recognizing the crucial role feature selection plays in model development. Another significant area of interest in wireless sensor networks is the development of energy-efficient solutions. Compressive sensing (CS) has emerged as a popular strategy among these. Many studies have explored the integration of clustering with compressive sensing to improve the overall network lifetime and efficiency.

4.5.1 TRADITIONAL MACHINE LEARNING-BASED INTRUSION DETECTION

An ML-based intrusion detection system (IDS), proposed by Singh et al. [13], employed Gaussian Process Regression (GPR) to accurately predict the k-barrier coverage probability while minimizing computational cost and time complexity. They used an analytical approach to select and extract the following six features: the number of nodes, sensing range (R_s), sensor-to-intruder velocity ratio, mobile-to-static node ratio, angle of the intrusion path, and required k. The authors illustrated that the proposed GPR-based models could outperform the benchmark support vector regression algorithm in terms of correlation coefficient, root mean square error, and complexity. On the other hand, Zhang et al. [37] proposed an ANN-based IDS that utilized the classification algorithm of kernel extreme learning machine (KELM). This model achieved better energy efficiency and significantly shortened detection time while guaranteeing high detection accuracy.

4.5.2 Deep Learning-based Intrusion Detection

Of late, deep learning (DL), a class of ML algorithms that uses multi-layer neural networks, has gained attention and application in IDS. Otomun et al. [38] investigated the feasibility of DL in IDS and compared a DL-based method with its adaptive ML-based counterpart. Via NS-3 simulations, they showcased that both methods achieved the same detection and accuracy rates; however, the ML-based IDS was more time-efficient than the DL-based solution regarding detection time. Nguyen and Kim [39] employed a convolutional neural network (CNN) model within their genetic algorithm (GA)-based intrusion detection system. Their approach introduced a hybrid learning methodology that combined a CNN deep feature extractor with a bagging classifier aimed at enhancing the overall classification performance. The proposed scheme demonstrates notable accuracy in large-scale intrusion detection tasks; nonetheless, it is noteworthy that this method exhibits a drawback in terms of significantly increased time consumption.

DL-based architectures featuring fully connected feed-forward ANN have been explored in [12, 40–42]. Singh et al. [12] predicted the number of barriers for fast intrusion detection, assuming that any two sensors can communicate with each other if the transmission range (R_c) of sensors is at least twice the sensing range (R_s) of sensors. In contrast, Amaran and Mohan [40] applied optimal multi-layer perceptron (OMLP) with the Dragonfly algorithm to detect intrusions and identify their class type. Muruganandam et al. [41], used a customized feed-forward ANN to precisely estimate the k-barrier count for effective intrusion detection and mitigation across a region of interest (RoI). They discovered that when evaluating a lot of barriers, the territory of the RoI was less important than other inputs to the ANN, such as the sensing range (R_s), the transmission area, and the number of sensors (N). The proposed model exhibited better detection accuracy when compared to GPR and Random Forest. However, the posited method is limited by only being able to process positive input characteristic values. The same problem and similar approach have been studied in [42], where the same four potential features were considered.

Gautami et al. [43] classified various types of attacks, including denial-of-service (DoS), remote to local (R2L), user to root (U2R), and probes, distinguishing them from normal network data. These same four attack classes were also studied by Hussain et al. [44]. The authors proposed two new algorithms in their intrusion-detection framework: an artificial bee colony-based feature selection algorithm and a new CNN model. This DL-driven framework was shown to be faster and more accurate than architectures based on ML models such as a support vector machine (SVM), decision tree (DT), and K-nearest neighbors (KNN). A similar performance comparison has also been demonstrated by Goyal et al. [45] in their DL-based IDS. They showed that the DL-powered framework achieved higher detection accuracy than DT and SVM.

4.5.3 Feature Selection-based Intrusion Detection

Intrusion detection systems that rely on machine learning algorithms, such as SVM, ANN, and KNN often suffer from reduced accuracy and intrusion misclassification due to suboptimal feature selection. To rectify this, several works [46–49] have focused on developing more effective feature selection strategies. Subbiah et al. [47] presented Boruta feature selection with a grid search random forest algorithm, demonstrating that improved feature selection significantly boosts detection accuracy. A similar outcome has been substantiated by Jiang et al. [46]; they have reduced the number of features used in their ML algorithm to achieve faster training efficiency, low memory usage, and higher accuracy. In another study, Nancy et al. [48] used a recursive feature selection algorithm to select the optimal number of features for classification and analysis. Plus, they integrated an extension of the DT algorithm using fuzzy temporal constraints for classifying the network traffic and users. Furthermore, CNN was employed to handle large volumes of data. The authors asserted that their model produced higher precision levels, increased packet delivery ratio and throughput, and reduced energy consumption. Likewise, Riyaz and Ganapathy [49] implemented a feature selection algorithm in conjunction with an existing CNN model in their intrusion detection system. They developed a new methodology for variable selection for the existing correlation coefficient variance-based feature selection algorithm, which is claimed to be more efficient in choosing the most important features for classification. This framework was shown to result in higher detection accuracy and shorter training and testing periods.

4.5.4　Modern Sensing Models

Recently, Zhu et al. [50] conducted a survey on modern soft-sensing modeling techniques. The study explored various approaches, including SVM, DL, GA, neural network (NN), fuzzy logic (FL), probabilistic latent variable models (PLVMs), and other effective methods to capture real-time variables in the sensor fermentation processes. Zroug et al. [51] proposed an ANN-based prediction framework for a number of crucial performance metrics of CSMA/CA protocol in WSNs. The proposed method is designed to handle performance predictions of massive networks comprising a large number of nodes, which is a limitation of the existing techniques. Another prediction model presented in [52] aimed to determine the degrees of coverage and sink connectivity in WSNs.

In many WSN applications, power consumption emerges as the principal constraint. Accordingly, a considerable body of researchers has studied this domain to proffer effective solutions. A compressive sensing (CS) technique is a contemporary technique used in many applications for reducing power consumption and extending network lifetime [53–56]. Nguyen et al. [53] applied CS in landslide monitoring events. The main idea of the paper is as follows: At first, in each node, the environment data is sensed by multiple sensors (e.g., soil moisture sensor, temperature sensor, and acceleration sensor). Next, to save the power consumption in the transmission process, instead of sending the whole sensed data, the data is first converted from the time domain to the frequency domain by using Fourier transform, and finally, the node sends a random amount of the corresponding Fourier coefficients to the gateway node. To save power consumption in various catastrophic scenarios, Yang et al. [55] proposed a compressed network coding-based distributed data storage scheme by exploiting the correlation of sensor readings and by reducing the total number of transmissions. The proposed approach used CS and network coding techniques. Wei and He [56] explored the big data P2P direct connection perception idea towards the spatial compressed sensing technology. The research results showed improved performance in locating objects, reducing the number of defected nodes in the networks, reducing node energy consumption, and enhancing calculation speed.

A number of studies focused on incorporating clustering with the compressive sensing technique for the betterment of the network lifetime [57–60]. Patil and Parveen [57] proposed Integrated CS-Clustering Mechanism (ICCM), which integrated clustering with CS for an energy-efficient WSN network. First, an optimized clustering was used to form an efficient clustering, and then each cluster head used an optimized CS mechanism while sending data to the base station. The evaluation results showed that ICCM outperformed the existing models in terms of energy consumption. Liu et al. [58] showed that a bottleneck effect is produced if the clustering method is integrated with the traditional data compression technique, resulting in an energy hole. Then, they came up with a hybrid CS method that integrates CS with hexagon-clustered WSN for effectively balancing the network load and extending the network lifetime.

Similarly, Yuan et al. [59] proposed a CS-based scheme for improving the network lifetime. In each cluster, m-dimensional data is formed using CS technology to ensure that the data can be recovered. After that, to send the m-dimensional data to the sink, the cluster head used a two-stage routing scheme to ensure that each node carries a reduced amount of time for achieving higher energy utilization. Also, to increase network lifetime and reduce the number of transmissions, Joan et al. [60] presented a distributed CS-based transmission scheme. In order to reduce the number of channel usage in a given time, the model selected only a few active nodes and relayed based on a cost function responsible for keeping the balance between the reconstruction error and energy consumption.

4.5.5 Computational Issues Related to Antenna Arrays (5G, mmWave, and THz)

Millimeter-wave (mmWave) and Multiple-Input, Multiple-Output (MIMO) technologies are pivotal in shaping the capabilities of 5G wireless networks [61]. MmWave, which operates at extremely high frequencies, offers vast bandwidths for data transmission, enabling unprecedented data rates and ultra-low latency. This spectrum range unlocks the potential for applications like augmented and virtual reality, high-definition video streaming, and IoT devices that demand high-speed connectivity [62]. However, mmWave signals are sensitive to obstacles and environmental conditions, making MIMO indispensable. MIMO leverages multiple antennas at both the transmitter and receiver to enhance signal quality, increase capacity, and mitigate interference. As we move into the realm of terahertz (THz) wireless networks, the importance of these technologies becomes even more pronounced. THz frequencies, ranging from 0.1 THz to 10 THz, offer orders of magnitude more bandwidth than mmWave, potentially enabling data rates that are several times faster than current 5G speeds. However, THz signals are highly susceptible to absorption and scattering, necessitating advanced MIMO techniques and beamforming to overcome these propagation challenges.

Antenna arrays are fundamental in advancing 5G and THz wireless networks, particularly in mmWave and MIMO technologies. In mmWave, antenna arrays harness the directional characteristics of mmWave signals, utilizing beamforming to focus and steer beams [63], thereby mitigating signal attenuation and enhancing reliability. This is essential for achieving the desired coverage and capacity in dense urban environments. Similarly, in MIMO, antenna arrays significantly boost data throughput by exploiting spatial diversity and multipath propagation. Multiple antennas at both ends of the communication link allow for the simultaneous transmission of multiple data streams, increasing spectral efficiency, reducing interference, enabling spatial multiplexing, and improving overall network performance. Furthermore, in THz wireless networks, antenna arrays can aid in beamforming and beam steering at THz frequencies, ensuring reliable connectivity.

While these next-generation wireless technologies offer significant benefits in terms of data rates and capacity, they also present several computational challenges related to antenna arrays. Some of these computational complexity issues are described below.

4.5.6 Beamforming

Beamforming in 5G and THz networks involves complex mathematical calculations to determine the phase and amplitude of signals across multiple antenna elements to form the desired beam. In mmWave massive MIMO systems, a large antenna array is employed for beamforming to counteract path losses through focused transmissions. In traditional massive MIMO setups, the conventional digital baseband beamforming (DB) necessitates a dedicated radio frequency (RF) chain for each antenna [64–66]. However, due to the substantial power consumption and cost associated with mixed-signal and RF components such as analog-to-digital converters (ADCs), digital-to-analog converters (DACs), data converters, and mixers [67], a shift towards hybrid beamforming, which operates in both the baseband and analog domains, has emerged.

The use of a large array of antenna elements is intended to boost capacity gains in massive MIMO systems. However, determining the optimal number of RF chains, which directly affects complexity and power consumption, is typically done considering practical constraints like the number of multiplexed streams, antenna elements, constant amplitude, and quantized phases of analog phase shifters [68].

Another notable challenge in this domain is real-time adaptation. According to the analysis presented in [69], there is no universal hybrid beamforming configuration that strikes the ideal balance between complexity and performance. Therefore, to achieve the best performance with hybrid beamforming, the configuration should adapt dynamically to the specific application and channel conditions [68]. Researchers have investigated different methods of adaptation, including neural networks [70], hardware solution [71, 72], and reinforcement learning [73].

4.5.7 CHANNEL ESTIMATION

Channel estimation involves determining the characteristics of the wireless channel between the transmitter and receiver, which is essential for beamforming, signal processing, and overall system performance. The complexity of channel estimation in mmWave and THz systems is influenced by several factors, including high antenna array density, narrow beamwidths, complex channel models, and extensive multipath propagation components. The performance and computational complexity of channel estimation methods for millimeter and terahertz frequency bands have been studied in [74–77].

Several channel estimation strategies, including Bayesian learning with the relevance vector machine, orthogonal matching pursuit, least absolute shrinkage and selection operator optimization, and least squares channel estimator were compared in [74]. The researchers found that although the Bayesian learning-based estimator offers higher accuracy, it comes at the expense of increased computational complexity. In [76], a low complexity channel estimation model for mmWave systems with large antenna arrays was studied. The authors found that, as the number of antennas increases, more measurements are necessary to approach mean rate performance similar to that of perfect estimation. Yang et al. [77] proposed a channel estimation framework for the mmWave lens MIMO system utilizing a low-complexity antenna selection method, which was shown to enhance transmission throughput.

A deep neural network-driven channel estimation scheme was proposed in [78]. They explored the use of mixed-resolution analog-to-digital converters (ADCs) to balance power consumption and system performance. By leveraging the sparsity of the mmWave channel, they framed beamspace channel estimation as a sparse signal recovery problem. They showed that the deep learning-based channel estimation approach outperformed the traditional method in the same setup.

4.5.8 PRECODING

Precoding techniques are used to pre-distort the transmitted signals to compensate for the channel's characteristics, and this requires real-time computation and adaptation to changing channel conditions [79]. Furthermore, finding the optimal combination of analog and digital beamforming weights is a complex optimization problem [80]. Algorithms like singular value decomposition (SVD) [81, 82] or alternating minimization are commonly used. The complexity depends on the system's dimensions, including the number of antennas and beams and the convergence criteria.

In hybrid beamforming, a digital baseband stage is responsible for precoding (at the transmitter) and combining (at the receiver) signals across multiple antennas [83]. The complexity is primarily determined by the number of antennas and the modulation scheme. For MIMO systems, it is proportional to the number of transmitter and receiver antennas. A comparison of a wide range of existing precoding methodologies can be found in [79].

Park et al. [83] introduced a closed-form solution for hybrid analog/digital precoding in frequency-selective mmWave systems that demonstrates the effectiveness of the precoding methods in achieving high spectral efficiency. A hybrid precoding technique was also proposed in [84], where the primary focus was minimizing energy consumption by utilizing a low-complexity algorithm. Deep learning-based hybrid precoding strategies were presented in [85] and [86], where a deep learning model was harnessed to improve the performance and reduce the computation time compared to conventional optimization and codebook-based techniques.

4.6 CONCLUSION

The chapter presents computational modeling and analysis of the fundamental characteristics of Wireless Sensor Networks (WSNs) for various applications including intrusion detection. It encompasses sensor deployments, sensor capabilities, and application requirements under various constraints. The WSN modeling and analysis allow quick simulation and performance evaluation and comparison for parameter changes in various aspects before the physical experimental setup or practical deployment, minimizing costs and mitigating risks. Additionally, it illustrates the impact of different models on network performance, discussing their strengths, weaknesses, and interrelationships. Moreover, the chapter reviews the recent notable trend in these endeavors, involving the application of machine learning (ML) and artificial neural networks (ANN) to enhance accuracy and efficiency for applications in realistic environments.

REFERENCES

1. Filippo Campagnaro, Fabian Steinmetz, and Bernd-Christian Renner. Survey on low-cost underwater sensor networks: From niche applications to everyday use. *Journal of Marine Science and Engineering*, 11(1):125, 2023.
2. R Alekya, Neelima Devi Boddeti, K Salomi Monica, Ramadoss Prabha, and Venkataramani Venkatesh. Iot based smart healthcare monitoring systems: A literature review. *European. Journal of Molecular and Clinical Medicine*, 7:2020, 2021.
3. Deepika Sharma, and Jaiteg Singh. A comparative analysis of healthcare monitoring systems using wsn. *ECS Transactions*, 107(1):8695, 2022.
4. Crispino S Abella, Salvo Bonina, Antonino Cucuccio, Salvatore D'Angelo, Gianluca Giustolisi, Alfio D Grasso, Antonio Imbruglia, Giorgio S Mauro, Giuseppe AM Nastasi, and Gaetano Palumbo, et al. Autonomous energy efficient wireless sensor network platform for home/office automation. *IEEE Sensors Journal*, 19(9):3501–3512, 2019.
5. Mamoona Majid, Shaista Habib, Abdul Rehman Javed, Muhammad Rizwan, Gautam Srivastava, Thippa Reddy Gadekallu, and Jerry Chun-Wei Lin. Applications of wireless sensor networks and internet of things frameworks in the industry revolution 4.0: A systematic literature review. *Sensors*, 22(6):2087, 2022.
6. Simon Kamm, Sushma Sri Veekati, Timo Müller, Nasser Jazdi, and Michael Weyrich. A survey on machine learning based analysis of heterogeneous data in industrial automation. *Computers in Industry*, 149:103930, 2023.
7. Himanshu Sharma, Ahteshamul Haque, and Frede Blaabjerg. Machine learning in wireless sensor networks for smart cities: A survey. *Electronics*, 10(9):1012, 2021.
8. Fadi Al-Turjman, and Joel Poncha Lemayian. Intelligence, security, and vehicular sensor networks in internet of things (IoT)-enabled smart-cities: An overview. *Computers & Electrical Engineering*, 87:106776, 2020.
9. Ala' Khalifeh, Khalid A Darabkh, Ahmad M Khasawneh, Issa Alqaisieh, Mohammad Salameh, Ahmed AlAbdala, Shams Alrubaye, Anwar Alassaf, Samer Al-HajAli, and Radi Al-Wardat, et al. Wireless sensor networks for smart cities: Network design, implementation and performance evaluation. *Electronics*, 10(2):218, 2021.
10. Xiaohui Li, and Andrey V Savkin. Networked unmanned aerial vehicles for surveillance and monitoring: A survey. *Future Internet*, 13(7):174, 2021.
11. Atif Ali, Yasir Khan Jadoon, Sabir Ali Changazi, and Muhammad Qasim. Military operations: Wireless sensor networks based applications to reinforce future battlefield command system. In *2020 IEEE 23rd International Multitopic Conference (INMIC)*, pp. 1–6. IEEE, 2020.

12. Abhilash Singh, J Amutha, Jaiprakash Nagar, and Sandeep Sharma. A deep learning approach to predict the number of k-barriers for intrusion detection over a circular region using wireless sensor networks. *Expert Systems with Applications*, 211:118588, 2023.

13. Abhilash Singh, Jaiprakash Nagar, Sandeep Sharma, and Vaibhav Kotiyal. A gaussian process regression approach to predict the k-barrier coverage probability for intrusion detection in wireless sensor networks. *Expert Systems with Applications*, 172:114603, 2021.

14. Rahul Priyadarshi, Bharat Gupta, and Amulya Anurag. Deployment techniques in wireless sensor networks: A survey, classification, challenges, and future research issues. *The Journal of Supercomputing*, 76:7333–7373, 2020.

15. Fengrong Han, Xinni Liu, Izzeldin Ibrahim Mohamed, Kamarul Hawari Ghazali, and Yue Zhao. A survey on deployment and coverage strategies in three-dimensional wireless sensor networks. In *Proceedings of the 2019 8th International Conference on Software and Computer Applications*, pp. 544–549, 2019.

16. Mohammed Farsi, Mostafa A Elhosseini, Mahmoud Badawy, Hesham Arafat Ali, and Hanaa Zain Eldin. Deployment techniques in wireless sensor networks, coverage and connectivity: A survey. *Ieee Access*, 7:28940–28954, 2019.

17. Mohammed Jouhari, Khalil Ibrahimi, Hamidou Tembine, and Jalel Ben-Othman. Underwater wireless sensor networks: A survey on enabling technologies, localization protocols, and internet of underwater things. *IEEE Access*, 7:96879–96899, 2019.

18. Bin Cao, Jianwei Zhao, Po Yang, Peng Yang, Xin Liu, and Yuan Zhang. 3-d deployment optimization for heterogeneous wireless directional sensor networks on smart city. *IEEE Transactions on Industrial Informatics*, 15(3):1798–1808, 2018.

19. M Premkumar, and TVP Sundararajan. Dldm: Deep learning-based defense mechanism for denial of service attacks in wireless sensor networks. *Microprocessors and Microsystems*, 79:103278, 2020.

20. Rahul Priyadarshi, Bharat Gupta, and Amulya Anurag. Wireless sensor networks deployment: A result oriented analysis. *Wireless Personal Communications*, 113:843–866, 2020.

21. Sumit Pundir, Mohammad Wazid, Devesh Pratap Singh, Ashok Kumar Das, Joel JPC Rodrigues, and Youngho Park. Intrusion detection protocols in wireless sensor networks integrated to internet of things deployment: Survey and future challenges. *IEEE Access*, 8:3343–3363, 2019.

22. Haejoon Jung, and In-Ho Lee. Secrecy performance analysis of analog cooperative beamforming in three-dimensional gaussian distributed wireless sensor networks. *IEEE Transactions on Wireless Communications*, 18(3):1860–1873, 2019.

23. Mustapha Reda Senouci, Abdelhamid Mellouk, and Amar Aissani. Random deployment of wireless sensor networks: A survey and approach. *International Journal of Ad Hoc and Ubiquitous Computing*, 15(1–3):133–146, 2014.

24. Wenliang Du, Jing Deng, Yunghsiang S Han, Shigang Chen, and Pramod K Varshney. A key management scheme for wireless sensor networks using deployment knowledge. In *IEEE INFOCOM 2004*, volume 1. IEEE, 2004.

25. Alberto Leon-Garcia. *Probability and Random Processes for Electrical Engineering*. Pearson Education India, 1994.

26. Brian Carter, and Rammohan Ragade. A probabilistic model for the deployment of sensors. In *2009 IEEE Sensors Applications Symposium*, pp. 7–12. IEEE, 2009.

27. Nasir Mehranbod, Masoud Soroush, Michael Piovoso, and Babatunde A Ogunnaike. Probabilistic model for sensor fault detection and identification. *AIChE Journal*, 49(7):1787–1802, 2003.

28. Riham Elhabyan, Wei Shi, and Marc St-Hilaire. Coverage protocols for wireless sensor networks: Review and future directions. *Journal of Communications and Networks*, 21(1):45–60, 2019.

29. Ya-Nan Zheng, Zhe Yu, Guoyong Mao, Yunyao Li, Dhanapal Pravarthana, Waqas Asghar, Yiwei Liu, Shaoxing Qu, Jie Shang, and Run-Wei Li. A wearable capacitive sensor based on ring/disk-shaped electrode and porous dielectric for noncontact healthcare monitoring. *Global Challenges*, 4(5):1900079, 2020.

30. Bin Cao, Jianwei Zhao, Zhihan Lv, and Xin Liu. 3d terrain multiobjective deployment optimization of heterogeneous directional sensor networks in security monitoring. *IEEE Transactions on Big Data*, 5(4):495–505, 2017.

31. Chuan Zhu, Chunlin Zheng, Lei Shu, and Guangjie Han. A survey on coverage and connectivity issues in wireless sensor networks. *Journal of Network and Computer Applications*, 35(2):619–632, 2012.

32. Fernando Ojeda, Diego Mendez, Arturo Fajardo, and Frank Ellinger. On wireless sensor network models: A cross-layer systematic review. *Journal of Sensor and Actuator Networks*, 12(4), 2023.

33. Alberto Elfes. Using occupancy grids for mobile robot perception and navigation. *Computer*, 22(6):46–57, 1989.

34. Sheldon M Ross. *Introduction to Probability Models*. Academic Press, 2014.

35. Ashraf Hossain, Subit Chakrabarti, and Prabir Kumar Biswas. Impact of sensing model on wireless sensor network coverage. *IET Wireless Sensor Systems*, 2(3):272–281, 2012.

36. Yun Wang, Weihuang Fu, and Dharma P Agrawal. Gaussian versus uniform distribution for intrusion detection in wireless sensor networks. *IEEE Transactions on Parallel and Distributed systems*, 24(2):342–355, 2012.

37. Wenjie Zhang, Dezhi Han, Kuan-Ching Li, and Francisco Isidro Massetto. Wireless sensor network intrusion detection system based on mk-elm. *Soft Computing*, 24:12361–12374, 2020.

38. Safa Otoum, Burak Kantarci, and Hussein T Mouftah. On the feasibility of deep learning in sensor network intrusion detection. *IEEE Networking Letters*, 1(2):68–71, 2019.

39. Minh Tuan Nguyen, and Kiseon Kim. Genetic convolutional neural network for intrusion detection systems. *Future Generation Computer Systems*, 113:418–427, 2020.

40. Sibi Amaran, and Madhan R Mohan. An optimal multilayer perceptron with dragonfly algorithm for intrusion detection in wireless sensor networks. In *2021 5th International Conference on Computing Methodologies and Communication (ICCMC)*, pp. 1–5. IEEE, 2021.

41. S Muruganandam, Rahul Joshi, P Suresh, N Balakrishna, Kakarla Hari Kishore, and SV Manikanthan. A deep learning based feed forward artificial neural network to predict the k-barriers for intrusion detection using a wireless sensor network. *Measurement: Sensors*, 25:100613, 2023.

42. Abhilash Singh, J Amutha, Jaiprakash Nagar, and Sandeep Sharma. A deep learning approach to predict the number of k-barriers for intrusion detection over a circular region using wireless sensor networks. *Expert Systems with Applications*, 211:118588, 2023.

43. A Gautami, J Shanthini, and S Karthik. A quasi-newton neural network based efficient intrusion detection system for wireless sensor network. *Computer Systems Science & Engineering*, 45(1), 2023.

44. Khurram Hussain, Yuanqing Xia, Ameer N Onaizah, Tayyab Manzoor, and Khurrum Jalil. Hybrid of woa-abc and proposed cnn for intrusion detection system in wireless sensor networks. *Optik*, 271:170145, 2022.

45. Aman Goyal, Sourav Mishra, and Vijay K Chaurasiya. Intrusion detection in wireless sensor networks using deep learning. In *2023 4th International Conference for Emerging Technology (INCET)*, pp. 1–13. IEEE, 2023.

46. Shuai Jiang, Juan Zhao, and Xiaolong Xu. Slgbm: An intrusion detection mechanism for wireless sensor networks in smart environments. *IEEE Access*, 8:169548–169558, 2020.

47. Sridevi Subbiah, Kalaiarasi Sonai Muthu Anbananthen, Saranya Thangaraj, Subarmaniam Kannan, and Deisy Chelliah. Intrusion detection technique in wireless sensor network using grid search random forest with boruta feature selection algorithm. *Journal of Communications and Networks*, 24(2):264–273, 2022.

48. Periasamy Nancy, Sannasy Muthurajkumar, Sannasi Ganapathy, SVN Santhosh Kumar, Munuswamy Selvi, and Kannan Arputharaj. Intrusion detection using dynamic feature selection and fuzzy temporal decision tree classification for wireless sensor networks. *IET Communications*, 14(5):888–895, 2020.

49. B Riyaz, and Sannasi Ganapathy. A deep learning approach for effective intrusion detection in wireless networks using cnn. *Soft Computing*, 24:17265–17278, 2020.

50. Xianglin Zhu, Khalil Ur Rehman, Bo Wang, and Muhammad Shahzad. Modern soft-sensing modeling methods for fermentation processes. *Sensors*, 20(6):1771, 2020.

51. Siham Zroug, Ikram Remadna, Laid Kahloul, Saber Benharzallah, and Sadek Labib Terrissa. Leveraging the power of machine learning for performance evaluation prediction in wireless sensor networks. In *2021 International Conference on Information Technology (ICIT)*, pp. 864–869, 2021.

52. Pakpoom Hoyingcharoen, and Wiklom Teerapabkajorndet. Expected probabilistic detection and sink connectivity in wireless sensor networks. *IEEE Sensors Journal*, 19(12):4480–4493, 2019.

53. Dinh-Chinh Nguyen, Tran Duc-Tan, and Duc-Nghia Tran. Application of compressed sensing in effective power consumption of wsn for landslide scenario. In *2015 Asia Pacific Conference on Multimedia and Broadcasting*, pp. 1–5, 2015.

54. Vivek Kumar Singh, Shekhar Verma, and Manish Kumar. Odecs: An on-demand explosion-based compressed sensing using random walks in wireless sensor networks. *IEEE Systems Journal*, 13(3):2466–2475, 2019.

55. Xianjun Yang, Xiaofeng Tao, Eryk Dutkiewicz, Xiaojing Huang, Y Jay Guo, and Qimei Cui. Energy-efficient distributed data storage for wireless sensor networks based on compressed sensing and network coding. *IEEE Transactions on Wireless Communications*, 12(10):5087–5099, 2013.

56. Pengcheng Wei, and Fangcheng He. The compressed sensing of wireless sensor networks based on internet of things. *IEEE Sensors Journal*, 21(22):25267–25273, 2021.

57. Nandini S Patil, and Asma Parveen. Integrated compressive sensing based clustering approach to improve network lifetime in WSN. In *2021 IEEE International Conference on Mobile Networks and Wireless Communications (ICMNWC)*, pp. 1–5, 2021.

58. Dan Liu, Qian Zhou, Zhi Zhang, and Baoling Liu. Cluster-based energy-efficient transmission using a new hybrid compressed sensing in WSN. In *2016 IEEE Conference on Computer Communications Workshops (INFOCOM WKSHPS)*, pp. 372–376, 2016.

59. Yicong Yuan, Wei Liu, Tian Wang, Qingyong Deng, Anfeng Liu, and Houbing Song. Compressive sensing-based clustering joint annular routing data gathering scheme for wireless sensor networks. *IEEE Access*, 7:114639–114658, 2019.

60. Joan Enric Barceló-Lladó, Antoni Morell, and Gonzalo Seco-Granados. Amplify-and-forward compressed sensing as an energy-efficient solution in wireless sensor networks. *IEEE Sensors Journal*, 14(5):1710–1719, 2014.

61. Robert W Heath, Nuria Gonzalez-Prelcic, Sundeep Rangan, Wonil Roh, and Akbar M Sayeed. An overview of signal processing techniques for millimeter wave mimo systems. *IEEE Journal of Selected Topics in Signal Processing*, 10(3):436–453, 2016.

62. Theodore S Rappaport, George R MacCartney, Mathew K Samimi, and Shu Sun. Wideband millimeter-wave propagation measurements and channel models for future wireless communication system design. *IEEE Transactions on Communications*, 63(9):3029–3056, 2015.

63. B.D. Van Veen and K.M. Buckley. Beamforming: A versatile approach to spatial filtering. *IEEE ASSP Magazine*, 5(2):4–24, 1988.

64. Fredrik Rusek, Daniel Persson, Buon Kiong Lau, Erik G Larsson, Thomas L Marzetta, Ove Edfors, and Fredrik Tufvesson. Scaling up mimo: Opportunities and challenges with very large arrays. *IEEE Signal Processing Magazine*, 30(1):40–60, 2012.

65. Erik G Larsson, Ove Edfors, Fredrik Tufvesson, and Thomas L Marzetta. Massive mimo for next generation wireless systems. *IEEE Communications Magazine*, 52(2):186–195, 2014.

66. Lu Lu, Geoffrey Ye Li, A Lee Swindlehurst, Alexei Ashikhmin, and Rui Zhang. An overview of massive mimo: Benefits and challenges. *IEEE Journal of Selected Topics in Signal Processing*, 8(5):742–758, 2014.

67. Irfan Ahmed, Hedi Khammari, and Adnan Shahid. Resource allocation for transmit hybrid beamforming in decoupled millimeter wave multiuser-mimo downlink. *IEEE Access*, 5:170–182, 2016.

68. Shajahan Kutty, and Debarati Sen. Beamforming for millimeter wave communications: An inclusive survey. *IEEE Communications Surveys & Tutorials*, 18(2):949–973, 2015.

69. Andreas F Molisch, Vishnu V Ratnam, Shengqian Han, Zheda Li, Sinh Le Hong Nguyen, Linsheng Li, and Katsuyuki Haneda. Hybrid beamforming for massive mimo: A survey. *IEEE Communications Magazine*, 55(9):134–141, 2017.

70. Muhammad Alrabeiah, Yu Zhang, and Ahmed Alkhateeb. Neural networks based beam codebooks: Learning mmwave massive mimo beams that adapt to deployment and hardware. *IEEE Transactions on Communications*, 70(6):3818–3833, 2022.

71. Suhasini Komarraju, and Abhijit Chatterjee. Self-aware mimo beamforming systems: dynamic adaptation to channel conditions and manufacturing variability. In *2022 Design, Automation & Test in Europe Conference & Exhibition (DATE)*, pp. 572–575. IEEE, 2022.

72. Alvaro Morales, Serguei Smirnov, Dmitri V Lioubtchenko, Joachim Oberhammer, Chigo Okonkwo, and Idelfonso Tafur Monroy. Photonic-based beamforming system for sub-thz wireless communications. In *2019 European Microwave Conference in Central Europe (EuMCE)*, pp. 253–256. IEEE, 2019.

73. Yu Zhang, Muhammad Alrabeiah, and Ahmed Alkhateeb. Reinforcement learning of beam codebooks in millimeter wave and terahertz mimo systems. *IEEE Transactions on Communications*, 70(2):904–919, 2021.

74. Visa Tapio, Mubarak Umar Aminu, Janne Lehtomäki, and Markku Juntti. Channel Estimation Algorithms for Hybrid Antenna Arrays: Performance and Complexity. In *2019 16th International Symposium on Wireless Communication Systems (ISWCS)*, pp. 293–297, 2019.

75. LIU Wei-Tao, Xia Zhong-Xi, Qian-Lin CHENG, Yang Dong-Lin, and Xiao-Fei Zhang. Channel estimation algorithm for mmwave communication with array antennas via reduced-dimension music. In *3rd Annual International Conference on Electronics, Electrical Engineering and Information Science (EEEIS 2017)*, pp. 299–305. Atlantis Press, 2017.

76. George C Alexandropoulos, and Symeon Chouvardas. Low complexity channel estimation for millimeter wave systems with hybrid a/d antenna processing. In *2016 IEEE Globecom Workshops (GC Wkshps)*, pp. 1–6, 2016.

77. Lu Yang, Yong Zeng, and Rui Zhang. Channel estimation for millimeter-wave mimo communications with lens antenna arrays. *IEEE Transactions on Vehicular Technology*, 67(4):3239–3251, 2017.

78. Rui Zhang, Weiqiang Tan, Wenliang Nie, Xianda Wu, and Ting Liu. Deep learning-based channel estimation for mmwave massive mimo systems in mixed-adc architecture. *Sensors*, 22(10):3938, 2022.

79. Mahmoud A Albreem, Alaa H Al Habbash, Ammar M Abu-Hudrouss, and Salama S Ikki. Overview of precoding techniques for massive MIMO. *IEEE Access*, 9:60764–60801, 2021.

80. Omar El Ayach, Sridhar Rajagopal, Shadi Abu-Surra, Zhouyue Pi, and Robert W Heath. Spatially sparse precoding in millimeter wave MIMO systems. *IEEE Transactions on Wireless Communications*, 13(3):1499–1513, 2014.

81. Quentin H Spencer, A Lee Swindlehurst, and Martin Haardt. Zero-forcing methods for downlink spatial multiplexing in multiuser mimo channels. *IEEE Transactions on Signal Processing*, 52(2):461–471, 2004.

82. Miroslav Waldecker, and Roman Marsalek. Impact of hardware imperfections and imperfect channel knowledge on svd-precoded MIMO. In *2022 32nd International Conference Radioelektronika (RADIOELEKTRONIKA)*, pp. 01–04. IEEE, 2022.

83. Sungwoo Park, Ahmed Alkhateeb, and Robert W Heath. Dynamic subarrays for hybrid precoding in wideband mmwave MIMO systems. *IEEE Transactions on Wireless Communications*, 16(5):2907–2920, 2017.

84. Xinyu Gao, Linglong Dai, Shuangfeng Han, I Chih-Lin, and Robert W Heath. Energy-efficient hybrid analog and digital precoding for mmwave MIMO systems with large antenna arrays. *IEEE Journal on Selected Areas in Communications*, 34(4):998–1009, 2016.

85. Ahmet M Elbir, and Anastasios K Papazafeiropoulos. Hybrid precoding for multiuser millimeter wave massive MIMO systems: A deep learning approach. *IEEE Transactions on Vehicular Technology*, 69(1):552–563, 2019.

86. Xiaofeng Li, and Ahmed Alkhateeb. Deep learning for direct hybrid precoding in millimeter wave massive MIMO systems. In *2019 53rd Asilomar Conference on Signals, Systems, and Computers*, pp. 800–805. IEEE, 2019.

5 Security Measures in Computational Modeling and Simulations

Daniel Dauda Wisdom, Olufunke Rebecca Vincent,
Abayomi-Alli Adebayo, Folorunso Olusegun, Isaac
Oniovosah Ayetuoma and Garba Alpha Baba

5.1 INTRODUCTION

A comprehensive review of security measures in computational modeling and simulations is presented. Computational modeling and simulations have emerged as indispensable tools across a wide range of scientific, engineering, and real-world applications; they offer the ability to mimic complex systems, predict outcomes, and gain insights that may be challenging or impossible to achieve through traditional experimentation alone [1]. This chapter introduction delves into the foundational concepts, significance, and diverse applications of computational modeling and simulations.

Computational modeling refers to the creation and manipulation of mathematical, statistical, or logical representations of systems, processes, or phenomena using computational methods [2]. Simulations, on the other hand, involve running these models over time to observe their behavior and make predictions or decisions. In essence, computational modeling is the construction of a virtual representation, while simulations are the execution of this representation. The adoption of computational modeling and simulations has revolutionized various fields due to several compelling reasons, namely: Complex Systems Understanding- Many real-world systems, such as weather patterns, ecological ecosystems, or financial markets, are too complex to fully comprehend through traditional methods. Computational models allow scientists and researchers to simplify these systems and gain insight into their behavior [3][4].

Conducting experiments or observations in the real world can be costly and time-consuming-Simulations provide a cost-effective means of testing hypotheses, optimizing processes, and conducting what-if scenarios. In fields like aerospace, healthcare, and engineering, simulating can potentially help identify as well as mitigate risks before they occur. This is especially critical for safety-critical applications [4]. Computational models can make predictions about future events or trends, ranging from weather forecasts to stock market behavior. These predictions inform decision-making in numerous domains.

Furthermore, computational modeling and simulations are essential for research and development efforts, enabling the design and testing of new materials, drugs, and technologies without the need for physical prototypes. Thus, the Applications of Computational Modeling and Simulations include the versatility of computational modeling and simulations, which is reflected in their wide-ranging applications, such as scientific research.

Computational models are used to study natural phenomena, from molecular interactions in biology to the behavior of celestial bodies in astronomy [5][7][8]. In addition, engineers employ

DOI: 10.1201/9781003457428-6

simulations to optimize the design of structures, products, and systems, ensuring safety and efficiency. As a result, simulations aid in drug discovery and disease modeling, as well as treatment planning, or medical device development. Financial institutions use simulations to assess risk, make investment decisions, and predict market trends.

Computational models help assess environmental impact, climate change, and conservation strategies. Simulations power video games, virtual reality experiences, and animated movies, providing immersive entertainment. Simulators are employed in training for fields such as aviation, medicine, and military operations, offering a safe and controlled learning environment [6][8]. Thus, computational modeling and simulations are transformative tools that have reshaped how we explore, understand, and interact with the world. They offer the potential to solve complex problems, optimize processes, and advance knowledge across an array of disciplines.

Hence, this chapter proposes Security Measures in Computational Modeling and Simulations. In addition, a proposed cyber-attacks model is presented, which significantly sheds light on potential threats. Finally, the chapter explores the secure development practices for simulation software, emphasizing the significance of secure software development life cycles (SDLC) in simulations. The chapter also explores the methodologies, technologies, and challenges that hinder the successful implementation of computational modeling and simulations aiming to provide readers with a seasoned understanding of security in computational modeling and simulations.

Thus, the chapter concludes with a focus on data privacy and compliance, addressing regulations such as General Data Protection Regulation and Health Insurance Portability and Accountability Act (GDPR and HIPAA) in simulation software development.

The chapter also extensively investigates Intrusion Detection and Prevention Systems (IDPS) in computational modeling environments, along with the development and execution of effective incident response and recovery strategies.

5.1.1 Contributions of the Chapter

The main contributions of Security Measures in Computational Modeling and Simulations are as follows: The chapter offers an introductory exploration of computational modeling and simulations foundational understanding, establishing a fundamental knowledge base for subsequent discussions on security measures. Then, a comprehensive review of existing literature was conducted, analyzing security frameworks, challenges, and advancements, setting the context for in-depth exploration in the field. The chapter highlights the increasing importance of secure simulations in critical decision-making processes across various fields, underlining their role in modern applications. The chapter thoroughly covers multifactor identification, verification, and authentication techniques, alongside the importance of securing data through encryption, access controls, and data integrity in simulations, with detailed examination of common threats targeting computational models and potential system vulnerabilities as well as providing insights into the risks associated with these systems.

Moreover, the chapter explores the intricacies of authentication and authorization mechanisms, emphasizing their roles in controlling access to computational modeling resources and achieving effective access management. The chapter examines network security for distributed simulations, discussing security protocols to protect data transmission and exploring secure file transfer protocols for maintaining the integrity of simulation data. In addition, the chapter highlights the benefits of IDS/IPS and various techniques for enhancing network security, including preventing eavesdropping and securing communication channels.

The chapter addresses security considerations in cloud environments and ensures data and resource security, reflecting the evolving landscape of computational modeling. The chapter emphasizes on real-time monitoring, analysis, and proposing a cyber-attacks model that aims at identifying potential threats. As well, the chapter offers secure development practices for simulation software, particularly in compliance with GDPR and HIPAA regulations, while effectively exploring Intrusion

Detection and Prevention Systems (IDPS), and it provides insights into effective incident response and recovery strategies in computational modeling environments, ensuring a proactive approach to handling security incidents. Finally, the chapter discusses ethical considerations and responsible modeling practices, emphasizing the ethical use of simulation data, thereby highlighting the importance of ethical standards in this domain.

5.1.2 CHAPTER ORGANIZATION

The rest of this chapter is structured as follows: Section 5.1 presents the Introduction. Subsection; 5.1.1 presents the contributions of this chapter while Subsection 5.1.2 presents a detailed structured organization of the rest of this chapter. Section 5.2 reviews related literature; Section 5.3 Computational Modeling; Simulations and Their Significance in Various Fields; Section 5.4 Computational Models for Critical Decision-Making Processes; Section 5.5 Technology and Non-Technology-Based Security Measures; Section 5.6, The Importance of Securing Data Used in Computational Models; Section 5.7 The Role of Authentication and Authorization Mechanisms in Controlling Access to Computational Modeling Resources; Section 5.8 Network Security for Distributed Simulations; Section 5.9 SSH File Transfer Protocol (SFTP) and Secure Copy Protocol (SCP): Secure File Transfer for Simulation Data; Section 5.10 Intrusion Detection and Prevention Systems (IDS/IPS): Enhancing Network Security; Section 5.11 Secure Cloud Computing for Simulations; Section; 5.12 Cyber Threat Intelligence for Model Security; Section 5.13 the Proposed Cyber Attacks Model; Section 5.14 Secure Development Practices for Simulation Software; Section 5.15 Intrusion Detection and Prevention Systems (IDPS) in Computational Modeling Environments; Section 5.16 Ethical Considerations and Responsible Modeling and 5.17 Concludes the review. Specifically, the chapter concludes with a discussion on ethical considerations and responsible modeling practices, ensuring the ethical use of simulation data.

5.2 RELATED WORK

This section presents a systemic and comprehensive literature review of related research work on security measures in computational modeling and simulations as follows.

The study [1] proposed a dynamic trust computation model called "Secured Trust" for securing communication in multi-agent systems. It aimed to address the challenges of evaluating trust and reputation in such systems, especially in the presence of malicious agents, and to distribute workload evenly among service providers. The objectives of the research were to develop a trust model that could effectively evaluate the trust and reputation of agents in multi-agent systems, particularly in the face of unpredictable behavior by malicious agents. Additionally, the research aimed to provide a load balancing algorithm for efficient workload distribution among service providers.

The study described a comprehensive mathematical model for evaluating trust, including factors related to trust computation. It also proposed a load balancing algorithm based on these factors. The research relied on simulation results to evaluate the effectiveness of the Secured Trust model. The proposed algorithm in the research is the "Secured Trust" model, which is a dynamic trust computation model for evaluating trust and reputation in multi-agent systems. It also introduced a load balancing algorithm to distribute workload among service providers efficiently. The research presented simulation results that indicated the Secured Trust model was more robust and effective in handling malicious agents' strategic behavior and was capable of balancing workload efficiently compared to existing trust models. However, the study lacks specific numerical data and quantitative results from the simulation to support its claims about the effectiveness of the Secured Trust model. In addition, the study does not explicitly discuss its limitations, but suggests that potential solutions to other threats in multi-agent systems, such as tampering of trust information and Sybil attacks, could be addressed by incorporating additional security measures. The research also highlights the need for authentication and access control systems to defend against certain types of attacks and suggests that future work could explore these areas.

The study [2] investigates Integrated Computational Materials Engineering (ICME), highlighting its challenges and opportunities. It emphasizes the necessity of merging bottom-up and top-down modeling with systems engineering and informatics to design materials for specific performance targets. The primary objectives include exploring computational and information aspects for designing materials with complex microstructures and identifying crucial yet underdeveloped elements to bolster ICME. However, the study does not outline a distinct research methodology; rather, it focuses on the significance of computational modeling while addressing uncertainties in representing physical phenomena for decision-based design. It does not propose specific algorithms but underscores the need for innovative approaches in areas like inverse design, knowledge systems, and process path design within ICME.

Although it does not offer numerical or experimental results, it provides a conceptual overview, discussing various ICME-related topics such as uncertainty quantification, multiscale modeling, and multi-level materials design. Despite lacking empirical data or case studies to support its propositions, the study emphasizes the criticality of addressing uncertainties, multiscale modeling issues, and integration with manufacturing and systems optimization in ICME. It stresses the importance of a unified, long-term vision among industry, academia, and government to propel ICME forward. Future work could involve refining computational tools and methods for ICME's advancement.

The research study [3] focuses on security threats and challenges within trust and reputation management across distributed systems. It aims to identify, analyze, and provide recommendations to counter security threats affecting trust and reputation models. The objectives involve describing critical threats that could undermine these models, proposing solutions to mitigate these risks, and serving as a guide for secure model design.

Although the study does not explicitly detail its methodology, it likely involves a review and analysis of existing trust and reputation models along with categorizing and discussing various security threats. Rather than proposing a specific algorithm, it concentrates on identifying threats and discussing potential solutions, lacking the introduction of new computational algorithms.

While it lacks empirical data, experiments, or case studies to bolster its claims, it analyzes security threats applicable to trust and reputation models, discussing these threats and suggesting ways to handle them. Acknowledging the absence of specific solutions, the study points toward future implementation and comparison of models in response to security threats. It suggests developing validation tools for these tests and aims to guide new designers of trust and reputation models.

The study [4] introduces the "Recommendations plus Reputations-based Trust Computational Model (RpR)" for the Social Internet of Things (SIoT). RpR aims to establish trustworthy relationships among SIoT objects by evaluating trust through recommendation and reputation parameters. It strives to accurately model trustworthiness in distributed SIoT environments by separating and defining the meanings of reputation and recommendations. The objectives encompass assessing object trustworthiness, investigating convergence and accuracy, and validating the model's efficacy through simulations.

The research methodology involves identifying reputation and recommendation meanings, numerically evaluating object trustworthiness, and simulating key properties. It develops a numerical model and a robust algorithm for trust estimation and merging recommendation and reputation estimates into the RpR model. This model demonstrates robustness in computing trust for numerous objects in minimal iterations, especially downgrading trust for unreliable objects over time. Comparisons with other ranking systems support its effectiveness.

However, the research acknowledges limitations, such as not incorporating knowledge to compute final trust scores due to the difficulty in numerically assessing subjective properties. It suggests the integration of all three trust metrics and improving methods to filter out poor recommenders and address decay factor effects. Additionally, it emphasizes the necessity of reliable estimation transmission in SIoT environments. Finally, RpR stands as a recommendation and reputation-based trust computation model for SIoT, praised for its effectiveness and robustness but identifying areas for future enhancement, including integrating all trust metrics, refining filtering techniques, managing decay factor impact, and ensuring reliable estimation transmission in SIoT.

The study [42] outlines a roadmap for multiscale materials modeling, emphasizing the pivotal role of computational modeling in modern materials science. It underscores how modeling and simulation significantly contribute to discovering new materials, understanding material behavior, and making predictive designs. The primary goal is to offer a comprehensive overview of computational modeling's impact on materials science, showcasing advancements in various disciplines like phase field methods and molecular dynamics. It addresses challenges in interdisciplinary research concerning complex materials and aims to shift from fundamental material understanding to quantitative approaches.

The research methodology involves reviewing and analyzing the broad impact of computational modeling in materials science, focusing on advances in phase field methods, molecular dynamics, and interdisciplinary research. It does not specify a particular algorithm but concentrates on the overarching impact of computational modeling and the field's challenges and opportunities. It lacks specific empirical results, focusing instead on insights into modeling advancements and their efficiency in simulation and data usage. Rather than presenting detailed research findings, it serves as a roadmap and overview of computational modeling's contributions. It does not delve into specific empirical studies, experiments, or numerical simulations. The study highlights the use of simulation and data resources for reproducibility and efficiency in materials science.

Furthermore, the study identifies challenges and opportunities in the field, notably the need for improved infrastructure supporting multiscale modeling workflows. It envisions a future where a seamless and automated cyber-ecosystem could streamline multiscale modeling, reducing the manual steps currently involved. Limitations involve the manual intervention required in existing workflows, while future work aims to create a user-independent cyber-ecosystem facilitating seamless tool and infrastructure connections.

The study [43] introduces the Predictive Capability Maturity Model (PCMM), designed to evaluate the maturity level of computational modeling and simulation efforts, particularly in handling complex, nonlinear partial differential equations (PDEs) or integro-differential equations. This research focuses on assessing predictive capabilities reliant on numerical solutions to PDEs and aims to evaluate six key elements contributing to modeling and simulation maturity.

The specific objectives involve assessing representation and geometric fidelity, physics and material model fidelity, code verification, solution verification, model validation, and uncertainty quantification and sensitivity analysis. The methodology entails reviewing past investigations, such as the Capability Maturity Model Integration and NASA's Technology Readiness Levels, to identify deficiencies in existing maturity assessment approaches. From this review, the authors propose the PCMM, detailing these six elements and their associated attributes to gauge the maturity level of modeling and simulation efforts in engineering applications. While the study does not outline a specific algorithm, the PCMM offers a structured assessment by evaluating each element across four descriptive levels (0 to 3) to determine an effort's maturity. It extensively explains the PCMM's components and application, providing a comprehensive overview but lacking empirical results or case studies demonstrating its use in real-world modeling efforts.

The study acknowledges that PCMM scores should not necessarily be aggregated but recognizes the potential need for compression. It suggests a summary method for score aggregation and outlines future work, including refining the PCMM, conducting more case studies, and addressing challenges associated with aggregating scores.

Finally, the study highlights that PCMM scores offer valuable information but are just one aspect to consider for decision-makers in engineering systems. It suggests potential for further development and integration of PCMM into decision-making processes, leaving room for advancement and broader implementation.

The study [44] advocates for integrating computational models of cognitive processes into cyber-security tools, experiments, and simulations to tackle human-centered decision-making in securing computational networks. Objectives encompass addressing multi-disciplinary cyber-security challenges by employing cognitive modeling techniques in adversarial reasoning, human factors, dynamic simulations, and training effectiveness research. The aim is to bolster studies in cyber epidemiology, cyber hygiene, and human cyber-security performance. The methodology involves building initial group-level models based on average tendencies, incorporating statistics like skill proficiencies, demographics, and cultural factors. These models can then be refined to individual profiles for more precise predictions, and they can evolve over time using techniques like model tracing and dynamic parameter fitting based on recorded behavior.

While the study does not specify a proposed algorithm, it underscores the use of cognitive models as embedded computational agents for simulating human interactions with networks and software. It outlines two primary use-cases: simulating human behavior within networks, predicting software effects, and using simulated users for training scenarios; and employing model tracing to track human cognitive states for improved predictions of decisions and biases. However, the study lacks empirical results or outcomes from applying this methodology, focusing more on describing potential applications and benefits in cyber-security research and application contexts. It does not offer concrete examples or case studies demonstrating the methodology's application or empirical validation of cognitive models in cyber-security scenarios. While limitations are not explicitly stated, the study suggests that cognitive models perform better when tailored to specific subpopulations. Future work is suggested to address limitations, conduct empirical studies, and explore further applications of cognitive models in cyber-security studies.

The study [45] proposes a comprehensive survey of computational trust and reputation models for open multi-agent systems, aiming to showcase the most prominent models in the literature. Its objective is to offer an extensive overview of existing trust and reputation models while considering various analysis dimensions. The research intends to highlight the proliferation of cognitive models and introduces an objective definition of trust centered on reliance decisions.

Using a survey approach, the study reviews existing surveys and introduces new analysis dimensions to classify models. It emphasizes Castelfranchi's idea as a basis for an objective definition of trust. However, it does not present a specific algorithm; rather, it surveys and categorizes existing models, particularly focusing on cognitive models. Results reveal sustained interest in trust and reputation models, especially noting the recent surge in cognitive models. The study underscores the advantages of cognitive-based modeling, such as representing mental states and aligning closer to human understanding.

Although lacking empirical validations or practical applications of surveyed models, the study recognizes limitations in some cognitive models that remain highly descriptive due to complexity. It suggests future research focus on implementable cognitive models. The study concludes by reintroducing the objective definition of trust and classifying models based on this notion, highlighting the necessity for decision-making procedures in trust models. However, it does not explicitly state limitations or propose specific future research directions.

The study [46] introduces a simulation modeling approach aimed at representing computer networks and intrusion detection systems (IDS) to efficiently simulate cyber-attack scenarios. Its goal is to generate IDS alerts for testing and evaluating cyber security systems, specifically information fusion systems. The key objectives involve creating a simulation model capable of producing IDS alerts and ground truth files based on predefined computer network setups and attacks.

The research revolves around developing an object-oriented Java simulation model, focusing on platform independence and user-friendliness. This model facilitates the generation of various attack scenarios, distinguishes between auto-attack generation and event simulation, and offers detailed configurations for networks and attack actions. Notably, the simulator encompasses features like defining services, ports/protocols, vulnerabilities, and connector attributes to enhance the fidelity of cyber-attack representation.

Although the study does not explicitly detail a proposed algorithm, it highlights the capabilities and features of the simulation model, emphasizing its object-oriented Java design for versatility and user-friendliness. Ongoing work involves continual validation and feature addition to refine the model's accuracy in representing cyber-attacks, although specific simulation results or validation outcomes are not presented. The study lacks explicit demonstration of the simulation model's effectiveness through use cases or experiments. While limitations are not explicitly mentioned, ongoing efforts center on continuous validation and feature enhancements to improve the model's accuracy in simulating cyber-attacks. Future work is expected to focus on further improvements, validation experiments, and addressing any identified limitations.

5.3 COMPUTATIONAL MODELING, SIMULATIONS AND THEIR SIGNIFICANCE

This section presents Computational Modeling, Simulations and Their Significance in Various Fields on Security Measures in Computational Modeling and Simulations as follows.

Computational modeling is a process that involves creating mathematical or computational representations of real-world systems, processes, or phenomena. These representations, often presented in the form of equations, algorithms, or computer programs, are designed to simulate and mimic the behavior of the target system. Computational models enable researchers, scientists, engineers, and decision-makers to gain insights, make predictions, and conduct experiments in a controlled, virtual environment. While Simulations are the execution of computational models over time to observe and analyze their behavior [8]. By inputting specific parameters and initial conditions into a computational model, simulations allow for the exploration of how a system will evolve or respond under various circumstances. Simulations provide a dynamic and interactive means to test hypotheses, make predictions, and study complex systems without the need for physical experimentation.

Computational modeling and simulations hold immense significance in a wide array of fields, revolutionizing the way research is conducted, decisions are made, and innovations are developed. Computational models enable scientists to study natural phenomena, ranging from subatomic particle interactions in physics to ecological dynamics in biology [8][9]. These models facilitate the exploration of complex systems, allowing researchers to formulate hypotheses and test theories. For examples climate models help scientists understand and predict climate change by simulating the interactions of various climate variables over time. While engineers utilize computational modeling and simulations to design, analyze, and optimize structures, products, and systems. These tools ensure that designs meet safety and efficiency requirements. For example: Finite element analysis (FEA) simulates the structural behavior of bridges, buildings, and mechanical components under different conditions, aiding in design refinement [10].

Computational modeling and simulations are instrumental in drug discovery, disease modeling, treatment planning, and medical device development. They enable personalized medicine and the exploration of biological processes. For example computational fluid dynamics (CFD) is used to model blood flow in arteries and assess the impact of stents or other interventions on vascular health.

Financial institutions and economists also employ simulations to analyze market trends, assess risk, and make investment decisions. These tools help in portfolio optimization and risk management. For example, Monte Carlo simulations are used to model financial scenarios, estimate potential returns, and assess the probability of different investment outcomes. In addition, computational models aid in environmental impact assessments, climate modeling, and conservation efforts. They help evaluate the consequences of various policies and interventions. For example, agent-based models simulate interactions between individual organisms or agents in ecosystems to study population dynamics and biodiversity.

Simulations are at the heart of video games, virtual reality (VR) experiences, and animated movies, providing immersive entertainment and realistic experiences. For example, flight simulators provide a realistic training environment for aspiring pilots to practice flying without real-world risks [11].

Finally, Simulators and virtual environments are used in training for professions such as aviation, medicine, and military operations, offering a safe and controlled learning environment. Medical simulators allow medical students and practitioners to practice surgical procedures and patient care in a risk-free setting. In essence, computational modeling and simulations have become indispensable tools that transcend disciplinary boundaries, empowering researchers, professionals, and decision-makers to gain deeper insights, optimize processes, and innovate across a diverse range of fields and applications. Their significance continues to grow as technology advances, enabling more accurate and sophisticated representations of real-world phenomena respectively.

5.4　COMPUTATIONAL MODELS FOR CRITICAL DECISION-MAKING PROCESSES

Computational models are increasingly pivotal in critical decision-making due to their ability to handle complex real-world systems, vast data, and uncertainty [12]. They offer precision, accuracy, and scenario analysis across diverse fields. These models simulate intricate interactions among variables that are challenging for manual analysis, providing comprehensive insights [13]. They process vast data quickly, minimizing human errors and enabling risk assessment. In finance, they aid risk assessment and portfolio management. Their speed and accuracy are crucial in engineering and design for iterative optimization, reducing time and costs [14].

In healthcare, they assist in diagnoses and treatment plans based on patient data. Industries such as, aerospace and governments use simulations to test safety-critical systems and policy impacts. They incorporate probabilistic elements for uncertain scenarios, identifying critical decision factors [15]. They reduce the need for physical prototypes in manufacturing and optimize resource allocation. Ethical considerations and regulatory compliance are integrated into decision-making, ensuring alignment with ethical principles and regulations in sectors like healthcare and finance. As technology advances, computational models are expected to play an even more expansive role in critical decision processes due to their capacity to handle complexity and provide timely insights, ultimately enhancing decision-making effectiveness across various domains. Figure 5.1 displays the Markov chain transition mode.

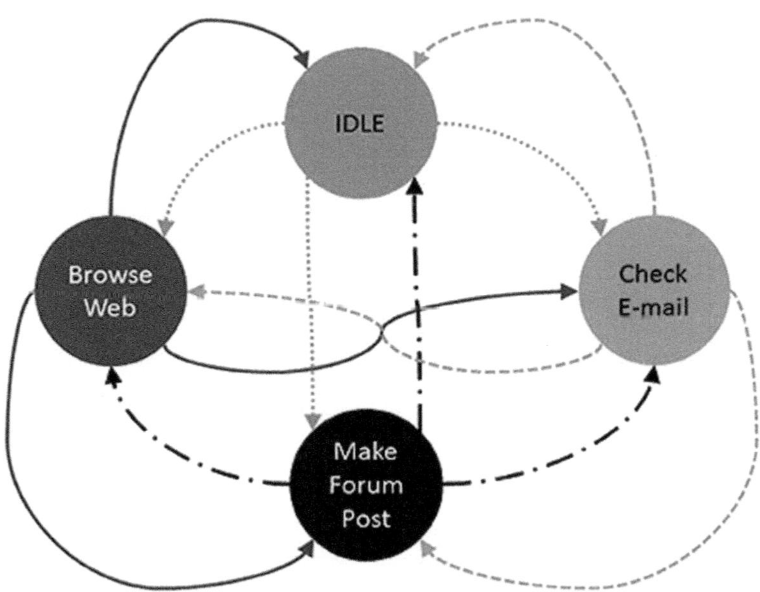

FIGURE 5.1　The Markov chain showing a four state transition mode of end users behavior.

5.5 TECHNOLOGY AND NON-TECHNOLOGY-BASED SECURITY MEASURES

This section explores the fusion of technology-based and non-technology-based security measures crucial for safeguarding assets, systems, and information against threats. Technology-based measures include Access Control Systems, employing biometrics and key cards, fortifying security in restricted areas [17]. Firewalls and Intrusion Detection Systems filter network traffic, thwarting unauthorized access and cyber threats. Encryption technologies such as SSL ensure data confidentiality during transmission. Surveillance Systems, Security Information and Event Management (SIEM) systems collect and analyze security-related data for threat detection and response [3][18]. Non-technology-based measures encompass Physical Security Measures such as fencing and locks, deterring unauthorized access and damage to assets. Security Policies and Procedures outline best practices, promoting compliance and awareness among employees. Employee Training and Awareness programs educate about security risks, reducing susceptibility to social engineering [20][21]. Security Guards, Physical Safes, and Locking Mechanisms offer tangible protection, while Crisis Management and Emergency Response Plans ensure coordinated responses to security incidents. Security Awareness Programs further reinforce a culture of vigilance and responsibility. The most effective security strategies blend both technology and non-technology measures, creating multi-layered defenses that adapt to diverse threats and organizational contexts. Selection of measures relies on asset nature, threat landscape, and organizational nuances to fortify security comprehensively [22][23].

5.5.1 MULTIFACTOR IDENTIFICATION, VERIFICATION, AND AUTHENTICATION TECHNIQUES

Multifactor identification, verification, and authentication (MFA) techniques are cybersecurity measures designed to enhance the security of digital systems, applications, and data by requiring users to provide multiple forms of evidence to prove their identity. MFA adds an extra layer of protection beyond traditional username and password authentication, making it significantly more difficult for unauthorized individuals to gain access.

Multifactor Identification (MFI)
Multifactor identification involves the use of two or more distinct factors to establish a user's identity. These factors typically fall into three categories:

Something You Know: This is typically a password or PIN, which is a knowledge-based factor. **Something You Have**: This includes physical items like a smart card, security token, or a mobile device, serving as a possession-based factor. **Something You Are**: Biometric information, such as fingerprint, retina scan, or facial recognition, constitutes an inherence-based factor.

Significance of Multifactor Identification:
It provides an additional layer of security by requiring attackers to compromise multiple factors.

It mitigates the risks associated with password-based attacks, such as brute force and credential stuffing.

It ensures a higher level of confidence in user identity [21].

Multifactor Verification (MFV)
Multifactor verification is the process of confirming the identity of a user who claims to be legitimate. It typically involves the presentation of multiple factors, often used in conjunction with multifactor identification. Verification mechanisms include: **One-Time Passwords (OTPs):** Users receive a temporary code via SMS, email, or a mobile app, which they must enter to verify their identity. **Push Notifications:** A notification is sent to the user's registered device, and they confirm or deny the login attempt. **Biometric Verification**: Users provide a biometric sample (e.g., fingerprint or facial scan), which is compared to a stored reference for verification.

Significance of Multifactor Verification:
It strengthens the authentication process by requiring users to validate their identity using a different factor than the one used for identification.

It enhances security against phishing attacks, as even if a password is compromised, the attacker still needs access to the second factor.

It provides real-time verification, adding an additional layer of protection [22].

Multifactor Authentication (MFA)
Multifactor authentication combines MFI and MFV to ensure that users are who they claim to be. Users must present multiple factors during the login process. MFA can include various combinations of factors, such as:

Two-Factor Authentication (2FA): Typically combines something the user knows (password) with something they have (a mobile app or token).

Three-Factor Authentication (3FA): Combines something the user knows (password), something they have (a mobile device), and something they are (biometric data).

Significance of Multifactor Authentication: Offers robust protection against unauthorized access by requiring multiple proofs of identity. It mitigates the risk of account compromise even if one factor is compromised. It also supports compliance with regulatory requirements for strong authentication in sensitive applications and industries [23].

5.5.2 Challenges and Considerations

Balancing security with user convenience is essential to encourage MFA adoption. Implementation: Properly integrating MFA into systems and applications is crucial for its effectiveness.

Backup Methods: Consideration should be given to providing alternative authentication methods in case primary factors are unavailable. Finally, multifactor identification, verification, and authentication techniques are vital components of modern cybersecurity strategies. They provide an effective means to protect digital assets and sensitive data by requiring users to provide multiple forms of evidence to prove their identity. As cyber threats continue to evolve, MFA remains a critical defense against unauthorized access and data breaches [16].

5.5.3 Data Security in Computational Modeling

Data security in computational modeling is a critical aspect of ensuring the confidentiality, integrity, and availability of sensitive information used in modeling and simulation processes. It encompasses a range of practices, policies, and technologies aimed at protecting data from unauthorized access, alterations, and breaches. This section presents the importance of data security in computational modeling and explores key considerations as follows.

1. **Protecting Sensitive Data**, such as, Data Privacy: Computational models often involve sensitive data, such as personal health records, financial information, or proprietary research data. Ensuring data privacy is essential to comply with regulations as GDPR or HIPAA. Research institutions and organizations invest significant resources in developing models. Protecting the intellectual property associated with these models is crucial [46].
2. **Securing Data Storage** such as Encryption: Data should be encrypted both in transit and at rest. Encryption ensures that even if unauthorized access occurs, the data remains unintelligible without the encryption key.

3. **Access Controls** Implement strict access controls to limit who can view, modify, or delete data. Role-based access control (RBAC) is often used to manage permissions. Regular Backups: Maintain secure and up-to-date data backups to prevent data loss in case of system failures or security incidents.
4. **Data Transfer and Sharing**, Secure Protocols Use secure communication protocols (e.g., HTTPS, SSH) when transferring data between systems. Avoid transmitting sensitive data in plain text. Data Masking and Anonymization: Before sharing data for research or collaboration, consider masking or anonymizing sensitive information to protect individual identities while maintaining data utility.
5. **Vulnerability Management**, Regular Patching: Keep software, operating systems, and modeling tools up to date with security patches to address known vulnerabilities.
6. **Penetration Testing**: Conduct regular security assessments to identify and remediate vulnerabilities in modeling systems [24].
7. **Insider Threat Mitigation**: Employee Training, Educate staff and researchers about data security best practices and the importance of protecting sensitive data.
8. **User Monitoring**: Implement monitoring systems to detect unusual or suspicious user activity within modeling environments.
9. **Regulatory Compliance** as Compliance Frameworks: this ensures that data security practices align with industry-specific regulatory frameworks, such as NIST for healthcare or financial services.
10. **Documentation**: Maintain thorough records of data handling and security procedures to demonstrate compliance with relevant regulations.
11. **Secure Collaboration** as Secure File Sharing: When collaborating on models or sharing results, always use secure file-sharing platforms with encryption and access controls.
12. **Data Access Agreements**: Establish clear agreements with collaborators or third parties regarding data handling, protection, and retention.
13. **Incident Response Plan**: Develop a robust incident response plan outlining the steps to take in case of a data security breach.
14. **Testing and Drills**: Regularly test and update the incident response plan through simulations and drills.
15. **Data Lifecycle Management** such as Data Retention Policies: Define data retention and disposal policies to ensure that data is only stored as long as necessary and is securely disposed of when no longer needed [37].
16. **Cloud Security** as Cloud Best Practices: when utilizing cloud-based modeling solutions follow cloud security best practices and consider encryption and access controls.
17. **External Data Sources**, Data Validation: when incorporating external data sources into models, validate the security practices of those sources to prevent data contamination or security risks.
18. **Continuous Monitoring** as Security Monitoring: Continuously monitor data access, system logs, and user activity for signs of unauthorized access or suspicious behavior.

In addition, data security is paramount in computational modeling to protect sensitive information, maintain trust, and ensure compliance with data protection regulations. A holistic approach to data security, involving policies, practices, and technology, is necessary to safeguard data throughout its lifecycle in modeling and simulation processes [25][26].

5.6 THE IMPORTANCE OF SECURING DATA USED IN COMPUTATIONAL MODELS

Securing data in computational models is critical due to its impact on security, privacy, and the reliability of outcomes. Key reasons include protecting sensitive information, ensuring legal compliance, and maintaining data integrity for accurate modeling outcomes [3]. Inaccurate data can lead to flawed decisions in crucial fields like healthcare and finance. Preventing data contamination and insider threats is crucial to safeguard against skewed results and potential catastrophic consequences. Additionally, data security safeguards intellectual property, upholds ethical obligations, and maintains trust and reputation. Breaches in critical infrastructure sectors could have national security implications. Ultimately, securing data used in computational models is both a technical cybersecurity concern and a fundamental responsibility, ensuring information protection, integrity of outcomes, and compliance with legal and ethical obligations. It is a critical factor in maintaining trust, mitigating risks, and ensuring reliability across various domains.

5.6.1 Explore Data Encryption, Access Controls, and Data Integrity in Simulations

Data encryption, access controls, and data integrity play vital roles in securing simulations, safeguarding sensitive information, authorizing appropriate data access, and maintaining reliable results [27]. Encryption employs algorithms like AES and RSA, using keys to transform and protect data in transit and at rest. It ensures data protection, compliance with regulations, and safe data sharing. Access controls, as RBAC and authentication methods, segment data access, prevent insider threats, and enable monitoring for accountability [28]. Data integrity measures, including hash functions, checksums, and digital signatures, ensure the reliability of simulation outcomes, protect against tampering, and enhance trust in results. These security components are crucial in fields like healthcare, finance, and engineering where simulations guide decision-making and compliance with data protection laws is essential [29][30].

5.6.2 Threats and Vulnerabilities in Computational Modeling

Computational modeling, while powerful, faces various threats and vulnerabilities that can compromise its accuracy and security. These encompass data quality issues such as contamination and manipulation, insider threats leading to data theft or manipulation, cybersecurity risks such as cyberattacks and data breaches, biases in models, validation challenges, hardware and software vulnerabilities, ethical considerations, inadequate documentation, and compliance risks [31][32]. To mitigate these, robust security measures like data encryption, access controls, and validation processes are essential. Ethical considerations and compliance with regulations are vital for responsible and reliable modeling outcomes. Organizations and researchers must prioritize these measures to ensure the integrity and reliability of computational modeling results.

5.6.3 Common Cybersecurity Threats Targeting Computational Modeling and Simulations

Computational modeling and simulations face an array of cybersecurity threats due to their critical roles in decision-making and industry applications. These threats include data tampering, insider threats, malware and ransomware, DDoS attacks, model poisoning, phishing, zero-day exploits, supply chain attacks, data interception, data leakage, and insider attacks on model training. These threats can lead to data manipulation, breaches, downtime, biased decisions, and compromised integrity of simulations. To counter these risks, organizations and researchers need robust cybersecurity measures such as access controls, encryption, regular software updates, employee training, and incident response plans. Additionally, ethical considerations and transparent, as well as responsible Artificial Intelligence (AI) practices are vital for ensuring the security and integrity of modeling and simulations [33].

5.6.4 Potential Vulnerabilities in Computational Modeling and Simulation Systems

Computational modeling and simulation systems harbor vulnerabilities that, if exploited, can compromise their integrity and security. These vulnerabilities include inadequate input validation allowing malicious data input, weak access controls granting unauthorized access or manipulation of simulations, insecure data storage and transmission risking interception, outdated software components with known vulnerabilities, insufficient logging hindering breach detection, lack of data validation leading to unreliable results, human error introducing system vulnerabilities, insider threats misusing privileged access, biases in models leading to unfair outcomes, risks from third-party integrations and supply chains, and ethical considerations posing reputational and legal risks [34][35]. To counter these vulnerabilities, organizations and researchers must enforce robust cybersecurity practices such as namely regular security assessments, access controls, data encryption, comprehensive logging and monitoring, and user training. Additionally, integrating ethical considerations and responsible Artificial Intelligence (AI) practices is crucial for ensuring fairness, transparency, and ethical behavior in modeling and simulation processes [36].

5.6.5 Authentication and Authorization

Authentication and authorization are fundamental concepts in cybersecurity and access control systems, ensuring that only authorized users have access to resources and data within organization systems. These two processes work together to establish trust, verify identities, and control access to sensitive information [36].

Authentication: is the process of verifying the identity of a user, device, or system entity trying to access a particular resource or system. It ensures that users are who they claim to be before granting access.
   Authentication processes has key components which are namely:

1. User Credentials: Authentication typically involves the use of credentials, such as usernames and passwords, to validate the identity of the user. Other factors, as smart cards, biometrics, or one-time passwords, can also be used [36].
2. Authentication Factors: Authentication can be based on one or more of the following factors: Something You Know: Knowledge-based factors, as passwords or PINs. Something You Have: Possession-based factors, like a smart card, security token or mobile device. Something You Are: Inherence-based factors, such as biometric data like fingerprints, retina scans, or facial recognition [37].

Authentication Process: 1. User Initiates Login: The process begins when a user initiates a login or access request to a system or resource. 2. Presentation of Credentials: The user provides the required credentials (e.g., username and password) or authentication factors.

3. Verification: The system compares the provided credentials or factors with stored data to verify the user's identity. 4. Access Granted/Denied: If the provided information matches the stored data, access is granted; otherwise, it is denied [38].

Authorization: Authorization is the process of determining what actions or resources an authenticated user or system entity is allowed to access or perform within a system. It defines the permissions and privileges granted to users after they have been authenticated.

Key Components of Authorization:

1. Access Control Policies: Authorization policies define which users or groups of users have access to specific resources, data, or functionalities.

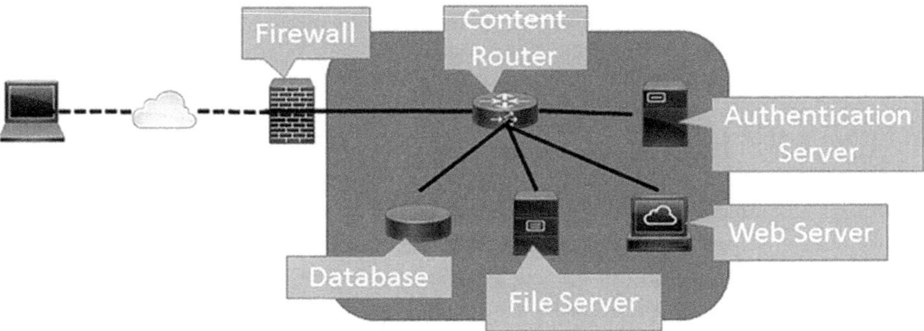

FIGURE 5.2   Traffic cascades.

2. Permissions: Permissions specify what actions users are allowed to perform once they are authenticated. They define read, write, execute, or administrative rights, for example. *Authorization Process*: 1. Authenticated User: After successful authentication, the user's identity is confirmed. 2. Access Request: The user requests access to a particular resource, data, or function within the system. 3. Authorization Check: The system evaluates the user's request against its access control policies and permissions. 4. Access Granted/Denied: Based on the evaluation, the system either grants or denies access to the requested resource or functionality [3][39][40].

5.6.6   RELATIONSHIP BETWEEN AUTHENTICATION AND AUTHORIZATION

Sequential Process: Authentication always precedes authorization. Users must be authenticated before the system can determine their authorization level. Authorization Based on Identity: Authorization decisions are made based on the authenticated identity of the user. Different users may have different authorization levels [41][42]. Granularity: Authorization can be highly granular, allowing organizations to define specific permissions for individual users or groups. Security: Proper authentication ensures that only legitimate users are granted access. Authorization ensures that legitimate users are given the appropriate level of access based on their roles and responsibilities. In addition, authentication and authorization are essential elements of access control and cybersecurity. They work together to verify identities and control access to resources, protecting sensitive information and systems from unauthorized access and potential security breaches. Properly implementing these processes is critical for maintaining the confidentiality, integrity, and availability of digital assets within organizations [4][43][44].

5.7   THE ROLE OF AUTHENTICATION AND AUTHORIZATION MECHANISMS IN CONTROLLING ACCESS TO COMPUTATIONAL MODELING RESOURCES

This section presents The Role of Authentication and Authorization Mechanisms in Controlling Access to Computational Modeling Resources on Security Measures in Computational Modeling and Simulations as follows.

Authentication and authorization mechanisms play a crucial role in controlling access to computational modeling resources, ensuring that only authorized users have the right level of access to these resources. The roles of Authentication in Controlling Access to Computational Modeling Resources are namely [45]:

1. Identity Verification: Authentication mechanisms verify the identity of users or system entities seeking access to computational modeling resources. Users are required to prove their identity through various factors, such as passwords, biometrics, or security tokens.
2. User Accountability: Authentication provides a means to trace actions back to specific individuals or entities. It ensures that any user interacting with the modeling system can be identified, which is crucial for auditing and accountability.
3. Preventing Unauthorized Access: Effective authentication prevents unauthorized users from gaining access to modeling resources. Only those with valid credentials or authentication factors can proceed further.
4. Secure User-Specific Configurations: Authentication allows organizations to establish user-specific configurations and privileges. Different users may have varying levels of access to modeling tools and data based on their roles and responsibilities [7][44][46].

5.7.1 AUTHORIZATION IN CONTROLLING ACCESS TO COMPUTATIONAL MODELING RESOURCES

Authorizations in Controlling Access to Computational Modeling Resources are achieved as follows:

1. Defining Access Levels: Authorization mechanisms define what actions or resources authenticated users can access within the computational modeling environment. This includes specifying read, write, execute, or administrative rights.
2. Role-Based Access Control (RBAC): RBAC is commonly used in modeling systems to assign roles and permissions to users. For example, researchers may have different permissions than administrators or data analysts. Authorization based on roles ensures that users only access what's necessary for their tasks.
3. Resource Protection: Authorization ensures that sensitive data used in modeling, proprietary algorithms, or intellectual properties are protected. Only users with the appropriate authorization can access and modify these resources.
4. Data Integrity: Authorization helps maintain data integrity by controlling who can modify or delete data within the modeling system. This is essential to prevent unauthorized alterations that could compromise the accuracy of models [47].

5.7.2 THE SYNERGY BETWEEN AUTHENTICATION AND AUTHORIZATION

1. Sequential Process: Authentication always precedes authorization. Users must first prove their identity through authentication before authorization mechanisms determine their level of access.
2. Granularity of Access: Authorization mechanisms can be highly granular, allowing organizations to specify permissions at a fine-grained level. This is particularly important in modeling, where different users may require different levels of access to data and computational resources.
3. Security Layers: The combination of authentication and authorization creates multiple layers of security. Even if an attacker manages to bypass authentication, they still need to pass through authorization checks to access critical modeling resources.
4. Compliance: Many industries have regulatory requirements that mandate strong authentication and proper authorization controls for sensitive data and models. Compliance with these regulations is essential in fields like healthcare, finance, and research [4][48].

Hence, authentication and authorization mechanisms work in tandem to ensure that only legitimate users have access to computational modeling resources and that their access is appropriate for their roles and responsibilities. This combination provides robust security, data protection, and accountability within the modeling environment, safeguarding sensitive information and the integrity of modeling outcomes.

5.7.3 Multi-Factor Authentication (MFA) and Role-Based Access Control (RBAC)

Multi-Factor Authentication (MFA) and Role-Based Access Control (RBAC) are two fundamental security mechanisms that play a pivotal role in strengthening access control and protecting digital assets within organizations. Multi-Factor Authentication (MFA): MFA is an authentication method that requires users to provide two or more authentication factors to gain access to a system, application, or resource. These factors fall into three main categories namely [48].

1. Something You Know: Knowledge-based factors include passwords, PINs, or answers to security questions. 2. Something You Have: Possession-based factors include mobile phones, smart cards, or security tokens. 3. Something You Are: Inherence-based factors involve biometrics, such as fingerprint scans, retina scans, or facial recognition.

Key Aspects of MFA: 1. Enhanced Security: MFA significantly enhances security by adding multiple layers of verification. Even if one factor is compromised, attackers would still need to overcome additional layers to gain access. 2. Mitigation of Password-Based Risks: MFA mitigates the risks associated with weak passwords or password reuse, as it requires an additional authentication factor. 3. Adaptability: Organizations can implement MFA across various systems and applications, providing a flexible and scalable security solution. 4. User Experience: MFA can be designed to balance security with user convenience, with options like biometric authentication on smartphones [3][49].

5.7.4 Role-Based Access Control (RBAC)

Role-Based Access Control (RBAC): is an access control model that assigns permissions and access rights to users based on their roles and responsibilities within an organization. RBAC revolves around the principle of granting the least privilege necessary for users to perform their tasks effectively. The Key Aspects of RBAC are: 1. Roles and Permissions: RBAC defines roles within an organization, such as "employee," "manager," or "administrator." Each role is associated with a specific set of permissions and access rights. 2. Access Control Policies: RBAC is governed by access control policies that dictate which roles can access specific resources or perform certain actions. These policies are typically managed centrally. 3. Scalability: RBAC scales well with organizations of varying sizes and structures. New users can be assigned roles and permissions easily, and changes can be made as organizational needs evolve. 4. Audit Trail: RBAC facilitates the creation of audit trails, making it easier to track who accessed what resources and when. This is vital for compliance and security monitoring.

5.7.5 The Synergy between MFA and RBAC

Multi-Factor Authentication (MFA) and Role-Based Access Control (RBAC) complement each other in access control and security namely: 1. Defense-in-Depth: MFA adds an extra layer of security to the authentication process, further safeguarding access to resources, while RBAC ensures that users, even with multiple factors of authentication, only have access to what's necessary for their roles. 2. Granularity: RBAC provides granularity in access control, defining permissions down to specific actions and resources. MFA adds an additional layer of granularity by requiring multiple factors, making it even harder for unauthorized access. 3. Adaptive Security: MFA can be adapted to different roles or risk levels. For example, more sensitive roles may require additional authentication factors, providing adaptive security based on RBAC roles. In addition, MFA and RBAC are essential components of a robust access control and security strategy. MFA strengthens authentication, while RBAC ensures that access permissions align with roles and responsibilities. The combination of these two mechanisms provides comprehensive protection against unauthorized access, data breaches, and other security threats [2][3][4][50].

5.8   NETWORK SECURITY FOR DISTRIBUTED SIMULATIONS

This section presents Network Security for Distributed Simulations on Security Measures in Computational Modeling and Simulations as follows. Network security is crucial in ensuring the reliability, integrity, and confidentiality of distributed simulations, which involve interconnected systems across different locations.

The key considerations for network security in distributed simulations include:

1. Secure Communication as Encryption: which Employ protocols as TLS/SSL to encrypt communication between simulation nodes, preventing interception and eavesdropping. Authentication: Implement strong authentication mechanisms, such as certificates or tokens, to verify the identity of participating nodes and prevent unauthorized access. Secure Channels: Establish mutually authenticated secure communication channels to prevent man-in-the-middle attacks [3].

2. Access Control such as Role-Based Access Control (RBAC): Implementing RBAC to restrict access based on users' roles, ensuring only authorized participants manipulate simulations. Access Monitoring: Maintain logs to detect unauthorized access attempts, facilitating forensic analysis in case of security incidents.

3. Checksums and Hashes: Use checksums and cryptographic hash functions to verify the integrity of simulation data during transmission and storage, preventing corruption and tampering. Digital Signatures: Apply digital signatures to ensure the authenticity and origin of simulation data.

4. Firewalls: Deploy firewalls to filter network traffic and block unauthorized access, allowing only necessary communication between simulation nodes. Intrusion Detection Systems (IDS): Utilize IDS to monitor network traffic for suspicious patterns, enabling real-time detection and response to potential security breaches [4][51].

5. Secure Data Storage using encryption, Encrypt stored simulation data to protect against unauthorized access, even in the event of compromised physical servers or storage media. Access Controls: Implement access controls on data repositories to restrict viewing, modification, or deletion of simulation data.

6. Redundancy: Implement network redundancy for high availability and fault tolerance, minimizing the impact of network disruptions or attacks. Failover Mechanisms: Prepare failover mechanisms to switch to backup resources in case of network failures or security incidents.

7. Patch Management such as Regular Updates: Keep all network components up to date with the latest security patches to address known vulnerabilities.

8. Disaster Recovery Plan: Develop a plan to restore network functionality and simulation data in the event of a security incident or data loss. Incident Response Plan: Create a plan outlining steps to identify, mitigate, and recover from security breaches, testing it periodically.

9. Secure Collaboration such as Secure File Sharing: Implement secure file-sharing mechanisms for collaborating on simulation data, utilizing secure file transfer protocols and access controls.

10. Security Training: Provide training on network security best practices, including password management, phishing awareness, and safe data handling.

Thus, by implementing these robust network security measures and maintaining continuous threat monitoring, organizations can enhance the security of their distributed simulation environments, ensuring the trustworthiness of results and safeguarding sensitive data [50][51].

5.8.1 Network Security Protocols and Measures to Protect Data Transmission in Distributed Simulations

Ensuring a secure transmission of data in distributed simulations is crucial for maintaining the integrity, confidentiality, and availability of simulation results and sensitive information. To achieve this, various network security protocols and measures should be implemented. Thus, this section will discuss systematically on these protocols and measures as follows.

1. Encryption Protocols as Transport Layer Security/Secure Sockets Layer (TLS/SSL): TLS and SSL are cryptographic protocols that provide secure communication over the Internet. They encrypt data during transmission, preventing eavesdropping and data interception. Implementing TLS/SSL ensures that data exchanged between simulation nodes is protected [3][52].
2. Virtual Private Networks (VPNs) as VPN Tunnels: VPNs create secure, encrypted tunnels over public networks, such as the Internet. Distributed simulations can benefit from VPNs to establish secure communication channels between geographically dispersed nodes.
3. Internet Protocol Security (IPsec), Data Authentication and Encryption: IPsec provides data authentication, integrity verification, and encryption at the IP layer. It is commonly used to secure communication between networked devices in distributed simulations.
4. Secure Shell (SSH) as Secure Remote Access: SSH provides secure, encrypted remote access to networked devices. It is essential for secure management and configuration of simulation nodes, ensuring that unauthorized access is prevented.
5. VPN and Tunneling Protocols as Layer 2 Tunneling Protocol (L2TP) and Point-to-Point Tunneling Protocol (PPTP): these protocols create secure tunnels for data transmission over public networks. They are commonly used in conjunction with VPNs for secure communication between distributed simulation nodes [4][53].

5.8.2 Secure File Transfer Protocols

Secure file transfer protocols are fundamental for the safe and encrypted transfer of sensitive data among devices, networks, or users. This section presents the vital secure file transfer protocols used, namely:

1. SSH File Transfer Protocol (SFTP). Encrypts data and authentication credentials using the SSH protocol for secure communication. Supports multiple authentication methods as password-based, public key, and certificate-based authentication. And is Versatile across various operating systems due to its platform independence.
2. FTP over SSL/TLS (FTPS): Uses SSL/TLS encryption to secure FTP sessions with options for explicit or implicit modes. As well as supports diverse authentication methods including username/password, X.509 certificates, and public keys.
   Employs ports 21 and 989/990 for implicit SSL/TLS and port 21 for explicit SSL/TLS FTPS.
3. HTTP Secure (HTTPS): Secure version of HTTP using SSL/TLS encryption for safe file transfers via web browsers. Utilizes various authentication methods such as basic, digest, and client certificate-based authentication.

4. Applicability Statement 2 (AS2): Designed for secure B2B transactions with S/MIME encryption and digital signatures. As well as Provides non-repudiation features, ensuring message authenticity cannot be denied by either party [9].
5. Secure Copy Protocol (SCP): Uses SSH for secure and encrypted file transfers, relying on public key or password-based authentication. Commonly utilized via command-line interface for quick and secure transfers between remote servers.
6. Web-based Distributed Authoring and Versioning (WebDAV): Allows secure data transfers over HTTPS for collaborative work, enabling remote server document editing. Supports file locking to prevent conflicts during simultaneous file edits.
7. Accelerated File Transfer Protocol (AFTP): is optimized for high-speed transfers, especially for large files, while incorporating encryption for data integrity.
8. Trivial File Transfer Protocol with Security Extension (TFTP-SEC) on the other hand enhances basic TFTP with features as authentication and encryption for secure bootstrapping and firmware upgrades. Thus, these protocols play a pivotal role in safeguarding data during transit and ensuring confidentiality. Choosing the appropriate protocol depends on specific security needs and use cases within an organization [7][8].

5.9   SSH FILE TRANSFER PROTOCOL (SFTP) AND SECURE COPY PROTOCOL (SCP): SECURE FILE TRANSFER FOR SIMULATION DATA

This section presents SSH File Transfer Protocol (SFTP) and Secure Copy Protocol (SCP): Secure File Transfer for Simulation Data on Security Measures in Computational Modeling and Simulations as follows. SFTP and SCP are two widely used protocols that provide secure and encrypted file transfer capabilities. They are particularly valuable when it comes to safeguarding the integrity and confidentiality of simulation data during transit between simulation nodes. SSH File Transfer Protocol (SFTP) such as Encryption and Security is a secure file transfer protocol that operates over SSH, ensuring that all data transferred between the client and the server is encrypted. It uses strong cryptographic algorithms to protect data confidentiality and integrity, making it resistant to eavesdropping and tampering. SFTP offers various authentication methods, including password-based authentication, public key authentication, and certificate-based authentication.

Public key authentication, in particular, enhances security by allowing secure and convenient access without the need for passwords [10]. SFTP is widely supported and can be used across different platforms and operating systems, making it versatile for various simulation node environments. It is typically implemented through SFTP client software and can be used in both interactive and automated (scripted) modes. SFTP is often preferred in scenarios where data security is paramount, such as the transfer of sensitive simulation results, research data, or proprietary algorithms. The encryption and authentication mechanisms make it suitable for protecting critical simulation data during transit [11].

Secure Copy Protocol (SCP): SCP is a straightforward and efficient protocol for securely copying files between systems. It is often used for quick and secure file transfers. SCP is based on the SSH protocol, which provides the necessary security for data transfer.

Just like SFTP, SCP uses SSH-based authentication methods, such as password-based and public key-based authentication which ensures that only authorized users can initiate secure file transfers. SCP is typically used from the command line, making it easy to use in both interactive and automated scripting scenarios. It is particularly useful for ad-hoc or one-time transfers. SCP is a reliable choice for securely moving files between simulation nodes within a trusted network or for transferring simulation data to remote storage locations. While efficient, it may not offer as many features or options for file management as SFTP [12].

5.9.1 Choosing Between SFTP and SCP for Simulation Data

The choice between SFTP and SCP largely depends on your specific simulation data transfer requirements. This section will provide some key considerations as follows namely.

Sensitivity of Data: If your simulation data is highly sensitive and must be protected with strong encryption and authentication, SFTP is a preferred choice. *Ease of Use*: SCP is often favored for its simplicity, making it suitable for quick and straightforward file transfers within trusted environments. *Automation*: If you need to automate file transfers between simulation nodes, both SFTP and SCP can be scripted, but the choice may depend on your scripting preferences. In addition, SFTP and SCP are valuable tools for ensuring secure file transfers in the context of simulation data. They provide encryption, authentication, and data integrity checks to protect sensitive information while it travels between nodes, contributing to the overall security and reliability of simulation processes. The choice between the two protocols should align with your specific security and operational needs [4].

5.10 INTRUSION DETECTION AND PREVENTION SYSTEMS (IDS/IPS): ENHANCING NETWORK SECURITY

This section presents Intrusion Detection and Prevention Systems (IDS/IPS): Enhancing Network Security as follows. Intrusion Detection Systems (IDS) and Intrusion Prevention Systems (IPS) are pivotal for network security, identifying and responding to potential security threats. (1) IDS Monitors network/system activities to spot suspicious behavior or security breaches without actively intervening. (2) IDS Analyzes network traffic, generates alerts upon detecting potential threats, and focuses on various types such as Network-Based IDS (NIDS), Host-Based IDS (HIDS), and Log-Based IDS. (3) IDS Notifies administrators about detected threats or anomalies but may sometimes generate false positive alerts. (4) IDS Operates in a passive mode, observing and reporting without taking direct action [12].

5.10.1 Intrusion Prevention System (IPS)

1. IPS extends beyond IDS by not only detecting threats but actively preventing or mitigating them.
2. IPS Inspects and blocks network traffic, responds to threats in real-time, and includes Network-Based IPS (NIPS) and Host-Based IPS (HIPS).
3. IPS immediately takes action upon detecting threats, blocking or modifying traffic to protect the network.
4. IPS utilizes signature-based detection for known attack patterns and behavioral-based detection for deviations from normal behavior.
5. IPS Enforces predefined security policies and rules to ensure compliance with security standards and regulations. Thus, IDS monitors for threats and sends alerts, whereas IPS goes further by actively preventing or mitigating those threats in real-time. Both systems are crucial for maintaining network security and mitigating potential risks [5].

5.10.2 Benefits of IDS/IPS

This section presents the importance of IDS/IPS systems in threat detection, incident response, and network security enhancement within computational modeling and simulations. The systems play a crucial role in detecting malware, intrusion attempts, and abnormal network behavior, allowing rapid response to security breaches. They also enforce security policies, provide detailed monitoring and reporting, and offer proactive protection by blocking or mitigating threats. To maximize their effectiveness, it is essential to configure, update, and manage these systems carefully. The passage emphasizes implementing IDS/IPS for monitoring, configuring firewalls for packet filtering, using access control lists, network segmentation, regular updates, user authentication, secure protocols, and security training for users and administrators. These measures collectively establish a robust security framework to safeguard both simulation data and the integrity of simulation results in distributed environments [44].

5.10.3 Techniques for Securing Communication Channels and Preventing Eavesdropping

This section outlines crucial methods to secure communication channels and prevent unauthorized access or eavesdropping, essential for maintaining confidentiality and integrity in data exchange. Here are the key techniques underlined namely:

1. Encryption: Implement end-to-end encryption using robust algorithms as Advanced Encryption Standard (AES) to protect data from interception. Secure key management is essential to safeguard encryption keys.
2. Virtual Private Networks (VPNs): Establish encrypted tunnels over public networks for secure data transmission, particularly through IPsec VPNs.
3. Secure Socket Layer (SSL) and Transport Layer Security (TLS): Employ SSL/TLS protocols to secure web-based communications, ensuring encryption, data integrity, and server authentication.
4. Secure Shell (SSH): Use SSH for encrypting data during remote access, file transfers, and command executions.
5. Public Key Infrastructure (PKI): Utilize PKI for issuing digital certificates, enabling secure authentication, and implementing digital signatures for message authenticity.
6. Perfect Forward Secrecy (PFS): Generate unique session keys for each session to prevent compromises from affecting past or future communications.
7. Diffie-Hellman Key Exchange: Securely negotiate encryption keys without transmitting them over the network.
8. Secure Voice and Video Communications: Encrypt VoIP and video calls to prevent eavesdropping on sensitive conversations.
9. Zero-Knowledge Protocols: Implement in select applications to ensure service providers have no access to user data, ensuring privacy.
10. Forward Secrecy: Rotate session keys regularly to limit the impact of potential compromises.
11. Secure Communication Protocols: Enforce HTTPS for secure browsing and use encrypted email protocols as S/MIME or PGP for end-to-end encryption.
12. Network Segmentation: Isolate sensitive data within networks to restrict access to authorized users and systems.
13. Continuous Monitoring: Deploy IDS/IPS to detect and respond to suspicious activities or unauthorized access attempts.
14. User Education: Train users to identify phishing and social engineering attacks that could lead to eavesdropping or data breaches.
15. Regular Updates: Ensure systems and software receives timely security patches to address vulnerabilities. Finally, implementing these measures will effectively strengthen communication security, safeguarding various forms of digital exchanges against unauthorized access or eavesdropping as well [46][54].

5.11 SECURE CLOUD COMPUTING FOR SIMULATIONS

This section emphasizes the significance of secure cloud computing practices in computational simulations, given the transformative impact of cloud technology on scalability, cost-efficiency, and accessibility. It delves into various security measures for ensuring the protection of sensitive simulation data and processes in cloud environments as follows. (1) Data Encryption: Encrypt data during transmission and storage within the cloud using secure protocols such as TLS/SSL which Employ at-rest encryption to safeguard stored data. (2) Identity and Access Management (IAM): Implement robust user authentication mechanisms, Role-Based Access Control (RBAC), and Multi-Factor Authentication (MFA) to control access. (3) Virtual Private Cloud (VPC): Establish isolated virtual networks (VPCs) to segregate simulation environments, using security groups to manage inbound/outbound traffic. (4) Data Governance and Compliance: Categorize and apply appropriate security controls to simulation data based on sensitivity. Ensure compliance with industry-specific regulations. (5) Regular Security Audits: Conduct vulnerability assessments, penetration testing and frequent scans to identify and address weaknesses in the cloud infrastructure. (6) Logging and Monitoring: Monitor cloud audit logs, employ Security Information and Event Management (SIEM) tools for centralized security event analysis. (7) Backup and Disaster Recovery: Regularly back up simulation data and devise a comprehensive disaster recovery plan to minimize data loss and downtime during unforeseen incidents. (8) Security as Code (IaC): Embed security controls into Infrastructure as Code (IaC) scripts to automate security configurations. (9) Secure Third-Party Services: Ensure third-party services or software meet security best practices and are regularly updated. (10) Education and Training: Educate teams on secure cloud practices, data handling, and reporting security incidents. (11) Patch Management: Keep software updated with the latest security patches to mitigate vulnerabilities. (12) Incident Response Plan: Develop and test an incident response plan to effectively address security breaches. Thus, implementing these measures is crucial for safeguarding sensitive data and ensuring the reliability of simulation results in cloud environments. Hence, a comprehensive security strategy, continuous monitoring, and user education are integral in mitigating risks and fortifying the security of cloud-based simulations [44][55].

5.11.1 CLOUD COMPUTING FOR SIMULATIONS AND ASSOCIATED SECURITY CONSIDERATIONS

This section highlights the advantages of employing cloud computing for conducting simulations in various domains namely. (1) Scalability: Cloud services offer on-demand access to scalable computing resources, allowing simulations to utilize clusters of virtual machines for computationally intensive tasks. (2) Resource Optimization: Users can efficiently allocate resources in the cloud, scaling them as required to optimize costs and prevent hardware underutilization. (3) Global Accessibility: Cloud services are accessible worldwide via the Internet, facilitating collaboration on simulations among teams and researchers without geographical constraints. (4) High-Performance Computing (HPC): Many cloud providers offer high-performance computing instances equipped with powerful GPUs and CPUs, ideal for simulations that demand high computational performance. (5) Data Management: Cloud storage solutions provide secure and scalable data storage options, ensuring easy accessibility and efficient backup of simulation data [44].

5.11.2 SECURITY CONSIDERATIONS ASSOCIATED WITH CLOUD-BASED SIMULATIONS

This section highlights crucial security considerations when utilizing cloud computing for simulations as follows.

1. Data Encryption: Encrypt data during transmission and storage using secure protocols as TLS/SSL to prevent unauthorized access.

2. Identity and Access Management (IAM): Employ robust user authentication mechanisms, Role-Based Access Control (RBAC), and Multi-Factor Authentication (MFA) to control access to simulation resources.

3. Virtual Private Cloud (VPC): Establish VPCs to isolate simulation environments, utilizing Security Groups to manage inbound and outbound traffic.

4. Data Governance and Compliance: Categorize data based on sensitivity and enforce appropriate security measures, ensuring compliance with industry-specific regulations for sensitive data.

5. Regular Security Audits: Conduct vulnerability assessments, penetration testing, and regular audits to identify and address weaknesses in the cloud infrastructure.

6. Logging and Monitoring: Enable cloud audit logs, implement Security Information and Event Management (SIEM) tools to track user activities, and detect security threats promptly [47].

7. Backup and Disaster Recovery: Regularly back up simulation data and develop a comprehensive disaster recovery plan to minimize data loss during unexpected incidents.

8. Security as Code (IaC): Automate security configurations by integrating security controls into Infrastructure as Code scripts.

9. Secure Third-Party Services: Ensure third-party services comply with security best practices and receive regular updates.

10. Education and Training: Educate teams on secure cloud computing practices, emphasizing data handling, sharing, and reporting security incidents.

11. Patch Management: Maintain regular updates for all software, including operating systems and applications, by installing the latest security patches.

12. Incident Response Plan: Develop and regularly test an incident response plan to manage security breaches effectively. Thus, adopting these practices is crucial for safeguarding sensitive data and ensuring the reliability of cloud-based simulations through a well-designed security strategy, continuous monitoring, and user education [56].

5.11.3 SECURING DATA AND RESOURCES IN CLOUD-BASED SIMULATIONS

Securing data and resources within cloud-based simulations is critical for preserving confidentiality, integrity, and compliance. Here are key practices to achieve this.

1. Data Encryption: Encrypt data during transmission and storage in the cloud using secure protocols and encryption mechanisms offered by cloud providers to protect data at rest.

2. Identity and Access Management (IAM): Implement robust user authentication methods as RBAC and MFA to control access. Limit privileges based on roles to prevent unauthorized entry.

3. Virtual Private Cloud (VPC) and Network Isolation: Create VPCs to isolate simulation environments and employ security groups or policies to regulate inbound/outbound traffic.

4. Data Governance and Compliance: Classify data, apply suitable security controls, and ensure compliance with regulations (e.g., HIPAA, GDPR) for sensitive data.

5. Regular Security Audits and Vulnerability Scans: Conduct routine assessments and scans to detect and address vulnerabilities, employing penetration testing to simulate attacks and fortify defenses.

6. Logging and Monitoring: Enable audit logs and SIEM tools to track activities, analyze security events, and set alerts for potential threats or deviations.

7. Backup and Disaster Recovery: Regularly back up simulation data securely and create a disaster recovery plan to minimize downtime during unexpected incidents.

8. Security as Code: Integrate security measures into IaC scripts to automate consistent application of security configurations.
9. Secure Third-Party Services: Ensure third-party services align with security standards and receive regular updates to address vulnerabilities.
10. Education and Training: Continuously educate teams on secure practices, data handling, and incident reporting to raise awareness and maintain security.
11. Patch Management: Regularly update all software to address vulnerabilities that could be exploited by attackers.
12. Incident Response Plan: Develop and regularly test an incident response plan to ensure a swift and effective response to security breaches. Hence, implementing these practices significantly bolsters the security of cloud-based simulations, safeguarding data and resources while ensuring the integrity and availability of simulation environments [47][56].

5.12 CYBER THREAT INTELLIGENCE FOR MODEL SECURITY

This section emphasizes the importance of Cyber Threat Intelligence (CTI) in safeguarding computational models and simulations through several crucial facets namely. (1) Understanding Threat Landscape: CTI offers insights into evolving threats, aiding organizations in anticipating risks and taking proactive measures. (2) Threat Actor Profiling: Identifies threat actors, their tactics, and capabilities, enabling tailored defenses against specific threats. (3) Early Warning System: Provides alerts and indicators of compromise for swift threat detection and response. (4) Vulnerability Assessment: Offers data on known vulnerabilities, aiding in prioritizing patch management and security updates. (5) Threat Intelligence Feeds: Real-time information on emerging threats, integrated into security systems for enhanced threat detection. (6) Threat Mitigation Strategies: Offers actionable intelligence to proactively defend against attacks and secure systems. (7) Incident Response Planning: Assists in developing response playbooks, detailing actions in the event of a security incident. (8) Intelligence Sharing: Collaboration within the CTI community enhances collective knowledge and response capabilities. (9) Risk Assessment and Prioritization: Helps evaluate and prioritize cyber risks, optimizing resource allocation. (10) Compliance and Reporting: Aids in meeting regulatory requirements, showcasing efforts to protect computational models. CTI, an indispensable element of model security, provides insights, profiles threats, and offers actionable intelligence, empowering organizations to defend against cyber threats effectively and respond swiftly to protect sensitive data and intellectual property [4][57].

5.12.1 THE IMPORTANCE OF CYBER THREAT INTELLIGENCE IN IDENTIFYING POTENTIAL THREATS TO COMPUTATIONAL MODELS

Cyber Threat Intelligence (CTI) is pivotal for protecting computational models due to several key reasons namely.

1. Early Threat Detection: Offers real-time insights into emerging threats, enabling proactive threat detection before attacks occur.
2. Understanding Threat Landscape: Provides comprehensive insights into evolving threats, helping accurately model and predict potential risks to computational models.
3. Threat Actor Profiling: Profiles threat actors, aiding in understanding their capabilities and likely targets, enabling anticipatory measures.
4. Tailored Defenses: Allows customizing security measures to counter specific threats, enhancing intrusion detection and access controls.

5. Vulnerability Assessment: Identifies known vulnerabilities, aiding in prioritizing patch management to mitigate potential model compromises.
6. Timely Mitigation: Provides actionable intelligence for proactive threat mitigation, such as blocking malicious IP addresses or strengthening authentication.
7. Incident Response Preparedness: Informs incident response plans, ensuring effective reactions to security incidents involving computational models.
8. Risk Management: Assists in evaluating and managing risks associated with models, allowing resource allocation to critical areas.
9. Compliance and Reporting: Enables compliance with regulatory standards by showcasing efforts in identifying and mitigating potential threats. Hence, CTI stands as an essential element in identifying threats to computational models, empowering organizations to proactively defend against evolving threats, safeguard critical assets, data, and intellectual property. In addition, it forms a fundamental part of modern cybersecurity strategies for simulation and modeling environments [1][58].

5.12.2 Monitoring and Analyzing Threats in Real Time

The section provides a comprehensive approach to real-time threat monitoring and analysis. The strategies are as follows: (1) Advanced SIEM Systems: Implement centralized systems to collect and analyze security events for anomaly detection and alert generation. (2) Baseline and Anomaly Detection: Establish normal behavior baselines and use anomaly detection to flag deviations, indicating potential threats. (3) IDS/IPS Systems: Employ systems that monitor network traffic for known attack patterns, generating alerts or taking automated action against threats. (4) Threat Intelligence Feeds: Utilize real-time threat feeds to enhance threat detection, incorporating information about known threats and vulnerabilities. (5) Behavioral Analytics: Analyze user and system behavior to identify irregular patterns that might signify insider threats or sophisticated attacks. (6) Network Traffic Monitoring: Continuously monitor network activities and perform packet analysis to detect unusual data flows or unauthorized access attempts. (7) User and Entity Behavior Analytics (UEBA): Focus on monitoring behaviors to identify potential insider threats or compromised accounts. (8) Real-Time Vulnerability Scanning: Employ tools to identify and prioritize systems with known vulnerabilities for immediate patching. (9) Automated Alerts: Configure alerts for critical security events, setting thresholds to reduce false positives. (10) Threat Hunting: Employ specialized teams to proactively search for threats that automated systems might miss. (11) Cloud Environment Monitoring: Extend real-time monitoring practices to cloud environments using cloud-native security tools. (12) Incident Response Plan: Develop and regularly practice an incident response plan outlining roles and actions for different security incidents. (13) Security Policy Updates: Regularly review and update security policies to align with the evolving threat landscape. (14) Collaboration and Communication: Establish clear communication channels and encourage collaboration among teams for swift response to security incidents. Therefore, implementing these strategies enables organizations to promptly detect and respond to security incidents, minimizing potential damage and data breaches by continuously monitoring and analyzing threats in real-time.

5.13 PROPOSED CYBER ATTACKS MODEL

This section introduces a proposed model outlining cyber-attacks within the context of security measures in computational modeling and simulations. The model primarily focuses on cyber assaults originating from Internet-based hackers, while acknowledging the potential for insider threats. The success of such attacks depends on both the hacker's skills and the network's weaknesses. Initially rooted in [46], the model categorizes hacker actions into stages corresponding to their capabilities within the network's framework, spanning from Stage 0 to Stage 9. Stage 0 involves reconnaissance activities directed at the external network to gather information, encompassing external machines accessible from the Internet (Stage 0 to Stage 4) and internal machines reachable through specific channels (Stage 5 to Stage 9).

Table 5.1 outlines typical hacker actions aligned with each attack stage. The attack process starts with targeting an organization's external machine and progresses through the external network until access to internal machines is achieved. Subsequently, infiltrating the internal network allows compromising internal machines, ultimately fulfilling the hacker's goal. The simulation model integrates automated and user-defined methods to generate cyber-attacks. The automated approach utilizes network specifications, connectivity, and a predefined template of attack stages to determine viable attack sequences based on network vulnerabilities and attacker capabilities.

A graph-based template guides this process, detailing feasible actions at different attack stages. Figure 5.3 depicts the current graph-based template used in the simulation model, with 'S_0' through 'S_9' representing Stages 0 to 9, respectively. The process involves utilizing an automated method depicted in Figure 5.3 to generate a specific multi-stage attack, following a structured attack

TABLE 5.1
Typical Hacker Actions in a Cyber Attack

Stage	Typical Actions
0	Recon. Footprinting
1	Intrusion User
2	Escalation Service
3	Intrusion Root
4	Goal DoS
5	Recon. Enumeration
6	Intrusion User
7	Escalation Service
8	Intrusion Root
9	Goal Pilfering

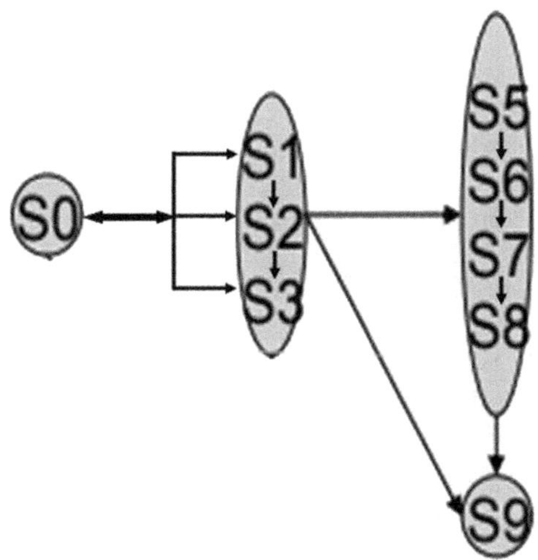

FIGURE 5.3   An attack strategy structured graph.

template and the specified network configuration. Users input details such as the target machine, attack objectives, and other attack-related parameters. The methodology operates in reverse through the network, starting with defining the attack target and finding a path outward that a hacker could exploit. It selects an attacker machine capable of communicating with the chosen target based on the network topology.

Upon determining the attack sequence for the current target, it identifies a new target – either a machine reachable by the current attacker or the current attacker itself. Choosing the attacker as the new target progresses the attack toward external machines, simulating how hackers breach a network. If the target is not the attacker, the process continues, determining another target using specific stages. However, if the target is an external machine, the logic generates an attacker IP address as attackers usually mask their actual IP addresses when attacking from the Internet. Subsequently, it determines the progression of the attack template for the external target using a different set of stages respectively [44][46].

The computer network is outlined, utilizing two primary elements: machines and connectors. The addition of subnets introduces a collective representation of machines. These modules visually depict the network and offer a logical means for users to input data concerning the network, such as whether a machine is accessible from the Internet. The interconnecting lines, displaying network connectivity, contribute to constructing a matrix indicating the network's topology, pivotal for generating attacks. Specific device details (e.g., IDS type) are stored as accessible variables tied to the device ID. Modifications to these devices are easily enacted by accessing a form through a double-click on their interface representation. Hackers navigate their way from the external network, aiming to access internal machines. Once infiltrated, internal machines become susceptible until the hacker achieves their objective.

5.14   SECURE DEVELOPMENT PRACTICES FOR SIMULATION SOFTWARE

The section outlines secure development practices crucial for safeguarding simulation software in computational modeling and simulations. The key practices include: (1) Threat Modeling: Identify potential threats and vulnerabilities specific to the simulation software's security requirements. (2) Security Requirements: Define clear security specifications aligned with industry standards and regulations. (3) Secure Coding Practices: Train developers in secure coding methods, utilizing trusted frameworks and libraries. (4) Code Review and Static Analysis: Regularly review and analyze code for vulnerabilities using automated tools. (5) Authentication and Authorization: Implement strong authentication and role-based access control (RBAC) mechanisms. (6) Data Protection: Encrypt sensitive data, validate and sanitize inputs to prevent attacks. (7) Secure Communication: Utilize secure protocols to safeguard data exchange between software components. (8) Error Handling: Implement effective error handling to prevent data leakage through error messages. (9) Input Validation: Validate user inputs at different layers to prevent various vulnerabilities. (10) Session Management: Ensure secure session practices to protect against session-related attacks. (11) Security Updates: Stay updated with security patches for all software components. (12) Third-Party Components: Assess and maintain the security of third-party components. (13) Logging and Monitoring: Implement comprehensive logging and monitoring for timely incident detection and response. (14) Secure Deployment: Configure production environments securely and manage configurations effectively. (15) Incident Response Plan: Develop and test an incident response plan to address security breaches. (16) User Training: Educate end-users on security best practices when using the software. (17) Compliance and Certification: Consider compliance with relevant security standards and certifications. (18) Secure DevOps (DevSecOps): Integrate security into the DevOps pipeline with automated testing and monitoring. Thus, implementing these practices helps minimize vulnerabilities, ensuring the software meets high security and reliability standards, fostering trust with users and stakeholders alike [10].

5.14.1 Developing Secure Simulation Software

To develop secure simulation software, a comprehensive approach to coding is vital, prioritizing security from the initial stages. Here's a breakdown of secure coding practices and methodologies namely. (1) Threat Modeling: Identify potential threats and vulnerabilities within the software, prioritizing mitigation based on their impact. (2) Security by Design: Integrate security into the software design phase, aligning functional and security requirements. (3) Input Validation: Validate user inputs rigorously to prevent common attacks such as SQL injection and cross-site scripting. (4) Avoid Hard-Coded Secrets: Refrain from embedding sensitive information directly into the source code; selection for secure configuration files. (5) Authentication and Authorization: Implement robust authentication and role-based access control mechanisms. (6) Secure Communication: Encrypt data during transit using secure protocols to prevent interception attacks. (7) Error Handling: Implement secure error handling, avoiding exposing sensitive information in error messages. (8) Buffer Overflow Prevention: Employ secure coding practices to prevent memory-related vulnerabilities. (9) Secure Data Storage: Encrypt stored data and employ robust key management practices. (10) Input/Output Sanitization: Sanitize inputs and outputs from external sources to prevent data corruption or malicious code execution. (11) Avoiding Code Injection: Safeguard against code injection attacks by validating and sanitizing data inputs.

(12) Secure Development Frameworks: Leverage trusted frameworks with built-in security controls. (13) Regular Code Reviews: Perform regular security-focused code reviews and utilize automated analysis tools. (14) Security Testing: Conduct penetration testing and vulnerability scanning to identify and address weaknesses. (15) Security Training: Train developers in secure coding and keep them updated on emerging threats and practices [19].

(16) Secure DevOps (DevSecOps): Integrate security into the DevOps pipeline for continuous security monitoring. (17) Regular Updates: Stay informed about security updates for all components and promptly apply patches. (18) Threat Modeling throughout Development: Continuously update threat models to account for evolving software features and changes.

By adhering to these practices and embedding security methodologies throughout the development process, the risk of security vulnerabilities in simulation software can be significantly reduced, fostering a more secure, reliable, and trustworthy software environment [4].

5.14.2 The Importance of Secure Software Development Life Cycles (SDLC) in Simulations

A secure Software Development Life Cycle (SDLC) is crucial for simulation software due to several key reasons which are: (1) Early Threat Mitigation: Security-focused SDLC identifies and addresses security threats early, saving costs and preventing issues from propagating. (2) Regulatory Compliance: Essential for meeting strict industry regulations regarding data security and privacy. (3) Data Protection: Protects sensitive simulation data from unauthorized access or breaches. (4) Intellectual Property Protection: Safeguards proprietary algorithms and trade secrets embedded in the software. (5) Maintaining Simulation Integrity: Ensures the reliability and accuracy of simulation results by preventing unauthorized alterations. (6) Reduced Vulnerability Surface: By implementing security practices, vulnerabilities are minimized, reducing the risk of breaches.

(7) Risk Management: Allows effective identification, prioritization, and mitigation of risks, minimizing potential security incidents.

(8) Enhanced Trust: Maintains user and stakeholder trust by ensuring the security and integrity of simulation results. (9) Cost Savings: Addressing security issues early reduces later development costs and minimizes financial impacts of incidents. (10) Cultural Emphasis on Security: Promotes a security-conscious mindset among developers, fostering better practices. (11) Incident Response Preparedness: Includes plans for swift reactions to security incidents. (12). Continuous Improvement: An iterative process that learns from past incidents to enhance security practices in subsequent cycles [4][19]. Hence, a secure SDLC is not just about protection; it is vital for reliability, compliance, trust-building, and responsible development of simulations. Integrating security across all development phases helps mitigate risks, comply with regulations, and build credibility with users and stakeholders, making it an essential part of simulation software development.

5.14.3 DATA PRIVACY AND COMPLIANCE IN SIMULATION SOFTWARE DEVELOPMENT

Developing simulation software involves critical considerations for data privacy and compliance, especially when handling sensitive information or adhering to regulatory standards. Here are key points namely: (1) Protecting Sensitive Data: Safeguard sensitive data processed, stored, or transmitted within simulation software to prevent unauthorized access or misuse. (2) Compliance with Regulations: Adhere to industry-specific and regional regulations such as Health Insurance Portability and Accountability Act (HIPAA), General Data Protection Regulation (GDPR), or Payment Card Industry Data Security Standard (PCI DSS) to avoid legal and financial implications. (3) Informed Consent: Obtain explicit consent when simulations involve personal data or human subjects, ensuring users are aware of data usage.

(4) Data Minimization: Collect and retain only necessary data, reducing the risk associated with handling excessive sensitive information. (5) Data Encryption: Encrypt sensitive data during transmission and storage using robust and updated encryption protocols. (6) Access Controls: Implement strong authentication mechanisms to limit access to authorized individuals. (7) Data Anonymization: Anonymize or pseudonymize data, especially for research simulations, to minimize exposure risks. (8) Data Retention Policies: Establish policies for data retention and disposal, deleting or anonymizing data no longer needed.

(9) Data Transfer Across Borders: Comply with regulations governing international data transfers if applicable to the simulation software. (10) Auditing and Logging: Employ comprehensive auditing mechanisms to track data access, essential for compliance and incident response. (11) Incident Response Plan: Develop a robust plan outlining steps to address data breaches or privacy incidents promptly. (12) Security Assessments: Conduct regular security assessments such as penetration testing to identify and rectify potential vulnerabilities. (13) Compliance Documentation: Maintain records showcasing adherence to data privacy regulations for audits and legal purposes. (14) User Education: Educate users about data privacy practices and how the software handles their data, fostering transparency.

(15) Privacy by Design: Incorporate privacy considerations into the software's design to make it inherently privacy-aware. (16) Privacy Impact Assessments: Conduct assessments to gauge potential privacy risks and take steps to mitigate them. (17) Legal and Ethical Considerations: Respect individuals' rights and privacy expectations, ensuring compliance with legal and ethical norms. (18) Ongoing Compliance Monitoring: Stay updated with evolving data privacy regulations and adapt software practices to maintain compliance. Hence, Data privacy and compliance are vital in simulation software development. Adhering to these practices not only mitigates risks but also establishes trust with users and stakeholders. Integrating privacy and compliance into the development process reflects responsible and ethical software practices [3].

5.14.4 DATA PRIVACY REGULATIONS

Data privacy regulations, notably General Data Protection Regulation (GDPR), and Health Insurance Portability and Accountability Act (HIPAA), hold immense importance in computational modeling and simulations, especially concerning personal or sensitive data. The significance are: (1) Protection of Personal Data: GDPR and HIPAA impose stringent safeguards on personal and healthcare information, ensuring their confidentiality in simulations involving patient records or user profiles. (2) Consent and Transparency: These regulations mandate explicit consent for data processing, crucial for simulations involving human subjects. Transparency ensures individuals understand data usage within simulations.

(3) Data Minimization: Emphasizing data minimization, GDPR and good privacy practices advocate collecting only essential data in simulations, reducing risks associated with handling sensitive information. (4) Data Security and Encryption: Both regulations stress robust data security, necessitating encryption for sensitive data at rest and in transit, preventing unauthorized access. (5) Access Controls and Authorization: Implementing access controls ensures only authorized entities access sensitive data within simulations, complying with GDPR and HIPAA mandates. (6) Data Retention Policies: Regulations prescribe limits on data retention; data not pertinent to simulation goals should be deleted or anonymized, reducing long-term privacy risks. (7) Privacy Impact Assessments (PIAs): Conducting PIAs helps assess and mitigate potential privacy risks in computational modeling, aligning simulations with privacy requirements. (8) Incident Response and Breach Notification: Regulations mandate prompt response and notification procedures for data breaches, vital in simulations to mitigate their impact on privacy. (9) Legal and Ethical Considerations: These regulations reinforce ethical and legal obligations during computational modeling, exhibiting a commitment to privacy rights. GDPR and HIPAA are integral in computational modeling, ensuring responsible data handling and protecting individuals' privacy. Compliance is not just a legal mandate but also showcases ethical conduct, commitment to privacy, and data security [19].

5.14.5 ENSURING DATA PRIVACY AND COMPLIANCE WITH REGULATIONS SUCH AS GDPR, HIPAA

Ensuring compliance with data protection laws such as General Data Protection Regulation (GDPR), Health Insurance Portability and Accountability Act (HIPAA), or other relevant regulations is crucial in computational modeling and simulations. a comprehensive guide is presented as follows. (1) Understand Regulations: Identify and understand the specific regulations applicable to your project, considering the type of data being used. (2) Data Classification: Categorize data based on sensitivity levels, requiring varied protection measures for different data types. (3) Consent and Transparency: Obtain explicit consent from individuals whose data is used in simulations. Communicate clearly about data usage and allow consent withdrawal. (4) Data Minimization: Collect only necessary data, aligning with the principle of minimal information gathering. (5) Security Measures: Implement robust security, including encryption, access controls, authentication, and regular updates to tackle emerging threats. (6) Access Controls: Restrict data access to authorized entities using role-based access controls, regularly reviewing access permissions. (7) Data Retention and Deletion: Establish clear data retention policies and ensure proper and irreversible data deletion when no longer required. (8) Privacy Impact Assessments (PIAs): Conduct PIAs to identify and mitigate potential privacy risks associated with simulations. (9) Incident Response Plan: Develop a comprehensive incident response plan to address data breaches promptly and effectively. (10) Compliance Documentation: Maintain detailed records demonstrating compliance with data privacy regulations, essential for audits and inquiries. (11) User Education: Educate simulation participants on data privacy best practices and their rights regarding their data. (12) Legal and Ethical Considerations: Ensure compliance with both legal and ethical standards, respecting privacy rights and transparency principles. (13) Continuous Monitoring and Auditing: Regularly monitor data practices and conduct audits to ensure ongoing compliance. (14) External Expertise: Consider seeking external legal or privacy expertise for alignment with current regulations and best practices. (15) International Data Transfers: Comply with cross-border data transfer regulations if the simulation involves international data transfers. (16) Encryption and Anonymization: Encrypt data and anonymize or pseudonymize where feasible to reduce privacy risks. Thus, Upholding data privacy and compliance demands a proactive and thorough approach, not only as a legal obligation but also an ethical responsibility. Integrating these considerations into computational modeling and simulations ensures sensitive information protection and builds trust with stakeholders [9].

5.15 INTRUSION DETECTION AND PREVENTION SYSTEMS (IDPS) IN COMPUTATIONAL MODELING ENVIRONMENTS

This section presents Intrusion Detection and Prevention Systems (IDPS) in Computational Modeling Environments as follows. In computational modeling environments, where sensitive data and complex simulations are conducted, the use of Intrusion Detection and Prevention Systems (IDPS) is crucial for safeguarding against various cyber threats. IDPS plays a vital role in monitoring and analyzing network and system activities to identify and respond to potential security incidents. This section presents how IDPS is utilized in protecting computational modeling environments namely. *Real-time Monitoring*: IDPS continuously monitors network traffic, system logs, and user activities in real-time. This helps in the early detection of suspicious or malicious behavior within the computational environment.

Anomaly Detection: IDPS employs anomaly detection techniques to establish a baseline of normal behavior within the modeling environment. Any deviation from this baseline is flagged as a potential intrusion, enabling swift response. *Signature-based Detection*: IDPS uses predefined signatures or patterns of known attacks to identify and prevent malicious activities. This includes recognizing common attack patterns or known malware signatures. *Incident Response*: Upon detecting a potential intrusion, IDPS initiates incident response mechanisms to mitigate the impact. This may involve isolating affected systems, blocking malicious traffic, or alerting security personnel [10].

5.15.1 TYPES OF ATTACKS IDPS CAN MITIGATE AND THEIR CONFIGURATION

In safeguarding computational modeling environments, Intrusion Detection and Prevention Systems (IDPS) play a crucial role in countering various cyber threats such as namely:

1. Malware and Ransomware: IDPS detects, prevents, and combats malware and ransomware by updating signatures regularly and employing heuristic analysis to identify new threats.
2. Denial-of-Service (DoS) and Distributed Denial-of-Service (DDoS) Attacks: IDPS identifies and mitigates DoS and DDoS attacks by monitoring network traffic patterns, detecting sudden spikes, and using filtering mechanisms to limit malicious traffic.
3. Unauthorized Access and Intrusions: Monitoring login attempts and user behavior, IDPS detects unauthorized access, setting alerts for failed logins and enforcing strong authentication.

4. SQL Injection and XSS Attacks: IDPS detects and prevents SQL injection and XSS attacks by analyzing input data for malicious patterns and sanitizing or blocking harmful inputs.
5. Zero-Day Exploits: Using behavior-based analysis, IDPS detects and prevents previously unknown or zero-day exploits by identifying abnormal system behavior.

Hence, Effective configuration of IDPS involves regular signature updates, minimizing false positives, configuring response actions based on threat severity, and integrating IDPS with other security solutions for comprehensive protection. IDPS ensures real-time monitoring, anomaly detection, and mitigation, but ongoing updates are vital for adapting to evolving security challenges [54].

5.15.2 Incident Response and Recovery for Simulations

In simulation environments, an incident response plan is crucial for several reasons:

1. Early Detection and Mitigation: The plan enables swift detection of anomalies and threats, allowing for quick mitigation. This minimizes the impact on ongoing simulations, preventing severe consequences like security breaches or system failures.
2. Preserving Simulation Integrity: Complex simulations rely on accurate data and models. The response plan ensures the integrity of simulations by identifying and isolating unauthorized access or disruptive activities that could compromise results.
3. Regulatory Compliance: Especially in sensitive fields, adherence to regulations is essential. The plan provides a structured approach to handle and report security incidents, ensuring compliance with data protection and security standards.
4. Minimizing Downtime: Simulations are resource-intensive; the plan aims to minimize downtime by swiftly restoring normal operations. This preserves project timelines and optimizes computational resources.
5. Protecting Sensitive Information: Given the involvement of sensitive data, the response plan includes measures to safeguard information. It promptly addresses unauthorized access or breaches, ensuring data protection [19][55].

5.15.3 Developing and Executing an Effective Incident Response and Recovery Strategy

The key steps for an effective incident response and recovery strategy in simulation environments are: (1) Risk Assessment: Identify specific risks and vulnerabilities in the simulation setup. (2) Incident Response Team: Form a dedicated team with clear roles and responsibilities. (3) Documentation: Develop comprehensive documentation, including response plans and contact lists. (4) Detection and Analysis: Use monitoring tools and intrusion detection systems for continuous monitoring and anomaly detection. (5) Incident Identification: Establish criteria to identify and classify incidents within the simulation context. (6) Containment and Eradication: Isolate affected systems and conduct a thorough analysis to eliminate vulnerabilities. (7) Recovery Strategy: Restore data from backups, reconfigure systems, and validate simulation integrity. (8) Post-Incident Review: Assess lessons learned, update documentation, and provide additional training for the response team.

(9) Continuous Improvement: Regularly test the response plan through simulations, and use feedback for ongoing improvements. Following these steps ensures an adaptive and proactive approach

to tackle security incidents in simulation environments, maintaining their resilience in the face of potential threats [56].

5.16 ETHICAL CONSIDERATIONS AND RESPONSIBLE MODELING

Ethical considerations in computational modeling and simulations are critical, especially in healthcare and financial domains such as. (1) Healthcare Simulations: Patient data usage requires ethical guidelines ensuring confidentiality, patient consent, and compliance with healthcare regulations. Addressing biases in medical data is vital for fair patient care. (2) Financial Simulations: Transparency and consent are crucial when using financial data such as customer transactions. Recognizing and rectifying biases in financial data ensures ethical decision-making and prevents discriminatory outcomes. (3) Transparency and Accountability: Ethical modeling demands transparency in healthcare and financial simulations. Clear communication about simulation purposes, methods, and impacts fosters accountability and trust. This transparency ensures responsible use of simulations in both domains. Thus, Adhering to these ethical guidelines helps maintain trust, integrity, and fairness in computational simulations across healthcare and financial sectors [13][57].

5.16.1 RESPONSIBLE MODELING PRACTICES AND ETHICAL USE OF SIMULATION DATA

Ethical modeling in healthcare and finance revolves around several key principles namely: (1) Data Accuracy and Validation: Ensuring the accuracy and quality of input data is crucial for reliable simulations. Ethical practices involve rigorous validation and verification processes to minimize errors in patient and financial data. (2) Security Measures: Protecting sensitive information is paramount. Ethical considerations demand robust security measures to safeguard healthcare and financial data from unauthorized access or cyber threats. (3) Cultural and Social Sensitivity: Ethical modeling accounts for cultural and social influences. This includes understanding diverse patient populations in healthcare and acknowledging social factors affecting economic decisions in financial simulations. (4) End-User Education: Educating users on interpreting and using simulation outcomes ethically is crucial. This empowers medical professionals and decision-makers to make informed and ethical choices based on simulation results. (5) Continuous Monitoring and Evaluation: Ongoing monitoring and evaluation of simulations help identify ethical concerns and unintended consequences. This allows for necessary adjustments to ensure responsible and ethical use of simulation data. By integrating these ethical considerations into computational modeling practices, simulations in healthcare and finance can positively impact decision-making while upholding values of privacy, fairness, transparency, and accountability throughout the modeling lifecycle respectively [58].

5.17 CONCLUSION AND FUTURE SCOPE

This chapter comprehensively explores security measures in computational modeling and simulations, covering technical nuances and ethical considerations. The chapter begins with foundational knowledge on these technologies, progresses through a systemic review of existing literature, and underscores the significance of computational models in decision-making processes. The chapter delves into securing access to computational resources, emphasizing the interplay between authentication and authorization methods. The chapter extensively covers securing data within models, addresses common cybersecurity threats and potential system vulnerabilities, and navigates through network security for distributed simulations. The chapter proposes a cyber-attacks model and examines security concerns in cloud-based simulations while stressing the importance of cyber threat intelligence. The chapter discusses secure software development practices, compliance with regulations, and real-time security measures like intrusion detection systems and incident response strategies. Ethical considerations conclude the chapter, providing readers with a well understanding of security in computational modeling and simulations. Future work will focus on developing a dynamic threat detection and adaptive security frameworks for real-time protection in simulation environments.

ACKNOWLEDGMENTS

This research was not funded. However, the authors wish to appreciate the reviewers and chief editor, who gave their time to constructively review this chapter, which has significantly improved it.

REFERENCES

[1] Das, A., & Islam, M. M. (2011). Secured trust: A dynamic trust computation model for secured communication in multi-agent systems. *IEEE*, 9(2), 261–274. DOI: 10.1109/TDSC.2011.57

[2] Panchal, J. H., Kalidindi, S. R., & McDowell, D. L. (2013). Key computational Modeling issues in *Integrated Computational Materials Engineering. Computer-Aided Design*, 45, 4–25. Elsevier Ltd. All rights reserved.

[3] Gómez Mármol, F., & Martínez Pérez, G. (2009). Security threats scenarios in trust and reputation models for distributed systems. *Computers & Security*, Elsevier Ltd. All rights reserved, 28, 545–556.

[4] Jayasinghe, U., Truong, N. B., Lee, G. M., & Um, T. (2016). RpR: A Trust Computation Model for Social Internet of Things. In *2016 International IEEE Conferences on Ubiquitous Intelligence & Computing, Advanced and Trusted Computing, Scalable Computing and Communications, Cloud and Big Data Computing, Internet of People, and Smart World Congress*, 18–21 July 2016, Toulouse, France, 2016, pp. 930–937, doi: 10.1109/UIC-ATC-ScalCom-CBDCom-IoP-SmartWorld.2016.0146.

[5] Ahmad, M. A., Wisdom, D. D., & Isaac, S. (2020). An empirical analysis of cybercrime trends and its impact on moral decadence among secondary school level students in Nigeria. In *The 26th iSTEAMS Bespoke Multidisciplinary Conference*, Accra, Ghana.

[6] Al-Besher, A., & Kumar, K. (2022). Use of artificial intelligence to enhance e-government services. *College of Computing and Informatics*, Saudi Electronic University, Riyadh, 11673, Saudi Arabia. DOI: 10.1016/j.measen.2022.100484

[7] Baracaldo, N., Chen, B., Ludwig, H., & Safavi, J, A. (2017). Mitigating poisoning attacks on machine learning models: A data provenance based approach. *Session: Defense against Poisoning*. AISec'17, November 3, 2017, Dallas, TX, USA.

[8] Alsammari, B. M. (2023). A framework for developing secure Internet of Medical Things: A comprehensive roadmap from an artificial intelligence perspective. *Journal of Theoretical and Applied Information Technology,* 101(4), 1455–1468.

[9] Brundage, M., Avin, S., Clark, J., Toner, H., Eckersley, P., Garfinkel, B., Dafoe, A., Scharre, P., Zeitzoff, T., Filar, B., Anderson, H., Roff, H., Allen, G. C., Steinhardt, J., Flynn, C., Ó hÉigeartaigh, S., Beard, S., Belfield, H., Farquhar, S., Lyle, C., Crootof, R., Evans, O., Page, M., Bryson, J., Yampolskiy, R., & Amodei, D. (2018). *The* Malicious Use of Artificial Intelligence: Forecasting, Prevention, and Mitigation. February, 2018.

[10] Kabir, M. H., Hasan, K. F., Hasan, M. K., & Ansari, K. (2021). Explainable artificial intelligence for smart city application: A secure and trusted platform. In: Ahmed, M., Islam, S.R., Anwar, A., Moustafa, N., Pathan, A.S.K. (eds) *Explainable Artificial Intelligence for Cyber Security. Studies in Computational Intelligence*, vol. 1025. pp 241–263. Springer, Cham. 2021. https://doi.org/10.1007/978-3-030-96630-0_11

[11] Telo, J. (2017). AI for enhanced healthcare security: An investigation of anomaly detection, predictive analytics, access control, threat intelligence, and incident response. *Journal of Advanced Analytics in Healthcare Management*, 1(1), 21–37.

[12] Nwakanma, C. I., Ahakonye, L. A. C., Njoku, J. N., Odirichukwu, J. C., Okolie, S. A., Uzondu, C., Nweke, C. C. N., & Kim, D.-S. (2023). Explainable Artificial Intelligence (XAI) for intrusion detection and mitigation in intelligent connected vehicles: A review. *Applied Sciences,* 13(3), 1252. www.mdpi.com/, https://doi.org/10.3390/app13031252

[13] Singh, S. K., Rathore, S., & Park, J. H. (2019). BlockIoTIntelligence: A blockchain-enabled intelligent IoT architecture with artificial intelligence. *Department of Computer Science and Engineering*, Seoul National University of Science and Technology (SeoulTech), Seoul Korea, Elsevier, 2019.

[14] Mustapha, A. M., Arogundade, O. T., Vincent, O. R., Adeniran, O. J., & Chen, X. (2017). A model-based business process compliance management architecture for SMSE towards effective adoption of cloud computing. In October 2017 *International Conference on Computing Networking and Informatics (ICCNI)* (pp. 1–6). IEEE.

[15] Wisdom, D. D., Tambuwal, A. Y., Chun, P. B., Adamu, H. K, & Ajayi, E. A. (2018). Enhanced model for computer viruses counter measures. *1st International Conference on Education and Development (ITED)*, Baze University, Abuja, Nigeria.

[16] Vetrivel, S. C., Maheswari, R., Saravanan, T. P. (2024). Industrial IOT: Security Threats and Counter Measures. In: Prasad, A., Singh, T. P., Dwivedi Sharma, S. (eds) *Communication Technologies and Security Challenges in IoT. Internet of Things.* pp. 403–425. Springer, Singapore. https://doi.org/10.1007/978-981-97-0052-3_20.

[17] Wisdom, D. D., Vincent, O. R., Igulu, K., Christian, A. U., Hyacinth, E. A., Baba, G. A., & Esther, O. O. (2023). *Cybersecurity in the Industry 4.0 and 5.0.* Taylor and Francis.

[18] Wisdom, D. D., Ajayi, E. A., Arinze, U. C., Idris, H., Bello, U. M., & Aladesote, O. I. (2021). An optimized TWIN battery resource management scheme in wireless networks, lecture notes in networks. *Proceedings of Sixth International Congress on Information and Communication Technology,* Vol. 217, Springer Nature, 2021. 978-981-16- 2101-7,511607

[19] Kelly, B. S., Quinn, C., Belton, N., Lawlor, A., Killeen, R. P., & Burrell, J. (2023). Cybersecurity considerations for radiology departments involved with artificial intelligence. *European Radiology.* https://doi.org/10.1007/s00330-023-09860-1.

[20] Kavitha, S., Bora, A., Naved, M., Raj, K. B., & Nadh Singh, B. R. (2021). An Internet of things for data security in cloud using artificial intelligence. *International Journal of Grid and Distributed Computing,* 14(1), 1257–1275.

[21] Ahsan, M., Nygard, K. E., Gomes, R., Chowdhury, M. M., Rifat, N., & Connolly, J. F. (2022). Cybersecurity Threats and Their Mitigation Approaches Using Machine Learning—A Review. *Journal of Cybersecurity and Privacy*, 2(3), 527–555.

[22] Jakka, G., Yathiraju, N., & Ansari, M. F. (2022). Artificial intelligence in terms of spotting malware and delivering cyber risk management. *Journal of Positive School Psychology*, 6(3), 6156–6165.

[23] Al-Besher, A., & Kumar, K. (2022). Use of artificial intelligence to enhance e-government services. *Measurement & Sensors*, 24(Suppl. C), 100484.

[24] Vijayakumar, K. P., Pradeep, K., Balasundaram, A., & Prusty, M. R. (2023). Enhanced Cyber Attack Detection Process for Internet of Health Things (IoHT) Devices Using Deep Neural Network. School of Computer Science and Engineering, Vellore Institute of Technology, Chennai 600127.

[25] Isaac, S., Wisdom, D. D., Ahmed, M. A., & Arinze, U. C. (2020). Battery-Life Management with an Efficient Sleep-Mode Power Saving Scheme (BM-ESPSS)‖ in IEEE 802.16e networks, *International Journal of Mechatronics, Electrical and Computer Technology (JMEC)*, 7(2).

[26] Li, B., Wu, Y., Song, J., Lu, R., Li, T., & Zhao, L. (2021). DeepFed: Federated deep learning for intrusion detection in industrial cyber–physical systems. *IEEE Transactions on Industrial Informatics,* 17(8), 5658–5667. DOI: 10.1109/TII.2021.3060480

[27] Radoglou-Grammatikis, P., Rompolos, K., Sarigiannidis, P., Argyriou, V., Lagkas, T., Sarigiannidis, A., Goudos, S., & Wan, S. (2022). Modeling, detecting, and mitigating threats against industrial healthcare systems: A combined software defined networking and reinforcement learning approach. *IEEE Transactions on Industrial Informatics,* 18(3), 2022.

[28] Meshram, C., Imoize, A. L., Aljaedi, A., Alharbi, A. R., Jamal, S. S., & Barve, S. K. (2021). An efficient electronic cash system based on certificateless group signcryption scheme using conformable chaotic maps. *Sensors*, 21(21), 7039. https://doi.org/10.3390/s21217039

[29] Wisdom, D. D., Saidu, I., Tambuwal, A. Y., Ahmad, M. A., Isaac, S., and Farouk, N. (2019). An Efficient Sleep- Window-Based Power saving Scheme (ESPSS) in IEEE 802.16e Networks. *15th International Conference on Electronics Computer and Computation ICECCO*, Abuja, Nigeria, 2019, pp. 1–6, doi: 10.1109/ICECCO48375.2019.9438963.

[30] Imoize, A. L., Irabor, D. O., Peter, G., & Chakraborty, C. (2022). Blockchain technology for secure COVID-19 pandemic data handling. In *Smart Health Technologies for the COVID-19 Pandemic: Internet of Medical Things Perspectives* (pp. 141–179). University of Lagos, University of KwaZulu-Natal, Birla Institute of Technology, Mesra. DOI: 10.1049/PBHE042E_ch6

[31] Imoize, L., Hemanth, J., Do, D.-T., & Sur, S. N. (Eds.). (2022). Explainable artificial intelligence in medical decision support systems. *Institution of Engineering and Technology.* https://doi.org/10.1049/PBHE050E

[32] Ramasamy, L. K., Khan K. P., F., Imoize, A. L., Ogbebor, J. O., Kadry, S., & Rho, S. (2021). Blockchain-Based wireless sensor networks for malicious node detection: A survey. *IEEE Access, 9,* 128765–128785. https://doi.org/10.1109/ACCESS.2021.3111923

[33] Sei, Y., Ohsuga, A., & Imoize, A. L. (2022). Statistical test with differential privacy for medical decision support systems. In L. Imoize, J. Hemanth, D.-T. Do, and S. N. Sur (Eds.), *Explainable Artificial Intelligence in Medical Decision Support Systems* (pp. 401–433). Institution of Engineering and Technology. https://doi.org/10.1049/PBHE050E_ch15

[34] Ayoade, O. B., Oladele, T. O., Imoize, A. L., Awotunde, J. B., Adeloye, A. J., Olorunyomi, S. O., Idowu, A. O. (2022). Explainable artificial intelligence (XAI) in medical decision systems (MDSSs): Healthcare systems perspective. In L. Imoize, J. Hemanth, D.-T. Do, and S. N. Sur (Eds.), *Explainable Artificial Intelligence in Medical Decision Support Systems* (pp. 1–43). Institution of Engineering and Technology. https://doi.org/10.1049/PBHE050E_ch1

[35] Kumar, R. L., Wang, Y., Poongodi, T., & Imoize, A. L. (2021). *Internet of Things, Artificial Intelligence and Blockchain Technology.* Springer International Publishing, Cham. https://doi.org/10.1007/978-3-030-74150-1

[36] Abikoye, O. C., Oladipupo, E. T., Imoize, A. L., Awotunde, J. B., Lee, C.-C., & Li, C.-T. (2023). Securing critical user information over the Internet of Medical Things platforms using a hybrid cryptography scheme. *Futur. Internet, 15*(3), 99. https://doi.org/10.3390/fi15030099

[37] Eneh, A. H., Wisdom, D. D., Wisdom, D. D., Arinze, U. C., Philibus, E., & Philibus, E. (2022). An Optimized Database Management System. *In Proceedings of the LASUSTECH 30th Multidisciplinary Innovation Conference* 2020, Series 30 Vol. 2 (pp. 25). Lagos, Nigeria. DOI: 10.22624/AIMS/iSTEAMS/LASUSTECH2022V30P25

[38] Wisdom, D. D., Vincent, O. R., Igulu, K. T., Arowolo, M. O., Hyacinth, E. A., Christian, A. U., & Baba, G. A. (2023). Mitigating cyber threats in healthcare systems: The role of artificial intelligence and machine learning. *Artificial Intelligence and Blockchain Technology in Modern Telehealth Systems, Institute of Engineering and Technology (IET).*

[39] Igulu, K. T., Wisdom, D. D., Arowolo, M. O., Singh, T. P. (2023). *Computational Intelligence in big data analytics. CI-Industry-4.0: Computational Intelligence in Industry 4.0 and 5.0 Applications.* Taylor and Francis group.

[40] Wisdom, D. D., Vincent, O. R., Oduntan, O. O., Igulu, K., & Garba, A. B. (2022). Enhanced cybersecurity framework for 5G-integrated oil and gas industry. *COMPUTOLOGY: Journal of Applied Computer Science and Intelligent Technologies, 2*(1), 8–22. Publisher: Journal Press India.

[41] Wisdom, D. D., Vincent, O. R., Igulu, K. T., Baba, G. A., Oduntan, O. E., Akinyemi, A. E., & Umar, B. (2023). *Cyber-threats in Healthcare Systems: The Role and Application of Artificial Intelligence and Machine Learning.* Nova Science Publishers, USA.

[42] Van der Giessen, E., Schultz, P. A., Bertin, N., Bulatov, V. V., Cai, W., Csányi, G., Foiles, S. M., Geers, M. G. D., González, C., Hütter, M., Kim, W. K., Kochmann, D. M., LLorca, J., Mattsson, A. E., Rottler, J., Shluger, A., Sills, R. B., Steinbach, I., Strachan, A., & Tadmor, E. B. (2020). Roadmap on multiscale materials modeling. *Modelling and Simulation in Materials Science and Engineering, 28*(4), 043001. https://doi.org/10.1088/1361-651X/ab7150

[43] Oberkampf, W. L., Pilch, M., & Trucano, T. G. (2007). *Predictive Capability Maturity Model for Computational Modeling and Simulation (SAND2007-5948).* Sandia National Laboratories. www.sandia.gov/ Publications/publications.html

[44] Veksler, V. D., Buchler, N., Hoffman, B. E., Cassenti, D. N., Sample, C., & Sugrim, S. (2021). Simulations in cyber-security: A review of cognitive modeling of network attackers, defenders, and users. *Simulation & Gaming, 52*(5), 623–647. https://doi.org/10.1177/1046878120983521

[45] Pinyol, I., & Sabater-Mir, J. (2013). Computational trust and reputation models for open multi-agent systems: A review. *Springer Intelligent Review*, 40(1), 1–25. https://doi.org/10.1007/s10462-011-9277-z

[46] Kuhl, M. E., & Sudit, M. (2007). Cyber-attack modeling and simulation for Network security analysis. In S. G. Henderson, B. Biller, M.-H. Hsieh, J. Shortle, J. D. Tew, & R. R. Barton (Eds.), *Proceedings of the 2007 Winter Simulation Conference*, IEEE. DOI: 10.1109/WSC.2007.4419720

[47] Sudit, M., Stotz, A, & Holender, M. (2005). Situational awareness of coordinated Cyber-attack. In *Proceedings of the International Society for Optical Engineering Conference*, Orlando, FL.

[48] Isabona, J., Imoize, A. L., Ojo, S., Karunwi, O., Kim, Y., Lee, C.-C., & Li, C.-T. (2022). *Development of a Multilayer Perceptron Neural Network for Optimal Predictive Modeling in Urban Microcellular Radio Environments*. MDPI. https://doi.org/10.3390/app12115713

[49] Wisdom, D. D., Christian, A. U., Hyacinth, E. A., Esther, O. O., Igulu, K., & Garba, A. B. (2022). IoT devices battery life energy management scheme. *COMPUTOLOGY: Journal of Applied Computer Science and Intelligent Technologies*, 2(2), 1–13. https://doi.org/10.17492/computology.v2i2.2201

[50] Wisdom, D. D., Igulu, K., Esther, O. O., Baba, G. A., Ahmad, A., & Sidi, A. (2021). A review of best practices and methods of mitigating cybersecurity risks in healthcare systems and a newly proposed algorithm. *Computology: Journal of Applied Computer Science and Intelligent Technologies*, 1(2), 27–36. DOI: 10.17492/computology.v1i2.2104

[51] Olaleye, T. O., & Vincent, O. R. (2020). A predictive model for students performance and risk level indicators using machine learning. In *2020 International Conference in Mathematics, Computer Engineering and Computer Science (ICMCECS)* (pp. 1–7). IEEE.

[52] Nubi, O. J., & Vincent, O. R. (2020). Virtual reality: A pedagogical model for Simulation based learning. In *2020 International Conference in Mathematics, Computer Engineering and Computer Science (ICMCECS)* (pp. 1–6). IEEE.

[53] Lawal, O. M., Vincent, O. R., Agboola, A. A. A., & Folorunso, O. (2021). *An* improved hybrid scheme for e-payment security using elliptic curve cryptography. *International Journal of Information Technology*, 13, 139–153.

[54] Sultan, F., Poylisher, A., Serban, C., Lee, R. C. J., & Chiang, C. J. (2012). "Timesync: Virtual time for scalable, high-fidelity hybrid network emulation," in *Military Communications Conference, 2012-Milcom 2012* (pp. 1–6). IEEE, Orlando, FL.

[55] Imoize, A. L., Balas, V. E., Solanki, V. K., Lee, C. C., & Obaidat, M. S. (Eds.). (2023). *Handbook of Security and Privacy of AI-Enabled Healthcare Systems and Internet of Medical Things* (1st ed.). CRC Press. DOI: 10.1201/9781003370321

[56] Awotunde, J. B., Folorunso, S. O., Imoize, A. L., Odunuga, J. O., Lee, C.-C., Li, C.- T., & Do, D.-T. (2023). An ensemble tree-based model for intrusion detection in industrial Internet of Things networks. *Applied Sciences*, 13(4), 2479. https://doi.org/10.3390/app13042479

[57] Meshram, C., Ibrahim, R. W., Meshram, S. G., Jamal, S. S., & Imoize, A. L. (2022). An efficient authentication with key agreement procedure using Mittag–Leffler–Chebyshev summation chaotic map under the multi-server architecture. *Springer Journal of Supercomputing*, 78, 4938–4959.

[58] Oladipupo, E. T., Abikoye, O. C., Imoize, A. L., Awotunde, J. B., Chang, T.-Y., Lee, C.-C., & Do, D.-T. (2023). An Efficient authenticated elliptic curve cryptography scheme for multicore wireless sensor networks. *IEEE Access*, 11, 1306–1323. doi: 10.1109/ACCESS.2022.3233632.

6 Computational Models for Training, Testing, and Validating Wireless Networks and Systems

Lateef Adesola Akinyemi, Olamide Peter Oshinuga,
Ernest Mnkandla, Mbuyu Sumbwanyambe,
Sunday Oladayo Oladejo, Stephen Obono Ekwe and
Adedotun Temitope Ajibare

6.1 INTRODUCTION

A computational model is a computer-based system built using algorithms [1]. It functions as a mathematical framework that empowers researchers to explore, examine, and forecast intricate and complex systems [2], such as wireless networks and related systems. Examples of such models include Random Forests, Deep Neural Networks, and Linear Regression.

Computational models rely heavily on data, necessitating extensive datasets for effective performance. These models require a training process before they can be employed to tackle complex tasks. During training, the models learn and establish connections within the large chunks of data they are being trained with, enhancing their suitability for practical applications such as decision-making, numerical prediction, and classification, as illustrated in Figure 6.1.

The training process of computational models is pivotal in improving their performance, ensuring their ability to provide reliable outcomes in real-world scenarios [3]. However, it is essential to highlight that these models operate exclusively within the boundaries of their training data. Consequently, their efficacy relies heavily on the quality and pertinence of the training data. Computational models are highly reliable and function effectively when the training data is sufficient. In instances where the model fails to deliver the desired outcomes, improvement can be achieved through additional rounds of data-driven training [4]. The adaptability and capacity for continuous learning from training data serve as critical advantages of computational models, empowering them to evolve and adjust to the dynamic realm of evolving challenges. Ultimately, computational models emerge as robust instruments for grasping, dissecting, and predicting complex systems.

In the field of wireless communication, the precise prediction of signal strength in wireless networks presents a paramount and ongoing challenge [5]. Signal strength is a critical metric that demands continuous monitoring, as it directly influences network performance and reliability. This challenge has widespread consequences for the functioning of wireless networks in various aspects.

In tackling the challenge, computational models emerge as potent tools, with machine learning and deep learning techniques standing out as particularly effective approaches. These models can accurately predict wireless network signal strength [6], offering robust solutions to the communication challenges encountered in the wireless domain. As technology continues its advancement, harnessing the potential of these computational models becomes ever more crucial for achieving

DOI: 10.1201/9781003457428-7

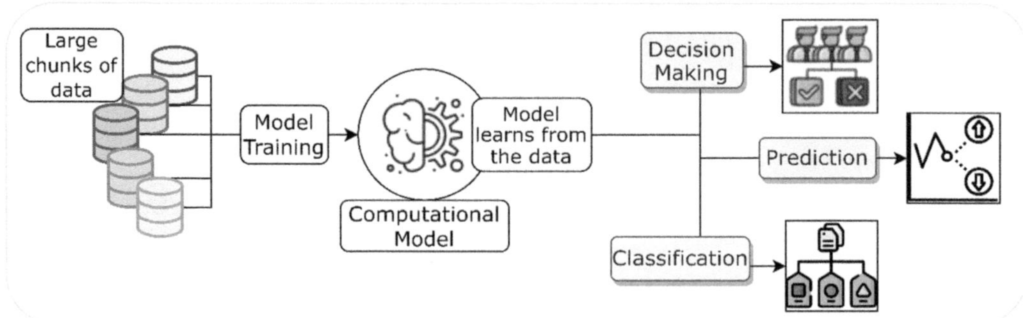

FIGURE 6.1 The training process of the computational models using large chunks of data.

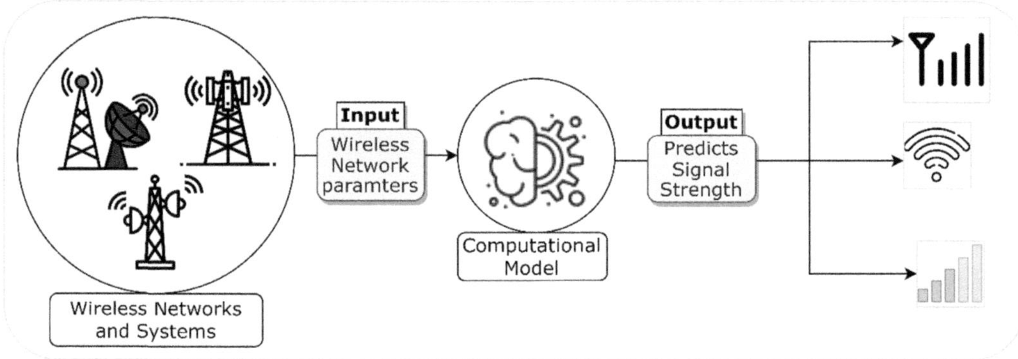

FIGURE 6.2 The prediction process of the employed computational models.

uninterrupted and reliable wireless networks. Accurate prediction of signal strength plays a pivotal role in optimizing network performance [7]. It ensures signal stability, reduces latency, enhances data transfer rates, and promotes the seamless and efficient operation of wireless networks. Additionally, it aids network operators in anticipating potential issues, simplifying the troubleshooting process.

This study aims to utilise computational models, such as Deep Neural Networks (DNN), and Random Forest (RF), among others, to predict wireless network signal strength. These models will be trained using a large dataset containing attributes of wireless network metrics and location-based properties, such as longitude, latitude, network latency, network type, and more. Through training, the models understand the relationship among the attributes in the data and are subsequently employed to predict the signal strength of the wireless network. As depicted in Figure 6.2, the computational model utilises wireless network parameters as inputs, resulting in the prediction of signal strength as the output. The quality of these predictions is strongly influenced by specific attribute values provided as input.

Validation of computational models plays a pivotal role in assessing the reliability and accuracy of their predictions [8]. This systematic process thoroughly compares the predicted and actual output values, effectively quantifying the performance of the model. Such validation is particularly crucial for ensuring the competency of computational models in generating accurate predictions. Commonly employed evaluation metrics, such as Mean Absolute Error (MAE) and Mean Square Error (MSE), are instrumental in this process. These metrics gauge the predictive accuracy of the model by measuring the disparity between actual and predicted signal strength generated by the

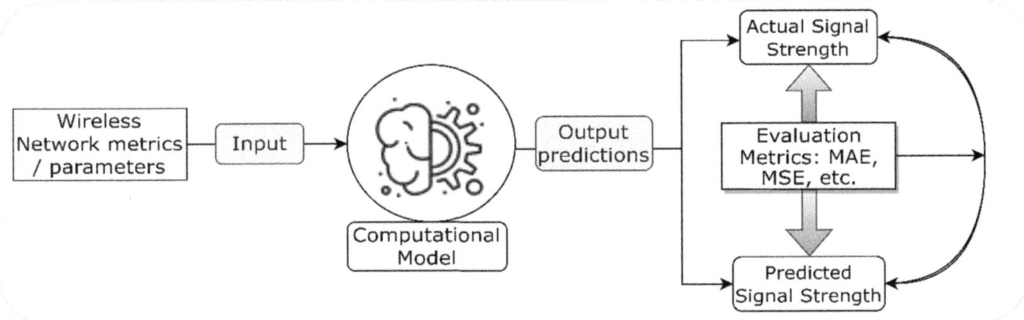

FIGURE 6.3 The validation process of the employed computational models.

computational model, as illustrated in Figure 6.3. The disparity provides valuable insights into the ability of the model to deliver reliable and precise predictions. Furthermore, the outcomes of model validation enable a comparative analysis of various computational models when applied to the same predictive task. This assists in identifying the most reliable model tailored to the specific predictive task at hand.

6.1.1 Key Contributions of the Chapter

This chapter makes the following significant contributions:

i. The chapter provides an explanation of computational models, their functioning, and the necessity of training for solving complex tasks.
ii. The chapter sheds light on the challenges in wireless communication arising from signal strength prediction and demonstrates how trained computational models can address these challenges by accurately predicting signal strength.
iii. The chapter elucidates the training process of computational models using a dataset comprising wireless network metrics and location-based properties. These models are subsequently applied to predict wireless network signal strength based on these properties.
iv. The chapter analyses the dataset in use to uncover the relationships between wireless network metrics and their impact on the prediction of wireless network signal strength.
v. The chapter emphasises the importance of computational model validation as a means to verify the efficacy of these models in predicting wireless network signal strength. This validation is achieved through evaluation metrics such as Mean Absolute Error and Mean Square Error.
vi. The chapter predicts the signal strength of wireless networks using computational models, including Deep Neural Networks (DNN), Random Forest, Linear Regression, and K-nearest neighbours.
vii. The chapter compares the predictive performance of the computational models and identifies which one is the most effective for the predictive task in this chapter.

6.1.2 Organisation of the Chapter

Section 6.1 discusses the introductory section and contributions of the book chapter on computational models for training, testing, and validating wireless systems and networks. Furthermore, Section 6.2 discusses the related works concerning training, testing and validation in wireless networks and

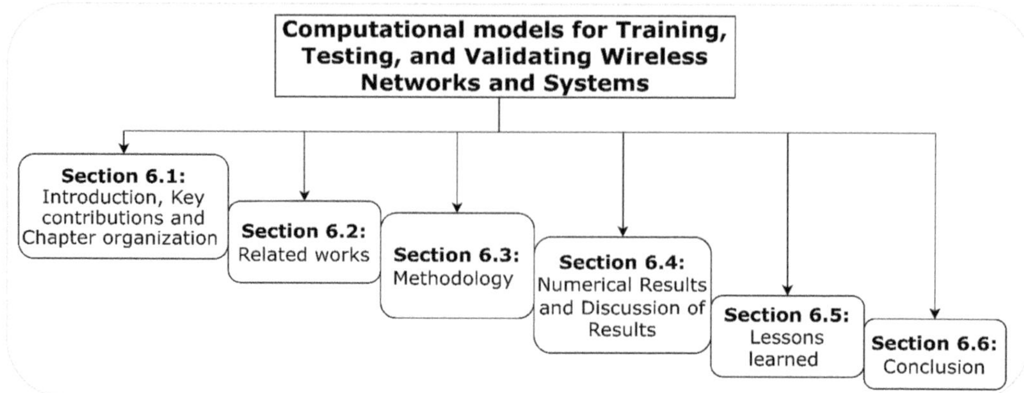

FIGURE 6.4   An organogram depicting the hierarchical arrangement of each section within the chapter.

systems, addressing gaps of related works, a summary of the reviewed works is equally presented in a tabular form stating the aim, and the results achieved. The methodology employed in this study is presented in Section 6.3. Numerical simulation results are presented and discussed in Section 6.4. Additionally, lessons that have been learnt thus far are summarily discussed in the computational model for training, testing and validating wireless networks and systems presented in Section 6.5. Section 6.6 completes the book chapter by making conclusions for the book chapter. The organisational structure of the chapter is depicted in Figure 6.4.

6.2   RELATED WORKS

This section reviews prior research, which focuses on the application of trained computational models, including machine learning, deep learning, and artificial intelligence, to address challenges in wireless communication and how the models undergo validation.

The study in [6] explored using machine learning models to predict wireless network signal strength. To improve the accuracy of predictions, the study incorporates feature engineering methods. While neural networks and linear models demonstrated effectiveness, the most successful among them was KNN, which achieved a lower MAE score of 0.65 compared to other models' performance.

The study in [9] employed computational models, incorporating machine learning algorithms like LR, RF, and SVM, to predict the received signal strength of a wireless network. Following the predictions, the models were validated using metrics such as MAE and RMSE. Subsequently, the prediction results of the models were compared. The study concluded that SVM outperformed the other algorithms in signal strength prediction, as indicated by its lower RMSE score than the other employed models.

The study conducted in [10] explored computational models comprising machine learning algorithms such as KNN, RF, DT, and others for signal strength prediction. It concludes that KNN with three neighbours outperformed other models for signal strength prediction.

The study in [11] utilised deep learning to predict the signal strength of a wireless network. The model's performance was validated using test data, resulting in an RSME score of 8.59

The study in [12] delves into the evaluation of computational models, incorporating various deep learning techniques, including RNN and CNN, to enhance routing accuracy and improve wireless network performance. Following validation and comparative analysis, it was determined that DL outperformed other ML methods, achieving an impressive 94% accuracy and a 32% reduction in loss.

The study in [13] introduces the DL approach, specifically using CNN for traffic detection. The study utilises a dataset comprising spectrum pictures in the time domain, which is employed for training with actual and artificial data. The approach achieved an impressive accuracy score of 96%. Furthermore, it demonstrated the capacity to distinguish between various transmission speeds with an accuracy rate of 87%.

The research in [14] introduces a computational model based on a DL approach for identifying cyberattacks in a network of wireless sensors. Specifically, a CNN-LSTM model was employed for this task. The model was trained using network data, leading to a substantial improvement in accuracy compared to conventional ML methods. Further model validation resulted in impressive training and validation accuracies of 96.02% and 95.08%, respectively.

The research in [15] presents a computational model incorporating a DL approach to forecast channel quality in wireless networks. It utilised DNN and LSTM for the predictions. The models were trained using data from Orange Senegal. Following the projections, an evaluation was conducted using the RMSE metric, and LSTM outperformed DNN with a lower RMSE score of 0.27, while DNN yielded a score of 0.28.

The research in [16] introduces computational models, CNN and MLP, to predict and classify traffic in wireless networks. These models were trained using network attributes and were validated, achieving an impressive accuracy score of 99%. Additionally, in the works of authors in [17–20], different artificial intelligence schemes and metaheuristic algorithms have been employed to solve wireless networks and systems in fifth- and sixth-generation environments using different testing functions and validation approaches.

The research discussed in [21] deals with the growing demand to address security challenges and safeguard the privacy worries of patients in devices and systems.

This study in [22] proposes an approach that incorporates snort and Zeek, enhanced with a machine learning (ML) model, to distinguish between legitimate network traffic and DDoS attack traffic.

The studies in [6, 9–11] have successfully harnessed trained computational models to predict signal strength in wireless networks, followed by validation to confirm the models' predictive prowess. Other research in [13–16] applied trained computational models to solve different complex tasks in wireless communication networks.

6.2.1 Addressing Research Gaps of Related Works

Firstly, in addition to prediction and validation, we delve deeper into the dataset through comprehensive data analysis. This analysis serves as a crucial exploration, uncovering the connections between essential network metrics and their direct impact on the signal strength of wireless networks. Furthermore, this study introduces novel approaches to enhance the accuracy and effectiveness of computational models, which were achieved by employing feature selection and advanced encoding techniques to refine the models. Finally, to comprehensively assess and compare the predictive capabilities of the various models, various metrics were utilized. These metrics allow us to quantify the performance of each model. This study extends beyond prediction and validation, venturing into data analysis, feature importance, encoding techniques, and model comparison.

6.2.2 Summary of Related Works

As mentioned above, the summary of the research relating to wireless networks is presented in Table 6.1, highlighting the computational models employed, the task done, and the summarised result.

TABLE 6.1
Summary of the Related Works

Study	Computational Models	Aim / Task	Result
[6]	Neural Networks, Linear Models and KNN	Prediction of signal strength.	KNN outperformed other models with an MAE score of 0.65.
[9]	LR, RF, and SVM	Prediction of signal strength.	SVM outperformed other models.
[10]	ANN, KNN, DT, RF and K-Means	Prediction of signal strength.	KNN with 3 neighbours outperformed other models.
[11]	Neural Network	Prediction of signal strength.	RMSE score of 8.59.
[12]	RNN and CNN	Enhancement of routing accuracy.	94% accuracy and 32% loss reduction.
[13]	CNN	Detection of traffic.	Accuracy rate and score of 87% and 96%.
[14]	CNN-LSTM	Identification of cyber-attacks.	Training and validation accuracies of 96.02% and 95.08%,
[15]	DNN and LSTM	Prediction of channel quality.	RMSE Sores: 0.27 for LSTM, 0.28 for DNN. LSTM outperformed.
[16]	CNN and MLP	Prediction and classification of traffic.	Accuracy score of 99%.

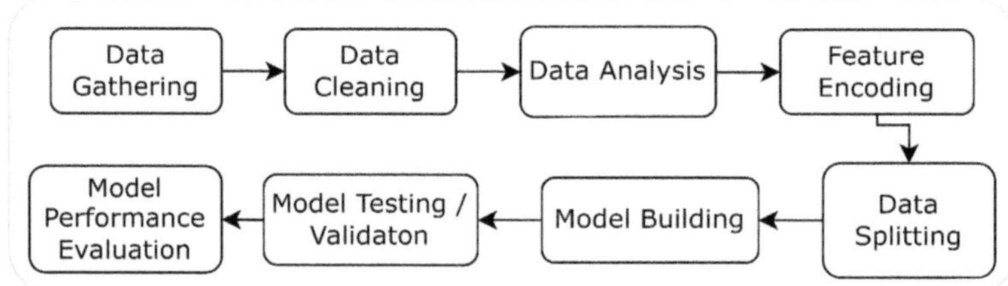

FIGURE 6.5 Showing the sequential flow of the methods employed.

6.3 METHODOLOGY OF THE STUDY

This section highlights and discusses the key method employed in the chapter to achieve its aim, which is the accurate prediction of wireless network signal strength using computational models.

The key methods employed in this study is summarized in Figure 6.5. It serves as a reference for understanding the sequential flow of these methodological approaches.

6.3.1 DATA GATHERING

The dataset encompasses a collection of 12 distinct attributes, representing a combination of wireless network metrics and location-based attributes, and comprises a total of 16,829 rows [23]. It encompasses attributes associated with wireless network signal metrics, including signal quality, latency, signal strength, and more. Furthermore, this dataset incorporates location-based attributes such as longitude and latitude. The target attribute of interest is the signal strength of the wireless network at a given timestamp, which is to be predicted. The attributes contained in the dataset are:

- *Timestamps*: This signifies the moment at which the signal measurements were documented, with a ten-minute gap between each time stamp.
- *Latitude and Longitude*: These attributes denote the geographical coordinates where network measurements were captured.

- *Network Latency*: This pertains to the delay in transmitting data from the origin to the target. It is quantified in milliseconds (ms).
- *Data Throughput*: This represents the pace at which data is sent or received through a network link. It is quantified in megabits per second (Mbps).
- *Network Type*: This denotes the technology or generation of the network in operation. It comprises categorical options such as 5G, 4G, LTE, and 3G.
- *Signal Strength*: This represents the quantification of the strength or potency of a wireless signal at a particular location and within a specified network category. Its measurement is expressed in decibels milliwatt (dBm). It is the target attribute which is to be predicted by the computational models.

6.3.2 Data Cleaning

This is done to remove any irregularities that may exist within the dataset, including empty cells, duplicate entries, and inconsistent attributes. Such irregularities can potentially affect the performance of the computational model when making predictions. Consequently, thorough data cleaning is performed to improve the efficiency and accuracy of the predictions generated by the computational models.

6.3.3 Data Analysis

This step is crucial because it aims to uncover the relationships between the attributes within the data, offering valuable insights into how these attributes affect the performance of wireless networks. In particular, the analysis reveals the connections between the Network Type attribute and other wireless network metrics. This aids in understanding the impact of specific attributes as they appear in the data and how they influence the prediction of wireless network signal strength.

6.3.4 Feature Encoding

Computational models exclusively operate with numeric data and necessitate the conversion of categorical variables into numerical format for practical training. This study employs Label Encoding as the chosen method for this transformation. Using label encoding, unique numerical labels are assigned to each category within the categorical variables, enabling the models to comprehend and process these variables. This process enhances the model's ability to make meaningful predictions regarding wireless network signal strength.

6.3.5 Data Splitting

The dataset is partitioned into two subsets: 80% designated as training data and 20% as testing data, as depicted in Figure 6.6. The training data is used to educate the computational models, enabling them to discern and learn the unique patterns within the data. Through the training process, the models develop an understanding of the relationships inherent in the wireless network metrics present in the dataset. The models are further employed to predict the wireless network signal strength. On the other hand, the testing data is used to validate the performance of the models in making predictions, ensuring the reliability and accuracy of the analytical results.

6.3.6 Model Building

This section involves the development of four distinct computational models, as depicted in Figure 6.6. These models are trained and subsequently employed to predict the signal strength of the wireless network.

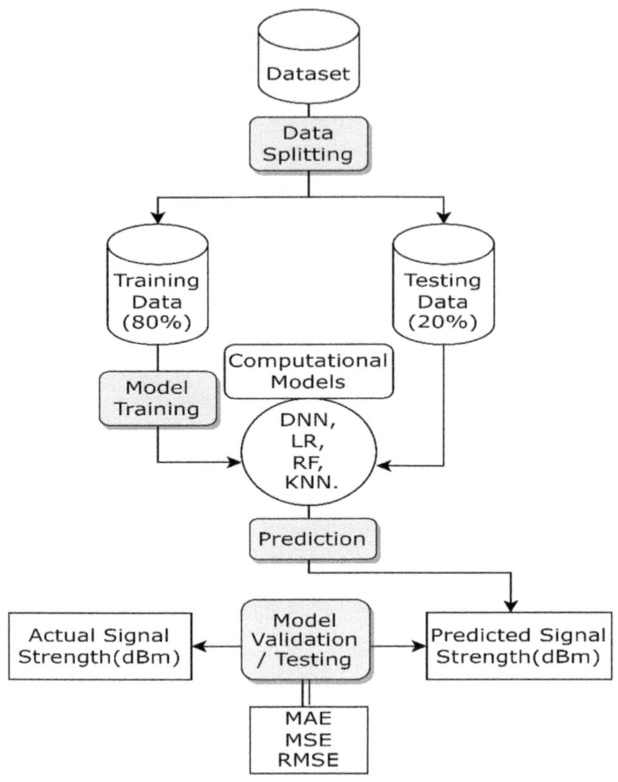

FIGURE 6.6 A step-by-step representation of the methods employed clearly and sequentially.

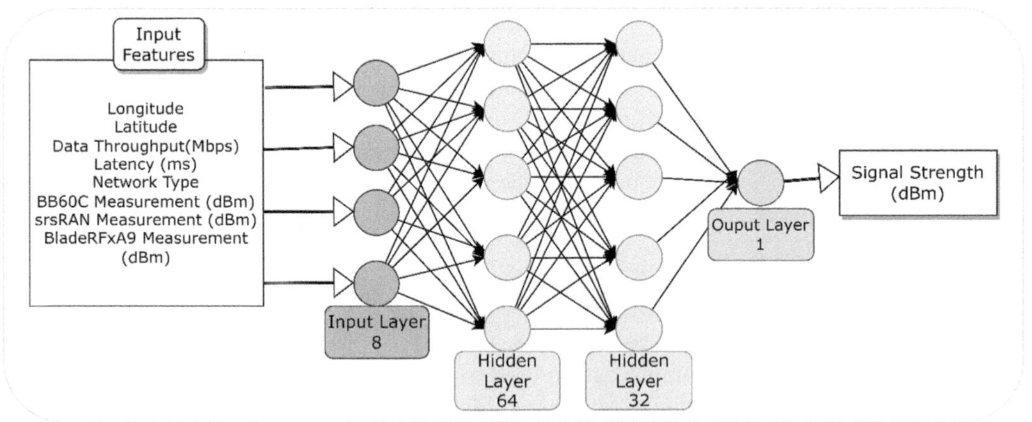

FIGURE 6.7 Architecture of the Deep Neural Network.

Deep Neural Network (DNN): This consists of three layers, which is the input, hidden and output layer. The architecture of the DNN is illustrated in Figure 6.7. It consists of a sequential arrangement of layers, starting with an input layer that receives eight input features. This is followed by two hidden layers, one with 64 neurons using ReLU (Rectified Linear Unit) activation and another with 32 neurons also using ReLU activation. The hidden layers are crucial for extracting complex patterns from the input data. The output layer comprises a single neuron responsible for predicting the continuous signal strength of the wireless network. This model employs the Adam optimiser and Mean Squared Error (MSE) loss function during training.

Linear Regression (LR): This statistical method investigates the relationship between input variables. It aims to identify the optimal line that effectively characterises the association between the input variables. This statistical technique employs methodologies to estimate the slope and intercept of the most suitable line for the model [24]. Polynomial features are introduced to improve the performance of the model.

Random Forest (RF): This involves constructing a multitude of decision trees, with "10 n_estimators" indicating the number of trees in the ensemble, and then amalgamating their predictions to generate a final prediction. These trees are built from subsets of the original dataset along with a selection of random variables [25].

K-Nearest Neighbours (KNN): This is used for predicting continuous numeric values. A value of K, representing the number of nearest neighbours considered, is set to 5. KNN works by finding the five training data points that are closest to a given input and then computing the predicted value as the average or weighted average of their corresponding target values [26].

6.3.7 MODEL VALIDATION/TESTING

This involves utilising various evaluation metrics to assess the effectiveness of the employed computational models in predicting the signal strength of the wireless network. It quantifies the difference between the actual signal strength of the testing data and the predicted signal strength generated by each model, as depicted initially in Figure 6.6. Additionally, these differences are visualised through validation plots, which illustrate a comparison between the actual and predicted signal strength. This approach verifies the accuracy of predictions made by each model and determines the most effective among them. To achieve this, the following evaluation metrics are employed:

Mean absolute error (MAE): This metric is employed to evaluate the accuracy of predictions made by computational models, specifically in predicting signal strength. It is calculated by summing the absolute differences between the actual and predicted signal strengths and then dividing this sum by the total number of data points, as shown in Equation 6.1. Lower MAE indicates superior prediction performance, signifying that the predictions made by the model closely align with the actual signal strength values. Conversely, a higher MAE suggests poorer model performance, indicating that the predictions by the model deviate more significantly from the actual signal strength values.

$$MAE = \frac{1}{N} \sum_{i=1}^{N} |y_i - y_a| \tag{6.1}$$

where y_i represents the predicted signal strength, y_u represents actual signal strength, and N represents the number of data points.

Mean squared error (MSE): This statistical method is used to validate predictions. It is calculated by summing up the squared difference between predicted and actual signal strength values and dividing it by the total number of data points, as shown in Equation 6.2. The lower the MSE value, the better the model is at prediction, while higher values imply that the model performs poorly.

$$MSE = \frac{1}{N} \sum_{i=1}^{N} |y_i - y_a|^2 \tag{6.2}$$

Root mean square error (RMSE): This is calculated as the square root of mean square error, as shown in Equation 6.3. It not only considers the average discrepancy between predicted and actual values but also factors in the impact of significant errors.

$$RMSE = \sqrt{\frac{1}{N}\sum_{i=1}^{N}\left|y_i - y_a\right|^2}$$

(6.3)

6.4 RESULTS AND DISCUSSION

This section visualizes, explains, and compares the results obtained in this study based on the analysis of the data used and the predictive outcomes of the computational models for wireless network signal strength prediction.

6.4.1 Data Analysis

This section presents the results obtained from the analysis of the utilized data. It uncovers the relationships between the attributes within the dataset, encompassing network performance metrics and the target attribute (signal strength).

6.4.1.1 Relationship between Network Type and the Target Attribute

This analysis entails calculating the average signal strength for four distinct network types and presenting the outcomes graphically. Figure 6.8 vividly depicts the average signal strength for each network type. It is evident from the visual representation that 5G boasts the highest average signal strength at -95 dBm, followed by 4G and LTE, while 3G trails with the lowest signal strength at -85 dBm. The difference in average signal strength between 4G and LTE is minimal. This analysis suggests that 5G outperforms other network types in terms of signal strength, hinting at potential advantages in terms of speed and reliability, while 4G and LTE exhibit similar average signal strengths.

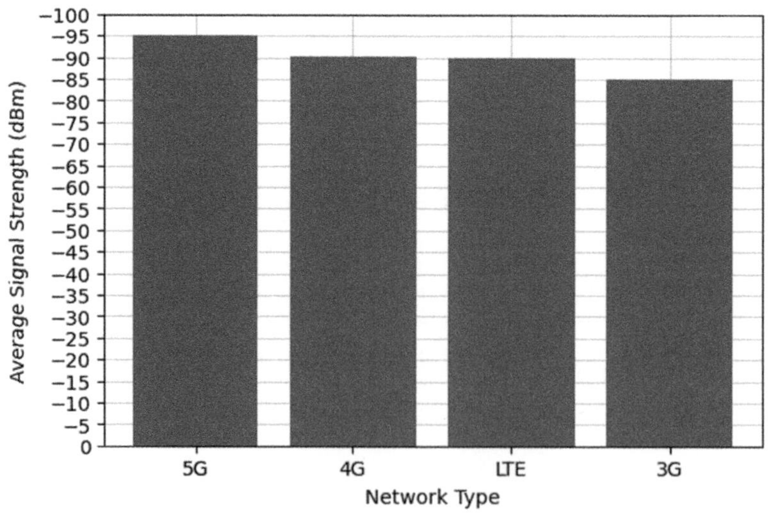

FIGURE 6.8 A bar chart representing the average signal strength of each network type in decibels milliwatt (dBm).

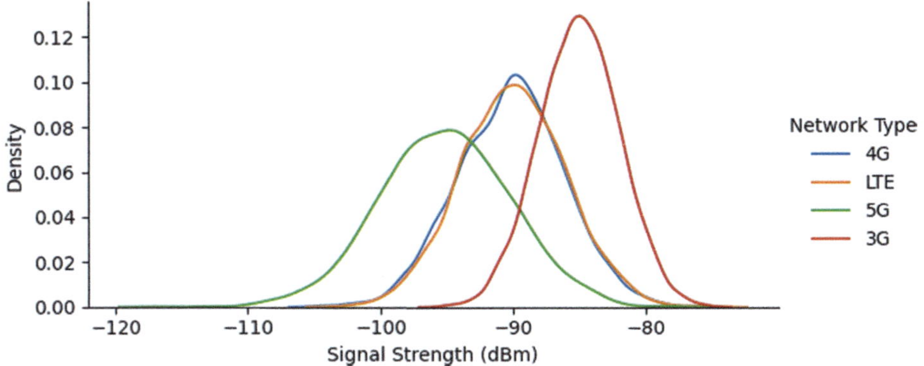

FIGURE 6.9 KDE plot showing the connection and distribution between the Network Types attributes and the Signal Strength of the wireless network.

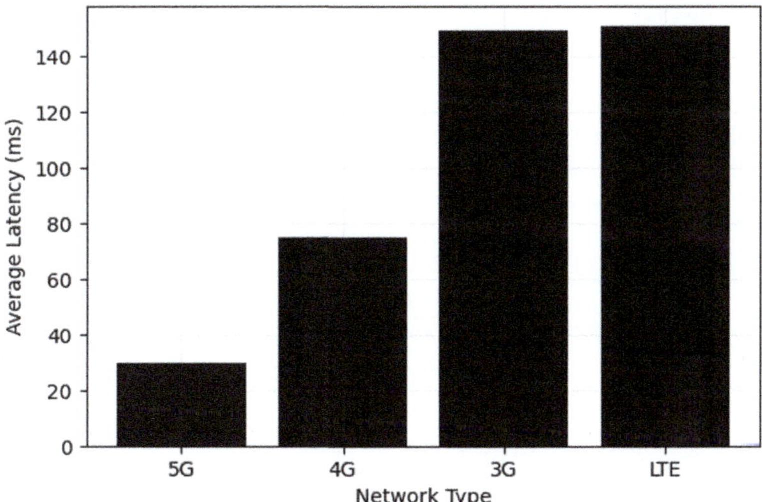

FIGURE 6.10 A bar chart representing the average latency of each network type.

Using a Kernel Density Estimation (KDE) plot, Figure 6.9 further illustrates the signal strength distribution for the four distinct network types. The plot reveals a distinct pattern, highlighting the unique signal strength range of 5G, spanning from -110 dBm to -95 dBm. In contrast, 4G and LTE share a similar range, extending from -90 dBm to -75 dBm, while 3G, with the lowest signal strength, spans from -75 dBm to -80 dBm. This visual representation underscores the diverse signal strength characteristics among network types, emphasising the robustness of 5G, the comparable performance of 4G and LTE, and the relative weakness of 3G. Such insights enhance the user experience for different network technologies.

6.4.1.2 Relationship between Network Latency and the Target Attribute

This analysis involves calculating the average latency for each network type and presenting the results visually, as depicted in Figure 6.10. In the figure, it is evident that 5G boasts the lowest average latency at 29 ms, signifying its exceptional speed and responsiveness. In contrast, 4G exhibits a latency of 75 ms, while LTE and 3G lag significantly with latencies of 150 ms and 149 ms, respectively. These findings emphasise the superior performance of 5G in terms of latency, making it the optimal choice for applications demanding low response times, such as real-time communication and online gaming while highlighting the need for improvements in LTE and 3G networks to provide faster and more responsive services.

6.4.2 Prediction of Wireless Network Signal Strength

This section computes and visualizes the results of computational models employed for the prediction of signal strength of the wireless network using scatter and line validation plots.

Scatter validation plot: This performs a comparison between the actual and predicted signal strength generated by each computational model using a scatter plot. A dotted red line is overlaid to represent the ideal scenario of perfect predictions. The performance of the employed models is assessed by examining the proximity of the scatter plot points to this dotted red line. Figures 6.11 – 6.14 display the outcomes yielded by Deep Neural Network, Linear Regression, Random Forest, and K-Nearest Neighbours.

Figures 6.11 and 6.12 reveal that both Linear Regression (LR) and Deep Neural Network (DNN) exhibited similar predictions that closely approached the red dotted line, indicating strong performance in predicting the signal strength of the wireless network. However, it is worth noting that LR outperformed DNN, as its predictions were even closer to the ideal line. Both LR and DNN consistently predicted values around -85 dBm, especially when the actual signal strength fell within the -97 dBm to -87 dBm range. However, they faced challenges in providing accurate predictions within this specific range. Despite this limitation, these models demonstrated generally strong performance. The superiority of LR over DNN is evident in its predictions, which align more closely with the dotted red line, symbolising nearly perfect predictions compared to DNN.

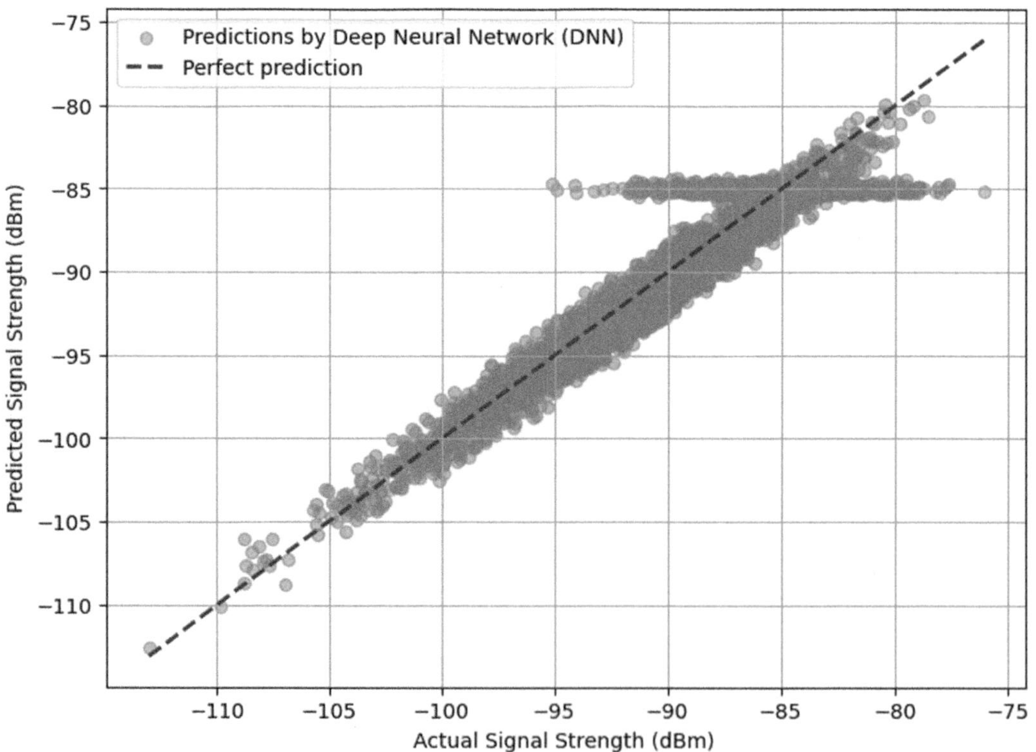

FIGURE 6.11   A scatter validation plot showing the predictions made by deep neural network.

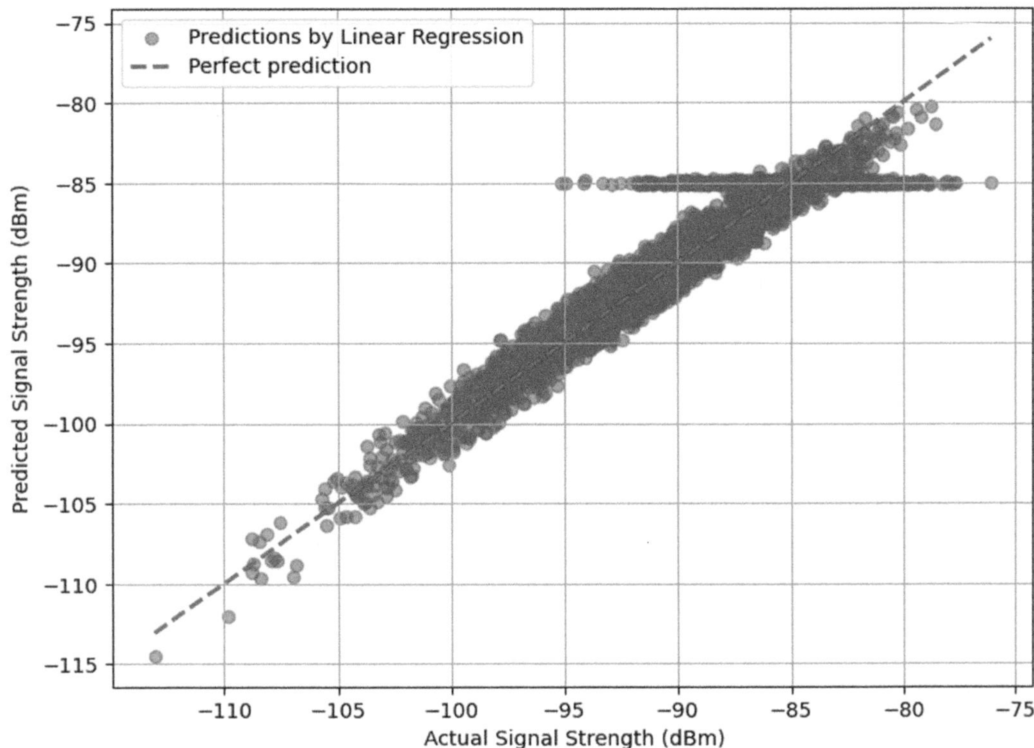

FIGURE 6.12   Scatter validation plot showing the predictions made by Linear Regression.

K-Nearest Neighbor (KNN) and Random Forest (RF) both exhibited strong performance in predicting signal strength, with their predictions showing a notable similarity, as illustrated in Figures 6.13 and 6.14. Their predictions were generally in proximity to the dotted red line, indicating their effectiveness. However, RF significantly outperformed KNN, as its scatter plot predictions were even closer to the red line, demonstrating higher accuracy. It's important to highlight that KNN encountered difficulties when predicting signal strength within the range of -115 dBm to -105 dBm, where its precision lagged behind other computational models. While KNN and RF performed well overall, RF predictions displayed a slightly superior closeness to the red line, underscoring its enhanced accuracy compared to KNN.

Line validation plot: This method is used to validate the model by visualising and comparing the actual and predicted signal strength produced by the computational models. In this approach, the actual and predicted signal strengths are represented using a vertical plot, where the x-axis displays the number of data points, and the y-axis represents the signal strength values. Figures 6.15–6.18 display the line validation plots for the first 500 data points generated by the employed computational models, where the actual and predicted signal strength are represented using a dot and cross symbol. This visualisation technique helps to discern the difference between the actual and predicted signal strength.

Figures 6.15–6.18 provide a granular and comprehensive visualisation of signal strength predictions made by each computational model for the initial 500 data points. It visibly depicts the similarity of predictions generated by Linear Regression (LR) and Deep Neural Network (DNN) as well as Random Forest (RF) and K-Nearest Neighbours (KNN). It is evident that LR, RF, and DNN demonstrated predictions that closely resemble the actual values, highlighting their capacity for accurate signal strength prediction.

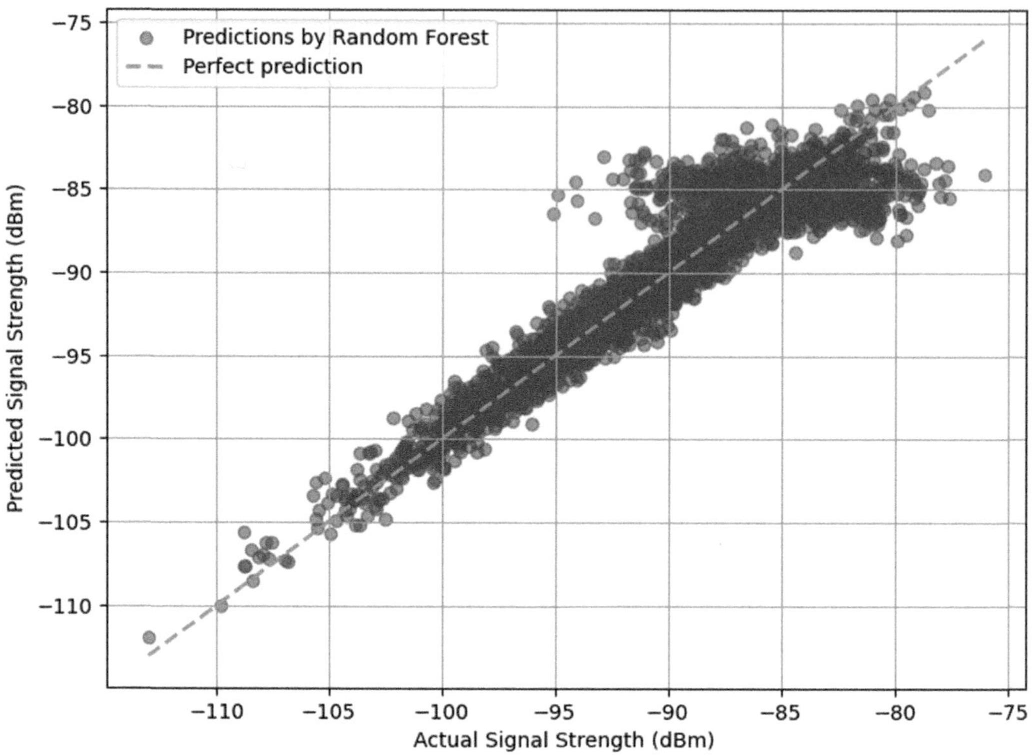

FIGURE 6.13 Scatter validation plot showing the predictions made by Random Forest.

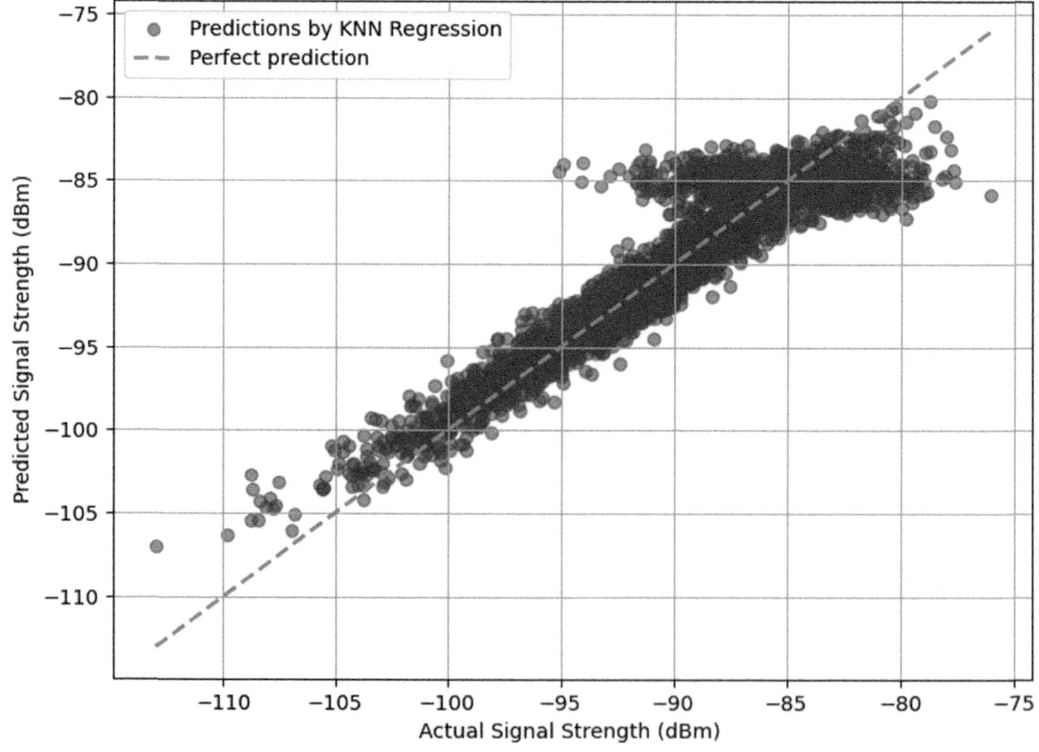

FIGURE 6.14 Scatter validation plot showing the predictions made by K-Nearest Neighbours.

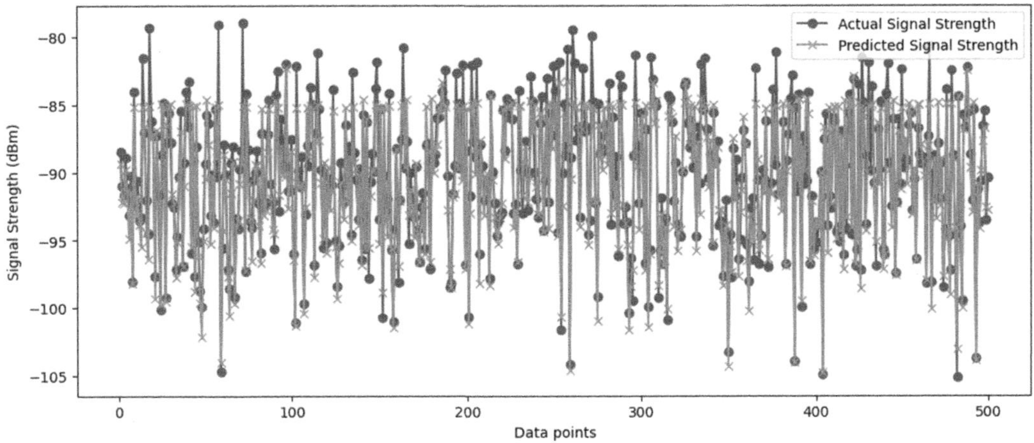

FIGURE 6.15 Line validation plot displaying the actual and predicted signal strength by the Deep Neural Network.

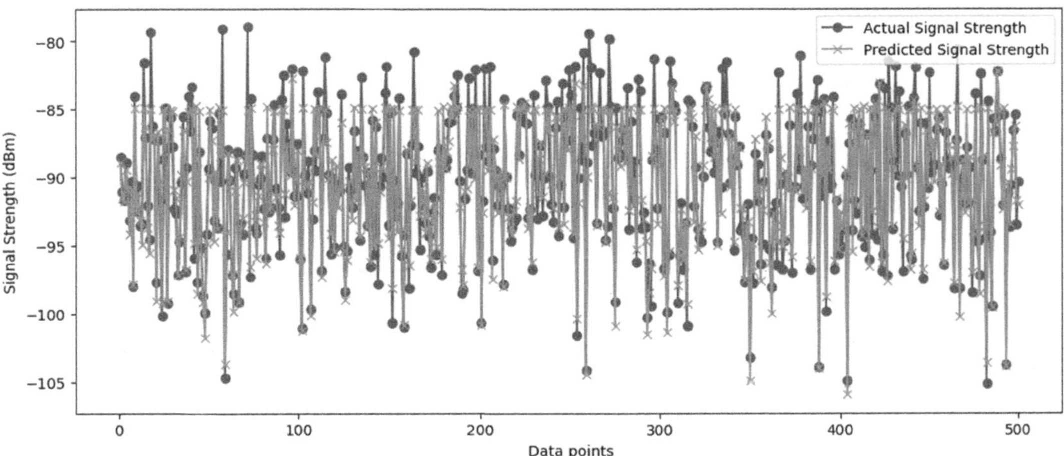

FIGURE 6.16 Line validation plot displaying the actual and predicted signal strength by Linear Regression.

6.4.3 EVALUATION METRICS SCORES

This section presents and compares the evaluation metrics scores of the models employed for the prediction of wireless network signal strength.

The evaluation metrics scores for each of the employed models are presented in Table 6.2 This is done to verify, measure, and compare the accuracy of predictions generated by the computational models. The evaluation metrics are Mean Absolute Error (MAE), Mean Square Error (MSE), and Root Mean Square Error (RMSE).

From the results presented in Table 6.2, Linear Regression (LR) achieved the lowest MAE, signifying that, on average, its predictions were the closest to the actual signal strength values. This indicates a relatively high level of accuracy in predicting signal strength. Additionally, LR exhibited lower MSE and RMSE values, suggesting that its predictions had low variability and were generally close to the true values. Both Deep Neural Network (DNN) and Random Forest (RF) demonstrated competitive performance, with similar MAE scores. They provided reasonably

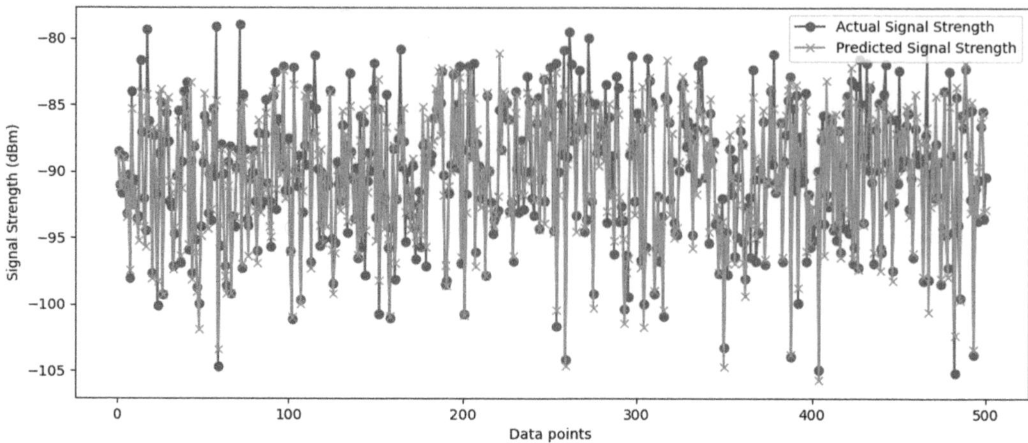

FIGURE 6.17   Line validation plot displaying the actual and predicted signal strength by Random Forest.

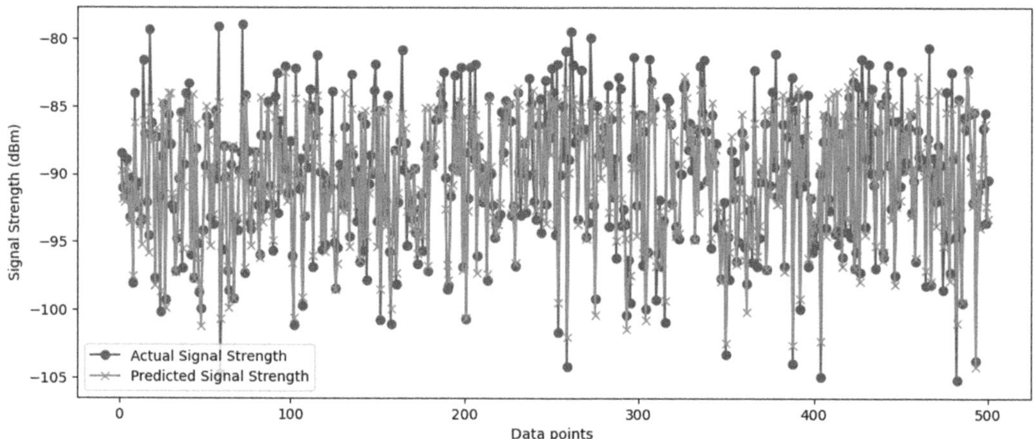

FIGURE 6.18   Line validation plot displaying the actual and predicted signal strength by K-Nearest Neighbours.

TABLE 6.2
Comparison of Evaluation Metrics Scores of the Computational Models

Computational models	MAE	MSE	RMSE
Deep Neural Network (DNN)	1.27	3.16	1.78
K-Nearest Neighbor (KNN)	1.35	3.60	1.90
Linear Regression (LR)	1.22	3.02	1.74
Random Forest (RF)	1.26	3.46	1.86

accurate predictions with a moderate level of variability, as indicated by their respective MSE and RMSE values. K-Nearest Neighbor (KNN) performed slightly less in terms of MAE compared to LR, DNN, and RF, implying a somewhat higher average prediction error. The MSE and RMSE values of KNN also indicated more variability in its predictions. In summary, while LR had the lowest MAE and offered precise predictions, DNN and RF also exhibited commendable performance.

6.5 LESSONS LEARNED

In this chapter, the following are the key lessons learned as summarised as follows:

- *Computational model training*: Computational models are initially devoid of knowledge and only become effective through data-driven training. Adequate training is paramount for these models to acquire the necessary knowledge and patterns from the data, enabling them to make accurate predictions and effectively address various tasks.
- *Computational model validation/testing*: This pertains to the validation and testing of the trained computational models. It involves assessing the performance and reliability of the models when applied to predictive tasks. By rigorously evaluating these models, we gain insights into their real-world effectiveness, ensuring they can be trusted to make accurate predictions.
- *Prediction of signal strength of wireless networks*: The prediction of signal strength in wireless networks emerges as a pivotal concern within the domain of wireless communication. Accurately predicting signal strength not only holds significant importance but also holds the potential to enhance overall network performance and streamline troubleshooting processes.

6.6 CONCLUSION

The accurate prediction of signal strength in wireless networks emerges as a paramount challenge in wireless communication networks. This challenge carries profound implications for network performance and reliability. By exploring computational models, including machine learning and deep learning techniques, we have demonstrated their capability to accurately predict wireless network signal strength. Computational models, such as Deep Neural Network, Linear Regression, and Random Forest have showcased their capacity to provide reliable and precise predictions. Crucially, we have underscored the pivotal role of validation in assessing the reliability and accuracy of the computational models. The systematic comparison of predicted and actual signal strength values using evaluation metrics like Mean Absolute Error and Mean Square Error has revealed the competence of the models in delivering dependable predictions. The findings of this study hold relevance for the field of wireless communication. Future work should focus on the integration of real-time data and the development of more efficient training processes, which will be critical for achieving even more accurate predictions.

REFERENCES

[1] J.B. Vancouver, and X. Li, "Computational Modeling," *Oxford Bibliographies Online Datasets*, Mar. 2018. doi: 10.1093/obo/9780199846740-0135

[2] V.K. Mago, and V. Dabbaghian, Eds., "Computational Models of Complex Systems," *Intelligent Systems Reference Library*, 2014. doi: 10.1007/978-3-319-01285-8

[3] V.K. Kodavalla, "Methodology for Determining Optimal Model and Training Data in Deep Learning," *2022 IEEE International Conference on Data Science and Information System (ICDSIS)*, Jul. 2022. doi: 10.1109/icdsis55133.2022.9916009

[4] L. Jian, J. Li, and S. Luo, "Exploiting Expertise Rules for Statistical Data-Driven Modeling," *IEEE Transactions on Industrial Electronics*, vol. 64, no. 11, pp. 8647–8656, Nov. 2017. doi: 10.1109/tie.2017.2703659

[5] S. Choudhary, A. Sharma, and K. Srivast, "Modeling and Optimisation of Mobile Signal Strength in Challenging Atmospheric Conditions," Mar. 2021. doi: 10.21203/rs.3.rs-295295/v1

[6] S. Mohammadjafari, S. Roginsky, E. Kavurmacioglu, M. Cevik, J. Ethier, and A. B. Bener, "Machine Learning-Based Radio Coverage Prediction in Urban Environments," *IEEE Transactions on Network and Service Management*, vol. 17, no. 4, pp. 2117–2130, Dec. 2020. doi: 10.1109/tnsm.2020.3035442

[7] N. Moraitis, L. Tsipi, and D. Vouyioukas, "Machine Learning-Based Methods for Path Loss Prediction in Urban Environment for LTE Networks," *2020 16th International Conference on Wireless and Mobile Computing, Networking and Communications (WiMob)*, Thessaloniki, Greece, 2020, pp. 1–6, doi: 10.1109/WiMob50308.2020.9253369.

[8] K.T. Hu, A. Urbina, and J. Mullins, "A Perspective on the Integration of Verification and Validation into the Decision Making Process," *Model Validation and Uncertainty Quantification*, vol. 3, pp. 265–273, 2015. doi: 10.1007/978-3-319-15224-0_28

[9] H. Panda, M. Das, and B. Sahu, "Received Signal Strength Prediction Model for Wireless Underground Sensor Networks Using Machine Learning Algorithms," *Journal of Information and Optimization Sciences*, vol. 43, no. 5, pp. 949–962, Jul. 2022. doi: 10.1080/02522667.2022.2092061

[10] R. Azoulay, E. Edery, Y. Haddad, and O. Rozenblit, "Machine Learning Techniques for Received Signal Strength Indicator Prediction," *Intelligent Data Analysis*, vol. 27, no. 4, pp. 1167–1184, Jul. 2023. doi: 10.3233/ida-226750

[11] M. Wu, C. Li, and C. Song, "Research on Wireless Intelligent Propagation Model Based on Deep Learning," *2020 IEEE 9th Joint International Information Technology and Artificial Intelligence Conference (ITAIC)*, Dec. 2020. doi: 10.1109/itaic49862.2020.9339171

[12] S.S. Al-Twalah, "Deep Learning Techniques for Accuracy Optimisation in Wireless Networks," 2020. http://paper.ijcsns.org/07_book/202003/20200322.pdf

[13] T. De Schepper, M. Camelo, J. Famaey, and S. Latré, "Traffic Classification at the Radio Spectrum Level Using Deep Learning Models Trained with Synthetic Data," *International Journal of Network Management*, vol. 30, no. 4, Feb. 2020. doi: 10.1002/nem.2100

[14] S. M. Naser, Y. H. Ali, and D. A.-J. OBE, "Deep Learning Model for Cyber-Attacks Detection Method in Wireless Sensor Networks," *Periodicals of Engineering and Natural Sciences (PEN)*, vol. 10, no. 2, pp. 251, Apr. 2022. doi: 10.21533/pen.v10i2.2838

[15] N. Diouf, M. Ndong, D. Diop, K. Talla, M. Sarr, and A. C. Beye, "Channel Quality Prediction in 5G LTE Small Cell Mobile Network Using Deep Learning," *2022 9th International Conference on Soft Computing & Machine Intelligence (ISCMI)*, Nov. 2022. doi: 10.1109/iscmi56532.2022.10068487

[16] P.H. Kumar, and T. Samanta, "Deep Learning Based Optimal Traffic Classification Model for Modern Wireless Networks," *2022 IEEE 19th India Council International Conference (INDICON)*, Nov. 2022. doi: 10.1109/indicon56171.2022.10039822

[17] A. Zappone, M. Di Renzo, and M. Debbah, "Wireless Networks Design in the Era of Deep Learning: Model-Based, AI-Based, or Both?" in *IEEE Transactions on Communications*, vol. 67, no. 10, pp. 7331–7376, Oct. 2019, doi: 10.1109/TCOMM.2019.2924010.

[18] M. F. Ahmad Fauzi, R. Nordin, N. F. Abdullah, and H. A. H. Alobaidy, "Mobile Network Coverage Prediction Based on Supervised Machine Learning Algorithms," in *IEEE Access*, vol. 10, pp. 55782–55793, 2022, doi: 10.1109/ACCESS.2022.3176619.

[19] F. Wilhelmi, M. Carrascosa, C. Cano, A. Jonsson, V. Ram and B. Bellalta, "Usage of Network Simulators in Machine-Learning-Assisted 5G/6G Networks," in *IEEE Wireless Communications*, vol. 28, no. 1, pp. 160–166, February 2021, doi: 10.1109/MWC.001.2000206.

[20] S.O. Oladejo, S.O. Ekwe, L.A. Akinyemi, and S.A. Mirjalili, "The Deep Sleep Optimiser: A Human-Based Metaheuristic Approach," *IEEE Access*, vol. 11, pp. 83639–83665, 2023.

[21] A.L. Imoize, V.E. Balas, V.K. Solanki, C.-C. Lee, and M.S. Obaidat, Eds., "Handbook of Security and Privacy of AI-Enabled Healthcare Systems and Internet of Medical Things," 1st ed., CRC Press, 2023. DOI: 10.1201/9781003370321

[22] M. AbdulRaheem, I.D. Oladipo, A.L. Imoize, J.B. Awotunde, C.C. Lee, G.B. Balogun, and J.O.Adeoti, "Machine Learning Assisted Snort and Zeek in Detecting DDoS Attacks in Software-Defined Networking," *International Journal of Information Technology*, pp. 1–17, 2023.

[23] "Cellular Network Analysis Dataset," *Kaggle*, Jun. 16, 2023. www.kaggle.com/datasets/suraj520/cellular-network-analysis-dataset

[24] M. Huang, "Theory and Implementation of Linear Regression," *2020 International Conference on Computer Vision, Image and Deep Learning (CVIDL), Chongqing*, China, 2020, pp. 210–217. doi: 10.1109/CVIDL51233.2020.00-99

[25] M. Aria, C. Cuccurullo, and A. Gnasso, "A Comparison among Interpretative Proposals for Random Forests," *Machine Learning with Applications*, vol. 6, pp. 100094, Dec. 2021. doi: 10.1016/j.mlwa.2021.100094

[26] M. Mailagaha Kumbure, and P. Luukka, "A Generalised Fuzzy K-Nearest Neighbor Regression Model Based on Minkowski Distance," *Granular Computing*, vol. 7, no. 3, pp. 657–671, Sep. 2021. doi: 10.1007/s41066-021-00288-w

7 Modeling and Simulation of Non-Lambertian Beams-Based Vehicular Visible Light Communications in 6G and Beyond

Jupeng Ding, Chih-Lin I, Jintao Wang and Jian Song

7.1 INTRODUCTION

For addressing the serious spectrum shortage and complex interference issues of the well-known vehicular radio frequency (RF) technology paradigm, vehicular visible light communication (VLC) is becoming a promising enabling technique for future 6G, and beyond vehicular wireless networks. Compared to the conventional radio frequency (RF) based vehicular connection technology, vehicular VLC is capable of offering impressive advantages, including unregulated and abundant optical spectrum resource, low-cost light front end, natural immunity to the conventional electromagnetic interference, provision of ubiquitous lighting function and high security performance [1–5].

Currently, due to the high directivity of vehicular VLC signal propagation, the limited coverage footprint and the unsatisfying vehicular link are viewed as the bottleneck issues that seriously hinder vehicular VLC commercialization. For enhancing the vehicular connection performance, a variety of communication enabling techniques and outdoor channel characterization are introduced to vehicular VLC system designs [6–9]. Particularly, the authors in [6] investigated how the selective combing receiver technique could improve the system performance of one typical vehicular VLC system. Moreover, the study in [7] presented one simultaneous light wave information and power transfer scheme for harvesting the energy for the infrastructure-to-vehicular VLC applications and showed the relationship between the achievable rate and the harvested energy amount. Furthermore, the authors in [8] proposed one optical intelligent reflecting surfaces aided beyond 5G vehicle-to-vehicle solution for road safety applications. In addition, the work in [9] analyses the interference effects from another vehicle in vehicle-to-vehicle VLC in order to identify the maximum distance between the vehicles could communicate, thereby minimizing the accident possibility on the highway scenario

Objectively, these vehicular VLC works usually assume that the concerned headlamps' LED-based transmitters match conventional Lambertian optical raidation pattern, which actually ignore the potential optical beam diversity and the inspired design degree of freedom in the emerging vehicular VLC domain. As a matter of fact, for the diverse vehicular applications, the sufficient fundamental measurement and modeling works have shown that the distinct non-Lambertian

DOI: 10.1201/9781003457428-8

optical beams could be elaborately designed and produced by the international LED components manufacturers [10–13]. In addition, these non-Lambertian optical beams have been actively introduced and analyzed in many research branches of developing VLC domain [12–20], including but not limited to coordinated coverage [15], multiple optical beam switching [17], channel modeling and characterization [12–13], secure optical wireless links [14, 16], access points design [20], and cells planning [18]. And the effects of distinct optical beam pattern and the potential performance characteristics have been estimated and illustrated for the above VLC branch directions and the related application scenarios.

7.1.1 KEY CONTRIBUTIONS OF THE CHAPTER

Nevertheless, up to now, to the best of our knowledge, the distinctive non-Lambertian optical beams, the potential available beam configurations, and the relevant performance characteristics are still waiting for fundamental investigation and numerical evaluation in the vehicular VLC research branch. Motivated by the discussions above, in this work, the typical non-Lambertian beams are adopted to construct the vehicular VLC transmission links and in addition are anticipated to provide more design and an optimized degree of freedom for the vehicular VLC technology.

7.1.2 CHAPTER ORGANIZATION

In this work, the Lambertian optical beam-based vehicular VLC are presented in Section 1.2. And the typical non-Lambertian optical beam-based vehicular VLC are investigated in Section 1.3. Numerical results and discussions evaluation are presented in Section 1.4. Finally, Section 1.5 concludes this chapter.

7.2 LAMBERTIAN OPTICAL BEAM-BASED VEHICULAR VISIBLE LIGHT COMMUNICATIONS

7.2.1 LAMBERTIAN OPTICAL BEAM PATTERN

In current vehicular VLC performance analysis, the LED sources usually follow a simple Lambertian optical beam pattern. This pattern assumption means that radiation intensity is just one cosine function of the viewing angle and could be given as [15–16, 21]:

$$I_{\text{Lam}}(\phi) = \frac{m_{\text{Lam}}+1}{2\pi}\cos^{m_{\text{Lam}}}(\phi) \qquad (7.1)$$

where m_{Lam} is the order of Lambertian optical beam pattern, and ϕ is the irradiance angle relative to the optical axis of the LED optical source. The Lambertian order m_{Lam} could be given by:

$$m_{\text{Lam}} = -\frac{\ln 2}{\ln\left(\cos\phi_{1/2}\right)}, \qquad (7.2)$$

where $\phi_{1/2}$ is the half power angle of the Lambertian beam [15–16]. Specifically, in Figure 7.1, the generalized 3D Lambertian light beam and the respective vehicular VLC application scenario are illustrated for clarity.

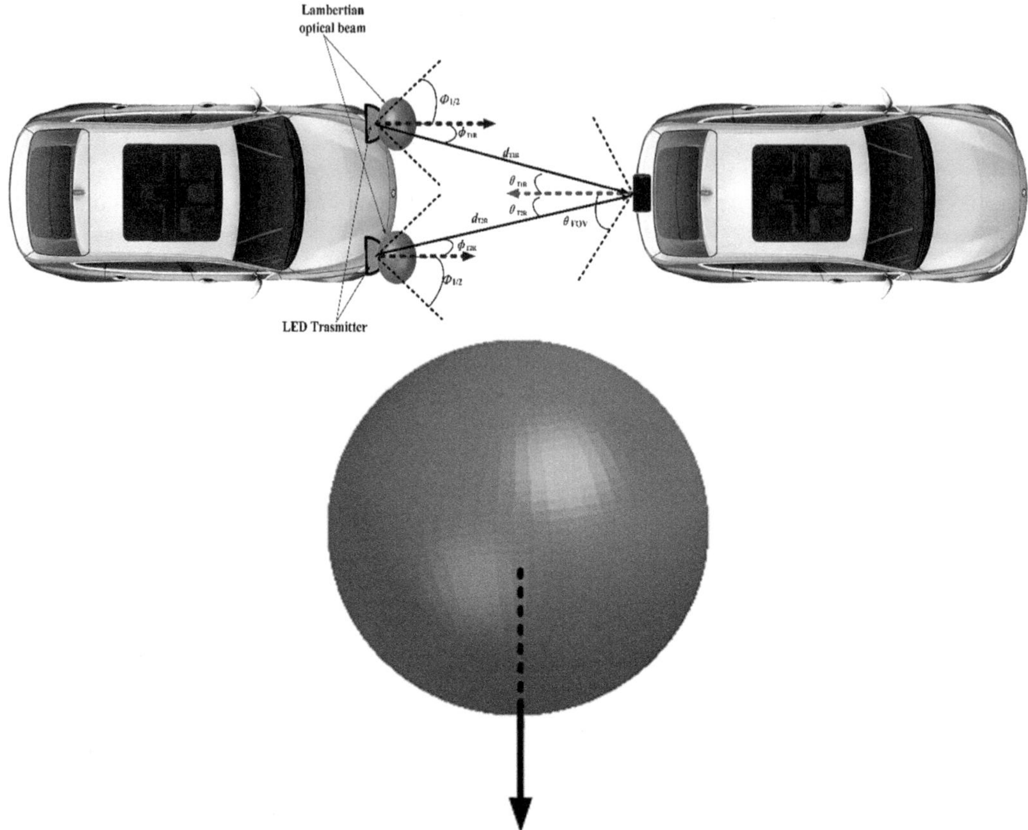

FIGURE 7.1 Schematic of Lambertian optical beam based vehicular visible light communications for 6G IoT network: (a) the respective application scenario, (b) the relevant 3D beam pattern.

7.2.2 LAMBERTIAN VVLC CHANNEL GAIN

For well-known Lambertian beam configuration, the vehicular VLC channel gain includes line of sight (LOS) and non-line of sight (NLOS) components. For simplified analysis, the NLOS components are usually ignored. Therefore, the respective vehicular VLC channel gain between the jth Lambertian optical source S_j^{Lam} and the optical receiver R could be given as [15–16, 21]:

$$H(S_j^{\mathrm{Lam}}, R) = \begin{cases} \dfrac{A_R}{d_j^{\,2}} I_{\mathrm{Lam}}(\phi)\cos(\theta)\dfrac{n^2}{\sin^2(\theta_{FOV})}, & 0 \le \theta_0 \le \theta_{FOV} \\ 0, & \theta_0 > \theta_{FOV} \end{cases} \tag{7.3}$$

where A_R denotes the physical area of photo detector (PD) of the optical receiver, d_j denotes the distance between the jth Lambertian optical source S_j^{Lam} and the PD of optical receiver R, $\dfrac{n^2}{\sin^2(\theta_{FOV})}$ denotes the optical concentrator gain at the receiver with internal refractive index n, θ is the incidence angle from the Lambertian optical source S^{Lam} and the optical receiver R, and θ_{FOV} is the field of view (FOV) at the optical receiver.

7.2.3 LAMBERTIAN VVLC PERFORMANCE METRICS

Following the work of Ding J, for convenience, the signal to noise ratio (SNR) of vehicular VLC could be calculated by [22]:

$$SNR_{\mathrm{Lam}} = \frac{(rI_{\mathrm{TX}})^2}{N_0 N_t^2}\left(\sum_{j=1}^{N_t} H(S_j^{\mathrm{Lam}}, R)\right)^2, \tag{7.4}$$

where r denotes the responsivity of PDs, I_{TX} denotes the emitted power of all optical source transmitters, N_0 denotes the noise power, $N_t=2$ denotes the amount of the optical source transmitter. In typical vehicular VLC environment, the shot noise is one dominant contributor to signal disturbance at the receiver. Then the noise power could be calculated as [22]:

$$N_0 = 2qI_{\mathrm{bg}}B + \frac{4K_b TB}{R_{\mathrm{f}}}, \tag{7.5}$$

where q is the electron charge in coulombs, I_{bg} is the current due to background light, B denotes the system modulation bandwidth, K_b is the Boltzmann constant, T is the absolute temperature and R_{f} is the feedback resistance of the transimpedance amplifier (TIA) [22].

7.3 NON-LAMBERTIAN OPTICAL BEAM-BASED VEHICULAR VISIBLE LIGHT COMMUNICATIONS

7.3.1 NON-LAMBERTIAN OPTICAL BEAM PATTERNS

According to the reported measurement results, the numerous non-Lambertian optical beam patterns have been numerically fitted for LED optical sources. In this work, without loss of generality, two typical non-Lambertian optical beam patterns are concerned in this work. Specifically, these optical beam patterns are modeled from the LUXEON Rebel LED and NSPW345CS LED respectively.

For the LUXEON Rebel non-Lambertian optical beam pattern, the spatial radiation intensity could be profiled by one sum of two Gaussian functions [11–12]:

$$I_{\mathrm{Rebel}}(\phi) = \sum_{i=1}^{N_1} g_{1i}^{\mathrm{Rebel}} \exp\left[-\ln 2\left(\frac{|\phi| - g_{2i}^{\mathrm{Rebel}}}{g_{3i}^{\mathrm{Rebel}}}\right)^2\right], \tag{7.6}$$

where the coefficient values of Gaussian functions are identified as $N_1=2$, $g_{11}^{\mathrm{Rebel}}=0.76$, $g_{21}^{\mathrm{Rebel}}=0°$, $g_{31}^{\mathrm{Rebel}}=29°$, $g_{12}^{\mathrm{Rebel}}=1.10$, $g_{22}^{\mathrm{Rebel}}=45°$, and $g_{32}^{\mathrm{Rebel}}=21°$. Correspondingly, the 3D display of LUXEON Rebel optical beam pattern and the respective vehicular VLC application scenario are shown in Figure 7.2. As for the case of NSPW345CS non-Lambertian optical beam pattern, the spatial radiation intensity should also be profiled by one sum of two Gaussian functions [11–12]:

$$I_{\mathrm{NSPW}}(\phi, \alpha) = \sum_{i=1}^{N_2} g_{1i}^{\mathrm{NSPW}} \exp\left[-(\ln 2)(|\phi| - g_{2i}^{\mathrm{NSPW}})^2\left(\frac{\cos^2\alpha}{(g_{3i}^{\mathrm{NSPW}})^2} + \frac{\sin^2\alpha}{(g_{4i}^{\mathrm{NSPW}})^2}\right)\right], \tag{7.7}$$

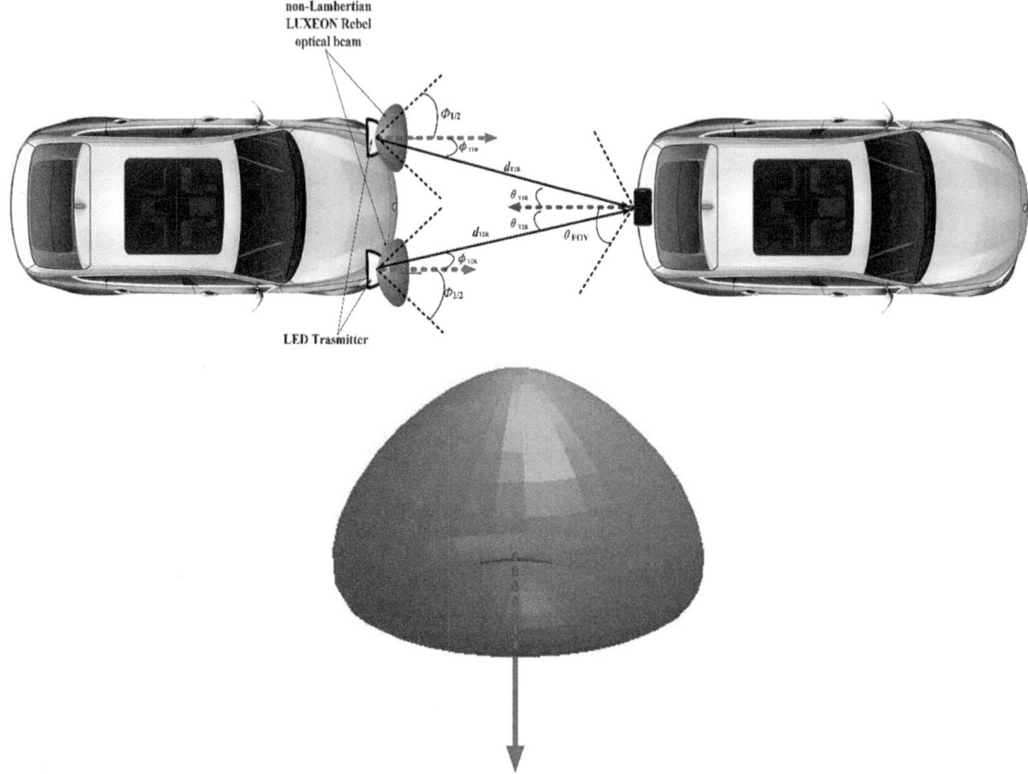

FIGURE 7.2 Schematic of LUXEON Rebel non-Lambertian optical beam based vehicular visible light communications for 6G IoT network: (a) the respective application scenario, (b) the relevant 3D beam pattern.

where α denotes the azimuth angle, the coefficient values of Gaussian functions are given as $N_2 = 2$, $g_{11}^{\text{NSPW}} = 0.13$, $g_{21}^{\text{NSPW}} = 45°$, $g_{31}^{\text{NSPW}} = g_{41}^{\text{NSPW}} = 18°$, $g_{12}^{\text{NSPW}} = 1$, $g_{22}^{\text{NSPW}} = 0$, $g_{32}^{\text{NSPW}} = 38°$, and $g_{42}^{\text{NSPW}} = 22°$. Similarly, the 3D display of NSPW345CS non-Lambertian optical beam pattern is shown in Figure 7.3.

7.3.2 Non-Lambertian VVLC Channel Gain

Apparently, the DC channel gain of non-Lambertian vehicular VLC is tightly relevant to the optical beam radiation pattern. Respectively, for the LUXEON Rebel beam case, the channel gain between the nth optical AP and the mth photo detector of receiver should be given by:

$$H(S_j^{\text{Rebel}}, R) = \begin{cases} \dfrac{A_R I_{\text{Rebel}}(\phi)}{P_{\text{normRebel}} d_j^2} \cos(\theta) \dfrac{n^2}{\sin^2(\theta_{\text{FOV}})}, & 0 \le \theta \le \theta_{\text{FOV}} \\ 0, & \theta > \theta_{\text{FOV}} \end{cases} \tag{7.8}$$

where $P_{\text{normRebel}}$ is the power normalization factor of this LUXEON Rebel non-Lambertian optical beam, to ensure that the whole radiated power equals 1W.

Similarly, for the NSPW345CS optical beam pattern case, the DC channel gain of non-Lambertian vehicular VLC should be replaced by:

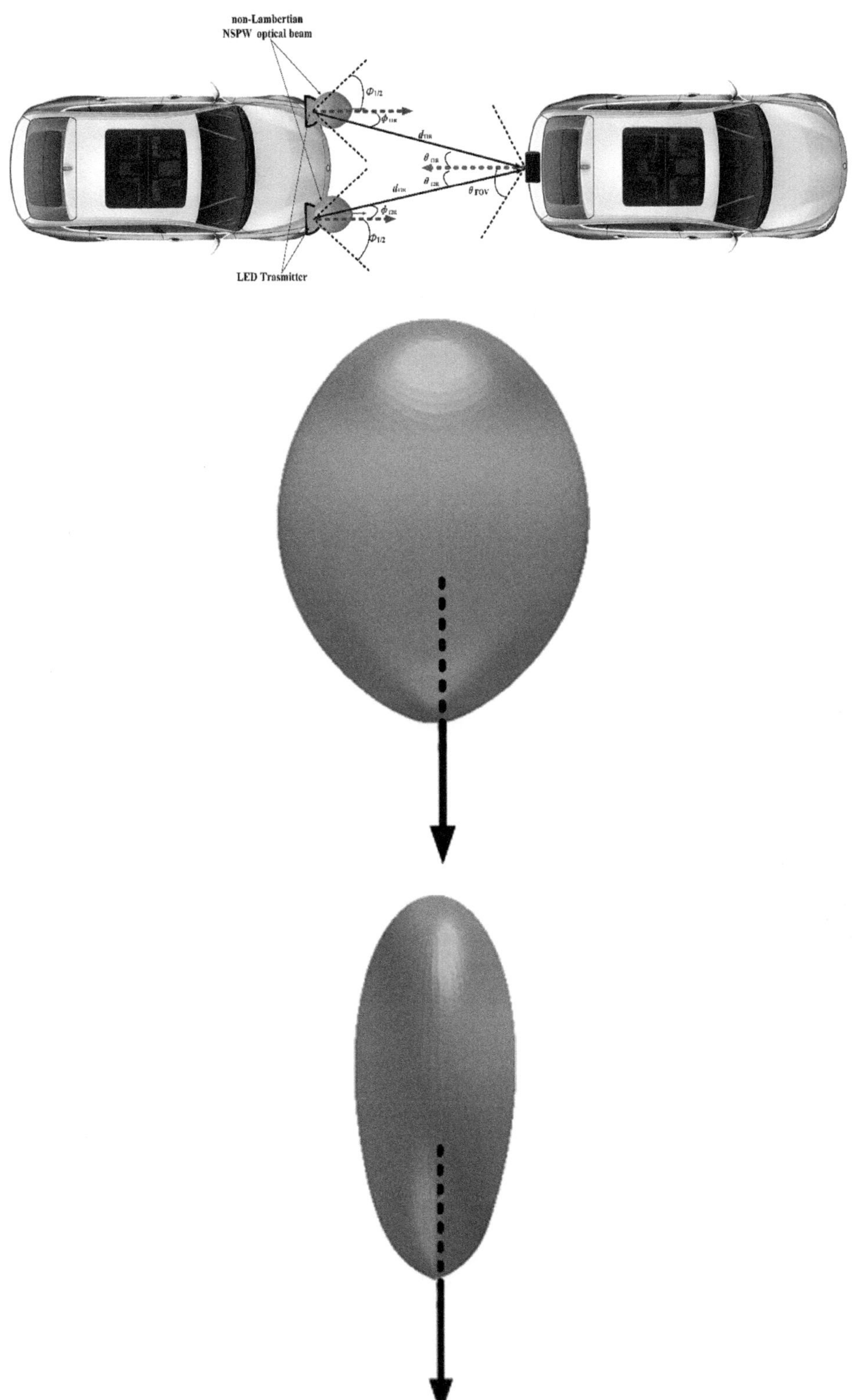

FIGURE 7.3 Schematic of asymmetric NSPW non-Lambertian optical beam based vehicular visible light communications for 6G IoT network: (a) the respective application scenario, (b) the relevant 3D beam pattern.

$$H(S_{j}^{\text{NSPW}}, R) = \begin{cases} \dfrac{A_R I_{\text{NSPW}}(\phi, \alpha)}{P_{\text{normNSPW}} d_j^{2}} \cos \theta_0 \dfrac{n^2}{\sin^2\left(\theta_{FOV}\right)} r, 0 \leq \theta_0 \leq \theta_{FOV} \\ \\ 0, \qquad\qquad\qquad\qquad\qquad\quad \theta_0 > \theta_{FOV} \end{cases}, \qquad (7.9)$$

where P_{normNSPW} is the power normalization factor of this NSPW345CS non-Lambertian optical beam, to ensure that the whole radiated power equal 1W.

7.3.3 NON-LAMBERTIAN VVLC PERFORMANCE METRICS

In the discussed LUXEON Rebel beam configuration, the vehicular VLC links are constructed between the two non-Lambertian optical sources with LUXEON Rebel beam pattern and the optical receiver located at the center of the rear of the vehicle ahead. In this situation, the related SNR at the VLC optical receiver could be given by:

$$SNR_{\text{Rebel}} = \frac{(rI_{\text{TX}})^2}{N_0 N_t^2} \left(\sum_{j=1}^{N_t} H(S_{j}^{\text{Rebel}}, R) \right)^2, \qquad (7.10)$$

Similarly, for the NSPW345CS optical beam configuration, the related SNR at the VLC optical receiver can be given by:

$$SNR_{\text{NSPW}} = \frac{(rI_{\text{TX}})^2}{N_0 N_t^2} \left(\sum_{j=1}^{N_t} H(S_{j}^{\text{NSPW}}, R) \right)^2, \qquad (7.11)$$

7.4 NUMERICAL RESULTS AND DISCUSSIONS

In the following section, for the typical vehicular VLC scenario, the SNR spatial distribution of the discussed Lambertian and two non-Lambertian vehicular VLC coverages are evaluated and compared. And for a fair comparison, all concerned light beam intensities are normalized within the following work. Furthermore, Table 7.1 summarizes the main parameters for this work [21–22].

As shown in Figure 7.4 (a), for conventional Lambertian optical beam configuration, the achievable SNR spatial dynamic range of this vehicular VLC coverage is about -17.62–75.33 dB while the average SNR of this conventional configuration is about 17.10 dB.

On the other side, for the LUXEON Rebel non-Lambertian optical beam configuration, the achievable SNR spatial dynamic range of this vehicular VLC coverage is degraded to about -18.47-73.89 dB while the average SNR of this non-Lambertian configuration is reduced to about 14.17 dB, as shown in Figure 7.4 (b). Compared to the above Lambertian vehicular VLC configuration, up to 2.93 dB average SNR loss is induced by this LUXEON Rebel vehicular VLC configuration.

Unlike the above LUXEON Rebel configuration, the vehicular VLC SNR spatial distribution of asymmetric NSPW non-Lambertian optical beam configuration renders better uniformity, as shown in Figure 7.4 (c): such that, the respective achievable SNR spatial dynamic range is changed to about -34.16-78.20 dB, while the average SNR of this asymmetric NSPW non-Lambertian configuration is increased to 21.10 dB. Correspondingly, the achieved average SNR gain is about 4.0 dB compared to benchmark Lambertian vehicular VLC configuration. Moreover, the respective maximum receiver SNR is 78.20 dB while up to 2.87 dB maximum vehicular VLC coverage SNR gain could be derived by this NSPW non-Lambertian optical beam configuration.

TABLE 7.1
Main Parameters Configuration

Parameter	Value
Emitted power of transmitter	1 W
LED optical sources spacing	1.20 m
Background light current	5100 μA
Absolute temperature	298 K
LED Lambertian index	1
Receiver field of view	90°
Physical area of PD	1 cm^2
Responsively of PD	0.53 A/W
Concentrator refractive index	1.54
Optical filter gain	1
LED Modulation bandwidth	20 MHz
Channel	LOS
Modulation	OOK-NRZ

For comparing the statistical characteristics of SNR distributions under the concerned three vehicular VLC optical beam configurations, the SNR cumulative distribution function (CDF) curves are described in Figure 7.5. For up to 80% receiver positions under Lambertian vehicular VLC configuration, the SNR is more than 8.17 dB while the counterpart of the LUXEON Rebel non-Lambertian configuration and the NSPW345CS non-Lambertian configuration is changed to 4.11 dB and 13.73 dB, respectively. Therefore, the proposed NSPW345CS non-Lambertian vehicular VLC configuration is capable of impressively improving the weak coverage domain of the typical vehicular VLC application scenario.

7.5 CONCLUSIONS AND FUTURE SCOPE

This work is motivated by the research limitation of the conventional Lambertian vehicular VLC technique schemes, which are unable to adapt to the diverse non-Lambertian vehicular VLC scenarios. In this work, the typical NSPW non-Lambertian LED transmitter is utilized to configure the vehicular VLC links. For the typical outdoor lane scenario, the SNR of 80% receiver positions is up to 13.73 dB, while the counterpart of the benchmark Lambertian vehicular VLC link is just 8.17 dB. Accordingly, this average SNR will be enhanced to 21.10, from the about 17.10 of the benchmark vehicular VLC link configuration. In the near future, the heterogeneous optical beam combination and cooperation methods will be proposed and investigated in the outdoor vehicular VLC domain for realizing performance benefits accordingly. In the near future work for 6G and beyond network, the exploration of diverse non-Lambertian vehicular VLC could be further extended to the customized vehicular beam forming, vehicular beam cooperation, vehicular dynamic beam configuration, vehicular beam switching, vehicular resource allocation, vehicular reconfigurable multiple-input multiple-output and other key techniques design.

ACKNOWLEDGEMENTS

This work was supported in part by the National Natural Science Foundation of China (Grants No. 62061043), Tianshan Cedar Project of Xinjiang Uygur Autonomous Region (Grants No. 2020XS27) and High-level Talents Introduction Project in Autonomous Region (Grants No. 042419004).

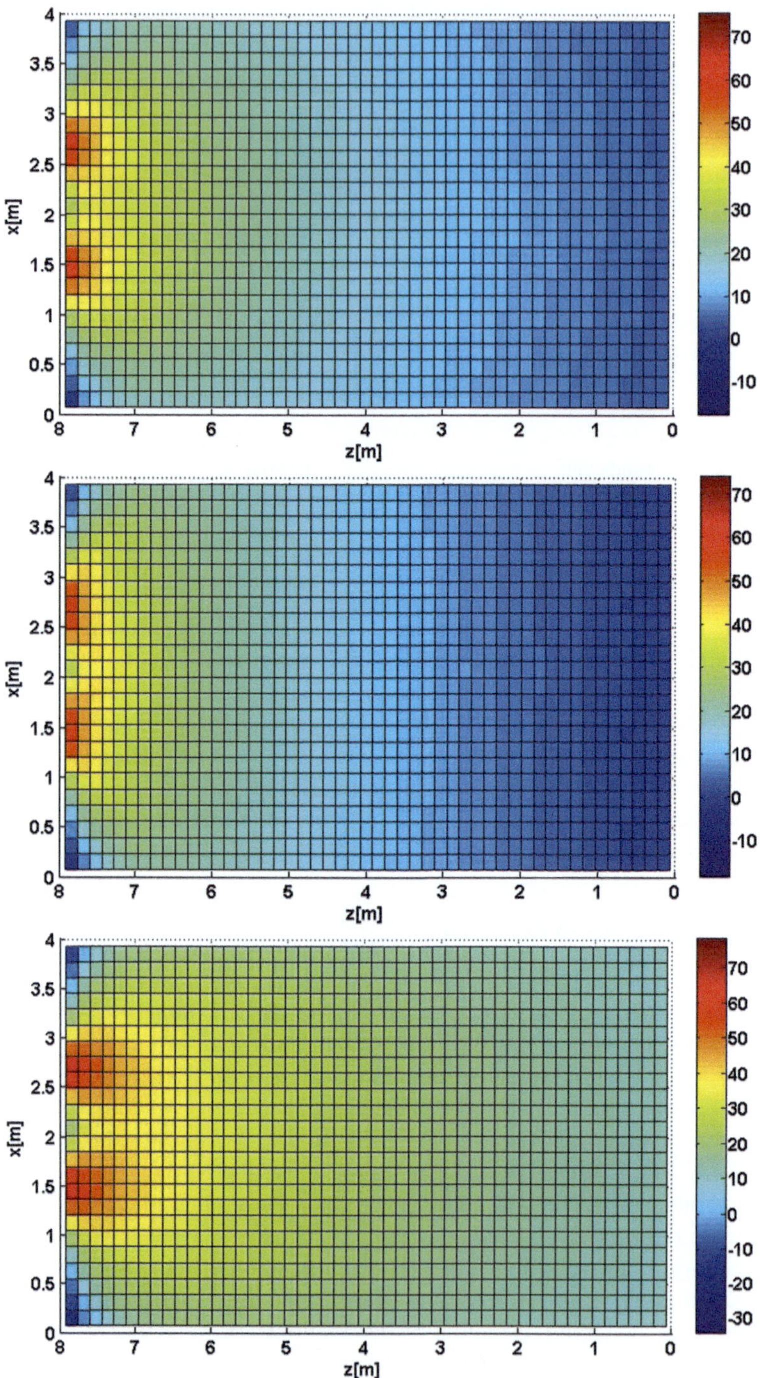

FIGURE 7.4 SNR spatial distribution of vehicular visible light communications for 6G IoT network in the case of (a) Lambertian optical beam configuration, (b) LUXEON Rebel non-Lambertian optical beam configuration and (c) asymmetric NSPW non-Lambertian optical beam configuration.

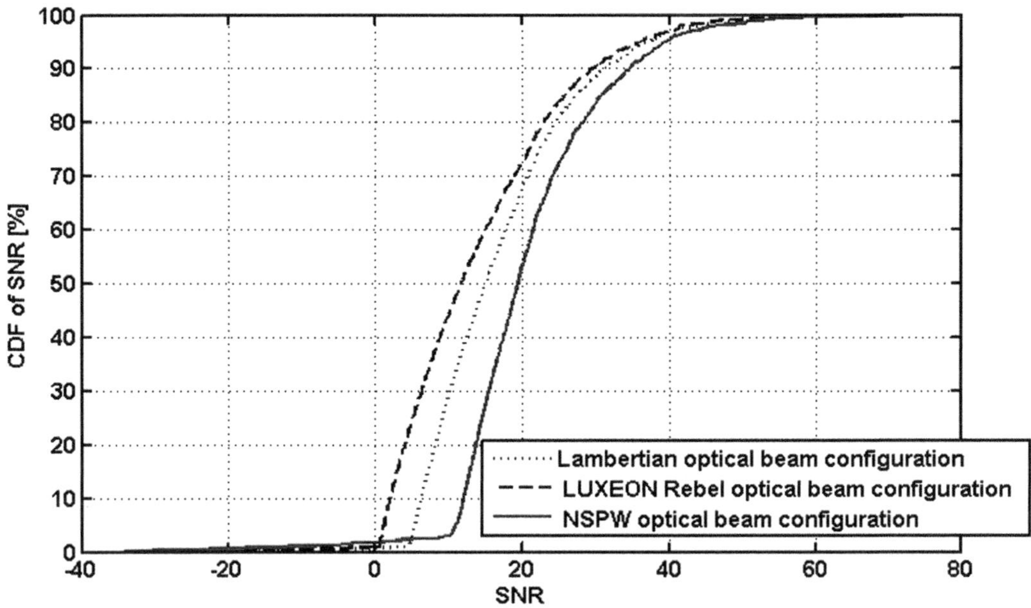

FIGURE 7.5 CDF curves comparison of SNR spatial distribution for three distinct optical beam configurations.

REFERENCES

[1] H. Mei, J. Ding, J. Zheng, X. Chen and W. Liu, "Overview of vehicle optical wireless communications," *IEEE Access*, vol. 8, pp. 173461–173480, 2020. doi: 10.1109/ACCESS.2020.3024854

[2] W. Liu, J. Ding, J. Zheng, X. Chen and C. -L. I, "Relay-assisted technology in optical wireless communications: A survey," *IEEE Access*, vol. 8, pp. 194384–194409, 2020. doi: 10.1109/ACCESS.2020.3031288

[3] S. U. Rehman, S. Ullah, P. Chong, S. Yongchareon and D. Komosny, "Visible light communication: A system perspective—overview and challenges," *Sensors*, vol. 19, no. 5, pp. 1–22, 2018.

[4] L. E. M. Matheus, A. B. Vieira, L. F. M. Vieira, M. A. M. Vieira and O. Gnawali. "Visible light communication: Concepts, applications and challenges," *IEEE Communications Surveys & Tutorials*, vol. 21, no. 4, pp. 3204–3237, Fourthquarter 2019.

[5] M. Z. Chowdhury, M. K. Hasan, M. Shahjalal, M. T. Hossan and Y. M. Jang, "Optical wireless hybrid networks: Trends, opportunities, challenges, and research directions," *IEEE Communications Surveys & Tutorials*, vol. 22, no. 2, pp. 930–966, Secondquarter 2020.

[6] Lee, I. E., M. L. Sim and F. W. L. Kung . "Performance enhancement of outdoor visible-light communication system using selective combining receiver." *Optoelectronics Iet*, vol. 3, no. 1, pp. 30–39, 2009.

[7] S. Refas, D. Acheli, S. Yahia and Y. Meraihi, "Energy harvesting based on SLIPT in I2V-VLC system," 2023 International Conference on Advances in Electronics, Control and Communication Systems (ICAECCS), BLIDA, Algeria, 2023, pp. 1–5. doi: 10.1109/ICAECCS56710.2023.10105093

[8] T. Pal, G. Singh, V. A. Bohara and A. Srivastava, "Optical IRS aided B5G V2V solution for road safety applications," *2022 IEEE International Conference on Advanced Networks and Telecommunications Systems (ANTS)*, Gandhinagar, Gujarat, India, 2022, pp. 169–174, doi: 10.1109/ANTS56424.2022.10227754.

[9] V. I. Kurnia, N. S. Tiyas, B. Pamukti and K. Sujatmoko, "Interference analysis for vehicle to vehicle communication using visible light communication," *2021 International Conference on Artificial Intelligence and Mechatronics Systems (AIMS)*, Bandung, Indonesia, 2021, pp. 1–5.

[10] S. Wang, K. Wang, F. Chen and S. Liu, "Design of primary optics for LED chip array in road lighting application," *Optics Express*, vol. 19, no. S4, pp. A716–A724, 2011.

[11] I. Moreno and C. Sun, "Modeling the radiation pattern of LEDs," *Optics Express*, vol. 16, no. 3, pp. 1808–1819, 2008.

[12] J. Ding and C. I, Z. Xu, "Indoor optical wireless channel characteristics with distinct source radiation patterns," *IEEE Photonics Journal*, vol. 8, no. 1, pp. 1–15, 2015.

[13] J. Ding, C. I, R. Xie, H. Lai and C. Zhang, "Actual radiation patterns-oriented non-deterministic optical wireless channel characterization," *Chinese Conference on Biometric Recognition*. Springer, Cham, 2018, pp. 517–527.

[14] J. Ding, C. L. I, J. Wang, H. Yang and L. Wang, "Secure optical wireless links with dynamic beam and diversity receiver," *2021 7th IEEE International Conference on Network Intelligence and Digital Content (IC-NIDC)*, Beijing, China, 2021, pp. 205–209. doi: 10.1109/IC-NIDC54101.2021.9660496

[15] J. Ding, C. -L. I, J. Wang, H. Yang and L. Wang, "Research on coordinated coverage of non-lambertian optical wireless communications," *2021 7th International Conference on Computer and Communications (ICCC)*, Chengdu, China, 2021, pp. 387–392. doi: 10.1109/ICCC54389.2021.9674276

[16] J. Ding, C. -L. I, J. Wang, H. Yang and L. Wang, "Multiple optical beam switching for physical layer security of visible light communications," *IEEE Photonics Journal*, vol. 14, no. 1, pp. 1–9, Feb. 2022, Art no. 7306209. doi: 10.1109/JPHOT.2021.3136233

[17] J. Ding, C. -L. I, J. Wang and H. Yang, "Optical beams switching-based coverage enhancement scheme for compact visible light communications," *Journal of Lightwave Technology*, vol. 40, no. 18, pp. 6139–6150, 15 Sept.15, 2022. doi: 10.1109/JLT.2022.3187785

[18] J. Ding and C. I, H. Zhang, "Cells planning of VLC networks using non-circular symmetric optical beam," *2019 IEEE International Conference on Communications (ICC)*, pp. 1–6, 2019.

[19] J. Ding, C. I, C. Zhang, B. Yu and H. Lai, "Evaluation of outdoor visible light communications links using actual LED street luminaries," *Chinese Conference on Biometric Recognition*. Springer, Cham, 2018, pp. 572–579.

[20] J. Ding, C. I, X. Chen and H. Lai, "Asymmetrical emission beams based visible light communication access points design," *2019 28th Wireless and Optical Communications Conference (WOCC)*, 2019, pp. 1–5.

[21] T. Komine and M. Nakagawa, "Fundamental analysis for visible-light communication system using LED lights," *IEEE Transactions on Consumer Electronics*, vol. 50, no. 1, pp. 100–107, 2004.

[22] Ding J, I C-L, Wang J, Yang H and Wang L. Performance comparison of repetition coding MIMO optical wireless communications with distinct light beams, *Sensors*, vol. 22, no. 3, pp. 1256, 2022. https://doi.org/10.3390/s22031256

8 Application of Computational Modeling in Electronics Devices

Solomon C. Nwaneri, Ibukun D. Adewale and Hannah C. Ugo

8.1 INTRODUCTION

Computational modeling is the scientific approach of using computers to simulate complex systems. Computational models enhance the understanding of the behaviour of various scientific and engineering systems. There are diverse applications of computational modelling, such as electronic device modeling, simulations in bioelectronics, assembly of electronic components on printed circuit boards, medical biotechnology and biological systems, geological studies, agricultural and resource economics [1–5]. The application of computational modeling in electronics devices, is our primary focus in this chapter.

The consumer electronics industry is arguably one of the largest technology industries globally with a market size of over 700 billion USD [6]. Electronic devices are generally developed from semiconductor materials such as silicon, germanium, selenium and gallium arsenide. In recent times, there has been increased use of graphene in the manufacture of electronic devices owing to its exceptional chemical and physical properties. Graphene is a 2-D, crystalline allotrope of carbon with overlapping valence and conduction band [7]. Based on its numerous properties, graphene has a wide range of electronic applications and biological engineering [8]. There are increasing requirements for the manufacture of extremely miniaturized electronic devices. Precision electronic devices are often useful in building small, flexible and user-friendly consumer electronics. There are various techniques for the fabrication of precision electronic devices. The application of a number of processes such as photolithography, has led to significant progress in high-precision manufacturing of miniaturized integrated circuits [9]. With recent innovations in the manufacture of flexible, thin and lightweight electronic components, the creation of more compact and portable consumer electronics is now common. The manufacture of these precision electronic devices cannot be achieved without computational modelling.

A computational model contains numerous variables that characterize the system being studied. Simulation is done by adjusting one or more variables while observing the outcomes. The use of computational modeling enhanced the conduct of several simulated experiments. This is also crucial to optimization of the performance of electronic devices [7]. Computational models can be classified into mechanistic and data-driven models. Mechanistic are dynamic, simulating changes in quantities of interest over time, and are generally represented by systems of non-linear ordinary or partial differential equations [10]. Data-driven models rely on experimental data for their computations. Such data-driven models are characterized by errors and uncertainty [11].

There are two basic types of computational mechanics modelling, namely: Computational Fluid Dynamics (CFD) and Computational Solid Mechanics (CSM) [1]. With Computational Fluid Dynamics, digital computers can be supported to perform numerical simulation of fluid flows based

DOI: 10.1201/9781003457428-9

on energy conservation, Newton's second law, and mass conservation [12, 13]. Computational Fluid Dynamics computations usually do not yield completely exact solutions despite recent improvements in its accuracy of prediction [14]. Also, Computational Fluid Dynamics tools can be used to model the thermo-fluidic behavior of a single electronic component that is mounted [15]. The modelling and simulation of electronic devices and circuits are carried out using several types of software such as CircuitMaker, CircuitLogix, Proteus, PSpice, Simulink, Simscape, and so forth.

8.1.1 CONTRIBUTIONS OF THE CHAPTER

The major contributions of this chapter are:

1. A detailed discussion of computational modeling and simulation of electronic devices.
2. Practical design, computation and simulation of electronic devices using MATLAB and LTSPICE software.
3. The chapter provides an understanding of the behaviour of Silicon Carbide MOSFETs under various conditions.
4. The chapter discusses various applications of wearable electronics and perform computational modeling and simulation of a wearable sensor.

8.1.2 CHAPTER ORGANIZATION

Section 8.2 describes recent developments in electronic systems modeling. In section 8.3, a review of related works on the applications of computational modeling and simulation in medical devices was done. Section 8.4 presents computational modeling and simulation of electronic devices. It includes the modeling and simulation of simple level 1 MOSFET circuits, silicon carbide MOSFETs and a wearable pulse sensor. While Section 8.5 and 8.6 presents the results for all the models and the discussion respectively, Section 8.7 summarizes the entire chapter.

8.2 ADVANCES IN OBJECT TRACKING SYSTEMS, WIRELESS MONITORING AND WEARABLE ELECTRONICS

8.2.1 ELECTRONICS SENSING AND OBJECT TRACKING SYSTEM

Advancements in electronics sensing and object tracking systems have enhanced the detection and real-life tracking of moving objects. Object tracking involves the monitoring in real-time of the location of an object. Electronics sensing and object-tracking systems have wide applications in visual surveillance systems such as the use of unmanned aerial vehicles for security purposes as it is tedious and expensive to use human operators [16].

The detection and tracking of moving objects in various forms of visual media is an essential aspect of diverse applications [17]. Machine learning has the ability to train existing data and automatically process and analyze unknown data. The use of machine learning in object tracking systems is capable of reducing the need for human intervention [18].

8.2.2 MODELING OF ELECTRONICS SYSTEMS FOR WIRELESS MONITORING

Electronics systems can help to monitor body parameters and diverse disease conditions in patients wirelessly. The use of cables to link sensors to external instruments can restrict its performance. Wireless monitoring is on the increase due to its ability to eliminate the usual discomfort caused by the use of wires. Wireless biological sensors have quite a number of benefits including their wearability, real-time and remote sensing [19].

In order to remotely monitor body parameters, sensing electrodes are often employed. For example, in [20], a miniaturized strain sensor was integrated with a wireless stent to monitor restenosis wirelessly. In some instances, the biosensors are integrated with antennas in order to share data wirelessly [21]. In recent times, these biosensors are not powered with batteries [22].

8.2.3 WEARABLE ELECTRONICS SYSTEMS MODELING AND SIMULATION

Wearable electronics systems are becoming the most integral part of human life. They offer robust functionality in life activities ranging from health monitoring to communication and entertainment. The use of wearable electronics involves the integration of sensors, data processing, storage and transmission circuitry [23]. With increased applications of wearable electronics in healthcare, concerted efforts should be channeled towards having wearable electronics with precision accuracy.

Wearable electronics systems are complex devices, comprised of both hardware and software components working together to perform diverse functions [24]. Sensors, that sense and monitor physiological parameters of the wearers are the key components of wearable electronics [25]. In addition to sensors, wearable electronics also include microcontrollers and low power batteries. Appropriate positioning of the sensors to ensure proper contact with the wearer is crucial to getting accurate measurements. They can be attached to clothing, mounted on the skin, worn directly on the body or implanted into the body [26]. The attachment of sensors to clothing is referred as electronic textiles (e-textiles). Smartwatches and health and fitness trackers are good examples of commercially available wearable electronics generally worn by consumers. In recent times, tattoo-based wearables have gained research interests as well as bio-fluidic based wearables [27, 28]. However, wearable electronics ought to be skin friendly and should not cause damage to the skin.

Wearable electronics system modeling and simulation help to reveal the possible performance of wearable sensors in real-life situations. Before performing modeling and simulation, it is imperative to identify relevant input parameters used to design the wearable electronics system. Models for each of the components in the wearable electronics system such as sensors, microcontrollers, and other accessories need to be created before simulating the system. Simulation is mostly performed using a simulation software. The MATLAB IoT Toolbox is one of the most common tools used by researchers for the modeling, simulation and analysis of wireless electronics systems [29].

8.3 RELATED WORKS

Previous studies related to the application of computational modeling in electronics devices were reviewed in this section. Also, studies that modeled both silicon MOSFETs and silicon carbide MOSFETs were also reviewed.

Zhao et al. [30] proposed a semi-analytical model utilizing the effects of edge bond relaxation in graphene nanoribbon field-effect transistors. They emphasized the need for an accurate calculation of the band structure, which is a necessary input for the model. Also, [31] used CircuitMaker for modeling and simulation of logic circuits. The application of the software in building logic elements for the synthesis of logic circuits as well as its use in teaching electronics was demonstrated.

One of the common approaches for modeling electronic devices is using artificial intelligence techniques, such as neural networks and fuzzy logic systems. These techniques can approximate complex non-linear relationships between input and output variables, offering advantages in speed, accuracy, and flexibility. In another study [32], the authors reviewed various electronic devices modeled using an adaptive neuro-fuzzy inference system (ANFIS), such as diodes, transistors, and thyristors. They showed that neuro-fuzzy inference system models can enhance the simulation speed, reliability, and time complexity of devices, and can be implemented in hardware using field programmable gate array (FPGA) technology.

In [33], the author proposed the modeling and hardware implementation of semiconductor circuit elements using artificial neural networks (ANN) and FPGA technology, focusing on MOSFETs. The author demonstrated that ANN models can optimize the architecture and parameters of semiconductor components and can be converted into very high-speed integrated circuit hardware description language (VHDL) for FPGA implementation. In [34], the authors developed a new method for modeling electronic devices using fuzzy logic systems, focusing on approximating non-linear curves. The authors presented a modeling principle whereby fuzzy logic patches approximate a three-dimensional surface to simplify complex device behaviors depending on two input variables and one output variable. They validated their fuzzy models for a rectifier diode and a bipolar transistor against SPICE simulations, emphasizing the fuzzy model's computational speed and ability to compute outputs for various input values in a single simulation step.

Belhamel et al. [35] introduced an innovative model-based design approach to create high-performance device driver models, particularly in the context of smart cities. The authors introduced an innovative approach to accurately model the operations of the STM32 (STMicroelectronics) analog converter peripheral using Simulink, enabling simulation and code generation for Low-Level STM32 firmware drivers. The study emphasized how model-based design can streamline the development of advanced smart applications by modeling various subsystems, including microcontrollers and peripherals, emphasizing the efficiency and high-quality product development achieved through this method.

Also, in [36], the authors investigated the simulation of double-gate MOSFETs using computational intelligence models, such as artificial neural networks and adaptive neuro fuzzy inference system (ANFIS). The study aimed to address the challenge of accurately and efficiently modeling the physics of quantum transport in nanoscale devices when traditional models become less effective due to quantum mechanical effects. The authors utilized 2-D numerical Non-Equilibrium Green's Function simulations of the DG MOSFET and employed ANN and ANFIS to relate input parameters to the output current. They demonstrated that both Artificial Neural Network and adaptive neuro fuzzy inference system models exhibit good agreement with the Non-Equilibrium Green's Function simulation.

Another approach for modeling electronic devices is using physical theories and numerical methods, such as drift-diffusion models, quantum transport models, Monte Carlo simulations, hydrodynamic models, multi-grid methods, and finite element methods. These methods can capture the detailed physics and mechanisms of device operation, fabrication, and design, facilitating early-stage analysis and development. Wachutka [37] studied the growing importance of problem-specific modeling and simulation tools in the advancement of power devices and microsystems. The author explained that these tools offer insights into fabrication processes, device operation, and design variants, facilitating early-stage analysis and development. The study addressed the need for modeling methodologies that consider various energy domains and ensure reliable validation and calibration of models for predictive simulations, especially for experiments near the safe operating limits.

Ray et al. [38] developed a Finite Element Method based device simulator tailored for high-voltage devices using silicon carbide technology. They emphasized the necessity for effective simulators in evolving post-CMOS electronics design and the implications of silicon carbide technology's properties for high-power devices. The proposed simulator was incorporated with drift-diffusion models, material parameters, and meshing mechanisms to analyze a silicon carbide-based PN junction diode. This simulator was validated against the TCAD Sentaurus tool, showcasing comparable results. Birajdar et al. [39] addressed the significance of numerical modeling and computer simulations in various scientific and engineering fields, highlighting the shift towards cost-effective and efficient computational modeling in solving complex problems across disciplines.

Furthermore, the authors in [40] investigated and compared methods for studying serial resonance in linear electrical circuits using both classic theoretical approaches and specialized software

(CircuitMaker). The research aimed to address the comparative analysis of these two methods, offering insights into their advantages and disadvantages. The authors in [41] provided a comprehensive exploration of computational electronics and semiconductor device modeling. The study covered various aspects of modeling, including MOSFET models, charge-based transistor models, mobility modeling, recombination statistics, and large-signal analysis. The authors carried out a literature review and simulation techniques to offer insights into semiconductor device physics and device modeling. They integrated extensive references and research contributions, focusing on quantum transport, Monte Carlo simulations, hydrodynamic models, and multi-grid methods.

A third approach for modeling electronic devices is using compact circuit models derived from detailed physical simulations or experimental data. These models can balance model complexity, accuracy, and generality, enabling efficient circuit simulation and design optimization. Also, the study in [42] reviewed a method for creating compact circuit models from detailed physical simulations of various devices. The method was based on the calibration of a simplified equivalent-circuit model on simulation results and could achieve an optimum balance of model complexity, accuracy, and generality. The study outlined a structured approach encompassing the modeling of traditional devices, optoelectronic systems, thermal effects, hot-carrier impacts, and quantum-mechanical phenomena.

A fourth approach for modeling electronic devices is using novel materials and structures, such as graphene, memristors, and single-photon avalanche diodes. These devices offer unique properties and capabilities for high-speed, low-power, and neuromorphic applications. However, they also pose new challenges and opportunities for modeling and simulation. Vargas-Bernal [7] discussed the potential applications of graphene in electronic devices, highlighting its exceptional properties for high-speed transistors. The study emphasized the importance of mathematical modeling to predict the electrical behavior of graphene-based devices and address defects. It also explained the use of graphene in radiofrequency electronics and combining it with other materials for new structures. The author concluded by outlining the need for modeling and research to harness graphene's potential in the field of electronics.

Zhao et al. [30] proposed a semi-analytical model utilizing the effects of edge bond relaxation in graphene nanoribbon field-effect transistors. They emphasized the need for an accurate calculation of the band structure, which is a necessary input for the model. Kalpachka [31] used CircuitMaker for the modeling and simulation of logic circuits. The application of the software in building logic elements and synthesis logic circuits as well as its use in teaching electronics was demonstrated. Also, in [43], three key semiconductor devices were modelled (Silicon-on-Insulator Four-Gate Transistor (G4 FET), Perimeter Gated Single-Photon Avalanche Diode (PG-SPAD), and Insulator-Metal Transistor with volatile memristance) with critical review of the functionalities, applications, and simulation of each device. The G4 FET's unique design and potential uses, the Single Photon Avalanche Diode's significance in single-photon detection, and Insulator-Metal Transistor's role in neuromorphic computing were highlighted. These devices have been rigorously simulated and integrated into commercial circuit simulators, showcasing their novel capabilities in various applications. The study underscores the importance of accurate compact models for these emerging devices, providing a comprehensive reference for further research in the semiconductor field. In another study [44], the modeling of silicon carbide MOSFET with particular emphasis on internal stray capacitances and inductances a correlation between the transient behavior of gate drive circuit and main power circuit were investigated.

Shamsir et al. [45] discussed semiconductor device modeling and simulation for electronic circuit design. The study covered various aspects of modeling, including MOSFET models, charge-based transistor models, mobility modeling, recombination statistics, and large-signal analysis. The authors delved into the historical evolution of device modeling methodologies, from classical models to modern multi-gate transistors and memristors. They explored the development of SPICE, the significance of compact models for circuit simulation, and advancements in quantum-level circuit/device modeling. Also, [46] investigated the effect of the threshold-voltage shift due to hysteresis effect on the switching performance of SiC MOSFETs. They also proposed a method to measure the threshold voltage shift caused by the hysteresis effect.

8.4 METHODOLOGY

This section describes the methodology used to implement this study. The approach adopted was the modeling and simulation of the device in modeling and simulation software. The two main software used in this study are MATLAB and LTSPICE electronic design automation application software. Three models were developed. The models are: (1) Level 1 MOSFET model (2) Silicon carbide MOSFET circuit.

8.4.1 COMPUTATIONAL MODELING OF METAL OXIDE FIELD EFFECT TRANSISTORS

Metal Oxide Field Effect Transistors are unipolar transistors known for their switching capabilities. MOSFETs have four main terminals: Source, drain, gate and body. CMOS integrated circuits are principally designed with MOSFETs. The nonlinear behaviour of MOSFETS in electronic circuits can be predicted by computational modeling [47].

In this section, a level 1 MOSFET model was implemented in MATLAB to enable us to predict the behavior of the drain source current of the MOSFET for various gate source voltage (V_{GS}) values. The model was developed based on the following parameters:

Threshold voltage, Vth;
Transconductance parameter, K = 0.01;
Output conductance parameter, λ;
Gate Source voltage, V_{GS};

The Gate Source voltage, V_{GS} was swept from 0 to 5 volts.
The drain source current for level 1 MOSFET model is given by:

$$I_{ds} = K\left(V_{GS} - V_{th}\right)^2 - \lambda\left(V_{GS}\right) \tag{8.1}$$

8.4.2 COMPUTATIONAL MODELING OF SILICON CARBIDE METAL OXIDE FIELD EFFECT TRANSISTORS

In recent times, the modeling of Silicon Carbide MOSFETs has received significant research interests because of their good electrical and thermal properties [44, 48, 49]. Silicon Carbide MOSFETs are known to have wide band gaps [50] and are well preferred for these properties in addition to their higher switching speeds and lower parasitic capacitances. However, one common feature of silicon carbide MOSFETs is the occurrence of hysteresis in the floating terminals of the device [51]. Threshold voltage hysteresis impairs the functionality of silicon carbide MOSFETs and could affect their reliability [52]. Also, the dynamic characteristics of the threshold voltage is affected more by the gate resistance and source inductance as compared to the gate inductance and drain inductance [53]. Hauck et al. [54] analyzed the performance of 4H-SiC n-MOSFETs of any geometry by simulating the drain current of the MOSFET.

In this chapter, modeling cond simulation are presented for NMOS SiC MOSFET similar to the model developed in [55]. The semiconductor device model, whose schematic is shown Figure 8.3, is an N-type SiC MOSFET that derives its parameters from the SCT20N120. This was developed in the LTSPICE electronic design automation application software making use of STMicroelectronics library to simulate the non-linear I-V characteristics of the MOSFET. Here, the Figure 8.1 shows that the drain-to-source voltage V_{ds} was made to vary between 0 and 80 volts in steps of 0.01 volts

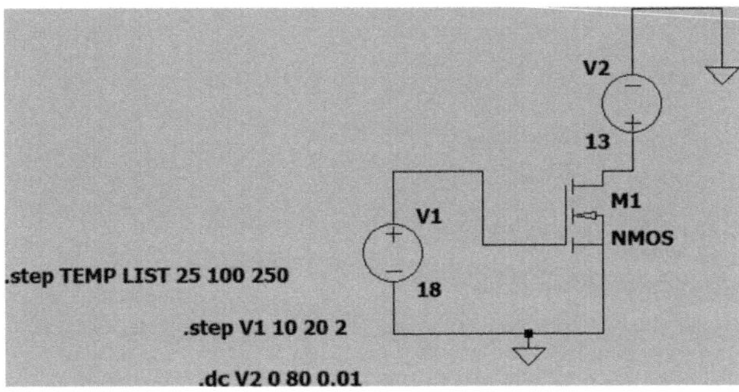

FIGURE 8.1 SiC NMOS MOSFET Circuit.

and a swept voltage of 10–20 volts in incremental steps of 2 volts was supplied to the gate. In addition, to explore the temperature dependency of the SiC MOSFET, the simulation was done at 25, 100 and 250 degrees Celsius respectively.

8.4.3 Computational Modeling of a Wearable Pulse Sensor

A computational model was performed in MATLAB for a simple wearable pulse sensor. The main function of a wearable pulse sensor is to measure the heart rate of its wearer. The pulse sensor produces a signal which corresponds to the pulsatile blood flow. A simple pulse signal was generated in the form of a sine wave. The basic signal parameters include time span, set between 0 and 60 seconds, and a time step set to 0.1. The key variables initialized include time, heart rate and pulse signal.

A sine wave to produce the pulse signal was generated based on equation 8.2:

$$\text{Pulse signal} = \sin(2\pi)\frac{\text{heart rate}}{60} \times \text{time} \tag{8.2}$$

The model was implemented for three different scenarios; normal pulse rate, bradycardia and tachycardia. In normal heat rate, the heart rate was fixed at 60 beats per minute. For the bradycardia scenario, the heart rate was fixed at 45 beats per minute. While for the tachycardia scenario, the heart rate was fixed for 150 beats per minute.

8.5 RESULTS

8.5.1 Results of Modeling and Simulation of Level 1 MOSFET

The results for the computational modeling of a level 1 MOSFET at various threshold voltage 0.5V, 1.5V and 1.5V are shown in Figure 8.2, 8.3, 8.4 and 8.5, respectively. All the three figures indicate a non-linear inversely proportional relationship between the drain source currents and the gate source voltage.

8.5.2 Results of Modeling and Simulation of Silicon Carbide MOSFET

i. Parameter Sensitivity Analysis

The drain-source current values at 25°C, 100°C and 250°C are shown in Figures 8.6, 8.7 and 8.8, respectively. The figures demonstrate the temperature dependency of the SiC MOSFET. The results revealed that the drain-to-source current (I_{DS}) values are proportional to the gate-to-source voltage,

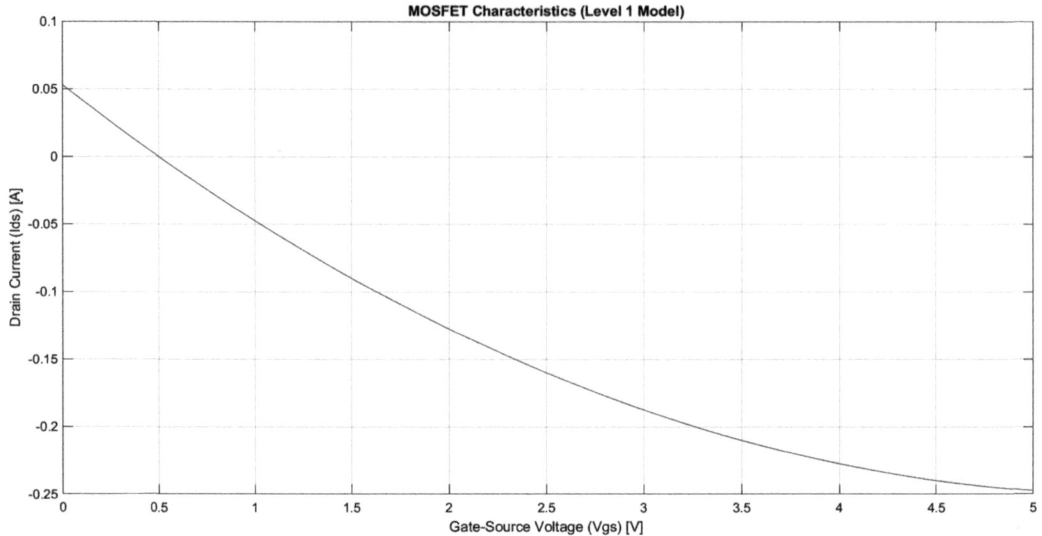

FIGURE 8.2 MOSFET Characteristics for Level 1 Model at Threshold voltage of 0.5.

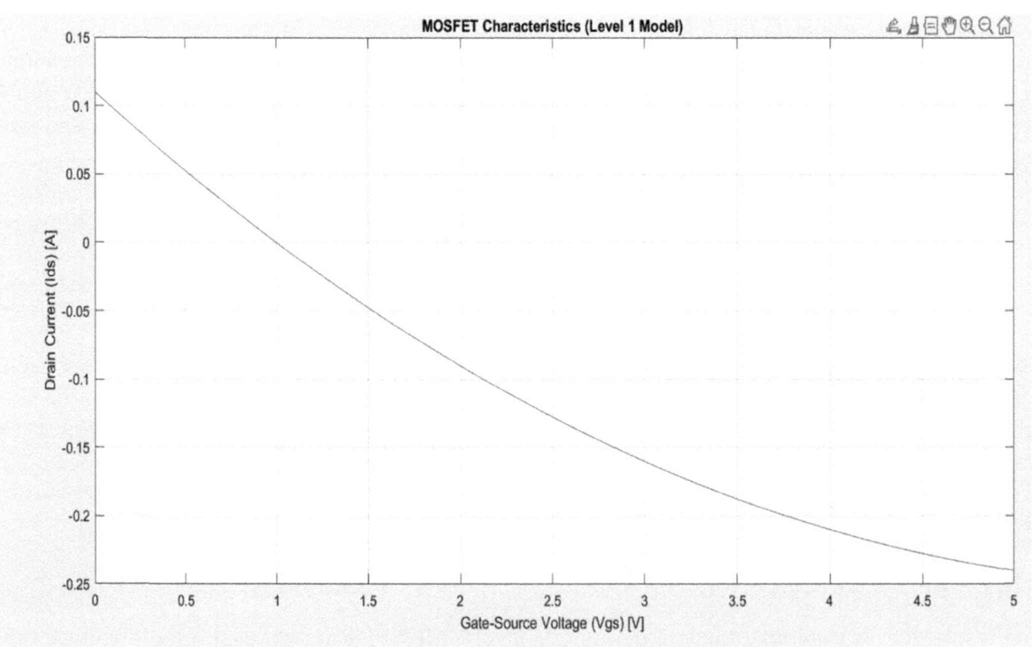

FIGURE 8.3 MOSFET Characteristics for Level 1 Model at Threshold voltage of 1V.

V_{GS} for the different input V_{GS} voltages (10V, 12V, 14V, 16V, 18V and 20V), represented by different coloured lines. The drain source current I_{DS} increased with increase in V_{GS}. The simulation results also revealed that a corresponding decrease in the drain source current I_{DS} with increased temperature. The maximum drain source current I_{DS} values at 25°C, 100°C and 250°C were 4.1 mA, 3 mA and 1.6 mA, respectively.

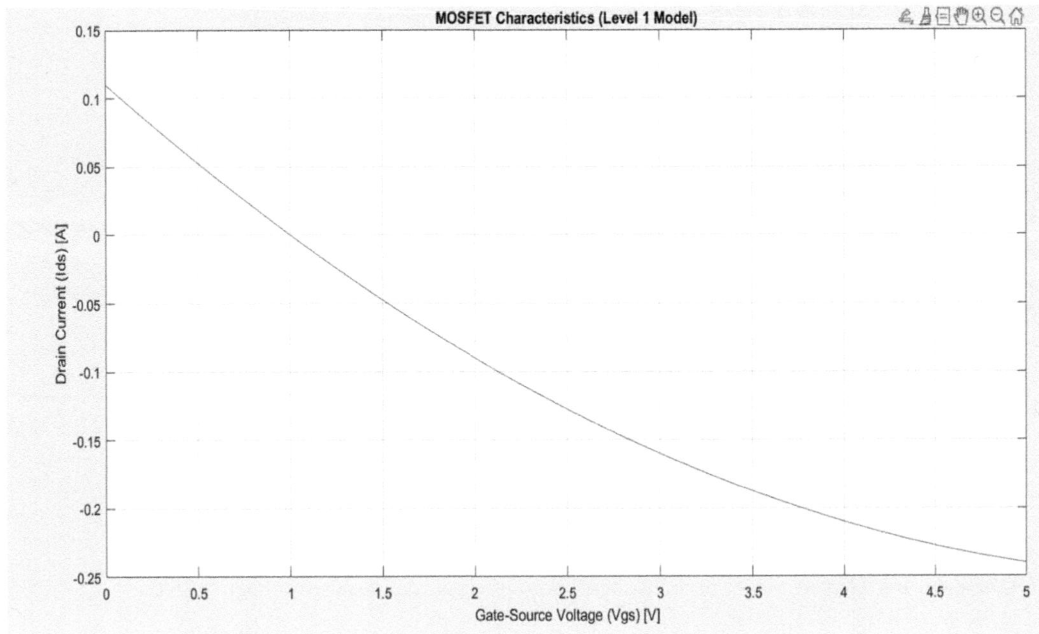

FIGURE 8.4   MOSFET Characteristics for Level 1 Model at Threshold voltage of 1.5V.

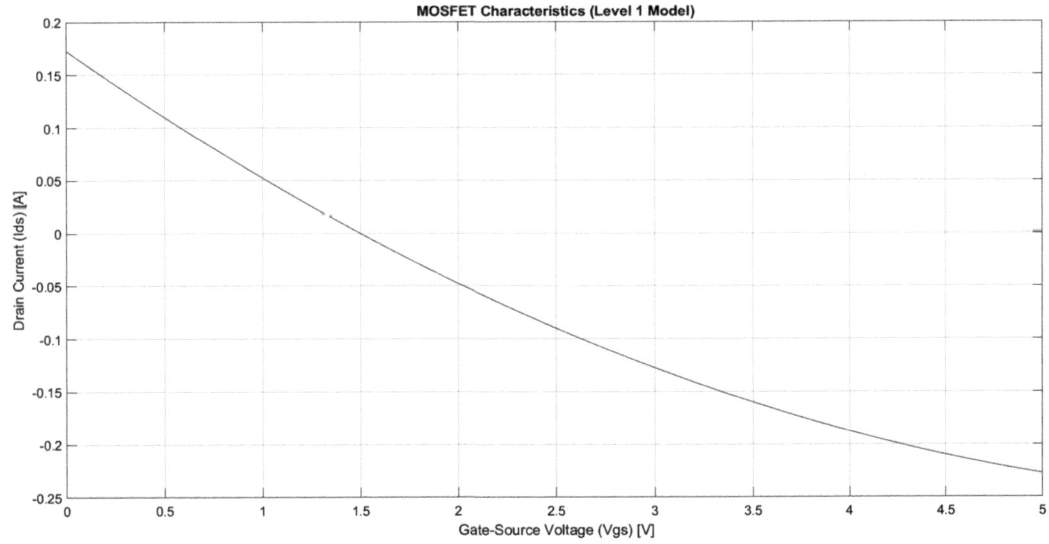

FIGURE 8.5   MOSFET Characteristics for Level 1 Model at Threshold voltage of 2V.

ii. Parameter Performance Analysis

The power dissipation of the SiC MOSFET is shown in Figure 8.9. The power dissipation was found to be proportional to the V_{GS} levels.

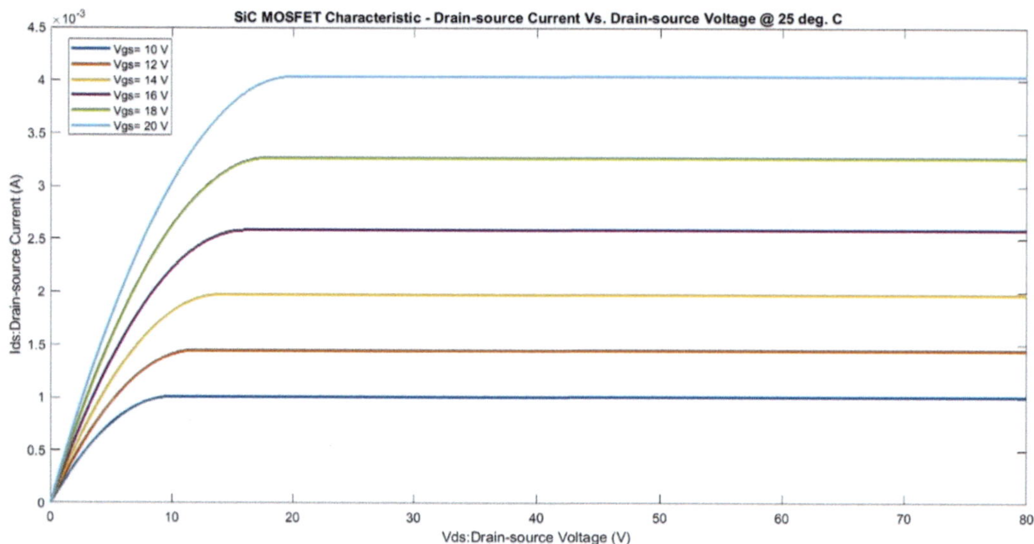

FIGURE 8.6 I–V Characteristics of drain-source current versus drain source voltage @25°C.

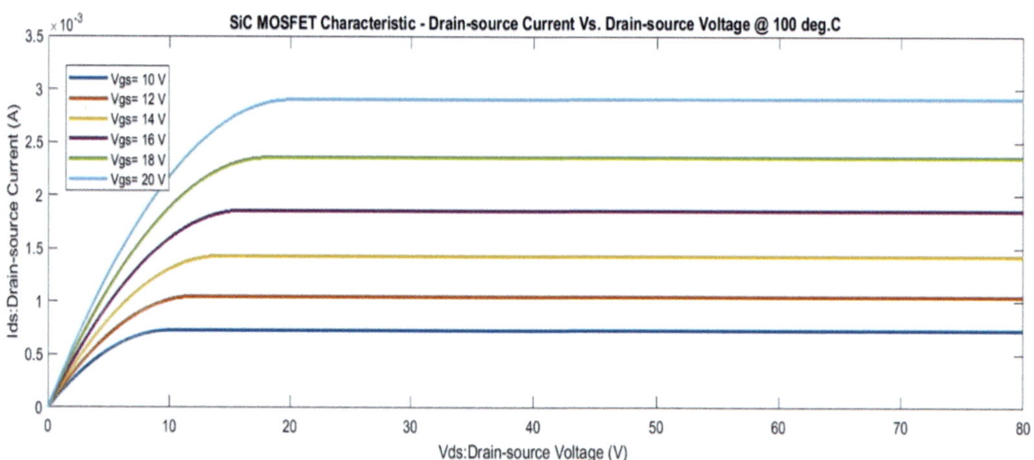

FIGURE 8.7 I–V Characteristics of drain-source current versus drain source voltage @100°C.

8.5.3 Simulation of Wearable Pulse Sensors

Figure 8.10 shows a graphical representation of the simulation of wearable pulse sensors with time against pulse signal. The pulse rate is measured per minute. Hence, the maximum time indicated on the X-axis is 60 seconds. The wearable pulse sensor can measure heart rate of patients while being worn by the patient.

Figure 8.11 shows the simulation of a wearable pulse sensor for tachycardia (heart rate higher than the normal heart rate of 100 beats per minute). Last, Figure 8.12 shows the simulation of wearable pulse sensor for Bradycardia.

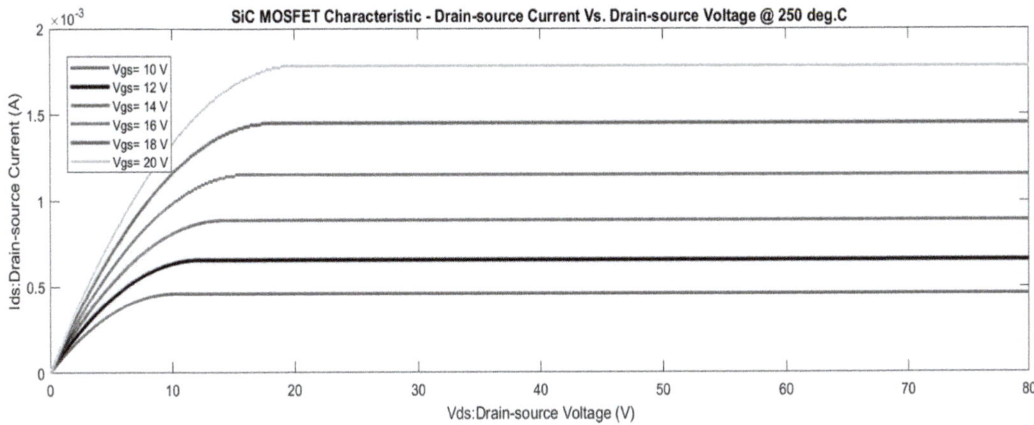

FIGURE 8.8   I–V Characteristics of drain-source current versus drain source voltage @250°C.

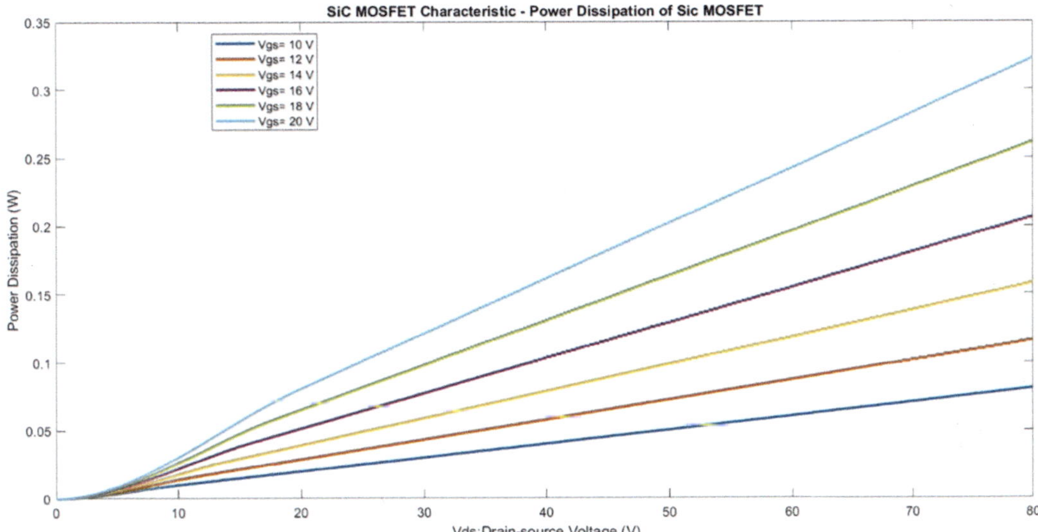

FIGURE 8.9   Power dissipation of the Sic MOSFET.

8.6   DISCUSSION

This section discusses the results of the computational modeling of the MOSFET, SiC MOSFET and the wearable pulse sensor models. The computational models developed have enhanced our understanding of the behaviour of these models. The peak drain current in the MOSFET model was found to increase with increased threshold voltage. The drain source currents were found to be inversely proportional to the gate source voltage. By increasing the gate source voltage (V_{GS}), there was a sharp decrease in the drain currents for the MOSFET models. Also, when the threshold voltage, V_{TH} is increased, it leads to higher drain current for the MOSFET leading to increased power consumption.

Similarly, our findings suggest that power dissipation in Silicon Carbide MOSFETs is proportional to the input gate to source voltage V_{GS}. Thus, more power is dissipated with increased gate to source voltage. This resultant proportionality confirms the fact that the power dissipation of the

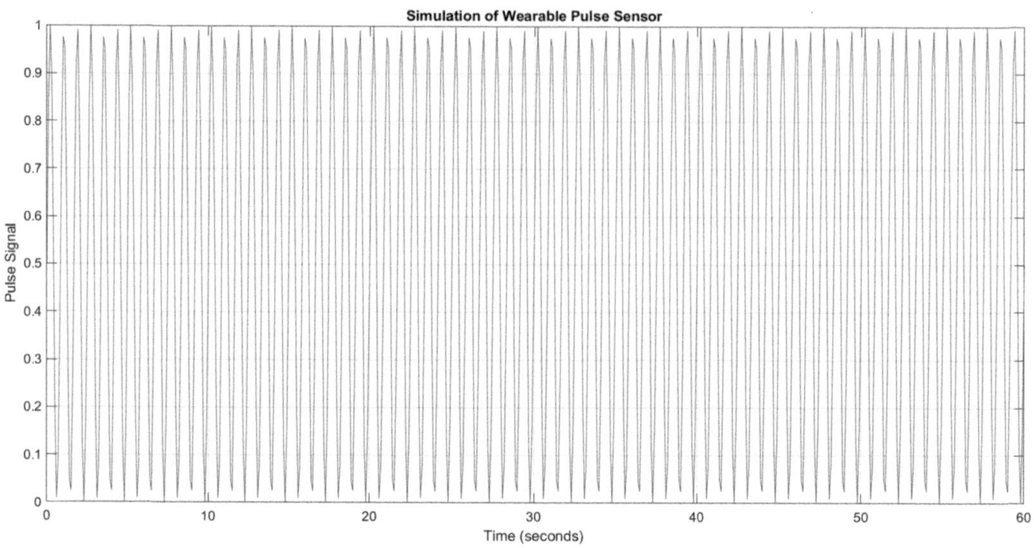

FIGURE 8.10 Simulation of wearable pulse sensor for normal pulse rate.

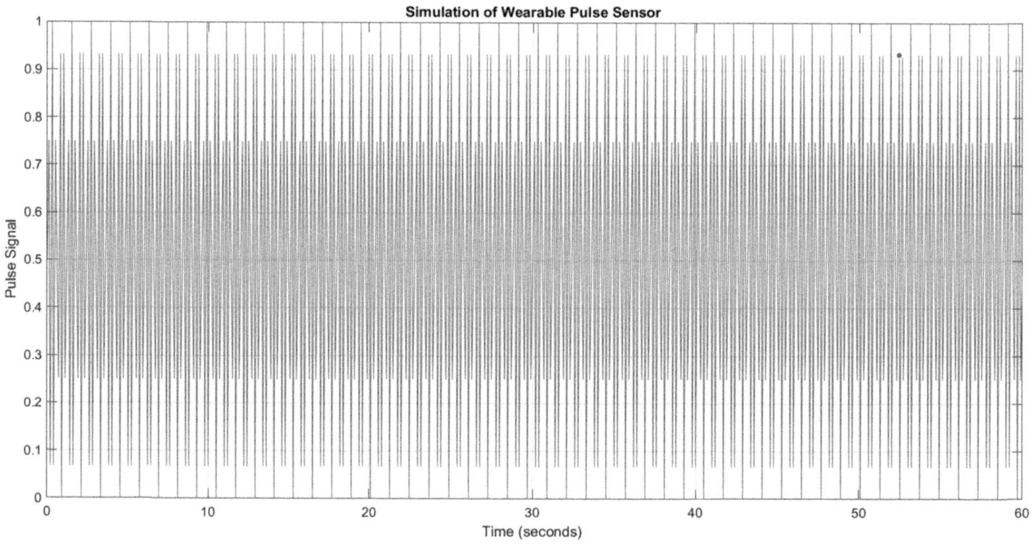

FIGURE 8.11 Simulation of wearable pulse sensor for tachycardia.

MOSFET is indeed a composite of two dissipations, that is, the switching power dissipation and the resistive power dissipation. The results are similar to those obtained in [55].

The modeling and simulation of a wearable pulse rate sensor is critical to its design. The modeling and simulation of normal heart rate, tachycardia and bradycardia revealed variations in the width of the waveforms generated. The width of the graph decreased with increased heart rates. In Bradycardia, where the heart rate is lower than 60 beats per minute, the width of the waveform was the widest. Conversely, in tachycardia, where the heart rate is higher than 100 beats per minute, the pulse signal had the thinnest width. The modeling of the wearable pulse sensor is limited by not considering the location and dimension of the sensor.

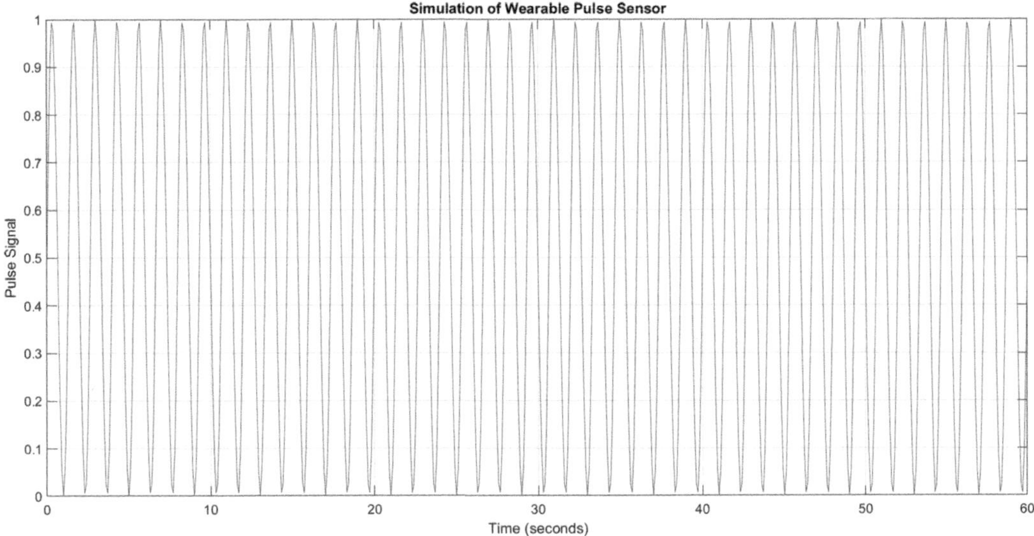

FIGURE 8.12   Simulation of Wearable Pulse Sensor for Bradycardia.

8.7 CONCLUSION

Computational modeling of electronic devices has enhanced the understanding of the behaviour of electronic devices. In this chapter, the various applications of computational modeling in electronic devices were discussed. Diverse studies on computational modeling of electronic devices were reviewed. These include: the application of artificial intelligence, such as neural networks and fuzzy logic systems in electronic device modeling, the use of compact circuit models to balance model complexity, modeling of novel materials and structures such as graphene. This chapter further described the modeling of Silicon Carbide MOSFETs and the effects of hysteresis on its threshold voltage. Finally, the concept of wearable electronics systems modeling and simulation and their behaviour in real-life situations was discussed. Hypothetical models for wearable pulse sensor were developed.

REFERENCES

[1] C. Bailey, H. Lu, and D. Wheeler, "Computational modeling techniques for reliability of electronic components on printed circuit boards". *Applied Numerical Mathematics*, vol. 40, pp. 101–107, 2002.

[2] H. Tang, K.C Tan, and Z. Yi, *Neural Networks: Computational Models and Applications.* New York: Springer, 2007.

[3] J.W. Jerome, B. Chini, M. Longaretti, and R. Sacco, "Computational modeling and simulation of complex systems in bio-electronics". *Journal of Computational Electronics*, vol. 7, pp. 10–13, 2008.

[4] J. Nolan, D. Parker, and G.C. van Kooten, "An overview of computational modeling in agricultural and resource economics". *Canadian Journal of Agricultural Economics*, vol. 57, pp. 417–429, 2009.

[5] E.C. Holland, P.N. Watton, and Y. Ventikos, "Biological Fluid Mechanics: Integrative and Multiscale Computational Modeling". In *Comprehensive Biotechnology* (Second Edition) M. Moo-Young, Ed, Massachusetts: Academic Press, 2011, pp. 203–216.

[6] Consumer Electronics Market (2023). Available: https://straitsresearch.com/report/consumer-electronic-market [accessed 27 September 2023]

[7] R. Vargas-Bernal, "Advances in computational modeling of electronic devices based on graphene". *IEEE Journal on Emerging and Selected Topics in Circuits and Systems*, vol. 5, no. 1, pp. 109–115, 2015.

[8] H. Chang, and H. Wu, "Graphene-based Nanomaterials: Synthesis, properties, and optical and optoelectronic applications". *Advanced Functional Materials*, vol. 23, pp. 1984–1997, 2013.

[9] H. Shim, S. Jang, and C. Yu, "High-resolution patterning of organic semiconductors toward industrialization of flexible organic electronics". *Matter*, vol. 5, no. 1, pp. 23–25, 2022.

[10] J. Heijman, H. Sutanto, H.J.G.M Crijns, S. Nattel, and N.A. Trayanova, "Computational models of atrial fibrillation: Achievements, challenges, and perspectives for improving clinical care". *Cardiovascular Research*, vol. 117, no. 7, pp. 1682–1699, June 2021.

[11] T. Kirchdoerfer, and M. Ortiz, "Data-driven computational mechanics". *Computer Methods in Applied Mechanics and Engineering*, vol. 304, no. 1, pp. 81–101, 2016.

[12] J.D. Anderson, and J. Wendt, *Computational Fluid Dynamics*. New York: McGraw-Hill, 1995.

[13] M.M. Bhatti, M. Marin, A. Zeeshan, and S.I. Abdelsalam, "Editorial: Recent trends in computational fluid dynamics". *Frontiers in Physics*, vol. 8, pp. 1–3, 2020.

[14] H. Hu, "Computational Fluid Dynamics". In *Fluid Mechanics* (Fifth Edition), P. K. Kundu, I. M. Cohen, and D.R. Dowling, Eds, Massachusetts: Academic Press, 2012, pp. 421–472.

[15] E. Monier-Vinard, B. Rogie, V. Bissuel, N. Laraqi, O. Daniel, and M. Kotelon, "State of the art of thermal characterization of electronic components using computational fluid dynamic tools". *International Journal of Numerical Methods for Heat & Fluid Flow*, vol. 27, no. 11, pp. 2433–2450, 2017.

[16] S. Kamate, and N. Yilmazer, "Application of object detection and tracking techniques for unmanned aerial vehicles". *Procedia Computer Science*, vol. 61, pp. 436–441, 2015.

[17] J.T. Zhou, J. Du, H. Zhu, X. Peng, Y. Liu, R.S.M. Goh, Anomaly Net: An anomaly detection network for video surveillance. *IEEE Transactions on Information Forensics and Security,* vol. 14, pp. 2537–2550, 2019.

[18] Z. Huang, Y. Liu, C. Zhan, C. Lin, W. Cai, and Y. Chen, "A novel group recommendation model with two-stage deep learning". In *IEEE Transactions on Systems, Man, and Cybernetics: Systems*, vol. 52, no. 9, pp. 5853–5864, Sept. 2022.

[19] Y. Cui, "Wireless biological electronic sensors". *Sensors*, vol. 17, no. 10, 2017. www.mdpi.com/1424-8220/17/10/2289 [accessed November 8, 2023]

[20] B. Rigo, A. Bateman, J. Lee, H. Kim, Y. Lee, L. Romero, Y.C. Jang, R. Herbert, W. Yeo, "Soft implantable printed bioelectronic system for wireless continuous monitoring of restenosis". *Biosensors and Bioelectronics*, vol. 241, pp. 115650. Available: www.sciencedirect.com/science/article/pii/S0956566323005924 [accessed November 8, 2023)

[21] N. Oyunbaatar, D. Kim, G. Prasad, Y. Jeong, and D. Lee, "Self-rollable polymer stent integrated with wireless pressure sensor for real-time monitoring of cardiovascular pressure". *Sensors and Actuators A: Physical*, vol. 346, pp. 113869, 2022. www.sciencedirect.com/science/article/abs/pii/S092442472 2005040 [access November 8, 2023]

[22] R. Herbert, H. Lim, B. Rigo, W. Yeo. "Fully implantable wireless batteryless vascular electronics with printed soft sensors for multiplex sensing of hemodynamics". *Science Advances*, vol. 8, no. 19, May 2022. Available: https://pubmed.ncbi.nlm.nih.gov/35544557/ [Accessed December 26, 2023]

[23] M. Catrysse, R. Puers, C. Hertleer, L. Van Langenhove, H. van Egmond, and D. Matthys, Towards the integration of textile sensors in a wireless monitoring suit. *Sensors and Actuators A: Physical*, vol. 114, no. 2–3, pp. 302–311, 2004.

[24] K. Guk, G. Han, J. Lim, K. Jeong, T. Kang, E.K. Lim, and J. Jung, "Evolution of wearable devices with real-time disease monitoring for personalized healthcare". *Nanomaterials (Basel)*, vol. 9, no. 6, pp. 813, May 29, 2019.

[25] R.L. Bulathsinghala, W. Ding, and R.D.I.G. Dharmasena, "Triboelectric nanogenerators for wearable sensing applications: A system level analysis". *Nano Energy*, vol. 116, pp. 108792, 2023. [accessed November 29, 2023]

[26] T. Zhang, N. Liu, J. Xu, Z. Liu, Y. Zhou, Y. Yang, S. Li, Y. Huang, and S. Jiang, "Flexible electronics for cardiovascular healthcare monitoring". *The Innovation*, vol. 4, no. 5, pp. 100485, 2023.

[27] T. Ha, J. Tran, S. Liu, H. Jang, H. Jeong, and R. Mitbander, et al., (2019, Jul), "A chest-laminated ultrathin and stretchable E-tattoo for the measurement of electrocardiogram, seismocardiogram, and cardiac time intervals". *Advanced Science*, vol. 6, no. 14, Available: https://pubmed.ncbi.nlm.nih.gov/31380208/ [Accessed December 28, 2023]

[28] S.M. Iqbal, I. Mahgoub, E. Du, M.A. Leavitt, and W. Asghar, Advances in healthcare wearable devices. *npj Flexible Electronics*, vol. 5, pp. 1–14, 2021.

[29] O.M. Mirza, H. Mujlid, H. Manoharan, S. Selvarajan, G. Srivastava, and M.A. Khan, "Mathematical framework for wearable devices in the Internet of Things using deep learning". *Diagnostics*, vol. 12, 2750, 2022. Available: www.mdpi.com/2075-4418/12/11/2750 [accessed November 9, 2023]

[30] P. Zhao, M. Choudhury, K. Mohanram, and J. Guo, "Computational model of edge effects in graphene nanoribbon transistors". *Nano Research*, vol. 1, no. 5, pp. 395–402, 2008.

[31] G. Kalpachka, "Computer modeling and simulations of logic circuits". *International Journal of Modern Education and Computer Science*, vol. 12, pp. 31–37, 2016.

[32] T.H. Nagrare, and A.A. Gurjar, A. A., "Adaptive neuro fuzzy inference system for modeling of electronics devices: A review". *2017 Innovations in Power and Advanced Computing Technologies (i-PACT)*, 2017-Janua, pp. 1–4, 2017.

[33] R. Tuntas, "The modeling and hardware implementation of semiconductor circuit elements by using ANN and FPGA". *Acta Physica Polonica A*, vol. 128, no. 2B, pp. 78–82, 2015.

[34] G. Oltean, and M. Gordan, "Electronics and telecommunications electronic devices modeling by fuzzy logic interpolator". *Acta Technica Napocensis* [Online], 2000. Available: www.bel.utcluj.ro/~goltean/articole/ATN2000_1.pdf [Accessed November 10, 2023]

[35] L. Belhamel, A. Buscarino, A. Cucuccio, L. Fortuna, and G. Rascona, "Model-based design for high performance devices oriented to advanced smart solutions". In *2020 6th IEEE Congress on Information Science and Technology (CiSt)*, Agadir – Essaouira, June 2020, pp. 442–447.

[36] M. Hayati, M. Seifi, and A. Rezaei, "The computational intelligence in simulation of DG MOSFET: Application to the simulation of the nanoscale CMOS circuit". *IEEE International Conference on Semiconductor Electronics, Proceedings, ICSE*, pp. 138–142, 2008.

[37] G. Wachutka, "The art of modeling and predictive simulation in power electronics and microsystems". In *Proceedings of the 20th International Conference on Mixed Design of Integrated Circuits and Systems*, MIXDES 2013.

[38] A. Ray, G. Kumar, S. Bordoloi, D.K. Sinha, P. Agarwal, and G. Trivedi, "FEM based device simulator for high voltage devices". *Communications in Computer and Information Science*, vol. 711, pp. 127–135, 2017.

[39] G. Birajdar, M. Ishaq, S. Kirkup, and B. Leye, (2020). Numerical Modeling and Computer Simulation. In D. M. Cvetković, and G. A. Birajdar (Eds.), *Numerical Modeling and Computer Simulation* (Issue June). IntechOpen. https://doi.org/10.5772/intechopen.77641

[40] V. Milovanski, and G. Kalpachka, "Computer modeling and simulations of processes in serial resonance". *Bulgarian Chemical Communications*, vol. 52, no. 2, pp. 225–228, 2020.

[41] D. Vasileska, and S.M. Goodnick, "Computational electronics". *Synthesis Lectures on Computational Electromagnetics*, vol. 6, no. 1, pp. 1–216, 2006.

[42] S. Luryi, and A. Pacelli, "Automatic generation of RF compact models from device simulation". *2002 International Conference on Modeling and Simulation of Microsystems – MSM 2002*, 2002, pp. 702–709.

[43] M. Sakib Hasan, M. Shamim Ara Shawkat, S. Amer, S. Kamrul Islam, N. McFarlane, and G. Rose, "Modeling Emerging Semiconductor Devices for Circuit Simulation". In *Computational Models in Engineering*, K. Volkov, Ed, London: IntechOpen, 2020, pp. 1–26.

[44] Y. Mukunoki, Y. Nakamura, K. Konno, T. Horiguchi, Y. Nakayama, A. Nishizawa, M. Kuzumoto, and H. Akagi "Modeling of a silicon-carbide MOSFET with focus on internal stray capacitances and inductances, and its verification". In *IEEE Transactions on Industry Applications*, vol. 54, no. 3, pp. 2588–2597, May–June 2018.

[45] S. Shamsir, M. Sakib Hasan, O. Hassan, P. Sarathi Paul, M. Razuan Hossain, and S. Islam, Semiconductor Device Modeling and Simulation for Electronic Circuit Design. In *Modeling and Simulation in Engineering – Selected Problems*, J. Valdman, Ed., London: IntechOpen, 2020, pp. 1–22.

[46] A. Huerner, P. Sochor, M. Feil, and R. Elpelt, "Influence of the threshold-voltage hysteresis on the switching properties of SiC MOSFETs". In *PCIM Europe digital days 2021; International Exhibition and Conference for Power Electronics, Intelligent Motion, Renewable Energy and Energy Management*, 2021, pp. 1–8.

[47] A. Grebennikov, and M.J. Franco, *Switchmode RF and Microwave Power Amplifiers* (Third Edition). Massachusetts: Academic Press, 2021.

[48] F. Loche-Moinet, L. Theolier, and E. Woirgard, Electro-thermo-mechanical modelling of a SiC MOSFET transistor under non-destructive short-circuit. *Microelectronics Reliability*, vol. 150, 2023. Available: www.sciencedirect.com/science/article/pii/S0026271423002433 [accessed November 30, 2023]

[49] L. Zhang, T. Dai, P.M. Gammon, V.A. Shah, P.A. Mawby, and M. Antoniou, Simulation Study of a 650V Hybrid-Channel SiC Trench MOSFET with improved on-state performance. *Power Electronic Devices and Components,* vol. 4, 2023. Available: www.journals.elsevier.com/power-electronic-devices-and-components [accessed November 30, 2023]

[50] I. Kwon, H. Kwon, and H. Cho, Development of high temperature operation silicon based MOSFET for harsh environment application. *Results in Physics*, vol. 11, pp. 475–481, 2018.

[51] I. Matacena, L. Maresca, M. Riccio, A. Irace, and G. Breglio, SiC MOSFETs capacitance study. e-Prime – Advances in Electrical Engineering. *Electronics and Energy*, vol. 5, 100251, pp. 1–10, 2023.

[52] D. Peters, T. Aichinger, T. Basler, G. Rescher, K. Puschkarsky, and H. Reisinger, "Investigation of threshold voltage stability of SiC MOSFETs". In *2018 IEEE 30th International Symposium on Power Semiconductor Devices and ICs (ISPSD)*, Chicago, IL, USA, 2018, pp. 40–43.

[53] Y. Cai, T. Sun, P. Sun, Z. Zhao, X. Li, H. Wang, Z. Chen, and B. Cao, "Influence of parasitic parameters on dynamic threshold voltage hysteresis of silicon carbide MOSFETs". *Journal of Power and Energy Systems*, vol. 9, no. 6, pp. 2251–2262, 2023.

[54] M. Hauck, J. Lehmeyer, G. Pobegen, H.B. Weberand, and M. Krieger, "An adapted method for analyzing 4H silicon carbide metal-oxide-semiconductor field-effect transistors". *Communications Physics*, vol. 2, no. 5, 2019. Available: www.nature.com/articles/s42005-018-0102-8#citeas. [accessed 28 December 2023].

[55] M. Alhalabi, A. Rashed, N. B. Iqbal, and A.A. Tarabsheh, "Modelling of SiC power MOSFET in Matlab, Simulink, and LTSpice". In *2018 IEEE International Conference on Environment and Electrical Engineering and 2018 IEEE Industrial and Commercial Power Systems Europe (EEEIC / I&CPS Europe)*, Palermo, Italy, 2018, pp. 1–6.

9 High-Speed Stream Ciphers for Wireless Communication Systems

Design and Simulation

Oleksandr Kuznetsov, Emanuele Frontoni,
Natalia Kryvinska, Mykola Mormul and Oleg Poplavskiy

9.1 INTRODUCTION

The advent of wireless technologies has revolutionized the way we communicate, bringing forth a plethora of applications that range from simple personal communications to complex industrial systems [1], [2]. With this technological evolution, the need for robust security mechanisms to protect against an ever-growing landscape of cyber threats has become paramount [3], [4]. High-speed stream ciphers emerge as a pivotal element in this context, offering a unique blend of rapid data encryption and strong security, making them ideal for the high-throughput requirements of modern wireless systems [5].

This chapter begins by setting the stage with an overview of the current state of wireless communication technologies. It provides a context for understanding the criticality of data security in wireless networks, highlighting the challenges posed by the diverse nature of wireless applications and the vulnerabilities inherent in wireless transmission. The discussion then shifts to the realm of stream ciphers, unraveling their significance in the wireless world. It delves into the principles that govern the design of these ciphers, shedding light on their structural intricacies, operational mechanisms, and the rationale behind their cryptographic strength.

The core of the chapter is dedicated to a detailed examination of various high-speed stream ciphers. It offers a comprehensive analysis of their design philosophies, algorithmic structures, and security features. This examination is not merely theoretical; it is complemented by rigorous computational modeling and simulation, providing empirical evidence of the performance and efficiency of these ciphers in real-world scenarios. The simulations are meticulously designed to mimic actual wireless communication environments, thereby offering valuable insights into the practical applicability of these cryptographic tools.

Furthermore, the chapter discusses the challenges and considerations involved in integrating these ciphers into existing wireless communication frameworks. It explores the trade-offs between security, speed, and resource utilization, crucial factors in the deployment of cryptographic solutions in wireless systems. The discussion extends to the future directions of stream cipher development, contemplating emerging trends, potential advancements, and the evolving landscape of wireless communication technologies.

In essence, this chapter is not just a technical discourse on stream ciphers; it is a narrative that intertwines the theoretical and practical aspects of wireless security. It aims to provide readers with

DOI: 10.1201/9781003457428-10

a holistic understanding of the application of high-speed stream ciphers in modern wireless systems, highlighting their indispensable role in securing the invisible yet omnipresent threads of wireless communication that bind the digital world.

9.1.1　Key Contributions of the Chapter

This chapter makes significant contributions to the field of cryptographic research, particularly in the context of wireless communication systems. First, it provides a comprehensive analysis of the fundamentals of stream ciphers, offering insights into their basic principles, functionality, and general models. This foundational knowledge is crucial for understanding the role of stream ciphers in securing wireless communications. Second, the chapter discusses recent advancements in stream ciphering, highlighting enhanced security protocols, efficiency improvements, and integration with advanced wireless technologies. These discussions reflect the latest trends and challenges in the field, contributing to the ongoing discourse on stream cipher development and application. Additionally, the chapter presents a detailed methodology for computational modeling and comparative analysis of stream ciphers, showcasing a systematic approach to evaluating their performance and efficiency. This methodological contribution is vital for researchers and practitioners in simulating and assessing the suitability of various cryptographic algorithms for wireless systems. Finally, the chapter offers a thorough discussion on the research findings, providing a critical analysis of the experimental results and their implications for wireless communication security.

9.1.2　Chapter Organization

The chapter is organized into several sections, each addressing a specific aspect of stream ciphers in wireless communication systems. Section 9.2 sets the stage by reviewing existing literature and research in the field, providing a background against which our research is positioned. In Section 9.3 the chapter delves into the core concepts of stream ciphers, including their basic principles, functionality, and general models. Section 9.4 outlines the approach taken for the empirical evaluation of stream ciphers, detailing the methods used to assess computational efficiency and the specific cryptographic algorithms examined. Section 9.5 presents the findings from the experimental analysis, including the results of long stream encryption, packet encryption, and IV and key setup performance testing. In Section 9.6 the chapter provides a comprehensive analysis of these results, discussing their implications and relevance in the broader context of wireless communication security. Finally, Section 9.7 synthesizes the key insights from the chapter, highlighting the contributions made to the field.

9.2　RELATED WORK

In the pursuit of advancing the field of wireless communication systems, particularly through the lens of high-speed stream ciphers, it is imperative to acknowledge and analyze the contributions of existing literature. This section provides a critical review of various scholarly works, each contributing uniquely to the domain of cryptographic security and computational modeling in wireless systems.

9.2.1　Literature Review

An et al. (2022) [6]: In their work on DCGAN-based symmetric encryption for end-to-end communication systems, An and colleagues propose a novel approach using deep convolutional generative adversarial networks. This study enhances the security of transmission in wireless systems, offering a fresh perspective on symmetric encryption and its integration with advanced neural network technologies.

Conrad, Misenar, and Feldman (2023) [7]: Their comprehensive chapter on Security Architecture and Engineering delves into key cryptographic concepts and their application in cloud and virtualization environments. This work is instrumental in understanding the broader scope of cryptographic mechanisms, including their modes of operation, in securing complex wireless systems.

Ghasemi and Babaie (2022) [8]: Focusing on RFID-based Internet of Things, their research introduces a lightweight secure authentication approach utilizing stream ciphering. This study is significant for its application of non-linear pseudo-random number generators in enhancing privacy and security in IoT environments.

Goswami and Trivedi (2023) [9]: Their work on FPGA implementation of modified SNOW 3G stream ciphers emphasizes the importance of hardware-efficient cryptographic solutions in wireless networks. This research contributes to the understanding of area, power, and efficiency considerations in cryptographic algorithm implementation.

Haseeb et al. (2022) [10]: The paper presents a secure mobile sensor network protocol for smart data gathering with cloud services. This research is pivotal in addressing security and efficiency in dynamic networks, particularly in the context of green systems and IoT.

Lima et al. (2022) [11]: Their exploration of event-based cryptography using the ChaCha20 stream cipher in automation networks of cyber-physical systems offers a novel approach to securing communication channels in discrete-event systems. This work is notable for its application of stream ciphering in automation networks without altering data size or structure.

Liu et al. (2022) [12]: The study introduces a novel pipelining encryption hardware system for 5G, focusing on the ZUC-256 stream cipher algorithm. This research is crucial for its contribution to enhancing the security and throughput in 5G wireless networks.

Noura et al. (2022) [13]: Their paper on LoRCA, a lightweight cipher scheme for IoV systems, addresses the need for efficient cryptographic solutions in the Internet of Vehicles. This work is significant for its balance between security level and performance, particularly in the context of resource-constrained environments.

Rudnytskyi et al. (2022) [14]: The article discusses cryptographic encoding in modern symmetric and asymmetric encryption, proposing a new approach to constructing asymmetric stream ciphers. This research offers valuable insights into the development of new cryptographic algorithms and the improvement of existing ones.

Sharma and Mehra (2023) [15]: Their systematic survey on secure communication in IoT-based UAV networks highlights the vulnerabilities and security solutions in this emerging field. This work is instrumental in understanding the security landscape of IoT-based UAV networks and the various cryptographic solutions applicable.

Sharma, Kumar, and Rai (2024) [16]: The paper introduces a new approach to implement stream cipher using memristor-based Chua's circuit for secure communication. This research contributes to the field by proposing a stream cipher scheme with higher complexity and simplicity, particularly relevant for secure communication systems.

Thabit et al. (2023) [17]: Their work on cryptography algorithms for enhancing IoT security provides a comprehensive analysis of cryptographic algorithms and their application in IoT. This survey is crucial for understanding the security challenges and cryptographic solutions in the rapidly growing IoT domain.

Wang, Wang, and Liu (2024) [18]: The study focuses on a key security measurement method of authentication based on mobile edge computing in urban rail transit communication networks. This research is significant for its approach to enhancing security in urban rail transit communication platforms using mobile edge computing.

9.2.2 Summary and Gap Identification

The reviewed literature collectively advances the understanding of cryptographic security in various wireless communication contexts, from IoT and UAV networks to 5G and urban rail transit systems. Each study contributes unique perspectives and solutions, addressing different aspects of security, efficiency, and implementation. However, a gap is identified in the comprehensive analysis and simulation of high-speed stream ciphers specifically tailored for advanced wireless communication systems. Our research fills this gap by providing an in-depth exploration of high-speed stream ciphers, their computational modeling, and simulation, thereby offering a holistic view of their application and efficacy in modern wireless communication systems.

9.3 FUNDAMENTALS OF STREAM CIPHERS

In the realm of wireless communication, the security of transmitted data is paramount. This section delves into the bedrock of stream ciphers, an encryption method integral to safeguarding data in motion. We aim to demystify the complexities of stream ciphers, laying a comprehensive foundation for understanding their critical role in wireless communication security. This exploration is not only academically enriching but also vital for practitioners and researchers striving to fortify the digital fortresses of modern wireless networks.

9.3.1 Basic Principles of Stream Ciphers

Stream ciphers are a subset of symmetric key ciphers where plaintext data is encrypted one symbol at a time with a pseudorandom cipher digit stream, known as the keystream [19, 20]. This method stands in contrast to block ciphers, which encrypt data in fixed-size blocks. The essence of stream ciphers lies in their ability to create an encrypted stream that closely mimics randomness, thereby ensuring the confidentiality of the transmitted data.

The core mechanism of a stream cipher revolves around the generation of a keystream and its subsequent use in the encryption process [5]. The keystream is typically as long as the plaintext message and is generated using a predetermined algorithm and a secret key [20]. The encryption process involves a bitwise exclusive OR (XOR) operation, where each bit of the plaintext is XORed with the corresponding bit of the keystream. This operation is elegantly simple yet effective, as the same process is applied for both encryption and decryption.

Figure 9.1 represents the key components and processes involved in stream cipher encryption and decryption. It includes the plaintext input, the stream cipher encryption process, the generation of ciphertext, and the subsequent decryption process to recover the original plaintext. The key stream generator, which is central to the operation of the stream cipher, receives inputs from the secret key and, optionally, an initialization vector. The diagram also highlights the encryption and decryption processes that are integral to the stream cipher's functionality.

The generation of the keystream is a critical aspect of stream ciphers [21]. The keystream must be unpredictable and should appear random to an outside observer. Any predictability in the keystream can lead to the cipher's compromise. Therefore, the algorithm used for keystream generation is designed to be complex and non-repetitive [22]. It often involves the use of linear feedback shift registers (LFSRs), nonlinear combinations of LFSRs, or more sophisticated mechanisms like nonlinear feedback shift registers (NFSRs) and cryptographic algorithms [23].

Stream ciphers require the sender and receiver to maintain synchronization, as the encryption and decryption processes depend on the alignment of the keystream with the plaintext. Loss of synchronization can lead to decryption errors and data corruption. This necessitates robust protocols for key management and distribution, ensuring that both parties maintain alignment throughout the communication process.

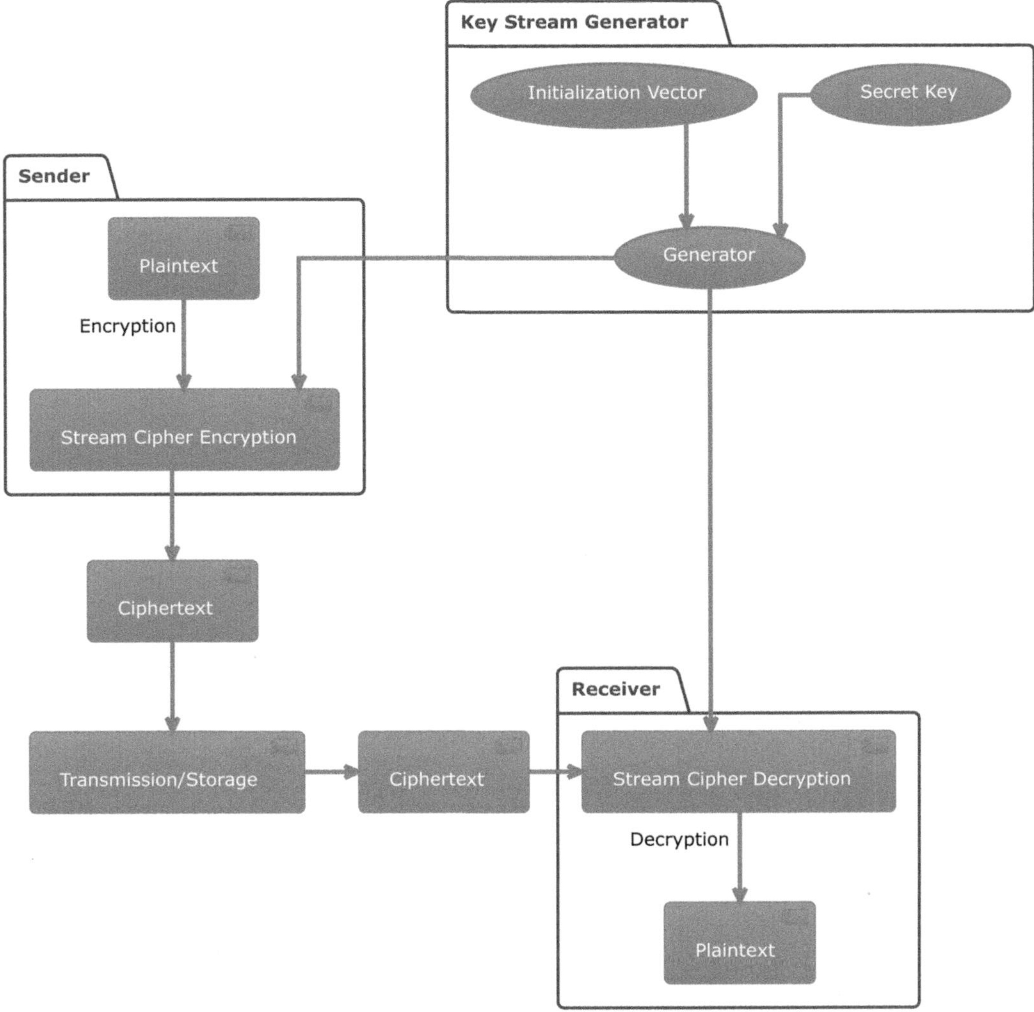

FIGURE 9.1 Processes Involved in Stream Cipher Encryption and Decryption.

Stream ciphers are particularly advantageous in wireless communication due to their efficiency in encrypting data of varying lengths and their low computational complexity. The bitwise XOR operation, fundamental to stream ciphers, is computationally less intensive compared to other cryptographic operations, making it ideal for high-speed data transmission in wireless networks. This efficiency is crucial in environments where processing power and battery life are limited, such as in mobile devices and IoT applications.

The security of a stream cipher is fundamentally tied to the properties of its keystream. Two key attributes are critical for security [20]:

- The keystream must have a long period, ensuring that the sequence does not repeat within the length of the encrypted message. This prevents attackers from exploiting patterns in the keystream.
- The keystream should closely resemble a truly random sequence, making it resistant to statistical analysis and cryptanalytic attacks. This randomness is essential to thwart attempts at predicting future bits of the keystream based on observations of past bits.

In summary, the basic principles of stream ciphers form the cornerstone of their application in securing wireless communication systems. Their design, centered on the generation of a robust and unpredictable keystream, coupled with the efficient XOR operation, strikes a balance between security and performance. As we progress further into the intricacies of stream ciphers, we will explore how these fundamental principles are ingeniously applied to protect the ever-expanding and dynamic domain of wireless communications. The subsequent sections will delve into the evolution of stream ciphers, their comparison with block ciphers, and their specific adaptations to meet the unique challenges of wireless communication environments. This comprehensive understanding is crucial for anyone engaged in the field of wireless security, be it as a developer, researcher, or practitioner. The insights gained here will not only deepen the appreciation of the elegance and complexity of stream ciphers but also highlight their indispensable role in the fabric of modern digital communication.

9.3.2 Description of Stream Cipher Functionality

Stream ciphers, a pivotal component in the domain of cryptographic systems, operate on the principle of stateful encryption, where the encryption of each symbol is dependent on a dynamically evolving state. This subsection aims to elucidate the intricate workings of stream ciphers, employing the theoretical framework of automata theory, which provides a robust mathematical foundation for understanding these complex systems.

Figure 9.2 shows the key elements and their relationships in the structure of a stream cipher [5], [24]. The figure shows the plaintext, ciphertext, keystream, and also includes key functions such as the state transition function and the output function, and their relationships with other elements. The functionality of stream ciphers, as elucidated through automata theory, reveals a complex interplay of mathematical and cryptographic principles [5, 24].

1. Alphabets and Spaces:
   - Plaintext Alphabet (X): Represents the set of all possible symbols in the plaintext.
   - Ciphertext Alphabet (Y): Denotes the set of all possible symbols in the ciphertext.
   - Keystream Alphabet (Z): Constitutes the set of symbols in the encryption sequence.
   - State Space (S): The internal states of the stream cipher, evolving with each encryption step.
   - Key Space (K): The set of all possible keys, chosen according to a probability distribution P_k.

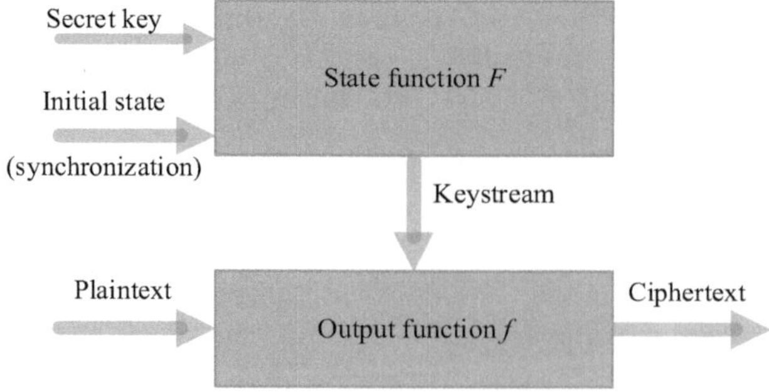

FIGURE 9.2 The General Structure of a Stream Cipher.

2. Symbol Notation:
    - x_i, y_i, z_i, and s_i symbolize the plaintext, ciphertext, keystream, and internal state symbols at time i, respectively.
    - The key k from K is selected based on p_k, typically following a uniform distribution, although non-random selection may be employed in certain scenarios.

The dynamic nature of their internal states, governed by meticulously designed state transition and output functions, forms the backbone of their security architecture [5, 24].

1. State Transition Function (F):
    - Described by $s_{i+1} = F(k, s_i, x_i)$, this function dictates the next state of the cipher based on the current state, key, and plaintext symbol.
    - This function is the cornerstone of the cipher's security, ensuring that the internal state evolves in a manner that is secure and unpredictable.
2. Output Function (f):
    - Defined by $y_i = f(k, s_i, x_i)$, it determines the ciphertext symbol based on the current state, key, and plaintext symbol.
    - In many stream ciphers, y_i is typically computed as $x_i + F(k_i, s_i)$, ensuring seamless and delay-free stream decryption.

The keystream, a critical output of these functions, must exhibit the highest degree of randomness and unpredictability to ensure the cipher's resilience against cryptographic attacks.

- The keystream is defined as the sequence $Z_i = F(k, s_i) : i = 0, 1, ...$, representing an extended key transformed into a pseudorandom sequence of potentially infinite length.
- The randomness, irreversibility, and unpredictability of the keystream are fundamental to the cipher's security. It must be as random as possible to resist cryptanalytic attacks and ensure robust encryption.

The design and analysis of stream ciphers, therefore, demand a deep understanding of both theoretical and practical aspects of cryptography. This includes a thorough grasp of probability theory for key selection, automata theory for state transitions, and a keen insight into potential vulnerabilities and attack vectors [19, 20].

- Randomness: The keystream's randomness is crucial in preventing attackers from predicting future keystream symbols based on past observations.
- Irreversibility: This property ensures that knowledge of the keystream does not facilitate the determination of the internal state or the secret key.
- Unpredictability: A keystream that is unpredictable, even if some internal states are compromised, upholds the cipher's integrity against sophisticated attacks.

Stream ciphers, with their inherent complexity and sophistication, stand as a testament to the ingenuity and depth of modern cryptographic practices. They are not just tools for securing data but are also embodiments of the elegant interplay between mathematics and computer science, aimed at safeguarding our digital communications in an increasingly interconnected world.

In the broader context of wireless communication systems, where speed and efficiency are paramount, the role of stream ciphers becomes even more pronounced. Their ability to encrypt data on-the-fly, adapting to the ever-changing landscape of wireless networks, makes them indispensable in our quest for secure and reliable digital communication.

9.3.3 GENERAL MODELS FOR STREAM CIPHERS

Stream ciphers, a fundamental aspect of cryptographic systems, are designed to provide secure communication by encrypting plaintext data into ciphertext using a keystream. This subsection aims to elaborate on the general models for stream ciphers, focusing on their structure, functionality, and the critical roles of keystream generators and output functions.

9.3.3.1 Keystream Generators

The structure of a stream cipher consists of a combination of a key stream generator and an output function. In the realm of cryptographic systems, key stream generators are pivotal components, and they can be broadly classified into two overarching categories: self-synchronizing generators and synchronous generators (Table 9.1) [5, 24].

- Self-Synchronizing Stream Ciphers, also known as asynchronous stream ciphers, possess a unique characteristic that sets them apart from their synchronous counterparts. In the event of a desynchronization between the cipher sequence generated by the receiving encryptor (decryptor) and the key stream of the transmitting encryptor, self-synchronizing generators have the capability to continue accurate decryption. This resilience is attributed to the design of these stream ciphers, where the function determining the next state of the cryptosystem incorporates elements from the previously generated ciphertext. This feature allows for a correction in the event of transmission errors or lost synchronization, enhancing the reliability of the cryptographic process in less-stable communication environments.
- Synchronous Stream Ciphers, on the other hand, operate on a different principle. In these systems, the key stream is generated independently of both the plaintext stream and the ciphertext stream. The initial state of the key stream generator in synchronous systems is typically a function of the secret key and, possibly, a randomized variable. The essence of a synchronous generator lies in its ability to transform a short random secret key into a long

TABLE 9.1
Comparative Analysis of Self-Synchronizing and Synchronous Stream Ciphers

Aspect	Self-Synchronizing Stream Ciphers	Synchronous Stream Ciphers
Synchronization	Can continue accurate decryption even if the cipher sequence desynchronizes from the key stream.	Require precise synchronization between sender and receiver for successful decryption.
Dependency	The next state of the cryptosystem is influenced by parts of the previously generated ciphertext.	Key stream generation is independent of both the plaintext and the ciphertext streams.
Error Correction	Capable of self-correction in case of transmission errors or synchronization issues.	Lack inherent error correction; desynchronization leads to decryption failure.
Initial State Dependency	The initial state often depends on a segment of the previously received ciphertext.	The initial state is generally a function of the secret key and, possibly, a randomized variable.
Key Stream Generation	The generation of key streams adapts based on the received ciphertext, providing dynamic synchronization.	Generates a pseudorandom sequence from a secret key, which remains constant throughout the communication session.
Use-Case Suitability	Better suited for environments with higher risk of transmission errors or synchronization issues.	Ideal for stable communication environments where precise synchronization can be maintained.

pseudorandom sequence. This sequence is then utilized for the encryption of messages, ensuring that the encryption process remains secure as long as the secret key is protected. The synchronous nature of these ciphers necessitates that the sender and receiver maintain exact alignment in their key stream generation to ensure successful decryption, making them more susceptible to synchronization issues than their self-synchronizing counterparts.

Synchronous Keystream Generators are finite-state machines characterized by their predictable and consistent generation of keystreams independent of the plaintext or ciphertext [5, 24].

- Initialization Function (Init): This function initializes the generator using a key K and an initialization vector IV. The IV is crucial for ensuring that each encryption instance is unique, preventing the reuse of the same key-IV pair.
- Next-State Function (Next): Responsible for transitioning the generator from its current state S_i to the next state S_{i+1}. This function is pivotal in maintaining the internal dynamics of the generator.
- Keystream Function (Strm): It outputs a keystream block Z_i from the current state S_i. The generator, upon initialization, enters a state $S_0 = \text{Init}(IV, K)$, and subsequently produces keystream blocks and updates its state as needed.

The predictability and consistency of synchronous generators make them suitable for environments where strict synchronization between the sender and receiver is maintained. However, they are vulnerable to certain types of attacks if the internal state is compromised.

Self-Synchronizing Keystream Generators rely solely on previous ciphertexts, the key, and the initialization vector. They operate in a stateless fashion, allowing for recovery from synchronization loss after receiving sufficient ciphertext blocks [5, 24]:

- Initialization Function (Init): Similar to synchronous generators, it initializes the generator but also outputs internal input S and a series of dummy ciphertext blocks.
- Keystream Function (Strm): This function takes the internal input S and a specified number of previous ciphertext blocks to produce a keystream block Z_i. The number of feedback blocks r and the functions Init and Strm are essential specifications for these generators.

These generators offer resilience against synchronization loss, making them ideal for scenarios where data transmission might be interrupted or unreliable. However, their reliance on previous ciphertexts can introduce latency in the decryption process.

The Table 9.1 delineates the fundamental differences and functionalities of self-synchronizing and synchronous stream ciphers, two pivotal categories in cryptographic systems. Self-synchronizing stream ciphers are characterized by their ability to maintain accurate decryption capabilities, even in the face of desynchronization, making them particularly suited for less-stable communication channels. They adaptively generate key streams based on the received ciphertext, thereby enabling a form of inherent error correction. In contrast, synchronous stream ciphers operate under the premise of maintaining a strict synchronization between the communicating parties. They generate a key stream independently of the plaintext and ciphertext, relying on the secrecy of the initial key and, potentially, a randomized variable. While offering security in stable environments, they are susceptible to decryption failures in the event of synchronization loss. The choice between these two types of stream ciphers is largely contingent upon the specific requirements of the communication environment, with self-synchronizing ciphers offering robustness in less predictable settings, and synchronous ciphers providing efficacy in controlled, stable scenarios.

9.3.3.2   Output Functions in Stream Ciphers

The output function in stream ciphers plays a crucial role in the encryption and decryption process. It is an algorithmic component that combines the keystream with the plaintext to produce ciphertext, and vice versa. The design of this function is critical for ensuring the security and efficiency of the cipher [5, 24].

1. The encryption and decryption process is formally described as follows:
   - Encryption: The output function, denoted as Out, takes a plaintext block P_i, a keystream block Z_i, and optionally, another input R, to produce a ciphertext block C_i. The process can be represented as $C_i = Out(P_i, Z_i, R)$.
   - Decryption: The inverse of the output function, Out^{-1}, is used for decryption. It takes a ciphertext block C_i, a keystream block Z_i, and the optional input R, to retrieve the original plaintext block P_i. The decryption process is represented as $P_i = Out^{-1}(C_i, Z_i, R)$.
2. Characteristics of the Output Function:
   - Invertibility: A key characteristic of the output function is its invertibility. For any given keystream block Z_i, plaintext block P_i, and other input R, it must be possible to recover the original plaintext, ensuring that $P_i = Out^{-1}(Out(P_i, Z_i, R), Z_i, R)$.
   - Security: The security of the output function is paramount. It must be designed to resist various cryptographic attacks, ensuring that the encryption process does not introduce vulnerabilities.
   - Efficiency: The function should be computationally efficient, allowing for quick encryption and decryption, which is especially important in real-time communication systems.
3. Types of Output Functions:
   - XOR Operation: A common and simple output function is the XOR operation, where the plaintext is XORed with the keystream. This method is popular due to its simplicity and effectiveness.
   - Modular Arithmetic: Another approach involves using modular arithmetic to combine the plaintext and keystream. This method can offer enhanced security compared to XOR, depending on the implementation.
   - Complex Functions: For higher security requirements, more complex functions can be employed. These may include cryptographic algorithms that provide additional layers of security, making the cipher more resilient to attacks.

Thus, the output function is a vital element in the design of a stream cipher. Its choice and implementation directly impact the overall security and performance of the cipher. As cryptographic techniques continue to evolve, the development of innovative and robust output functions remains a key area of research in the field of cryptography.

Table 9.2 provides an in-depth look at three types of output functions: XOR Operation, Modular Arithmetic, and Complex Functions.

This table serves as a comprehensive guide to understanding the various output functions used in stream ciphers, offering insights into their operational mechanisms, security implications, and efficiency considerations. It is an essential resource for anyone looking to grasp the nuances of cryptographic techniques in stream ciphers. This comparison elucidates the trade-offs between simplicity and security in cryptographic methods, highlighting the importance of selecting the appropriate output function based on the specific requirements and constraints of the cryptographic system in question.

TABLE 9.2
Comparison of Output Functions in Stream Ciphers

Function Type	XOR Operation	Modular Arithmetic	Complex Functions
Description	Simple method where plaintext is XORed with the keystream.	Uses modular arithmetic for combining plaintext and keystream.	Involves advanced cryptographic algorithms for added security.
Encryption Process	$Ci = P_i \oplus Z_i$	$Ci = (P_i + Z_i) \bmod n$	Varies based on algorithm
Decryption Process	$Pi = C_i \oplus Z_i$	$Pi = (C_i - Z_i) \bmod n$	Inverse operation of encryption
Key Characteristics	Invertible – Easy to implement	Invertible – More complex than XOR	Highly secure – Tailored for specific needs
Security Level	Moderate; vulnerable if keystream is compromised	Higher; offers enhanced security	Very High; resilient to various attacks
Efficiency	High; minimal computational resources	Moderate; more computational resources than XOR	- Lower; generally less efficient than simpler methods

9.3.4 Recent Advancements in Stream Ciphering for Wireless Communication Systems

This subsection provides an exploration into the latest advancements in stream ciphering, particularly focusing on their application in advanced wireless communication systems. The rapid evolution of wireless technology necessitates robust and efficient cryptographic solutions to ensure data security and integrity. Stream ciphers, known for their speed and efficiency, have undergone significant transformations to meet these emerging challenges.

9.3.4.1 Enhanced Security Protocols in Stream Ciphering for Advanced Wireless Communication Systems

In the rapidly evolving landscape of wireless communication, the need for enhanced security protocols in stream ciphering is more critical than ever [7, 19]. The aim is to address the growing concerns over data security and privacy in an era marked by increasingly sophisticated cyber threats.

Figure 9.3 serves as a visual representation of the multifaceted approach required to fortify stream ciphering in the context of modern wireless communication systems. This diagram is not merely a collection of terms but a structured depiction of the intricate relationships and dependencies among various components of stream ciphering security.

This visual representation serves not only as an educational tool but also as a strategic guide for researchers, practitioners, and policymakers in the field of cybersecurity and wireless communications.

1. Adaptive Encryption Algorithms
   - Dynamic Parameter Adjustment: Modern stream ciphers are being developed with the capability to adapt their encryption parameters in real-time. This adaptability is crucial in responding to varying network conditions and threat levels, ensuring optimal security and performance.
   - Context-Aware Encryption: Discussion on encryption algorithms that adjust their behavior based on the context of the communication, such as user location, device type, and network traffic patterns.

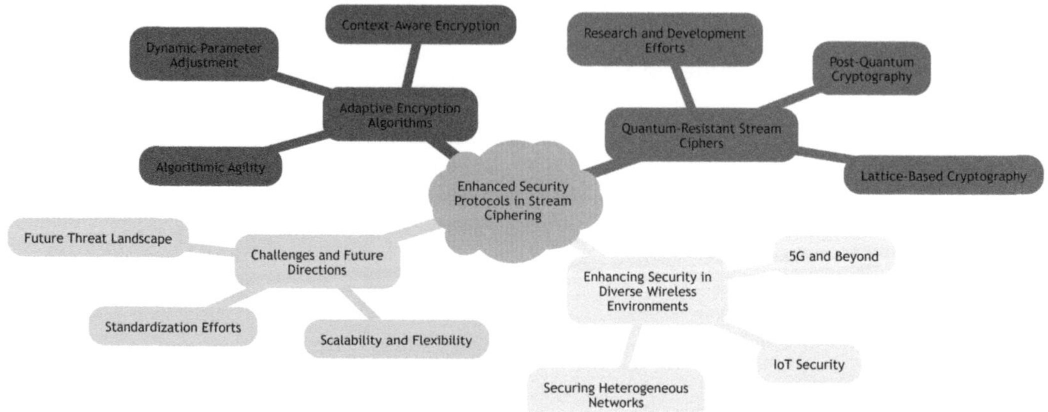

FIGURE 9.3 Enhanced Security Protocols in Stream Ciphering for Advanced Wireless Communication Systems.

- Algorithmic Agility: Exploration of the concept of algorithmic agility in stream ciphers, which allows for seamless switching between different cryptographic algorithms based on real-time assessments of threat landscapes.
2. Quantum-Resistant Stream Ciphers
   - Post-Quantum Cryptography: With the advent of quantum computing, traditional cryptographic algorithms face the risk of becoming obsolete. This section discusses the development of quantum-resistant stream ciphers designed to withstand attacks from quantum computers.
   - Lattice-Based Cryptography: An overview of lattice-based cryptographic approaches, which are gaining attention as promising candidates for quantum-resistant stream ciphers.
   - Research and Development Efforts: Insights into ongoing research initiatives and collaborations aimed at developing robust quantum-resistant cryptographic solutions for wireless communication systems.
3. Enhancing Security in Diverse Wireless Environments
   - 5G and Beyond: Addressing the unique security challenges posed by 5G networks, including higher data rates, lower latency, and increased connection density.
   - IoT Security: Examining the integration of advanced stream ciphers in IoT devices, focusing on the balance between security and resource constraints.
   - Securing Heterogeneous Networks: Strategies for implementing enhanced stream ciphers in heterogeneous networks, encompassing various wireless technologies and standards.
4. Challenges and Future Directions
   - Scalability and Flexibility: Discussing the challenges in scaling these advanced cryptographic solutions while maintaining flexibility across diverse network architectures.
   - Standardization Efforts: Highlighting the importance of global standardization in stream ciphering protocols to ensure interoperability and widespread adoption.
   - Future Threat Landscape: Speculating on future cybersecurity threats and the role of enhanced stream ciphers in mitigating these risks.

Thus, by addressing the unique challenges and opportunities presented by modern wireless networks, this subsection provides a comprehensive view of the state-of-the-art in stream ciphering security.

9.3.4.2 Efficiency Improvements for High-Speed Wireless Networks

In the domain of high-speed wireless networks, efficiency is not just a desirable attribute but a critical necessity [7, 19]. The focus is on innovations that streamline ciphering processes without compromising security, ensuring that the rapid pace of wireless communications is maintained.

Figure 9.4 shows four primary categories, each representing a key aspect of efficiency improvements:

1. Low-Latency Cipher Designs [5]:
   - Design Principles: Highlights the foundational principles guiding the development of low-latency cipher designs, crucial for real-time data transmission.
   - Innovative Algorithms: Focuses on the creation of new algorithms that are both efficient and secure, tailored for high-speed networks.
   - Case Studies: Provides real-world examples and applications of these innovative cipher designs, demonstrating their impact on network performance.
2. Resource-Efficient Implementations [2]:
   - Optimization Techniques: Discusses various methods employed to optimize cipher algorithms for better resource management.
   - Balancing Security and Efficiency: Explores the delicate balance between maintaining robust security and achieving high efficiency.
   - Emerging Technologies: Looks into how emerging technologies like edge computing contribute to the efficiency of stream ciphers.
3. Integration with Advanced Wireless Technologies [2, 12]:
   - 5G Network Adaptations: Details the specific adaptations of stream ciphers for the unique requirements of 5G networks.
   - IoT and M2M Communications: Examines the integration of efficient stream ciphers in IoT and M2M communications.
   - Cross-Layer Optimization: Investigates the potential benefits of a cross-layer design approach in stream cipher optimization.
4. Challenges and Future Directions [1, 15]:
   - Scalability Issues: Addresses the challenges in scaling these efficient cipher solutions in rapidly expanding network infrastructures.
   - Standardization and Interoperability: Highlights the importance of standardization for ensuring widespread compatibility and use.

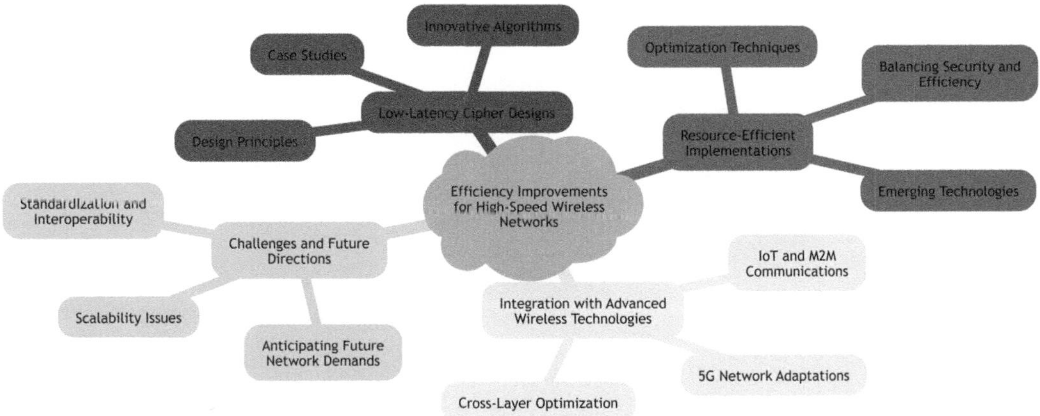

FIGURE 9.4 Efficiency Improvements for High-Speed Wireless Networks.

- Anticipating Future Network Demands: Discusses the need to design ciphers that can adapt to future network developments and demands.

Thus, Figure 9.4 serves as an effective tool for understanding the complex and multifaceted nature of efficiency improvements in stream ciphering, particularly in the context of advanced wireless communication systems. It visually represents the interconnectedness of various components, providing a clear and concise overview of the topic.

9.3.4.3 Integration with Advanced Wireless Technologies

The integration of stream ciphers with advanced wireless technologies is a pivotal aspect of modern cybersecurity [7, 19]. The focus is on the harmonization of cryptographic robustness with the unique characteristics and demands of these emerging technologies.

The detailed exploration of the integration of stream ciphers with advanced wireless technologies provides a comprehensive view of the current trends, challenges, and future directions in this crucial area of cybersecurity (Figure 9.5).

1. 5G Network Adaptations [1, 12]
   - Tailoring to 5G Specifications: Delve into how stream ciphers are being customized to align with the high-speed, low-latency, and massive connectivity features of 5G networks. This includes adapting to the network's architecture, such as small cell deployments and network slicing.
   - Enhanced Data Protection: Discuss the importance of stream ciphers in protecting the vast data flows characteristic of 5G networks, from user data to critical infrastructure communications.
   - Support for 5G Use Cases: Explore how stream ciphers are being optimized for various 5G use cases, including ultra-reliable low-latency communications (URLLC), massive machine-type communications (mMTC), and enhanced mobile broadband (eMBB).
2. IoT and M2M Communications [15, 17]
   - Resource-Constrained Environments: Address the challenges of implementing stream ciphers in IoT devices, which often have limited computational power and energy resources. Focus on lightweight cryptographic solutions that provide adequate security without overburdening these devices.

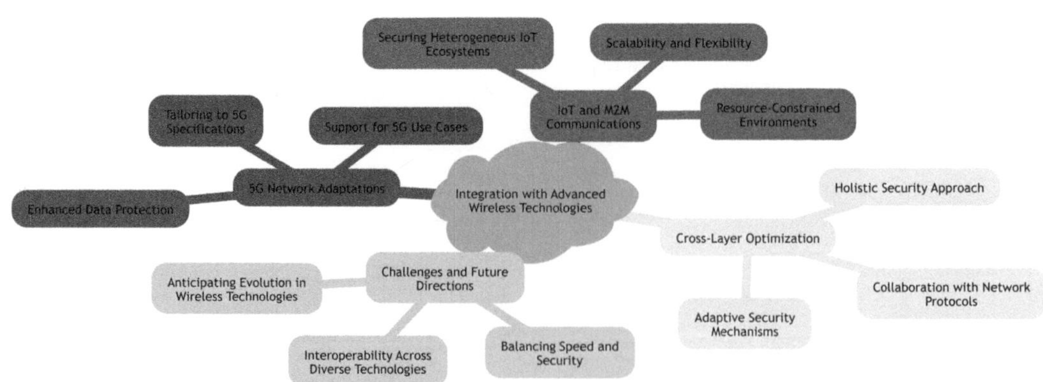

FIGURE 9.5 Integration with Advanced Wireless Technologies.

- Scalability and Flexibility: Discuss the scalability of stream ciphers to support the vast number of devices in IoT networks and their flexibility to adapt to different communication protocols and standards.
- Securing Heterogeneous IoT Ecosystems: Examine strategies for securing diverse IoT ecosystems, encompassing a wide range of devices and applications, from smart homes to industrial IoT (IIoT).

3. Cross-Layer Optimization [7]
   - Holistic Security Approach: Introduce the concept of cross-layer design in stream ciphering, where insights from different network layers (physical, data link, network, etc.) are utilized to enhance security and efficiency.
   - Adaptive Security Mechanisms: Explore adaptive security mechanisms that can dynamically adjust encryption parameters based on network conditions, traffic types, and threat levels.
   - Collaboration with Network Protocols: Highlight the importance of stream ciphers working in tandem with network protocols to ensure end-to-end security in wireless communications.

4. Challenges and Future Directions [2]
   - Interoperability Across Diverse Technologies: Discuss the challenges in ensuring that stream ciphers are interoperable across various wireless technologies, including legacy systems and emerging standards.
   - Balancing Speed and Security: Address the ongoing challenge of balancing high-speed data transmission with robust security in stream ciphering, particularly in high-throughput environments like 5G.
   - Anticipating Evolution in Wireless Technologies: Consider the need for stream ciphers to be adaptable and forward-compatible with future advancements in wireless technology, ensuring long-term relevance and effectiveness.

It underscores the intricate balance required between maintaining robust security measures and achieving the high efficiency necessary for modern wireless communication systems. The advancements in adapting stream ciphers for 5G, IoT, and cross-layer optimization reflect a concerted effort to address the unique demands of these technologies. As the digital landscape evolves, this integration will play a pivotal role in shaping the future of secure wireless communication. The continuous innovation in this field is not only a response to current technological demands but also a proactive approach to future challenges, ensuring that wireless networks remain secure, reliable, and efficient in the face of ever-increasing data volumes and evolving cyber threats.

9.3.4.4 Standardization and Compliance in Stream Ciphering for Advanced Wireless Communication Systems

In the rapidly evolving landscape of wireless communication, standardization and compliance play a pivotal role in ensuring the security and interoperability of stream ciphering technologies [7, 19]. This subsection delves into the complexities of developing and adhering to global security standards, focusing on their impact on stream ciphering within advanced wireless communication systems.

Table 9.3 serves as a comprehensive guide to understanding the various aspects of standardization and compliance in stream ciphering, offering insights into their operational mechanisms, challenges, and future considerations.

This comparison elucidates the complexities and necessities of establishing and adhering to global security standards in the realm of stream ciphering. It highlights the importance of harmonizing security protocols, ensuring interoperability, setting security benchmarks, and the collaborative efforts required to achieve these goals. Table 9.3 also reflects on the challenges and future directions,

TABLE 9.3
Key Aspects of Standardization and Compliance in Stream Ciphering for Advanced Wireless Communication Systems

Aspect	Description	Challenges	Key Organizations	Future Directions
Harmonizing Security Protocols	Establishing universal security standards for stream ciphers across various technologies and regions.	Keeping pace with rapidly evolving standards and technologies.	ITU, IEEE, IETF	Developing forward-thinking standards that anticipate technological advances.
Ensuring Interoperability	Facilitating seamless communication among diverse wireless systems through standardized stream ciphers.	Balancing the rigidity of compliance with the need for flexibility and innovation.	3GPP, ETSI	Accommodating regional variations while maintaining global standards.
Benchmarking Security Measures	Setting minimum security thresholds for stream ciphers to ensure robust protection.	Resource and cost implications, especially for smaller entities.	NIST, ISO/IEC	Adapting benchmarks to emerging threats and technologies.
Collaborative Efforts	Joint initiatives among standardization bodies to develop cohesive security standards.	Aligning different organizational goals and methodologies.	GSMA, ANSI	Enhancing collaboration for comprehensive and universal standards.
Global vs. Regional Standards	Balancing the development of global standards with regional specificities in wireless communication.	Navigating differing regional regulations and requirements.	Regional regulatory bodies	Striking a balance between universal and region-specific standards.

emphasizing the need for forward-thinking and adaptable standards that can cater to both global and regional requirements in the ever-evolving landscape of wireless communication technologies.

9.4 METHODOLOGY FOR COMPUTATIONAL MODELING AND COMPARATIVE ANALYSIS OF STREAM CIPHERS

In the intricate world of cryptographic systems, the efficiency and security of stream ciphers are of paramount importance. This section delves into the methodologies employed for computational modeling and comparative analysis of stream ciphers, focusing on their computational efficiency. The aim is to provide a comprehensive framework for evaluating and benchmarking various cryptographic algorithms, thereby facilitating informed decisions in the selection and deployment of these ciphers in advanced wireless communication systems.

9.4.1 METHODOLOGY FOR ASSESSING COMPUTATIONAL EFFICIENCY

The evaluation of computational efficiency in stream ciphers is a multifaceted process, encompassing various performance criteria. The methodology adopted here is inspired by the framework provided by eSTREAM (www.ecrypt.eu.org/stream/perf/), which outlines key performance measures for stream ciphers [25, 26]. This approach is designed to reflect the diverse implementation properties and application scenarios of stream ciphers.

1. Encryption Rate for Long Streams:
   - Objective: This measure focuses on the encryption rate for long data streams, a scenario where stream ciphers typically hold an advantage over block ciphers.
   - Testing Framework: The encryption rate is assessed by encrypting a long stream in chunks, approximately 4KB each, using the ECRYPT_encrypt_blocks function. The metric used is encryption speed in cycles/byte, calculated based on the number of bytes encrypted in a fixed time frame (250 µsec).
   - Significance: This criterion is often the most crucial in applications where large volumes of data need to be encrypted efficiently.
2. Packet Encryption Rate:
   - Objective: To determine the efficiency of stream ciphers in encrypting packets of varying lengths, particularly in comparison to block ciphers.
   - Methodology: The packet encryption rate is measured using the ECRYPT_encrypt_packet function on packets of different sizes (40, 576, and 1500 bytes), representative of typical internet traffic. This test includes separate IV setup for each packet and, if applicable, MAC finalization for authenticated encryption.
   - Relevance: This measure is critical in applications with diverse packet sizes, highlighting the efficiency of stream ciphers in smaller packet encryption scenarios.
3. Key and IV Setup:
   - Objective: To evaluate the efficiency of key and IV setup processes in stream ciphers.
   - Approach: The efficiency of the key setup (ECRYPT_keysetup) and IV setup (ECRYPT_ivsetup) is measured separately.
   - Context: While this test is generally considered less critical, it provides valuable insights into the operational efficiency of stream ciphers, especially in scenarios where frequent key or IV changes are required.

The methodology for assessing the computational efficiency of stream ciphers is not just a set of empirical tests but a conceptual framework that reflects the practical realities of cryptographic applications. It recognizes that stream ciphers are deployed in a variety of contexts, each with its unique efficiency demands. By adopting a multi-dimensional approach to performance evaluation, this methodology offers a holistic view of a cipher's operational characteristics.

1. Adaptation to Real-World Scenarios:
   - The methodology is designed to mimic real-world usage scenarios of stream ciphers, from encrypting lengthy data streams to handling small, frequent data packets.
   - This approach ensures that the performance metrics obtained are not only theoretically sound but also practically relevant.
2. Balancing Speed and Security:
   - A critical aspect of this methodology is the balance it strikes between encryption speed and cryptographic security.
   - It provides a platform to evaluate whether increased efficiency compromises the cipher's security or whether robust security features impede performance.
3. Future-Proofing the Evaluation:
   - The methodology is structured to be adaptable to future advancements in stream ciphering and wireless communication technologies.
   - It allows for the incorporation of new performance measures as the landscape of cryptographic applications evolves.

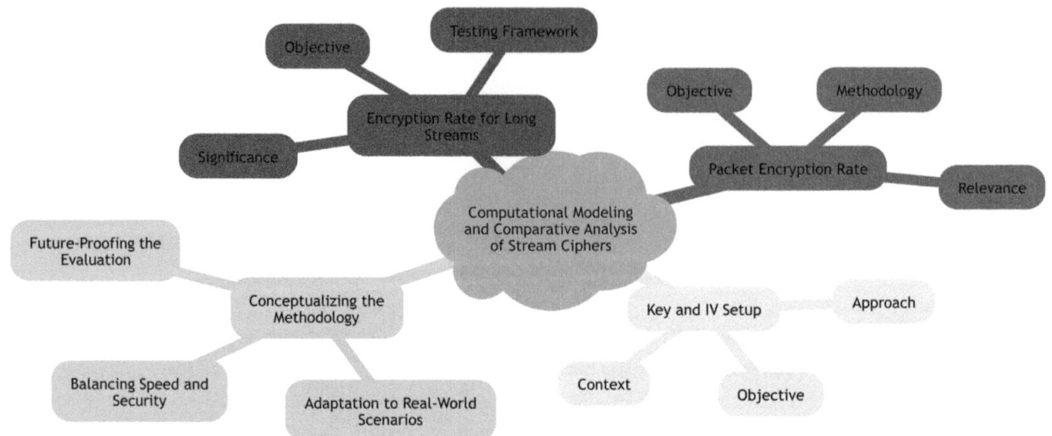

FIGURE 9.6   Computational Modeling and Comparative Analysis of Stream Ciphers.

Figure 9.6 provides a structured visualization of the key aspects involved in the computational modeling and comparative analysis of stream ciphers. Each node and sub-node is placed to illustrate the interconnected nature of these topics, emphasizing how each aspect contributes to a holistic understanding of stream cipher performance.

This detailed exploration of the methodology for assessing computational efficiency in stream ciphers provides a nuanced understanding of the various performance criteria and their relevance in real-world applications. It reflects a deep integration of theoretical knowledge and practical considerations, essential for the effective evaluation and advancement of cryptographic technologies in the field of wireless communication.

9.4.2   CRYPTOGRAPHIC ALGORITHMS IN COMPUTATIONAL MODELING AND SIMULATION OF ADVANCED WIRELESS COMMUNICATION SYSTEMS

In the intricate field of computational modeling and simulation for advanced wireless communication systems, the evaluation and analysis of cryptographic algorithms present a unique set of challenges and complexities. This section delves into the structural, design, and security aspects of various cryptographic algorithms, each playing a pivotal role in ensuring the integrity and confidentiality of data in wireless networks. The focus is on understanding the nuances of these algorithms, their design principles, security features, and operational parameters, all within the context of advanced computational modeling and simulation.

9.4.2.1   Mickey [27, 28]

- Structure and Design: Mickey is a stream cipher that is part of the eSTREAM portfolio. It is designed with a focus on hardware efficiency and simplicity. The cipher operates using two linear feedback shift registers (LFSRs) with dynamic, irregular clocking, which enhances its security against linear and differential cryptanalysis.
- Security Aspects: Mickey's security is rooted in its complex clocking mechanism, which makes it resistant to common attacks such as time-memory trade-off attacks. The cipher's structure is optimized to prevent predictability and ensure a high degree of randomness in its keystream.
- Parameters and Characteristics: Mickey operates with an 80-bit key and a 40-bit initialization vector (IV). Its compact design and efficient use of resources make it suitable for environments with limited computational capabilities, such as IoT devices in wireless networks.

- Context in Wireless Systems: In the realm of advanced wireless communication systems, Mickey's efficiency and security profile make it a viable option for securing data transmission in systems where hardware resources and power consumption are critical considerations.

9.4.2.2 Decim [29]

- Structure and Design: Decim is a hardware-oriented stream cipher, also part of the eSTREAM portfolio, designed for ultra-low power consumption and minimal hardware footprint. It employs an irregularly clocked LFSR and a non-linear filtering function, providing a balance between simplicity and security.
- Security Aspects: The cipher's design includes built-in countermeasures against several forms of cryptanalysis, including correlation attacks and algebraic attacks. Decim's irregular clocking mechanism plays a crucial role in enhancing its resistance to various cryptographic attacks.
- Parameters and Characteristics: Decim operates with an 80-bit key and a variable-length IV. Its design is particularly focused on applications where power efficiency is paramount, making it ideal for battery-operated wireless devices.
- Context in Wireless Systems: In the context of computational modeling and simulation of advanced wireless communication systems, Decim stands out for its application in scenarios where energy efficiency is as crucial as security, such as in sensor networks and wearable technology.

9.4.2.3 Strumok-256 [30, 31]

- Structure and Design: Strumok-256 is a high-performance stream cipher known for its exceptional speed and security. It is designed with a focus on providing robust encryption with minimal computational overhead. The cipher utilizes a complex arrangement of shift registers and non-linear combining functions to generate its keystream.
- Security Aspects: The strength of Strumok-256 lies in its large key size and sophisticated internal structure, which offers strong resistance against various forms of cryptanalysis, including brute-force attacks and statistical analysis. Its design ensures a high degree of randomness and unpredictability in the keystream.
- Parameters and Characteristics: Operating with a 256-bit key, Strumok-256 is tailored for scenarios requiring a higher level of security. Its efficient algorithm makes it suitable for high-speed data processing, a critical requirement in advanced wireless communication systems.
- Context in Wireless Systems: In the context of computational modeling and simulation for wireless systems, Strumok-256's blend of speed and security makes it an ideal choice for protecting high-volume data transmissions, such as in broadband wireless networks and secure cloud communications.

9.4.2.4 Strumok-512 [30, 31]

- Structure and Design: Building on the foundation of Strumok-256, Strumok-512 is an advanced stream cipher that offers even higher security. It features a similar design to its predecessor but with an extended key size and enhanced internal mechanisms for keystream generation.
- Security Aspects: With a 512-bit key, Strumok-512 provides an ultra-high level of security, making it virtually impervious to brute-force attacks. Its complex internal structure further fortifies it against sophisticated cryptanalytic techniques.
- Parameters and Characteristics: Strumok-512's design is optimized for environments where maximum security is paramount. Despite its enhanced security features, it maintains a high level of efficiency in data processing.
- Context in Wireless Systems: Strumok-512 is particularly relevant in wireless communication systems where the highest security standards are required, such as in military communications, government networks, and critical infrastructure control systems.

9.4.2.5 Trivium [32–34]

- Structure and Design: Trivium is a stream cipher known for its simplicity and high efficiency. It is part of the eSTREAM portfolio and is designed to be particularly efficient in hardware implementations. The cipher is based on a unique combination of three shift registers with non-linear feedback.
- Security Aspects: Trivium's security is derived from its minimalistic yet effective design, which provides a high level of security while maintaining simplicity. It is resistant to known forms of cryptanalysis, including linear and differential attacks.
- Parameters and Characteristics: Trivium operates with an 80-bit key and an 80-bit IV, making it a versatile option for various applications. Its design allows for fast data encryption and decryption, crucial for real-time communication systems.
- Context in Wireless Systems: In advanced wireless communication systems, Trivium's balance of simplicity, speed, and security makes it a practical choice for securing data in systems where both performance and security are essential, such as in IoT devices and high-speed wireless networks.

9.4.2.6 Sosemanuk [35, 36]

- Structure and Design: Sosemanuk is a stream cipher that combines the best elements of the Snow 2.0 cipher and the Serpent block cipher. It is part of the eSTREAM portfolio and is designed for high-speed software implementations. The cipher uses a series of linear feedback shift registers (LFSRs) and a non-linear filtering function derived from Serpent.
- Security Aspects: Sosemanuk offers strong resistance against known cryptanalytic attacks, including linear and differential cryptanalysis. Its security is bolstered by the complex interaction between its LFSR-based stream generation and the Serpent-inspired filtering function.
- Parameters and Characteristics: Operating with a 128-bit to 256-bit key and a 128-bit IV, Sosemanuk is tailored for applications requiring both speed and security. Its efficient algorithm makes it suitable for high-throughput data processing in wireless systems.
- Context in Wireless Systems: In the realm of advanced wireless communication, Sosemanuk's blend of speed and security makes it an ideal choice for encrypting data in real-time applications such as streaming services and secure voice communications.

9.4.2.7 Snow 2.0 [9, 37, 38]

- Structure and Design: Snow 2.0, an evolution of the earlier Snow cipher, is a stream cipher known for its high performance in both hardware and software implementations. It features a linear feedback shift register (LFSR) and a finite state machine (FSM), which work together to produce a highly secure keystream.
- Security Aspects: The cipher's security is enhanced by its complex internal state and the use of both linear and non-linear components in keystream generation. Snow 2.0 is resistant to a variety of attacks, including time-memory trade-off and algebraic attacks.
- Parameters and Characteristics: Snow 2.0 operates with either a 128-bit or 256-bit key, along with a 128-bit IV. Its design allows for fast data processing, making it suitable for high-speed wireless communication applications.
- Context in Wireless Systems: Snow 2.0's efficiency and robust security profile make it well-suited for use in wireless communication systems where high data rates and strong encryption are required, such as in 4G and 5G networks.

9.4.2.8 Salsa20 [39–41]

- Structure and Design: Salsa20 is a stream cipher known for its simplicity and high speed. Designed by Daniel J. Bernstein, it stands out for its unique 'hash-like' computation to generate its keystream, consisting of 20 rounds of simple, fast operations.

- Security Aspects: Salsa20 offers a high level of security, with resistance to known attacks such as differential cryptanalysis. Its design ensures a high degree of randomness and unpredictability in the keystream.
- Parameters and Characteristics: Salsa20 can operate with either a 128-bit or 256-bit key, along with a 64-bit IV. Its speed and efficiency make it suitable for a wide range of applications, from low-power IoT devices to high-speed network communications.
- Context in Wireless Systems: In advanced wireless communication systems, Salsa20's speed and security make it a valuable tool for securing data transmission. Its adaptability to different platforms and efficiency in processing make it suitable for various wireless applications, including secure messaging and VPNs.

9.4.2.9 Rabbit [42, 43]

- Structure and Design: Rabbit is a high-speed stream cipher known for its efficiency in software implementations. It operates using a combination of linear and non-linear components in its internal state, which consists of eight 32-bit registers and eight 32-bit counters.
- Security Aspects: Rabbit's security is derived from its complex internal state dynamics and fast mixing of state elements, making it resistant to common cryptanalytic attacks, including brute force and statistical attacks.
- Parameters and Characteristics: Rabbit uses a 128-bit key and a 64-bit IV. Its design allows for rapid data processing, making it suitable for applications requiring high throughput.
- Context in Wireless Systems: In wireless communication systems, Rabbit is ideal for high-speed data transmission scenarios such as streaming media, offering a balance of speed and security.

9.4.2.10 HC-128 and HC-256 [44, 45]

- Structure and Design: HC-128 and HC-256 are stream ciphers from the eSTREAM portfolio, designed for high performance in software. HC-128 operates with a 128-bit key, while HC-256 uses a 256-bit key. Both ciphers use a combination of linear and non-linear operations over a large internal state for keystream generation.
- Security Aspects: These ciphers offer strong security, with HC-256 providing an enhanced level of protection suitable for highly sensitive data. They are resistant to various attacks, including time-memory trade-off and differential cryptanalysis.
- Parameters and Characteristics: HC-128 and HC-256 are known for their high throughput and minimal memory requirements, making them efficient for a wide range of applications.
- Context in Wireless Systems: HC-128 is suitable for general-purpose wireless applications, while HC-256 is ideal for scenarios demanding higher security levels, such as secure corporate networks and critical infrastructure communications.

9.4.2.11 AES-128 and AES-256 [46, 47]

- Structure and Design: AES (Advanced Encryption Standard) is a widely-used block cipher with variants based on key sizes. AES-128 uses a 128-bit key, while AES-256 employs a 256-bit key. Both operate on a block size of 128 bits and use a series of transformations for encryption and decryption.
- Security Aspects: AES is renowned for its strong security profile, making it resistant to the most advanced cryptanalytic attacks. AES-256 offers an even higher security level due to its longer key size.
- Parameters and Characteristics: While inherently a block cipher, AES can be used in various stream cipher modes, such as CTR (Counter Mode), to encrypt data streams efficiently.
- Context in Wireless Systems: AES is versatile and robust, making it suitable for a wide range of wireless applications, from securing Wi-Fi networks to encrypting data in mobile communications and IoT devices.

TABLE 9.4
Comparative Overview of Cryptographic Algorithms in Wireless Systems

Algorithm	Key Size (bits)	IV Size (bits)	Structure	Security Aspects	Suitability for Wireless Systems
Mickey	80	40	Two LFSRs with dynamic clocking	Resistant to time-memory trade-off attacks	Ideal for constrained environments
Decim	80	Variable	Irregularly clocked LFSR	Resistant to correlation and algebraic attacks	Optimized for ultra-low power consumption
Strumok-256	256	256	Shift registers with non-linear functions	High resistance to brute-force attacks	Suitable for high-speed data transmission
Strumok-512	512	256	Enhanced Strumok-256 design	Ultra-high security level	Ideal for highly sensitive data protection
Trivium	80	80	Three shift registers with non-linear feedback	Resistant to linear and differential attacks	Balances simplicity, speed, and security
Sosemanuk	128-256	128	LFSRs with Serpent-based filtering	Strong resistance against cryptanalysis	Efficient for real-time applications
Snow2.0-128	128	128	LFSR and FSM	Resistant to time-memory trade-off and algebraic attacks	High performance in hardware and software
Snow2.0-256	256	128	Enhanced Snow2.0 design	Higher security level	Suitable for high data rate wireless systems
Salsa20	128-256	64	'Hash-like' computation	High randomness and unpredictability	Versatile for various wireless applications
Rabbit	128	64	Registers and counters based	Fast mixing of state elements	Ideal for high-speed data transmission
HC-128	128	128	Large internal state with linear and non-linear operations	Strong security	Efficient for general-purpose wireless applications
HC-256	256	256	Enhanced HC-128 design	Enhanced security level	Suitable for secure corporate networks
AES-128	128	--	Block cipher with series of transformations	Robust security	Versatile in stream cipher modes
AES-256	256	--	Extended key size AES variant	Higher security level	Ideal for securing sensitive data

Table 9.4 provides a detailed comparative overview of various cryptographic algorithms, highlighting their key features, structural designs, security aspects, and suitability for wireless communication systems:

- Key and IV Sizes: The table outlines the key and initialization vector (IV) sizes for each algorithm, indicating their potential cryptographic strength. Larger key sizes generally imply higher security levels, as seen in algorithms like Strumok-512 and AES-256.
- Structural Complexity: The structural design of each algorithm is summarized, providing insights into their operational mechanisms. This includes the use of linear feedback shift registers (LFSRs), finite state machines (FSMs), and non-linear functions, each contributing to the cipher's overall security and efficiency.
- Security Aspects: Table 9.4 highlights the security features of each algorithm, including their resistance to various cryptanalytic attacks. This information is crucial for understanding the robustness of each cipher in protecting wireless communications.
- Suitability for Wireless Systems: Each algorithm's suitability for different wireless communication scenarios is assessed, considering factors such as speed, resource efficiency, and security requirements. For instance, algorithms like Sosemanuk and Rabbit are noted for their efficiency in real-time applications, while HC-256 and AES-256 are more suited for environments requiring enhanced security.

The comparative analysis presented in this table is instrumental in our research on Computational Modeling and Simulation of Advanced Wireless Communication Systems. It underscores the importance of carefully evaluating and selecting cryptographic algorithms that not only provide robust security but also align with the specific operational demands of wireless communication systems.

9.5 RESULTS OF THE COMPUTATIONAL EXPERIMENT

9.5.1 RESULTS OF THE COMPUTATIONAL EXPERIMENT ON LONG STREAM ENCRYPTION

This section presents the findings from a comprehensive computational experiment focused on the encryption of long data streams using various cryptographic primitives. The experiment's primary objective was to evaluate the performance of different ciphers in encrypting 1GB of data, a scenario where stream ciphers typically exhibit significant advantages. The results are crucial for understanding the efficiency of these ciphers in real-world applications, particularly in wireless communication systems where large volumes of data are frequently transmitted.

The experiment involved generating 1GB of random data, followed by the random generation of a key and the initialization of cipher states using an initialization vector (IV). The encryption process for the 1GB data was then conducted, and the performance of each cipher was measured in terms of cycles per byte and Mbps. The testing environment was standardized using an Intel Core i5-7200U CPU @ 2.50 GHz 2.70 GHz processor.

The results, summarized in Table 9.5, provide insights into the efficiency of various cryptographic primitives in encrypting long data streams:

- High-Efficiency Ciphers: Strumok-256 and Strumok-512 stand out with their exceptional performance, achieving the highest Mbps rates with minimal cycles per byte. Their efficiency in processing large data volumes makes them highly suitable for applications requiring rapid data transmission without compromising security.
- Balanced Performance: Ciphers like HC-128, Snow2.0 (both 128 and 256 variants), and Sosemanuk demonstrate a balanced performance, offering high Mbps rates with relatively low cycles per byte. These ciphers are well-suited for general-purpose use in wireless systems where both speed and security are important.

TABLE 9.5
Encrypted 1 Gbytes Performance Results

Encryption Algorithms	Cycles/Byte	Mbps
Mickey	316	68
Decim	1545	14
Strumok-256	1	12840
Strumok-512	1	12861
Trivium	6	3785
Sosemanuk	4	4865
Snow2.0-128	3	8037
Snow2.0-256	3	8025
Salsa20	8	2690
Rabbit	6	3743
HC-128	2	10857
HC-256	6	3861
AES-128	9	2470
AES-256	13	1231

- Specialized Ciphers: Mickey and Decim, while not as fast as some other ciphers, have their unique advantages in specific scenarios. Mickey's moderate speed is offset by its hardware efficiency, making it ideal for constrained environments. Decim's lower Mbps rate is a trade-off for its design, which is optimized for ultra-low power consumption in IoT devices.
- AES Variants: AES-128 and AES-256, while slower in comparison to some stream ciphers, offer robust security. Their ability to operate in fast stream cipher modes makes them versatile for a wide range of wireless applications, from consumer devices to enterprise-level systems.
- Considerations for Wireless Systems: The results underscore the importance of selecting appropriate ciphers based on specific requirements of wireless communication systems. Factors such as data volume, transmission speed, hardware constraints, and security requirements must be considered to choose the most suitable cipher.

Figure 9.7 displays the number of computational cycles required per byte for various encryption algorithms. Lower values indicate higher efficiency in terms of the computational cycles needed per byte of data. For instance, Strumok-256 and Strumok-512 are the most efficient in terms of cycles per byte, suggesting they require fewer computational resources.

Figure 9.8 illustrates the encryption speed in Mbps for different algorithms. Higher values represent faster data processing speeds. Here, it is evident that Strumok-256 and Strumok-512 provide the highest encryption speeds, indicating their superiority in processing large volumes of data quickly.

Both charts offer a comparative view of the encryption algorithms in terms of their computational efficiency and data processing speed, highlighting the performance differences across various encryption methods.

The findings from this computational experiment provide valuable insights into the performance of various cryptographic ciphers in encrypting long data streams. The results highlight the diversity and specialization within the field of cryptography, emphasizing the need to carefully consider the unique attributes of each cipher when deploying them in wireless communication systems. As the demand for secure and efficient data transmission continues to grow, these ciphers will play a crucial role in ensuring the integrity and confidentiality of communications in an increasingly connected world.

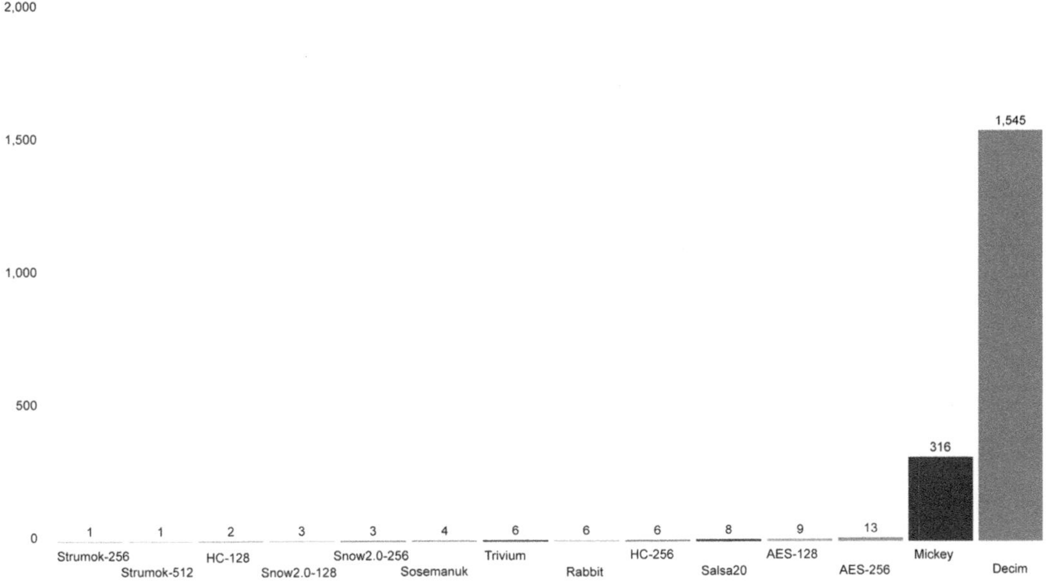

FIGURE 9.7 Cycles per Byte for Different Encryption Algorithms.

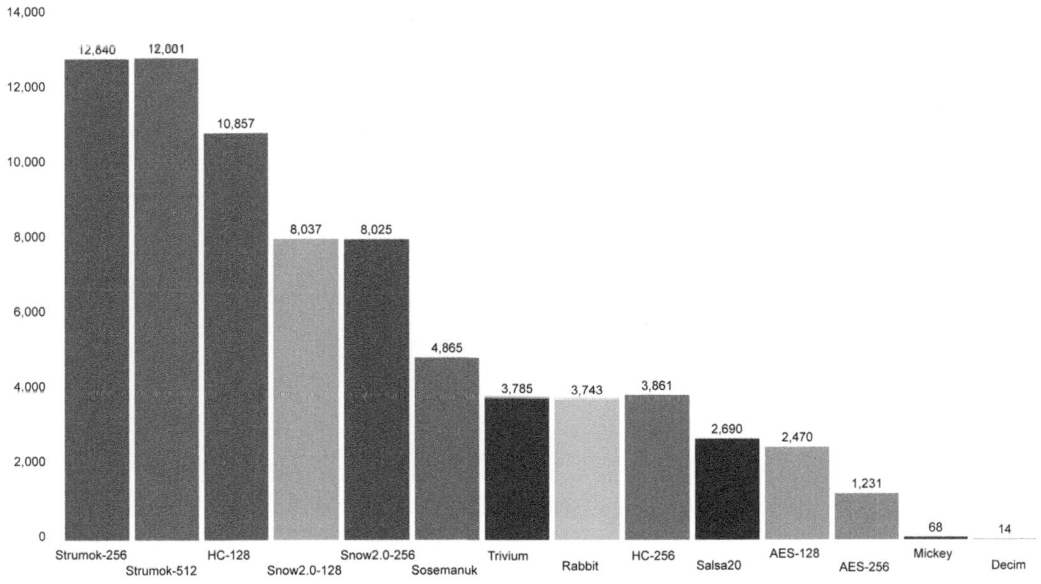

FIGURE 9.8 Encryption Speed in Mbps for Different Algorithms.

9.5.2 Results of the Computational Experiment on Packet Encryption

This subsection presents the results of a computational experiment designed to evaluate the performance of various cryptographic primitives in encrypting short data packets. The experiment's focus was on understanding how different ciphers handle the encryption of small data blocks, a scenario critical in wireless communication systems where data packets are often of limited size. The results provide insights into the efficiency and suitability of each cipher for packet-based data transmission.

The experiment involved encrypting data blocks of different sizes (40 bytes, 576 bytes, and 1500 bytes) under multiple keys. The performance of each cipher was measured in terms of cycles per byte and Mbps, providing a comprehensive view of their efficiency in handling small packets. The testing environment utilized an Intel Core i5-7200U CPU @ 2.50 GHz 2.70 GHz processor.

The results, summarized in Tables 9.6, 9.7, and 9.8, highlight the performance of each cryptographic primitive in encrypting short data packets:

TABLE 9.6
Encrypted 350 Blocks of 40 Bytes

Encryption Algorithms	Cycles/Byte	Mbps
Mickey	753	29
Decim	2302	9
Strumok-256	17	1273
Strumok-512	17	1244
Trivium	27	794
Sosemanuk	16	1318
Snow2.0-128	13	1647
Snow2.0-256	13	1698
Salsa20	12	1867
Rabbit	19	1155
HC-128	302	72
HC-256	2407	9
AES-128	13	1697
AES-256	18	1231

TABLE 9.7
Encrypted 120 Blocks of 576 Bytes

Encryption Algorithms	Cycles/Byte	Mbps
Mickey	349	62
Decim	1623	13
Strumok-256	3	8253
Strumok-512	3	8132
Trivium	7	2989
Sosemanuk	4	4851
Snow2.0-128	3	6662
Snow2.0-256	3	6662
Salsa20	8	2711
Rabbit	6	3351
HC-128	23	957
HC-256	196	111
AES-128	9	2436
AES-256	13	1728

TABLE 9.8
Encrypted 50 Blocks of 1500 Bytes

Encryption Algorithms	Cycles/Byte	Mbps
Mickey	330	66
Decim	1598	14
Strumok-256	2	10714
Strumok-512	2	10909
Trivium	6	3448
Sosemanuk	4	5769
Snow2.0-128	3	7317
Snow2.0-256	3	7407
Salsa20	8	2643
Rabbit	6	3571
HC-128	10	2198
HC-256	83	263
AES-128	9	2449
AES-256	13	1724

- High-Efficiency Ciphers: Strumok-256 and Strumok-512 demonstrate remarkable efficiency, especially in handling larger packets (576 and 1500 bytes), showcasing their potential for high-speed data transmission in wireless systems.
- Balanced Performance: Ciphers like Salsa20, Snow2.0, and Sosemanuk exhibit a balanced performance across different packet sizes, indicating their versatility in various wireless communication scenarios.
- Specialized Ciphers: Mickey and Decim, while slower, offer unique advantages in specific contexts, such as environments with stringent power or computational resource limitations.
- AES Variants: AES-128 and AES-256 maintain consistent performance across different packet sizes, reinforcing their suitability for secure wireless communications, especially when used in stream cipher modes.

Figure 9.9 provides a comparative view of the computational efficiency of various encryption algorithms when encrypting different block sizes. It clearly shows how the number of cycles per byte varies not only across different algorithms but also with the change in block size. Algorithms with lower cycles per byte are more efficient in terms of computational resources needed per byte of data.

Figure 9.10 provides a clear comparison of the encryption speeds in Mbps for various algorithms when processing different block sizes. It illustrates how the speed in Mbps varies not only among different algorithms but also changes with the size of the data blocks being encrypted. Algorithms with higher Mbps values are faster in processing data, which is crucial for applications requiring high-speed data encryption.

The findings from this computational experiment provide valuable insights into the packet encryption capabilities of various cryptographic ciphers. The results highlight the importance of selecting appropriate ciphers based on the specific requirements of wireless communication systems, considering factors such as packet size, transmission speed, and security needs. As wireless technologies continue to evolve, these ciphers will play a crucial role in ensuring the secure and efficient transmission of data in wireless networks.

9.5.3 Results of IV and Key Setup Performance Testing

This subsection presents the findings from a computational experiment designed to evaluate the performance of various cryptographic algorithms during the setup of initialization vectors (IV) and keys.

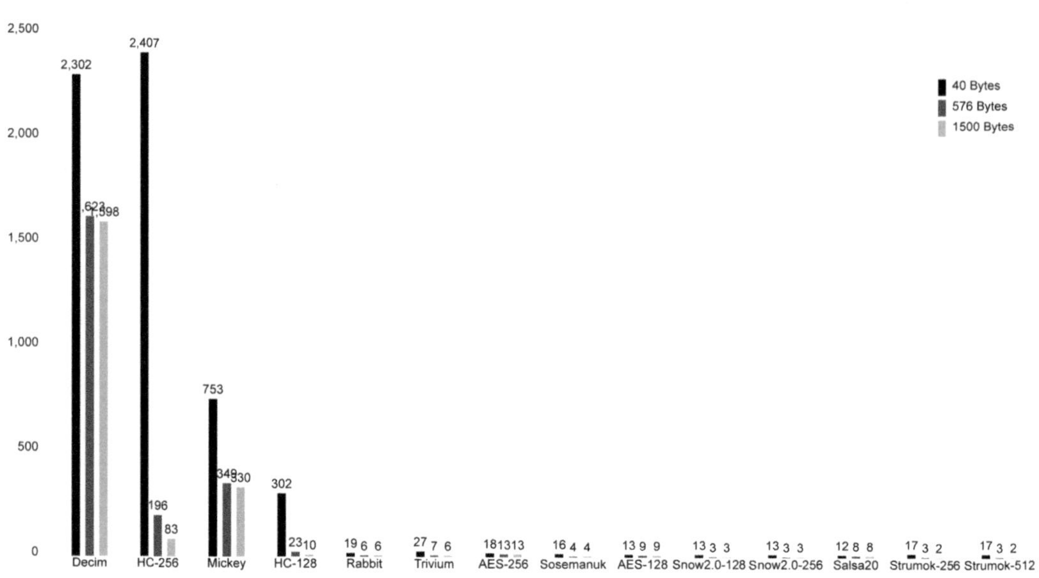

This chart shows the cycles per byte for different encryption algorithms across various block sizes (40 Bytes, 576 Bytes, 1500 Bytes). The data is ordered in ascending order of cycles per byte.

FIGURE 9.9   Cycles per Byte when Encrypting Different Block Sizes.

The experiment's focus was on understanding how different ciphers handle the setup process, a critical aspect in scenarios where frequent re-keying or re-initialization is required, such as in dynamic wireless communication systems. The results provide insights into the efficiency of each cipher in this crucial preparatory phase.

The experiment involved conducting 500 IV setups under multiple keys for each cipher. The performance of each cipher was measured in terms of cycles per setup and setups per second, providing a comprehensive view of their efficiency in the setup phase. The testing environment utilized an Intel Core i5-7200U CPU @ 2.50 GHz 2.70 GHz processor.

The results, summarized in Table 9.9, highlight the performance of each cryptographic primitive during the IV and key setup process:

- High-Efficiency Ciphers: Strumok-256 and Strumok-512 demonstrate exceptional efficiency in the setup process, with minimal cycles per setup and high setups per second. This efficiency is crucial in applications requiring frequent re-initializations.
- Balanced Performance: Ciphers like Sosemanuk, Snow2.0, and Rabbit exhibit a balanced performance, indicating their suitability for systems where moderate setup times are acceptable.
- Specialized Ciphers: Mickey and Decim, while slower in setup, offer unique advantages in specific contexts, such as environments where setup time is less critical.
- AES Variants: AES-128 and AES-256 show remarkably low cycles per setup, indicating their efficiency in the setup process. However, the setups per second metric is not applicable due to the extremely low cycle count.

The findings from this computational experiment provide valuable insights into the IV and key setup capabilities of various cryptographic ciphers. The results highlight the importance of considering

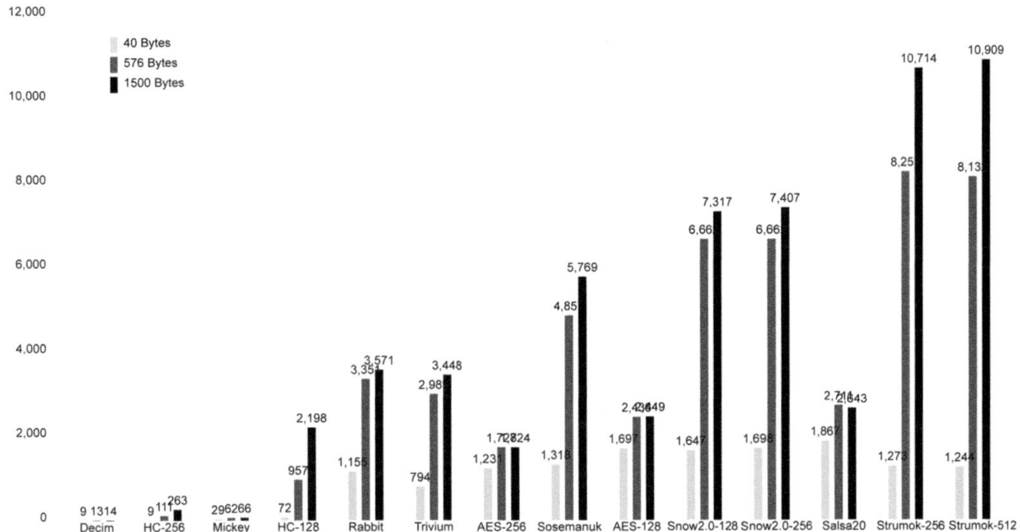

This chart shows the encryption speed in Mbps for different encryption algorithms across various block sizes (40 Bytes, 576 Bytes, 1500 Bytes). The data is ordered in ascending order of Mbps.

FIGURE 9.10 Mbps when Encrypting Different Block Sizes.

TABLE 9.9
IV and Key Setup Performance Results

Encryption Algorithms	Cycles/Setup	Setups/Sec
Mickey	17102	158579
Decim	35157	77148
Strumok-256	411	6578947
Strumok-512	401	6140350
Trivium	847	3505128
Sosemanuk	430	6250000
Snow2.0-128	324	8333333
Snow2.0-256	327	8333333
Salsa20	1	500000000
Rabbit	341	7936507
HC-128	11867	228519
HC-256	94981	28552
AES-128	0.24	--
AES-256	0.28	--

setup efficiency in selecting appropriate ciphers for wireless communication systems, especially in scenarios requiring dynamic key management. As wireless technologies continue to evolve, these ciphers will play a crucial role in ensuring the secure and efficient operation of wireless networks.

9.6 DISCUSSION ON THE RESEARCH FINDINGS

This discussion section delves into the comprehensive analysis of the results obtained from our extensive research on cryptographic algorithms within the scope of Computational Modeling and Simulation of Advanced Wireless Communication Systems. The focus is on dissecting the outcomes of computational modeling of ciphers and the experimental results, interpreting their implications in the broader context of wireless system security.

9.6.1 COMPUTATIONAL MODELING OF CRYPTOGRAPHIC ALGORITHMS

- Design and Structural Analysis: The computational modeling provided deep insights into the structural intricacies and design principles of various cryptographic algorithms. Algorithms like Strumok-256 and Strumok-512 demonstrated a high level of sophistication in their design, catering to the needs of high-security wireless applications. The simplicity and efficiency of Trivium and Sosemanuk highlighted their suitability for real-time wireless communication scenarios.
- Security Implications: The security analysis of these algorithms revealed a diverse range of resistance to cryptanalytic attacks. For instance, the robustness of AES-256 against advanced attacks makes it a reliable choice for securing sensitive data, while the agility of Rabbit and Salsa20 offers a balance between speed and security, crucial for high-speed wireless networks.
- Contextual Relevance: Each algorithm's features were evaluated in the context of their applicability to wireless systems. The findings underscored the importance of selecting ciphers that align with specific operational demands, such as power efficiency in IoT devices or high throughput in broadband communications.

9.6.2 RESULTS OF EXPERIMENTAL INVESTIGATIONS

- Long Stream Encryption Performance: The experiments on long stream encryption revealed significant differences in the performance of various ciphers. Strumok-256 and Strumok-512 emerged as frontrunners in terms of speed, making them ideal for applications involving large-volume data transmission. The results also highlighted the trade-offs between speed and security, as seen in the relatively slower performance of AES variants.
- Short Packet Encryption Efficiency: The encryption of short data packets showcased the adaptability of ciphers like Salsa20 and Snow2.0, which performed efficiently across different packet sizes. This adaptability is crucial in wireless networks where data packet sizes can vary significantly.
- IV and Key Setup Performance: The setup times for IVs and keys were critical in scenarios requiring frequent re-initializations. Ciphers like Salsa20 demonstrated exceptional efficiency in setup times, indicating their suitability for dynamic wireless environments where quick reconfigurations are common.

9.6.3 DISCUSSION IN THE CONTEXT OF ADVANCED WIRELESS SYSTEMS

- Implications for Wireless Security: The research findings have profound implications for the security of advanced wireless communication systems. The selection of appropriate cryptographic algorithms based on specific system requirements can significantly enhance the overall security posture of wireless networks.

- Future Directions: The results also pave the way for future research, particularly in the area of statistical security analysis, which will further refine our understanding of the suitability of these ciphers in evolving wireless communication scenarios.
- Integration Challenges: Integrating these cryptographic solutions into existing and emerging wireless technologies presents its own set of challenges, including compatibility, scalability, and performance optimization.

9.6.4 FUTURE SCOPE

- Quantum-Resistant Cryptography: With the advent of quantum computing, future research will focus on developing quantum-resistant cryptographic algorithms to safeguard wireless systems against quantum attacks.
- AI and Machine Learning Integration: Exploring the integration of artificial intelligence and machine learning in cryptographic systems could lead to adaptive security mechanisms that evolve in response to emerging threats.
- In-Depth Statistical Evaluations: Future studies will delve deeper into statistical security analysis, examining the randomness and unpredictability of cryptographic algorithms under various scenarios.
- Real-World Data Simulations: Implementing simulations with real-world data patterns will provide more accurate insights into the performance and security of cryptographic algorithms in practical wireless communication systems.
- Cross-Disciplinary Approaches: Combining insights from fields like network engineering, data science, and cybersecurity will enrich the understanding and development of cryptographic solutions for complex wireless systems.
- Standardization and Compliance: Ongoing research will also focus on aligning cryptographic algorithms with emerging global standards and regulatory requirements, ensuring widespread compatibility and adoption.

9.7 CONCLUSION

As we conclude this comprehensive exploration into the realm of cryptographic algorithms and their application in advanced wireless communication systems, it becomes evident that the field of cryptography is not just about securing data but also about adapting to the ever-evolving landscape of wireless technology. The intricate analysis of various cryptographic primitives, from the well-established AES variants to the specialized stream ciphers like Strumok and Salsa20, has illuminated the diverse requirements and challenges inherent in securing modern wireless communications.

The computational modeling and experimental results presented in this research offer a nuanced understanding of each algorithm's strengths and limitations. The high-speed performance of Strumok-256 and Strumok-512 in long stream encryption, the adaptability of Salsa20 and Snow2.0 in handling short data packets, and the efficiency of ciphers like Sosemanuk and Rabbit in IV and key setups, all underscore the importance of context in cryptographic selection. These findings are not just numbers and metrics; they represent the delicate balance between security, efficiency, and practicality that is essential in the wireless domain.

Moreover, this research has highlighted the significance of considering the operational environment when deploying cryptographic solutions. The choice of a cipher can profoundly impact the security and efficiency of a wireless system, whether it is a high-speed broadband network, a power-constrained IoT device, or a secure corporate communication channel. The adaptability of these cryptographic tools to different scenarios is a testament to the ingenuity and foresight of their design.

As we look towards the future, the insights gained from this research will undoubtedly guide the development and implementation of cryptographic solutions in wireless systems. The challenges of integrating these solutions into existing and emerging technologies, while optimizing for compatibility, scalability, and performance, will continue to drive innovation in the field. The journey of ensuring secure and efficient wireless communications is ongoing, and the role of cryptography in this journey remains as crucial as ever. This research not only contributes to the current understanding of cryptographic applications in wireless systems but also lays the groundwork for future advancements in this vital field.

ACKNOWLEDGEMENTS

- This project has received funding from the European Union's Horizon 2020 research and innovation programme under the Marie Skłodowska-Curie grant agreement No. 101007820 – TRUST. This publication reflects only the author's view and the REA is not responsible for any use that may be made of the information it contains.
- This research was funded by the European Union – NextGenerationEU under the Italian Ministry of University and Research (MIUR), National Innovation Ecosystem grant ECS00000041-VITALITY-CUP D83C22000710005.

REFERENCES

[1] C.-X. Wang *et al.*, "On the Road to 6G: Visions, requirements, key technologies, and testbeds," *IEEE Communications Surveys & Tutorials*, vol. 25, no. 2, pp. 905–974, 2023. doi: 10.1109/COMST.2023.3249835

[2] M. H. Alsharif, A. Jahid, R. Kannadasan, and M.-K. Kim, "Unleashing the potential of sixth generation (6G) wireless networks in smart energy grid management: A comprehensive review," *Energy Reports*, vol. 11, pp. 1376–1398, Jun. 2024. doi: 10.1016/j.egyr.2024.01.011

[3] V. Kampourakis, V. Gkioulos, and S. Katsikas, "A systematic literature review on wireless security testbeds in the cyber-physical realm," *Computers & Security*, vol. 133, pp. 103383, Oct. 2023. doi: 10.1016/j.cose.2023.103383

[4] K. Cao *et al.*, "Physical-layer security for intelligent-reflecting-surface-aided wireless-powered communication systems," *IEEE Internet of Things Journal*, vol. 10, no. 20, pp. 18097–18110, Oct. 2023. doi: 10.1109/JIOT.2023.3278238

[5] A. A. Kuznetsov, O. V. Potii, N. A. Poluyanenko, Y. I. Gorbenko, and N. Kryvinska, Stream Ciphers in Modern Real-time IT Systems. in *Studies in Systems, Decision and Control*. Cham: Springer Nature, 2022. doi: 10.1007/978-3-030-79770-6

[6] Y. An, M. Wang, L. Chen, and Z. Ji, "DCGAN-based symmetric encryption end-to-end communication systems," *AEU – International Journal of Electronics and Communications*, vol. 154, pp. 154297, Sep. 2022. doi: 10.1016/j.aeue.2022.154297

[7] E. Conrad, S. Misenar, and J. Feldman, "Chapter 4 – Domain 3: Security Architecture and Engineering," in *CISSP® Study Guide* (Fourth Edition), E. Conrad, S. Misenar, and J. Feldman, Eds., Syngress, 2023, pp. 107–223. doi: 10.1016/B978-0-443-18734-6.00007-6

[8] F. Ghasemi, and S. Babaie, "A lightweight secure authentication approach based on stream ciphering for RFID-based Internet of Things," *Computers and Electrical Engineering*, vol. 102, pp. 108288, Sep. 2022. doi: 10.1016/j.compeleceng.2022.108288

[9] S. S. P. Goswami, and G. Trivedi, "FPGA implementation of modified SNOW 3G stream ciphers using fast and resource efficient substitution box," *IEEE Embedded Systems Letters*, vol. 15, no. 4, pp. 238–241, Dec. 2023. doi: 10.1109/LES.2023.3298743

[10] K. Haseeb, Z. Jan, F. A. Alzahrani, and G. Jeon, "A secure mobile wireless sensor networks based protocol for smart data gathering with cloud," *Computers & Electrical Engineering*, vol. 97, pp. 107584, Jan. 2022. doi: 10.1016/j.compeleceng.2021.107584

[11] P. M. Lima, C. K. P. da Silva, C. M. de Farias, L. K. Carvalho, and M. V. Moreira, "Event-based cryptography for automation networks of cyber-physical systems using the stream cipher ChaCha20," *IFAC-PapersOnLine*, vol. 55, no. 28, pp. 58–65, Jan. 2022. doi: 10.1016/j.ifacol.2022.10.324

[12] Y. Liu, Z. Shen, S. Fang, and Y. Wang, "A novel pipelining encryption hardware system with high throughput and high integration for 5G," *China Communications*, vol. 19, no. 6, pp. 1–10, Jun. 2022. doi: 10.23919/JCC.2022.06.001

[13] H. N. Noura, O. Salman, R. Couturier, and A. Chehab, "LoRCA: Lightweight round block and stream cipher algorithms for IoV systems," *Vehicular Communications*, vol. 34, pp. 100416, Apr. 2022. doi: 10.1016/j.vehcom.2021.100416

[14] V. Rudnytskyi, O. Korchenko, N. Lada, R. Ziubina, L. Wieclaw, and L. Hamera, "Cryptographic encoding in modern symmetric and asymmetric encryption," *Procedia Computer Science*, vol. 207, pp. 54–63, Jan. 2022. doi: 10.1016/j.procs.2022.09.037

[15] J. Sharma, and P. S. Mehra, "Secure communication in IOT-based UAV networks: A systematic survey," *Internet of Things*, vol. 23, pp. 100883, Oct. 2023. doi: 10.1016/j.iot.2023.100883

[16] N. Sharma, P. Kumar, and S. K. Rai, "A new approach to implement stream cipher using memristor-based Chua's circuit for secure communication," *AEU – International Journal of Electronics and Communications*, vol. 175, pp. 155105, Feb. 2024. doi: 10.1016/j.aeue.2023.155105

[17] F. Thabit, O. Can, A. O. Aljahdali, G. H. Al-Gaphari, and H. A. Alkhzaimi, "Cryptography algorithms for enhancing IoT Security," *Internet of Things*, vol. 22, pp. 100759, Jul. 2023. doi: 10.1016/j.iot.2023.100759

[18] Y. Wang, Z. Wang, and X. Liu, "Key security measurement method of authentication based on mobile edge computing in urban rail transit communication network," *Computer Communications*, vol. 215, pp. 140–149, Feb. 2024. doi: 10.1016/j.comcom.2023.12.020

[19] Y. Qian, F. Ye, and H.-H. Chen, "Cryptographic Techniques," in *Security in Wireless Communication Networks*, IEEE, 2022, pp. 51–76. doi: 10.1002/9781119244400.ch4

[20] A. J. Menezes, P. C. van Oorschot, S. A. Vanstone, P. C. van Oorschot, and S. A. Vanstone, *Handbook of Applied Cryptography*. CRC Press, 2018. doi: 10.1201/9780429466335

[21] A. A. Kuznetsov, O. V. Potii, N. A. Poluyanenko, Y. I. Gorbenko, and N. Kryvinska, "Analysis of synchronous stream cryptoconversions," in *Stream Ciphers in Modern Real-time IT Systems: Analysis, Design and Comparative Studies*, A. A. Kuznetsov, O. V. Potii, N. A. Poluyanenko, Y. I. Gorbenko, and N. Kryvinska, Eds., in *Studies in Systems, Decision and Control*, Cham: Springer International Publishing, 2022, pp. 47–64. doi: 10.1007/978-3-030-79770-6_4

[22] S. Rubinstein-Salzedo, Cryptography. in *Springer Undergraduate Mathematics Series*. Cham: Springer International Publishing, 2018. doi: 10.1007/978-3-319-94818-8

[23] A. A. Kuznetsov, O. V. Potii, N. A. Poluyanenko, Y. I. Gorbenko, and N. Kryvinska, "Research of Second-Order Properties of NLFSR. Comparative Analysis of M-NLFSR and M-LFSR," in *Stream Ciphers in Modern Real-time IT Systems: Analysis, Design and Comparative Studies*, A. A. Kuznetsov, O. V. Potii, N. A. Poluyanenko, Y. I. Gorbenko and N. Kryvinska, Eds., in *Studies in Systems, Decision and Control*. Cham: Springer International Publishing, 2022, pp. 419–465. doi: 10.1007/978-3-030-79770-6_15

[24] "ISO/IEC 18033-4:2011 Information technology — Security techniques — Encryption algorithms—Part 4: Stream ciphers." Accessed: Jan. 18, 2024. [Online]. Available: www.iso.org/standard/54532.html

[25] "eSTREAM Optimized code HOWTO." Accessed: Jan. 18, 2024. [Online]. Available: www.ecrypt.eu.org/stream/perf/

[26] "Home," E-Stream. Accessed: Jan. 18, 2024. [Online]. Available: https://e-stream.it/

[27] S. Babbage, and M. Dodd, "The MICKEY Stream Ciphers," in *New Stream Cipher Designs: The eSTREAM Finalists*, M. Robshaw and O. Billet, Eds., in Lecture Notes in Computer Science. Berlin, Heidelberg: Springer, 2008, pp. 191–209. doi: 10.1007/978-3-540-68351-3_15

[28] "The eSTREAM portfolio page." Accessed: Jan. 18, 2024. [Online]. Available: www.ecrypt.eu.org/stream/e2-mickey.html

[29] C. Berbain *et al.*, "Decimv2," in *New Stream Cipher Designs: The eSTREAM Finalists*, M. Robshaw and O. Billet, Eds., in Lecture Notes in Computer Science. Berlin, Heidelberg: Springer, 2008, pp. 140–151. doi: 10.1007/978-3-540-68351-3_11

[30] A. A. Kuznetsov, O. V. Potii, N. A. Poluyanenko, Y. I. Gorbenko, and N. Kryvinska, "Stream Symmetric Cipher 'Strumok,'" in *Stream Ciphers in Modern Real-time IT Systems: Analysis, Design and Comparative Studies*, A. A. Kuznetsov, O. V. Potii, N. A. Poluyanenko, Y. I. Gorbenko, and N.

Kryvinska, Eds., in Studies in Systems, Decision and Control. Cham: Springer International Publishing, 2022, pp. 467–516. doi: 10.1007/978-3-030-79770-6_16

[31] I. Gorbenko, A. Kuznetsov, Y. Gorbenko, S. Vdovenko, V. Tymchenko, and M. Lutsenko, "Studies on statistical analysis and performance evaluation for some stream ciphers," *International Journal of Computing*, vol. 18, no. 1, Art. no. 1, Mar. 2019.

[32] Y. Tian, G. Chen, and J. Li, "On the design of Trivium." 2009. Accessed: Jan. 18, 2024. [Online]. Available: https://eprint.iacr.org/2009/431

[33] C. De Cannière, "Trivium: A Stream Cipher Construction Inspired by Block Cipher Design Principles," in *Information Security*, S. K. Katsikas, J. López, M. Backes, S. Gritzalis, and B. Preneel, Eds., in Lecture Notes in Computer Science, Berlin, Heidelberg: Springer, 2006, pp. 171–186. doi: 10.1007/11836810_13

[34] "The eSTREAM portfolio page." Accessed: Jan. 18, 2024. [Online]. Available: www.ecrypt.eu.org/stream/e2-trivium.html

[35] C. Berbain *et al.*, "Sosemanuk, a Fast Software-Oriented Stream Cipher," in *New Stream Cipher Designs: The eSTREAM Finalists*, M. Robshaw and O. Billet, Eds., in Lecture Notes in Computer Science. Berlin, Heidelberg: Springer, 2008, pp. 98–118. doi: 10.1007/978-3-540-68351-3_9.

[36] "The eSTREAM portfolio page." Accessed: Jan. 18, 2024. [Online]. Available: www.ecrypt.eu.org/stream/e2-sosemanuk.html

[37] P. Ekdahl, T. Johansson, A. Maximov, and J. Yang, "A new SNOW stream cipher called SNOW-V." 2018. Accessed: Jan. 18, 2024. [Online]. Available: https://eprint.iacr.org/2018/1143

[38] A. N. Alekseychuk, S. M. Koniushok, and M. V. Poremskyi, "Security evaluation for snow 2.0-like stream ciphers against correlation attacks over extension fields." 2019. Accessed: Jan. 18, 2024. [Online]. Available: https://eprint.iacr.org/2019/287

[39] "The eSTREAM portfolio page." Accessed: Jan. 18, 2024. [Online]. Available: www.ecrypt.eu.org/stream/e2-salsa20.html

[40] D. J. Bernstein, "The Salsa20 Family of Stream Ciphers," in *New Stream Cipher Designs: The eSTREAM Finalists*, M. Robshaw and O. Billet, Eds., in Lecture Notes in Computer Science. Berlin, Heidelberg: Springer, 2008, pp. 84–97. doi: 10.1007/978-3-540-68351-3_8

[41] T. R. Campbell, "Daence: Salsa20 and ChaCha in Deterministic Authenticated Encryption with no noNCEnse." 2020. Accessed: Jan. 18, 2024. [Online]. Available: https://eprint.iacr.org/2020/067

[42] M. Boesgaard, T. Pedersen, M. Vesterager, and E. Zenner, "The rabbit stream cipher – design and security analysis." 2004. Accessed: Jan. 18, 2024. [Online]. Available: https://eprint.iacr.org/2004/291

[43] "The eSTREAM portfolio page." Accessed: Jan. 18, 2024. [Online]. Available: www.ecrypt.eu.org/stream/e2-rabbit.html

[44] H. Wu, "A New Stream Cipher HC-256," in *Fast Software Encryption*, B. Roy and W. Meier, Eds., in Lecture Notes in Computer Science. Berlin, Heidelberg: Springer, 2004, pp. 226–244. doi: 10.1007/978-3-540-25937-4_15

[45] H. Wu, "A new stream cipher HC-256." 2004. Accessed: Jan. 18, 2024. [Online]. Available: https://eprint.iacr.org/2004/092

[46] J. Daemen, and V. Rijmen, "Specification of Rijndael," in *The Design of Rijndael: The Advanced Encryption Standard (AES)*, J. Daemen and V. Rijmen, Eds., in *Information Security and Cryptography*. Berlin, Heidelberg: Springer, 2020, pp. 31–51. doi: 10.1007/978-3-662-60769-5_3

[47] J. Daemen, and V. Rijmen, "Rijndael/AES," in *Encyclopedia of Cryptography and Security*, H. C. A. van Tilborg, Ed., Boston, MA: Springer US, 2005, pp. 520–524. doi: 10.1007/0-387-23483-7_358

10 Legal Frameworks Regulating Computational Models in Wireless Communication Systems

Abubakar Aliyu, Emeka Ogbuju, Taiwo Kolajo, Agbotiname Lucky Imoize, Ovye John Abari, Maryam Sufiyanu Masari and Francisca Oladipo

10.1 INTRODUCTION

The quick adoption of computational models has ushered in a new era of connectivity and technological development in the dynamic field of wireless communication systems. These computational models have revolutionised communication, but they have also raised a host of difficult problems that overlap with compliance, ethics, and a need for conflict resolution [1]. This chapter embarks on a journey through the intricate legal frameworks that govern these computational models in wireless communication networks, placing a heavy emphasis on their ethical components, transparency imperatives, compliance requirements, and dispute resolution methods. In a modern digitally connected world, wireless communication technologies are vital to modern society because they enable fast worldwide connectivity [2]. The complex interaction between computational models and wireless communication technologies is at the core of this connectedness [3]. These computational models serve as the directors of the orchestra in our digital symphony, arranging the distribution of resources, enhancing network efficiency, and meeting our various communication needs [4]. This symphony does, however, contain certain dissonant notes that present us with significant ethical conundrums, complex legal issues, and the requirement for skilled conflict-resolution techniques [5].

In wireless communication systems, the introduction of computational models has raised significant ethical questions [6]. These models are trusted to make judgements that affect our daily lives, and they are powered by sophisticated algorithms [7]. The fairness of these models, their possible biases, and their broader societal ramifications are under the spotlight as we navigate this digital environment [8]. Making sure computational models are used ethically in wireless communications requires a close study of their decision-making procedures, data handling procedures, and adherence to inclusivity and fairness standards [9]. It is crucial for these models to operate transparently so that stakeholders and users may understand the reasoning behind their choices and identify any hidden bias in the algorithms [10]. The ethical issues that computational models raise develop along with their capabilities [11]. These algorithms increasingly make decisions that impact the lives of individuals and their well-being beyond basic data processing [12]. Concerns over justice, transparency, and potential biases in these computer models weigh heavily on the conscience of the digital age [13]. We will examine how the ethical ramifications of these models are intricately woven throughout the cellular communications industry. In order to shed light on these models inner workings and make

DOI: 10.1201/9781003457428-11

sure that their decision-making processes are accessible for examination, criticism, and development, we will delve into the necessity of transparency.

In addition to ethical concerns, there have been significant changes to the legal framework controlling computational models in wireless communication networks [14]. As these models spread throughout our connected world, strict compliance regulations have been put in place to protect user rights, data privacy, and fair competition [15]. Companies engaged in the creation and application of these models must juggle a maze of rules, from intellectual property legislation to data protection requirements [16]. The difficulties of cross-border data transfers and the variances in worldwide regulatory standards further complicate compliance efforts. The ethical objectives of computational models must be turned into practical, responsible practices within this legal framework [17]. The legal frameworks governing computational models in wireless communication systems must change swiftly to keep up with the technological landscape [18]. Businesses engaging in this field must negotiate a complicated web of international and local rules, data protection restrictions, and intellectual property rights. As we attempt to balance the strict constraints of the legal environment with the ethical goals of computational models, the intersection of ethics and compliance becomes especially painful [19]. The sections that follow will take us on a tour through this legal environment as we examine the mechanisms that reconcile ethics and the law and decode compliance standards.

Wireless communication network interconnectedness unavoidably results in disputes over resources, spectrum use, interference control, and service prioritisation [20]. These systems' computational models, which are embedded at their core, have the ability to reduce disputes by using adaptive algorithms, dynamic resource allocation, and wise decision-making [21, 22]. Conflicts naturally arise due to the dynamic and linked nature of wireless communication systems, including interference, resource competition, and divergent service expectations [23]. The cognitive engines of these systems, computational models, have the capacity to effectively traverse these tensions through adaptive algorithms and wise decision-making [24]. Our tale will illustrate how computational models are made to put an end to conflict and preserve the harmony of network operations by peeling back the layers of these conflict resolution techniques. Conflict resolution appears to be the key component in this area, ensuring flawless service continuity and resource optimisation [25]. In this chapter, we examine the methods and strategies used by these models to resolve disputes amicably, preserve network services, and maximise resource efficiency. Conflict resolution becomes more and more important as modern wireless communication systems get more complicated since it is essential to maintain the reliability and effectiveness of these systems.

The understanding and creation of legal frameworks governing computational models in wireless communication networks have greatly benefited from the substantial contributions made by this chapter. First, it gives readers a basic awareness of the regulatory environment by providing a thorough introduction to the current legal frameworks relevant to computational models in wireless communication. The chapter provides an organised overview of relevant literature, highlighting current research gaps and areas of disagreement. It emphasises the significance of national laws and regulations controlling computational models in wireless communication systems while critically analysing them using a strong approach. This global assessment covers international agreements and regulations and offers a comprehensive view of the transparency and ethical concerns surrounding computational models in this field. The chapter also examines dispute resolution techniques in the context of computational models for wireless communication systems and investigates the legal ramifications of noncompliance. The extensive examination yielded useful insights for policymakers, scholars, and industry practitioners, culminating in well-informed recommendations. As a whole, this chapter is an invaluable resource that offers a thorough analysis of the legal nuances related to computational models in wireless communication systems as well as a research and regulatory development path.

10.1.1 KEY CONTRIBUTIONS

The main contributions of this chapter, "Legal Frameworks Regulating Computational Models in Wireless Communication Systems," are as follows:

a) The legal foundations controlling computational models in wireless communication networks are thoroughly examined in Section 10.1 of this chapter. For scholars, professionals, and legislators looking for a comprehensive grasp of the moral and legal implications of this crucial nexus between regulation and technology, it is an invaluable resource.

b) A thorough analysis of the body of current research provides the fundamental information on which our investigation is based. This chapter explores earlier research to provide an understanding of the present status of legal considerations for wireless communication computational models, as discussed in Section 10.2 of this chapter.

c) The research technique utilised in this study is comprehensive, providing insight into the methodical approach utilised to examine and comprehend the complex terrain of legal systems. Section 10.3 discusses in detail the technique used in the chapter.

d) One important contribution of this chapter, as discussed in Section 10.4, is a thorough examination of national laws and regulations, which provides insight into how other nations handle the legal issues of computational models in wireless communication.

e) In Section 10.5, the chapter oscillates its emphasis on international agreements and standards, offering a perceptive analysis of the state of the world and cooperative initiatives to standardise computational models with legal elements.

f) International computational models for wireless communication are examined in Section 10.7 in light of ethical issues and the need for openness. The ethical aspects of legal frameworks and their intersections are covered in this chapter.

g) The chapter explains the legal ramifications of deploying and using computational models in wireless communication systems, as well as the requirements for compliance.

h) A vital component is an investigation of computer model techniques for resolving conflicts. This chapter provides insightful information on conflict mitigation techniques for wireless communication networks in Section 10.8.

In essence, by offering a comprehensive assessment of the regulatory framework, ethical issues, compliance challenges, and tools for conflict resolution, the chapter contributes to the scholarly discussion on the legal aspects of wireless communication. It gives readers the information and awareness they need to successfully negotiate the shifting boundaries between law and technology in this exciting sector. The chapter serves as a stimulus for additional study and the formulation of public policy in this important area.

10.1.2 CHAPTER ORGANIZATION

The fundamental ideas controlling legal frameworks in wireless communication systems are explored in Section 10.1 of this chapter. We explore the complex web of regulations and guidelines that direct the implementation of computational models in wireless communication systems. Section 10.2 guides us through the corpus of existing knowledge on the topic and provides details on earlier research work. Here, we explore the landscape of connected works, spotting gaps and new patterns that serve as the foundation for our investigation. Section 10.3 describes our selected methodology and research approach, revealing the research procedure. We go over the techniques used to gather and analyse data. Section 10.4 focuses on a thorough analysis of national regulations and how they affect computational models. We explore the world of global standards in Section 10.5 by looking at international conventions and agreements. These international efforts at harmonisation are crucial in determining the compliance environment. The crucial facets of ethics and transparency in the context of computational models in wireless communication are introduced in Section 10.6. We discuss fairness, data privacy, ethical issues, and systems for assuring openness. Moving on to compliance, Section 10.7 explores the nuances of compliance and its legal repercussions. We look at the effects of non-compliance and emphasise the best methods for guaranteeing that laws are followed. The challenging task of conflict resolution within computational models for wireless communication is tackled in Section 10.8. In an effort to preserve peace in the wireless communication environment, we talk about conflict resolution techniques. Section 10.9, where we discuss the lessons we learned from our exploration, summarises the knowledge and experiences gained throughout the chapter. The stakeholders in the field can greatly benefit from knowing these lessons. Recommendations that come out of the chapter are covered in Section 10.10. In this chapter, we present guiding principles for business procedures, policy considerations, future research, and the development of moral and open wireless communication systems. Finally, we end our chapter with concluding thoughts in Section 10.11. Here, we provide a summary of the main ideas, consider the future, and share our closing thoughts on the topic of legal frameworks governing computational models in wireless communication systems.

10.2   REVIEW OF RELATED WORKS

The symbiotic relationship between technological breakthroughs and legal frameworks is both complex and crucial in the ever-changing world of wireless communication technologies [26]. The governance and laws that support the existence of computational models are essential to ensuring their smooth operation [27], from 5G networks to the Internet of Things (IoT) and beyond [28]. This chapter sets out on a thorough exploration of the legal framework that governs and regulates the field of computational models used in wireless communication systems. The complex nature of this sector necessitates a deep comprehension of the regulatory, administrative, and policy frameworks that govern it [29]. We begin by carefully examining the relevant literature, which is summarised in Table 10.1 and forms the cornerstone of our study. We intend to shed light on the many aspects of legal governance relevant to computational models by reviewing the body of literature, which will contextualise our views on compliance, ethics, and conflict resolution.

We explore the literature to learn about the complex legal issues of spectrum allocation, data privacy, cybersecurity, and the growing problems caused by cutting-edge technologies like edge computing and 6G networks. In addition, we scrutinise the international agreements and standards that define the limits and potential of wireless communication networks in a global context.

After a thorough examination of related literature, it is plainly obvious that investigating the legal frameworks governing computational models in wireless communication systems is not only a worthwhile academic endeavour but also a vital necessity. The body of literature already in existence, which consists of legal analyses, comparative studies, and empirical research, emphasises the profound interaction between law and technology in this dynamic field. The wealth of work reviewed here confirms, first and foremost, that the governance of computational models is an inherently complex and varied endeavour. A legal system that is both adaptive and foresighted is required due to the exponential growth of wireless technology and the quickening speed of invention. It is becoming more and more clear that our legal frameworks must change concurrently to account for these disruptive forces as we approach the 6G era and observe the growing influence of edge computing, IoT ecosystems, and blockchain-driven networks. The review emphasises this challenges worldwide scope. In order to enable smooth interoperability and worldwide connection, wireless communication systems transcend geographical boundaries, necessitating harmonised international agreements and standards. The requirement for international cooperation and alignment is underscored by the existence of numerous regulatory approaches in various nations, regions, and jurisdictions.

TABLE 10.1
Summary of the Reviews

References	Description of Work Done	Limitations
[30]	Discusses spectrum allocation, security, and IoT implications as it examines the legislative foundations for 5G networks.	The analysis may not cover the entire range of computational models used in wireless communication systems because it is primarily focused on 5G.
[31]	Examines the effects of the GDPR on the protection of user data as well as data privacy laws controlling wireless communications.	Concentrate primarily on data privacy and ignores more general legal issues relating to computational models in wireless systems.
[32]	Analyses the effects of spectrum management regulations in different nations on wireless technology by doing a comparative examination of these laws.	Limited to spectrum management; little attention is paid to other legal facets of wireless communication models.
[33]	Examines how AI, IoT, and wireless networks are converging, examining regulatory issues and how they affect innovation.	IoT and AI are mostly discussed; possible incomplete coverage of all computational models used in wireless communication systems.
[34]	Examines laws protecting consumer rights in wireless services, focusing on service quality and dispute settlement.	Consumer rights are the primary focus; spectrum allotment and cybersecurity are not thoroughly covered.
[35]	Highlights cybersecurity precautions and potential dangers while analysing security standards and laws unique to 5G networks.	Restricted to 5G security issues; other wireless technologies spectrum management or privacy restrictions are not included.
[36]	Discusses the worldwide consequences of reviews of international agreements governing spectrum distribution and coordination across boundaries.	Focuses on international agreements; may not cover in-depth particular national laws and regulations.
[37]	Examines the legal implications of edge computing for wireless systems, especially with regard to data security and privacy.	Mostly focused on edge computing; may not fully encompass additional computational models or more general wireless communication regulatory concerns.
[38]	Investigates antitrust laws and their application in the wireless communication sector, addressing problems with competition.	Antitrust-related issues take precedence; cybersecurity, spectrum allocation, and other regulatory issues are not fully addressed.
[39]	Examines how national laws affect the implementation of AI-driven wireless networks, including legal difficulties and prospects.	Focuses on AI-driven networks in particular; other computational models may have additional legal consequences that aren't discussed in this research.
[40]	Examines how national regulations affect the security of IoT devices in wireless networks, putting a focus on compliance and data protection.	Focuses on IoT security; additional study may be needed for wider areas of computational models and spectrum management.
[41]	Investigates the legal framework for spectrum sharing in wireless communication networks, looking at how it may be affected by 5G and other developments.	Focus is primarily on spectrum sharing; other legal issues, including data protection, may not be thoroughly covered.

(continued)

TABLE 10.1 (Continued)
Summary of the Reviews

References	Description of Work Done	Limitations
[42]	Focuses on data integrity as it investigates the legal and regulatory issues related to implementing wireless networks based on blockchain.	Focuses on blockchain-based networks; spectrum management and other computational models may have different legal considerations.
[43]	Discusses regulatory obstacles as it examines the effects of national regulations on the deployment of Wi-Fi 6 technology in wireless networks.	Adoption of Wi-Fi 6 is the major topic; it might not cover other computational models or more general legal issues in the wireless communication arena.
[44]	Discusses spectrum allotment and coordination for worldwide coverage as it looks into the legal elements of satellite communication rules.	Focuses on satellite communication; other wireless technology-specific rules may be very different.
[45]	Focuses on safety and privacy concerns while analysing the legal framework controlling drone communication in wireless networks.	Focuses on drone communication; different legal restrictions may apply to alternative computational models and broader spectrum management.
[46]	Explores the legal issues surrounding the deployment of 6G wireless networks, emphasising spectrum shortage and implications for global standards.	Emphasises mostly on 6G networks; other computational models and technologies legal environments may differ.
[47]	Discusses the significance that regulatory sandboxes have in testing and deploying new technologies while examining the legal elements of wireless innovation.	Concentrates on regulatory sandboxes; more comprehensive regulatory frameworks for computational models may require different considerations.
[48]	Examines the regulatory environment for network slicing in 5G wireless networks, putting a focus on virtualization and service differentiation.	With a primary focus on network slicing, this article may not fully address the special legal issues that other computational models and regulatory factors may present.
[49]	Analyses how national laws may affect the creation of wireless networks in smart cities, addressing concerns about privacy and data governance.	Addresses smart city networks; separate investigation of other computational models and regulatory issues may be necessary.

However, the most notable finding of our research would be that the current legal system has gaps and contradictions. These gaps take the form of ethical conundrums around data privacy and surveillance, compliance issues that limit technological advancement, and difficulties with conflict resolution in this intricate ecosystem. Our analysis will try to fill up these gaps by offering clarification and solutions. The assessment essentially supports the urgent necessity for a thorough and multifaceted examination of the legal frameworks controlling computational models in wireless communication systems. It emphasises the necessity of tackling the difficulties and opportunities brought about by evolving technology while ensuring that moral considerations, compliance standards, and dispute resolution procedures are deeply and carefully woven into our legal system. This review acts as our compass, directing us through the complex landscape of law and technology as we go into the core of our investigation, which covers ethics, compliance, and dispute resolution. It confirms the importance of our work, not just in terms of research but also in terms of societal improvement and the appropriate development of wireless communication systems. As a result, there is a pressing need for the examination of legal frameworks, with a focus on the potential and difficulties presented by evolving technologies as well as the necessity of addressing ethical, compliance, and dispute resolution issues within this intricate field.

10.3   REVIEW APPROACH

A thorough and effective methodology is used in this chapter, "Legal Frameworks Regulating Computational Models in Wireless Communication Systems," which incorporates legal analysis and review. This approach comprises a thorough analysis and study of the existing statutes, rules, and case law relevant to the regulation of computational models in wireless communication systems. To comprehend the exact legal provisions, their interpretations, and how they are used in practice requires extensive legal research. The topic of this chapter, which is largely concerned with the legal concerns of regulating computational models, is perfectly aligned with this method. As a result, readers will have a firm understanding of the laws and regulations controlling wireless communication systems. It enables us to examine and dissect the legal system. The benefits of this strategy include providing a thorough and authoritative assessment of the legal environment and enabling a thorough examination of how diverse computational models and their applications are impacted by legal provisions. We can provide a complete analysis of the legal frameworks guiding computational models in wireless communication by using the Legal Analysis and Review approach.

10.4   ANALYSIS OF NATIONAL LAWS AND REGULATIONS CONCERNING COMPUTATIONAL MODELS

Analysing national laws and regulations concerning computational models, particularly in the context of wireless communication systems, requires a comprehensive understanding of the legal landscape in different countries [49]. Table 10.2 provides the general framework for such analysis.

This analysis will change based on the nations or locations you are researching as well as the specific legal topics you are looking into. To provide accurate and insightful research, a comprehensive and current assessment of each legal authority is essential [50].

10.5   EXAMINATION OF INTERNATIONAL AGREEMENTS AND STANDARDS GOVERNING THESE MODELS

Understanding how these technologies are governed and standardised globally requires an investigation of international agreements and standards controlling computational models in wireless communication systems. This section provides a thorough explanation of this examination.

a) International Agreements

International agreements are treaties or conventions that many nations have signed to control particular parts of computational modelling and wireless communication. These contracts may deal with a variety of issues, such as spectrum distribution, data privacy, and cybersecurity [51].

   i. *Spectrum Allocation*

     Radio frequency spectrum is allotted for different wireless communication services via international agreements, as those made by the International Telecommunication Union (ITU). These agreements aid in preventing cross-border interference between various wireless services [52].

   ii. *Data Privacy*

     Standards for the protection of personal data are established by agreements like the European Union General Data Protection Regulation (GDPR). Businesses handling the data of EU people around the world are impacted by GDPR extraterritorial reach [53].

   iii. *Cybersecurity*

     Wireless communication system security is being improved via international cybersecurity agreements. For instance, a global convention called the Budapest Convention on Cybercrime addresses cybercrime, including assaults on wireless networks [54].

TABLE 10.2
Analysis of National Laws

Framework	Description
Identification of Jurisdictions	Determine which particular nations or areas you wish to study first. The laws and rules governing computational models in wireless communication may vary depending on the jurisdiction.
Scope of Analysis	Establish the parameters of your analysis. Are you concentrating on a particular component of computational models (such as privacy, security, spectrum allocation) or a thorough analysis of all pertinent laws?
Data Collection	Gather the pertinent legal papers for each jurisdiction that deal with computational models in wireless communication, such as statutes, regulations, and court rulings.
Categorization	Sort the gathered legal papers into groups according to their topics. For instance, categories may include ethical considerations, spectrum management, privacy, and security.
Key Legal Principles	List and briefly describe the main legal principles and provisions that fall under each heading. Definitions, rights and obligations, and enforcement procedures are all part of this.
Comparative Analysis	Comparatively, evaluate the legal systems in various jurisdictions. Compare and contrast how different countries approach computational models in wireless communication, noting any patterns.
Compliance Requirements	Find out what regulations must be followed by organisations that operate in the wireless communication industry. This may involve requirements for licencing, data protection, security measures, and reporting.
Ethical Considerations	Determine whether the legal frameworks specifically address ethical issues associated with computational models. Look for clauses that address justice, openness, and responsibility.
Enforcement and Penalties	Examine the enforcement measures in place, such as regulatory organisations, fines for noncompliance, and procedures for resolving disputes.
Case Studies	Include case studies or illustrations of how these rules and legislation have actually been put into practice. Give examples of significant legal decisions or enforcement actions.
International Agreements	Consider any international treaties or agreements that influence the national legal framework for wireless communication as they may have a significant influence on domestic laws.
Emerging Trends	Determine any new legal developments or proposed laws that might have an effect on the control of computational models used in wireless communication.
Stakeholder Perspectives	To give a thorough study, take into account the opinions of many stakeholders, including industry associations, advocacy groups, and legal professionals.
Recommendations	Consider making suggestions for enhancing or harmonising legal frameworks based on your findings, especially in areas where there are gaps or discrepancies.
Conclusion	Summarise the main conclusions of your research, focusing on the most important legal considerations and their consequences for computational wireless communication models in the chosen jurisdictions.

b) International Standards

International standards are technological guidelines that guarantee compatibility and interoperability between various computational models and wireless communication technologies [55]. Organisations like the International Electrotechnical Commission (IEC) [56], the Institute of Electrical and Electronics Engineers (IEEE) [57], and the International Organisation for Standardisation (ISO) [58] are responsible for creating these standards [57].

  i. *Wireless Communication Protocols*
The technical requirements for wireless communication protocols are defined by standards like IEEE 802.11 (Wi-Fi) and 3GPP (3rd Generation Partnership Project) [59]. These standards allow for easy communication between devices made by various vendors [58].

  ii. *Security Standards*
The application of security measures in wireless communication networks is governed by cybersecurity standards created by organisations like ISO (such as ISO/IEC 27001) [60]. Networks and user data are protected by adherence to these standards [61].

  iii. *Data Interchange Standards*
The information sharing between various computational models is facilitated in wireless communication by data interchange standards like Extensible Markup Language (XML) and JavaScript Object Notation (JSON) [62]. These guidelines guarantee that data can be interpreted and processed consistently [45].

c) Harmonization of Regulations

Harmonising laws across borders is made possible by international agreements and standards [63]. Because wireless communication systems and computational models frequently function internationally and cross international borders, this harmonisation is crucial [64]. Harmonised regulations aid in reducing disputes and difficulties related to compliance [65].

d) Challenges and Considerations

  i. *Jurisdictional Issues*
Due to jurisdictional concerns, determining which international agreements and standards apply can be complicated [66]. Different nations may adhere to international agreements to different degrees [67].

  ii. *Emerging Technologies*
It is difficult to adapt current international agreements and standards to new technologies because of the rapid developments in computational models and wireless communication [66]. To address new dangers and possibilities, regulations must change [68].

  iii. *Enforcement*
Because voluntary adherence by countries and organisations is frequently required, ensuring conformity with international agreements and standards can be difficult Different enforcement strategies may include trade sanctions or diplomatic attempts [67].

e) International Agreements

  i. *ITU Radiocommunication Sector (ITU-R)*
Under this section, the ITU is responsible for managing the global radio frequency spectrum and satellite orbits [69]. The World Radiocommunication Conferences (WRC) are held by the ITU-R to allocate spectrum resources and handle topics including satellite coordination and 5G spectrum distribution [70].

  ii. *United Nations Treaties*
The United Nations is responsible for establishing several international accords, for instance, the usage of the electromagnetic spectrum in maritime areas is governed by the United Nations Convention on the Law of the Sea (UNCLOS) [71].

  iii. *Bilateral and Multilateral Agreements*
In addition to international accords, many nations also sign bilateral and multilateral agreements to control wireless communication within particular geographical areas or for

particular purposes. These can include trade agreements containing telecommunications-related clauses [72].

f) Role of International Standards

i. *Interoperability*
International standards guarantee that systems and equipment created by various manufacturers can function together seamlessly [73]. For instance, the 3GPP standard permits 5G and other mobile networks to operate globally [74].

ii. *Quality and Reliability*
The effectiveness and dependability of wireless communication networks are addressed by standards [75]. For instance, the ISO 9001 standard establishes requirements for quality management systems that can be used in telecommunications [76].

iii. *Cybersecurity*
Information security management systems can be established, put into practice, maintained, and improved through time using a framework provided by international standards like ISO/IEC 27001. The protection of computational models in wireless systems depends on these standards [76].

g) Regulatory Bodies

i. *World Trade Organization (WTO)*
The WTO regulates international trade laws between countries by affecting the regulatory environment, its agreements controlling the sale of telecommunications services have an indirect impact on computational models [66].

ii. *Internet Engineering Task Force (IETF)*
The IETF develops and advocates voluntary Internet standards, but it is not a regulatory agency these standards, together with RFCs (Request for Comments), are essential for the operation of computational models for wireless Internet connection [77].

h) Balancing Sovereignty and Globalization

i. One of the main difficulties is balancing globalisation with sovereign nations that frequently desire control over their telecommunications infrastructure to safeguard their interests [78]. However, due to the Internet and wireless communication networks, and a lack of national boundaries, international cooperation is necessary [79].

i) Privacy and Data Protection

i. Concerns of privacy relating to computational models in wireless communication are addressed through international agreements and standards [66]. A notable example is GDPR in the EU. Many nations adhere to the standards outlined in the OECD Guidelines on the Protection of Privacy and Transborder Flows of Personal Data [53].

j) Emerging Technologies and IoT

i. International agreements and standards must change as the Internet of Things (IoT) develops to solve the particular problems brought on by a large number of interconnected devices [80]. This concerns matters like data governance, security, and spectrum allocation [5].

k) Enforcement and Dispute Resolution
  i. It might be difficult to uphold international agreements and norms [81]. When nations or organisations disobey, disputes may occur [82]. For the global telecommunications industry to remain in good order, mechanisms for resolving such disagreements are crucial [83]. These mechanisms are frequently diplomatic or arbitration-based [47].

l) Evolving Legal Frameworks
  Computational model regulation is governed by a variety of laws that are always changing to keep up with new developments in technology [84]. To address new difficulties, international organisations and governments continually review and revise agreements and standards [47].

It is a dynamic and difficult process to evaluate international agreements and standards controlling computational models in wireless communication networks. This process encompasses the interaction of geopolitical, technical, and legal aspects. Governments, companies, and individuals involved in wireless communications must be updated about these agreements and standards to maintain compliance, security, and interoperability in a fast-changing international environment.

10.6 ETHICS AND TRANSPARENCY OF EMERGING COMPUTATIONAL MODELS IN WIRELESS COMMUNICATIONS

An important field of research called "Ethics and Transparency of Emerging Computational Models in Wireless Communications" looks at the ethical issues and transparency needs connected to the expanding usage of computational models in wireless communication systems [14]. These models, which frequently use machine learning and artificial intelligence, are playing an increasingly important role in wireless network design, management, and operation [30]. Building trust, defending user rights, and promoting ethical innovation all depend on ensuring moral conduct and openness in their creation and use [54].

  a) *Data Privacy and Ethics*
    i. *Data Collection and Consent*
       Wireless communications computational models frequently rely on enormous volumes of user data obtaining informed consent for data gathering and ensuring that user privacy rights are honoured are ethical considerations [85].
    ii. *Data Anonymization and Protection*
       Techniques used by ethical models include data anonymization, unauthorised access, and data breaches [86].
  b) *Fairness and Bias Mitigation*
    i. *Bias Identification and Correction*
       It's possible for computational models to unintentionally reinforce biases found in their training data. To ensure that all users are treated fairly, ethical models are created to recognise and reduce these biases [18].
    ii. *Algorithmic Fairness*
       When using algorithms to make judgements, ethical models prioritise fairness, especially if those findings can affect the capacity of someone to receive assistance or other factors [87].
  c) *Transparency and Explainability*
    i. *Algorithm Transparency*
       To ensure that consumers and stakeholders can understand how decisions are made, especially when those decisions directly affect them, ethical models work to make algorithms transparent [88].

ii. Explainable AI (XAI)

Complex models are made interpretable using XAI approaches, allowing users to understand the rationale behind model outputs [89]. This openness encourages trust [90].

d) *Accountability*

   i. *Clear Responsibility*

Ethical models clearly define who is responsible for the model behaviour [91]. Mechanisms are in place to identify and address the accountable parties if models malfunction or have unintended consequences [92].

   ii. *Regulatory Compliance*

Ethical models comply with legal and regulatory obligations, making sure they fulfil standards and rules of behaviour unique to their business [93].

e) *Security and Trust*

   i. *Data Security*

To safeguard user data from breaches or unauthorised access, ethical models include strong data security measures [94].

   ii. *Trustworthiness*

The reliability of computational models and the wireless networks in which they function is increased by transparency and moral conduct [95].

f) *User Education and Informed Consent*

   i. *User Awareness*

User education is a top priority for ethical models, ensuring that people are informed of how their data are utilised and the effects of model-driven decisions [96].

   ii. *Informed Consent*

Users are given clear and thorough information, enabling them to make knowledgeable decisions about data sharing and involvement in services that employ computational models [97].

g) *Regulatory Compliance*

   i. *Data Protection Regulations*

Ethical models comply with data protection laws, such as the GDPR in Europe, to ensure that they uphold user rights and legal obligations for data processing [53].

   ii. *Industry-Specific Regulations*

Ethical models must adhere to industry-specific standards in fields like healthcare and finance to safeguard private data and ensure responsible use [89].

h) *Ethical Considerations in Model Development*

   i. *Ethical by Design*

Fairness, accountability, and openness principles are incorporated into ethical models during the design phase [91].

   ii. *Ethics Committees*

Some businesses set up ethics committees or boards to oversee and direct the moral creation and application of computational models [98].

Emerging computational models for wireless communications must take into account a variety of ethical and transparent factors, including data privacy, fairness, accountability, and regulatory compliance. These guidelines are necessary to establish confidence, safeguard user rights, and make sure that the employment of computational models in wireless networks complies with moral and societal norms. Integrating these moral and open practices is crucial for responsible innovation in wireless communication systems as computational models develop further.

10.7 COMPLIANCE AND LEGAL IMPLICATIONS OF NEW COMPUTATIONAL MODELS

In the digital age, "Compliance and Legal Implications of New Computational Models" is a crucial and developing subject [99]. Powered by artificial intelligence (AI), machine learning (ML), and data analytics, computational models are permeating a wide range of industries, from finance to healthcare and beyond [98]. These models have a lot of potential, but their adoption also raises challenging regulatory and legal questions. The present chapter explores the legal implications and compliance obligations related to these new computational models. A paradigm shift brought forth by the digital transformation of the twenty-first century [100] is characterised by the dominance of computational models [101]. These models are revolutionising sectors and changing judgements since they are supported by AI and data-driven algorithms [88]. They hold the potential for tremendous efficiency and improvements thanks to their capacity for processing enormous volumes of data, identifying patterns, and making predictions [102]. But as the significance of computational models grows, so does the demand for strong compliance frameworks and a keen understanding of the consequences they have for the law [103].

10.7.1 COMPLIANCE IN THE AGE OF COMPUTATIONAL MODELS

A complex network of compliance regulations is in place to address the introduction of new computational models [104]. Organisations using these models must navigate a complicated terrain of rules and regulations, as depicted in Figure 10.1, from data privacy legislation like the GDPR in Europe to sector-specific obligations like HIPAA in healthcare [105]. A multifaceted strategy is required for compliance [106].

The diagram illustrates the four (4) key elements and considerations in computational modelling for wireless communication systems that are required for compliance with regulations.

a) *Data Privacy and Security Compliance*
   The moral and legal use of data is the cornerstone of compliance in this situation [33]. To secure sensitive information, organisations must make sure they have the required consent, have data protection procedures in place, and maintain data security [31].

b) *Transparency and Accountability*
   New regulations require algorithmic decision-making to be transparent [107]. Organisations are required by law to disclose the workings of these models, especially in cases where they impact the rights of individuals or their ability to receive services [108].

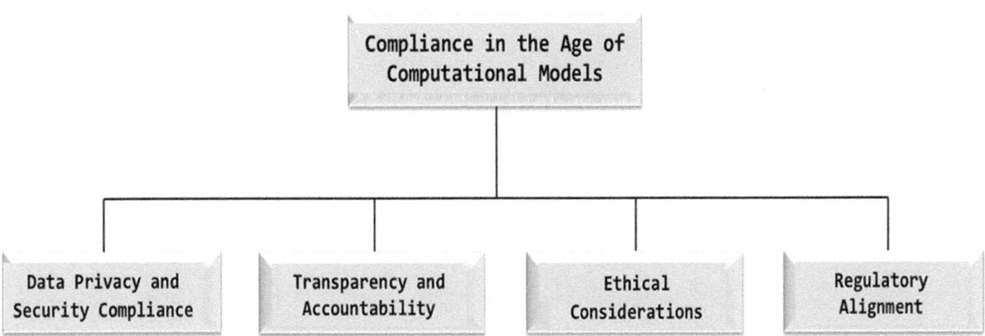

FIGURE 10.1 Regulatory Compliance of Computational Modeling.

c) Ethical Considerations

Compliance encompasses ethical considerations in addition to legal requirements [109]. The ethical ramifications of AI and ML models must be considered by organisations, especially in areas like bias reduction and fairness [110].

d) Regulatory Alignment

Organisations must adhere to legislation relevant to their industry, depending on the sector [111]. For instance, Basel III standards must be followed in the financial sector [112], whereas Software as a Medical Device (SaMD) rules from the FDA are crucial in the healthcare sector [113].

10.7.2 Legal Implications of Non-Compliance

Violations of the regulatory system governing computational models may have serious repercussions [114]. These legal repercussions are shown in Figure 10.2.

In the context of the legal frameworks governing wireless communication systems, this diagram highlights the dangers and repercussions of non-compliance.

a) *Fines and Penalties*

Regulators have the authority to penalise noncompliance severely, which can hurt the financial stability of an organisation [115].

b) *Reputation Damage*

The image of a company can be harmed by non-compliance [116], which also erodes confidence with partners, clients, and stakeholders [117].

c) *Legal Action*

Non-compliant models may result in legal action from those impacted by them, which could cost a lot of money and result in damages [118].

d) *0 Disruption*

Regulatory actions may interfere with business operations, necessitating resources for correction [119].

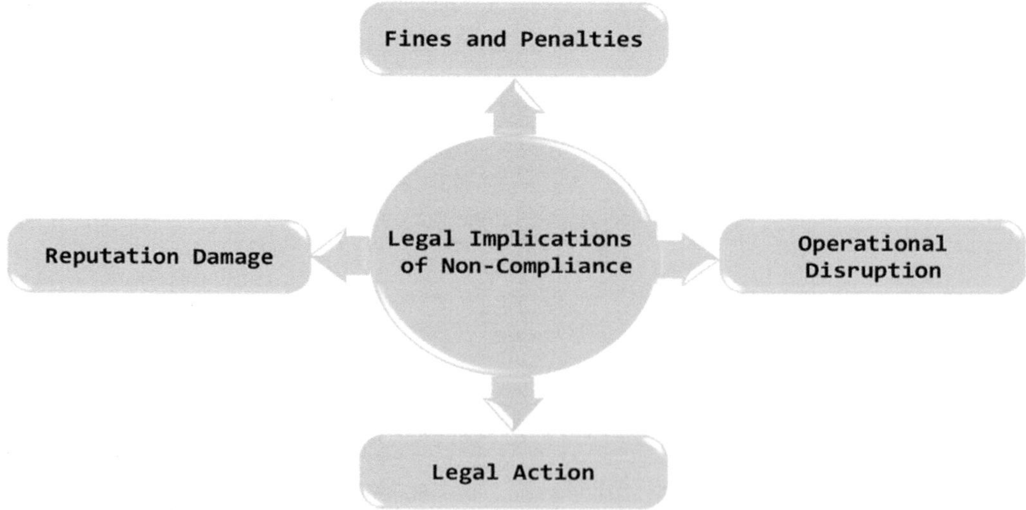

FIGURE 10.2   Risk and Consequences of Non-Compliance.

10.7.3 COMPLIANCE CHALLENGES

A proactive approach is necessary to successfully navigate the compliance and regulatory environment around computational models. Figure 10.3 depicts the compliance challenges in computational models.

An illustrated picture entitled *Compliance Difficulties* depicts the complex legal and regulatory hurdles that arise in the context of wireless communication systems and computational models.

a) *Compliance by Design*
   Organisations should incorporate compliance factors from the very beginning of the design and development of computational models [32].
b) *Continuous Monitoring*
   To maintain continuing compliance, model performance, data inputs, and results must be regularly monitored [120].
c) *Data Governance*
   To assist with compliance efforts, put in place strong data governance practices that cover data quality, provenance, and lineage [32].
d) *Education and Training*
   Give employees the knowledge and abilities they need to successfully negotiate the regulatory landscape [98].
e) *Legal Expertise*
   Engage a lawyer with experience in AI and data protection to advise you on compliance issues [121].

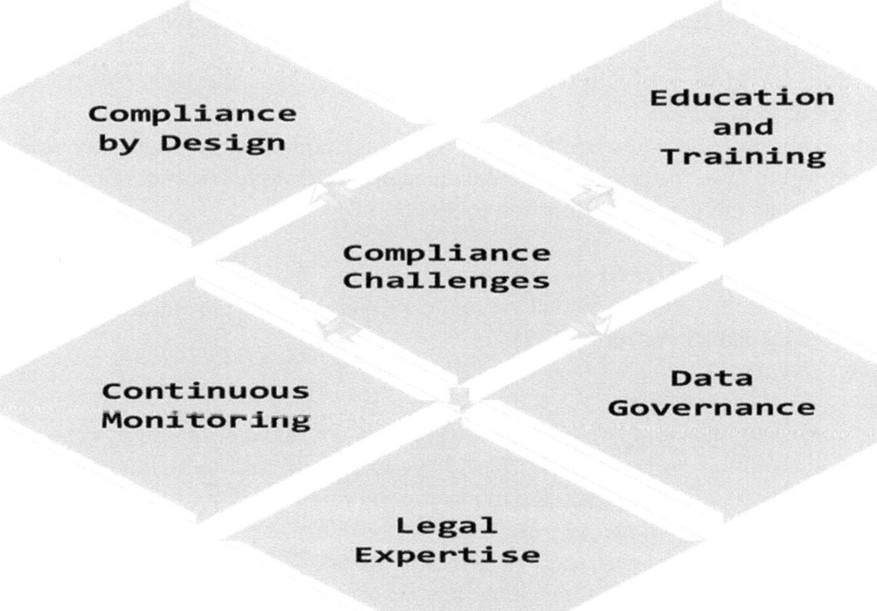

FIGURE 10.3   Compliance Difficulties.

The weight of legal duties and the necessity of compliance both increase as new computational models are adopted [122]. Companies that successfully negotiate this terrain will not only reduce legal risks but also establish themselves as accountable custodians of information and technology [123]. The path forward is fraught with difficulties, but it is also paved with chances for innovation and advancement, steered by a persistent dedication to compliance and moral behaviour in the field of new computational models [122].

10.7.4 IMPLICATIONS OF NON-COMPLIANCE

The consequences of breaching legal frameworks computational model regulation in wireless communication systems is complex and affects many parties, including enterprises, users, government agencies, and the larger technology community. The main ramifications include;

1. *Legal Consequences*
   a) *Fines and Penalties*
   Legal measures, including fines and penalties issued by regulatory bodies, are frequently the result of non-compliance. Depending on the type and degree of the infraction, these penalties could range in severity [124].
   b) *Legal Proceedings*
   Long-lasting, resource-intensive judicial actions might include businesses. Legal conflicts have the potential to damage the financial standing of a company and its reputation [125].
2. *Damage to Reputation*
   a) *Loss of Trust*
   Credibility among partners, consumers, and users can be damaged by noncompliance. Customers may stop trusting a service if they believe that their data is not treated responsibly or that it does not adhere to regulations [126].
   b) *Brand Damage*
   One important consequence is damage to reputation. The image of a business may be damaged by bad press and media attention to legal matters, making it challenging to win back the confidence of clients [127].
3. *Data Breaches and Security Risks*
   a) *Increased Vulnerability*
   Vulnerabilities in systems could result from noncompliance with security requirements, which would raise the possibility of data breaches. This may result in sensitive data compromise, data theft, and unauthorised access [128].
   b) *Regulatory Reporting*
   Organisations are required by numerous regulatory frameworks to quickly report data breaches. Failure to comply with these reporting requirements may intensify legal problems and raise regulatory concerns [129].
4. *Economic Impact*
   a) *Operational Disruptions*
   Violating compliance may require modifying current procedures and systems. This may cause interruptions to operations, cause downtime, and result in higher cleanup expenses [130].
   b) *Market Access Restrictions*
   Market access may be restricted by non-compliance, especially in areas with stringent regulatory requirements. In some areas, businesses may be prohibited from providing specific services or goods [129].

    5. *Regulatory Scrutiny and Oversight*
      a) *Increased Audits*
        Regulatory agencies might conduct more audits and inquiries into the activities of businesses that do not comply. The increased scrutiny may put more strain on staff and resources [131].
      b) *Stricter Regulations*
        Regulations may become more stringent as a result of non-compliance, which would affect the whole business. This may make the regulatory environment more difficult for all parties involved [132].
    6. *Loss of Business Opportunities*
      a) *Contractual Challenges*
        Violations of contractual agreements with partners and clients could result from non-compliance. This may lead to the dissolution of partnerships and commercial prospects [133].
      b) *Exclusion from Bidding*
        Non-compliance can keep companies out of profitable commercial prospects in sectors where participation in bids or tenders requires compliance [133].
    7. *Long-term Strategic Impact*
      a) *Strategic Reassessment*
        Organisations may be forced to reevaluate their long-term strategy if they fail to comply. This could entail adopting new technologies, modifying company strategies, or forming strategic alliances to comply with legal obligations [133].
      b) *Innovation Constraints*
        The adoption of cutting-edge techniques and developing technology may be hampered by noncompliance. To guarantee conformity to regulatory norms, businesses might have to give up on some achievements [134].
    8. *Increased Monitoring and Reporting Obligations*
      a) *Enhanced Reporting Requirements*
        Non-compliance usually leads to increased obligations for reporting and monitoring from regulatory organisations. This could result in more administrative labour for an organisation [126].
      b) *Regular Audits*
        More frequent audits may be applied to organisations to guarantee continuous compliance. This ongoing inspection may need a substantial amount of paperwork work and resources [131].

Violations of legal frameworks controlling computational models in wireless communication networks can have far-reaching effects on reputation, security, economic viability, and long-term strategic planning, in addition to legal ramifications. To avoid potential risks and maintain sustainable and responsible business practices, organisations need to proactively handle compliance needs as regulatory frameworks continue to grow.

10.8 CONFLICT RESOLUTION IN COMPUTATIONAL MODELS FOR WIRELESS COMMUNICATION SYSTEMS

In computational models for wireless communication systems, conflict resolution is a fundamental and complex process aimed at resolving disagreements, conflicting requests, and resource allocation issues that develop within the intricate framework of wireless networks [135]. Multiple parties that are vying for limited network resources, including devices, users, and services, may be involved in these conflicts [136]. The efficient and equitable use of these resources is ensured through effective conflict resolution, which also reduces interference and improves the overall performance and dependability of wireless communication networks [137]. A more thorough description of dispute resolution in this situation is provided in this section.

a) *Resource Allocation Conflicts*

Resource allocation is one of the main sources of contention in wireless communication systems [138]. Resources like spectrum, bandwidth, and computational power are scarce in wireless networks. There may be disagreements over how to allocate these resources if multiple devices and services simultaneously want access to them [139].

Conflict Resolution Mechanisms

Algorithms for dynamic resource allocation, such as those used in cognitive radio systems, allocate resources based on current demand. Mechanisms for bargaining and bidding can also be used to distribute resources among competing parties.

[140]

b) *Interference Management*

The quality of communication can be severely reduced by wireless device interference [141]. When many devices attempt to broadcast data concurrently on the same frequency or in close vicinity, conflicts may occur [142].

Conflict Resolution Mechanisms

By altering transmission strength, frequency hopping, or channel selection, interference mitigation algorithms and protocols seek to identify and address conflicts [141]. For instance, cognitive radio systems can detect the radio environment and change frequencies to lessen interference.

[143]

c) *Quality of Service (QoS) Conflicts*

Many different types of services with various QoS needs are frequently hosted by wireless communication networks [133]. These criteria may conflict, for example, when low-latency services vie with high-throughput applications for network resources [144, 145].

Conflict Resolution Mechanisms

Quality of service (QoS) management mechanisms establish traffic priority based on the particular needs of each service, guaranteeing that conflicting demands are fairly balanced.

[146]

d) *Security and Privacy Conflicts*

In wireless communication, data security and privacy are of utmost importance [147]. When models handle sensitive user data without the necessary safeguards, conflicts may occur that could compromise security or violate user privacy [148].

Conflict Resolution Mechanisms

For these conflicts to be resolved, strong access control, authentication, and encryption procedures are required. Ensuring GDPR compliance, as well as other data privacy laws, is essential.

[147]

e) Regulatory Compliance Conflicts

Systems for wireless communication are governed by several standards and regulations. When models and network activities do not follow these guidelines, conflicts may arise [149].

Conflict Resolution Mechanisms

Mechanisms for auditing and monitoring compliance make sure that all operations adhere to statutory and regulatory standards [150]. This can entail ongoing observation and reporting to regulatory bodies.

[151]

f) Cross-Stakeholder Collaboration

Network operators, service providers, and regulatory authorities may communicate and work together more easily thanks to collaborative platforms and industry standards [152]. Effective dispute resolution requires such cooperation [23].

Conflict Resolution Frameworks

Industry-wide standards and best practices establishment encourages consistent conflict resolution strategies and collaboration.

[22]

g) Real-time Conflict Detection

For a speedy resolution, quick conflict detection is essential [21]. To detect conflicts as they occur, systems should have real-time monitoring and detection capabilities [24].

Conflict Resolution Frameworks

AI-driven models and predictive analytics can foresee and identify prospective disputes, enabling proactive resolution.

[22]

h) Automation and Orchestration

Particularly in expansive and dynamic wireless networks, automation is crucial to conflict resolution [153]. Without human intervention, automated systems can quickly change resource allocation to resolve disagreements [154].

Conflict Resolution Frameworks

Based on preset policies and algorithms, network orchestration and automation frameworks enable dynamic resource allocation and dispute resolution.

[153]

Resource allocation, interference control, QoS optimisation, security, regulatory compliance, and cooperative efforts are just a few of the many facets of the multidimensional process that go into solving conflicts in computational models for wireless communication systems [155]. To ensure the seamless functioning of wireless networks, reduce disruptions, and maximise resource usage, effective conflict resolution techniques and frameworks are crucial. These methods are essential for handling the complex and dynamic problems that develop in the constantly changing wireless communication environment.

10.9 LESSONS LEARNED, RESEARCH DIRECTION AND RECOMMENDATIONS

In this section, we summarise the key findings from our investigation into the legal frameworks governing computational models in wireless communication systems, provide direction for future research endeavours, and make useful suggestions for industry stakeholders. Through the wealth of insights we have gained from our voyage, we have been able to distil important lessons, imagine future research possibilities, and offer practical recommendations to help us go further.

10.9.1 Lessons Learned

This chapter contains several insightful lessons that can be learned from it. The key lessons learned are as follows;

a) The chapter emphasises how closely connected legal systems and developing technologies are. It shows us that the law must change to keep up with technological developments to maintain the effectiveness of regulations.
b) It emphasises how crucial it is to incorporate ethical concerns into regulatory structures. Lessons learnt include the requirement to give priority to moral behaviour, data protection, and responsible technology development to foster confidence and safeguard user rights.
c) The chapter places a strong emphasis on how ubiquitous wireless communication networks are. Lessons learned here include the value of international cooperation and regulatory harmonisation to promote interoperability and address cross-border issues.
d) It reaffirms the necessity of strong cybersecurity protections inside regulated environments. The key takeaway is that, to protect vital infrastructure, regulatory compliance must include cybersecurity as a core component.
e) The chapter explains the necessity of strict data privacy laws to safeguard user data. Lessons include the value of openness, permission, and data security procedures in a world of data-driven wireless communication.
f) The chapter examines methods for resolving disputes. Lessons include the importance of using other dispute resolution procedures like mediation and arbitration to resolve disputes quickly.

The lessons learned from this chapter emphasise how the legal frameworks for wireless communication systems are dynamic and complex. The significance of strong cybersecurity and data privacy regulations is emphasised, along with adaptability, ethical considerations, international collaboration, and globalisation. Policymakers, industry participants, scholars, and anyone else interested in the interaction of law and technology in this important area will find these lessons to be useful.

10.9.2 Research Direction

Future research directions emerge, building on the insights obtained. Examining how developing technologies like edge computing and 5G may affect the legal frameworks that are in place is one notable way to go. Comprehending how these technologies interact with rules will be essential for well-informed policymaking. Furthermore, a thorough examination of the sociocultural implications and ethical aspects of computational models in wireless communication is necessary. Another interesting line of inquiry is to investigate new dispute resolution mechanisms in these models. Because the area is dynamic, it is necessary to continuously investigate the efficacy of current frameworks and identify any potential regulatory loopholes.

10.9.3 Recommendations

For policymakers, industry participants, and researchers, this chapter offered the following crucial suggestions. Taking into account the information and focus areas covered in the chapter, here are some suggestions.

a) Authorities and regulatory organisations should set up processes for routine changes to the laws controlling wireless communication technologies. Given the speed at which technology is developing, these frameworks should continue to be flexible and responsive to innovations.

b) Adapt legal systems to ethical issues. Consistent with public norms and expectations, make sure that rules emphasise responsible and ethical practices in technology development, data processing, and privacy protection.

c) Encourage international cooperation and standard-setting harmonisation in the wireless communication industry. This might improve interoperability, simplify things for international operators, and encourage a coordinated strategy for tackling shared problems.

d) To protect user rights and personal information in wireless communication systems, data privacy laws should be strengthened. As privacy expectations change, encourage transparency in data collection, usage, and permission processes.

e) Enforce stringent cybersecurity regulations within the bounds of the law. Impose strict security requirements on wireless communication system operators to protect against cyber threats and ensure the resilience of critical infrastructure.

f) Encourage the use of alternative dispute resolution procedures for resolving disputes in the wireless communication industry, such as arbitration and mediation. This can hasten the resolution of disputes and lessen legal conflicts.

g) To evaluate the effectiveness and efficiency of legal systems, institute recurring regulatory audits. These audits can spot areas for development and make sure that rules are still applicable.

h) Establish monitoring and reporting measures to monitor adherence to legal requirements. Regulators and the public can evaluate industry compliance with regulations with the aid of transparent reporting.

These suggestions support the overarching objective of maintaining the efficacy, adaptability, and favourable environment for responsible innovation of the legislative frameworks governing wireless communication systems. They discuss a variety of topics, demonstrating the complexity of this important field, ranging from ethics and data protection to cybersecurity and regulatory compliance.

10.10 CONCLUSION

This chapter has shown that the relationship between the law and the evolving field of wireless communication is both complex and essential. To understand the legal environment of wireless communication, we first looked at the international agreements and standards that support worldwide connectivity. These frameworks act as the foundation upon which the industry constructs its innovations, from the harmonising efforts of the ITU to the harmonious symphony of global standards. Transparency and ethics have become the compass points that point the way to responsible technological advancement. The moral necessity of upholding privacy and human dignity in the era of AI and IoT cannot be overemphasized. On our journey, compliance and legal ramifications were major concerns. The importance of industry stakeholders adhering to legal and regulatory demands was made clear by the compliance journey, which started with data protection and ended with cybersecurity. The legal repercussions of non-compliance, including severe penalties and reputational harm, highlighted the significant risks. Mechanisms for resolving conflicts paved the way for settling disputes inside the sector. These processes, which range from negotiation and mediation to arbitration and litigation, serve as bridges to agreements and justice compasses. We have a thorough understanding of the legal frameworks that weave the fabric of wireless communication as we pull the curtains in this chapter. These regulations are the foundation of our interconnected world; they go beyond simple legal requirements. The relationship between law and wireless communication will never be broken, even in a future that promises 6G, widespread IoT use, and AI-driven connectivity. Compliance is a must, ethical considerations are indisputable, and conflict resolution serves as a bridge to advancement on this road. In this environment, adaptation is essential, transparency is the basis for trust, and stakeholder discussion determines the course of action. This chapter demonstrates how crucial these frameworks are to determining how connectivity and innovation will develop in the future.

REFERENCES

[1] I. Lebovka, B. H. Mele, X. Liu, A. Zakieva and T. Schlamp "Computational modeling of cambium activity provides a regulatory framework for simulating radial plant growth," *Life*, vol. 12, 2023, pp. e66627.

[2] M. Theissen, L. Kern, T. Hartmann and E. Clausen "Use-case-oriented evaluation of wireless communication technologies for advanced underground mining operations," *Sensors*, vol. 23, no. 7, 2023, pp. 3537.

[3] D. J. Teece, "Profiting from innovation in the digital economy: Enabling technologies, standards, and licensing models in the wireless world," *Research Policy*, vol. 47, no. 8, pp. 1367–1387, 2018.

[4] S. Wang, X. Jiang, and M. B. Khaskheli, "The role of technology in the digital economy's sustainable development of Hainan Free Trade Port and genetic testing: Cloud computing and digital law," *Sustainability*, vol. 16, no. 14, p. 6025, 2024.

[5] S. Lehtilä, A. Alén and P. Korpisaari "Spectrum regulation and frequency allocation in the context of a smart city–using the regulatory approach in Finland as an example," *Information & Communications Technology Law*, 2023, pp. 1–15.

[6] T. Kohno, Y. Acar, and W. Loh, "Ethical frameworks and computer security trolley problems: Foundations for conversations," in *32nd USENIX Security Symposium (USENIX Security 23)*, 2023, pp. 5145–5162.

[7] J. Zhang and Z. M. Zhang, "Ethics and governance of trustworthy medical artificial intelligence," *BMC Medical Informatics and Decision Making*, vol. 23, no. 1, p. 7, 2023.

[8] A. Singhal, N. Neveditsin, H. Tanveer, and V. Mago, "Toward fairness, accountability, transparency, and ethics in AI for social media and health care: Scoping review," *JMIR Medical Informatics*, vol. 12, no. 1, p. e50048, 2024.

[9] R. Gibson, E. Bon and A. Römmele "Operationalizing data-driven campaigning: Designing a new tool for mapping and guiding regulatory intervention," *Policy Studies*, vol. 45, no. 2, 2023, pp. 1–17.

[10] A. Havolli and M. Fetaji, "Improving Radio Network Planning and Design in Next-Generation Mobile Networks Using AI and ML Algorithms," in *2023 12th Mediterranean Conference on Embedded Computing (MECO)*, June 2023, pp. 1–5.

[11] A. Parvez, R. S. Kartha, M. Tiwari and V. K. Singh "Integration of Blockchain Technology for Enhanced Wireless Networks," in *2023 International Conference on Inventive Computation Technologies (ICICT)*, pp. 1251–1253, April 2023.

[12] V. Wylde, E. Prakash, C. Hewage, and J. Platts, "Ethical challenges in the use of digital technologies: AI and big data," in Montasari, R., Carpenter, V., Masys, A.J. (eds), *Digital Transformation in Policing: The Promise, Perils and Solutions*, Cham: Springer International Publishing, 2023, pp. 33–58. https://doi.org/10.1007/978-3-031-09691-4_3

[13] J. Mahilraj M. Pandian and M. Subbiah "Evaluation of the Robustness, Transparency, Reliability and Safety of AI Systems," in *2023 9th International Conference on Advanced Computing and Communication Systems (ICACCS)*, pp. 2526–2535, March 2023.

[14] F. L. Shults and W. J. Wildman "The Ethics of Computer Modeling and Simulation," in *2018 Winter Simulation Conference (WSC)*, pp. 4069–4083, December 2018,.

[15] Z. Zhang, J. Zhang, Q. Luo and C. H. Chou "A biorealistic computational model unfolds human-like compliant properties for control of hand prosthesis," *IEEE Open Journal of Engineering in Medicine and Biology*, vol. 3, 2022, pp. 150–161.

[16] P. M. Salmon, G. J. M. Read, J. Thompson and S. Mclean "Computational modelling and systems ergonomics: A system dynamics model of drink driving-related trauma prevention," *Ergonomics*, vol. 63, no. 8, 2020, pp. 965–980.

[17] M. Mitchell, S. Wu, A. Zaldivar and P. Barnes "Model Cards for Model Reporting," in *Proceedings of the Conference on Fairness, Accountability, and Transparency*, pp. 220–229, January 2019.

[18] J. M. Durán and K. R. Jongsma, "Who is afraid of black box algorithms? On the epistemological and ethical basis of trust in medical AI," *Journal of Medical Ethics*, vol. 47, no. 5, 2021, pp. 329–335.

[19] I. Brown and C. T. Marsden, *Regulating code: Good Governance and better Regulation in the Information Age.* MIT Press, 2023.

[20] J. Moysen and M. Garcia-lozano "Conflict resolution in mobile networks: A self-coordination framework based on non-dominated solutions and machine learning for data analytics [application notes]," *IEEE Computational Intelligence Magazine*, vol. 13, no. 2, 2018, pp. 52–64.

[21] L. Pallottino, V. G. Scordio, A. Bicchi and E. Frazzoli, "Decentralized cooperative policy for conflict resolution in multivehicle systems," *IEEE Transactions on Robotics*, vol. 23, no. 6, 2007, pp. 1170–1183.

[22] A. Banerjee, S. S. Mwanje and G. Carle, "Game Theoretic Conflict Resolution Mechanism for Cognitive Autonomous Networks," in *2020 International Symposium on Performance Evaluation of Computer and Telecommunication Systems (SPECTS)*, pp. 1–8, July 2020.

[23] Y. Zhao, S. Zhou, H. Ding, Q. Liu, Z. Yang and C. Li "Research on time division multi-channel p-Persistent CSMA based on binary trees conflict resolution mechanism," *Journal of Advances in Computer Networks*, vol. 4, no. 2, 2016.

[24] W. Jacak and K. Pröll, "Conflict Resolution in Multiagent Systems Based on Wireless Sensor Networks," in *Computer Aided Systems Theory-EUROCAST 2009: 12th International Conference, Las Palmas de Gran Canaria, Spain, February 15-20, 2009, Revised Selected Papers 12*, pp. 753–760, Springer Berlin Heidelberg, 2009.

[25] I. David and E. Syriani, "Real-time collaborative multi-level modeling by conflict-free replicated data types," *Software and Systems Modeling*, vol. 22, no. 4, pp. 1131–1150, 2023.

[26] S. L. Ng, "Striking the balance: Accounting regulatory compliance and standards in Fintech," in Naz, F., Karim, S. (Eds), *Safeguarding Financial Data in the Digital Age*, Hershey, Pennsylvania, USA: IGI Global, 2024, pp. 214–237.

[27] E. T. Inau, R. nalugala, W. M. Nandwa and F. Obwanda "FAIR Equivalency, regulatory framework and adoption potential of FAIR Guidelines in health in Kenya," *Data Intelligence*, vol. 4, no. 4, pp. 852–866, 2022.

[28] U. Pagallo, M. Durante, S. Monteleone, "What is new with the Internet of Things in privacy and data protection? Four legal challenges on sharing and control in IoT," in Leenes, R., van Brakel, R., Gutwirth, S., De Hert, P. (eds), *Data Protection and Privacy: (In)visibilities and Infrastructures*, in Law, Governance and Technology Series, vol. 36, Cham: Springer, 2017. https://doi.org/10.1007/978-3z-319-50796-5_3

[29] R. B. L. Dixon, "A principled governance for emerging AI regimes: Lessons from China, the European Union, and the United States," *AI and Ethics*, vol. 3, no. 3, pp. 793–810, 2023.

[30] E. S. Chen and M. J. Walker, "Legal and ethical implications of 5G technology deployment," *International Journal of Law and Information Technology*, vol. 28, no. 1, 2020.

[31] M. A. Rodriguez and C. M. Gonzalez, "Ensuring data privacy in wireless sensor networks: A regulatory perspective," *Journal of Wireless Networks and Security*, vol. 6, no. 2, 2019.

[32] A. Mitra, "Adapting the regulation of spectrum and telecom networks to 5G technology – A cross country analysis," in *International Working Conference on Transfer and Diffusion of IT*, Cham: Springer Nature Switzerland, 2023, pp. 107–125.

[33] S. Rizvi, S. Campbell and K. Alden, "Why compliance is needed for Internet of Things?," *2020 International Conference on Software Security and Assurance* (ICSSA), Altoona, PA, USA, 2020, pp. 66–71. doi: 10.1109/ICSSA51305.2020.00019.

[34] M. D. Parker and L. M. Hayes, "Conflict resolution in 5G networks: A review of techniques and challenges," *International Journal of 5G and Wireless Communications*, vol. 10, no. 4, 2021.

[35] K. S. Turner and R. A. Wilson, "Compliance and certification of wireless devices: Challenges and solutions," *Wireless Networks*, vol. 24, no. 7, 2018.

[36] R. E. White and C. S. Hill, "Legal aspects of spectrum sharing in cognitive radio networks," *Telecommunications Policy*, vol. 44, no. 5, 2020.

[37] J. Barnett, "Antitrust mercantilism: The strategic devaluation of intellectual property rights in wireless markets," *Berkeley Technology Law Journal*, Forthcoming, USC CLASS Research Paper No. CLASS23-2, 2023.

[38] B. T. Clark and M. L. Turner, "Ensuring fairness in resource allocation for wireless communication systems," *Wireless Networks*, vol. 26, no. 4, 2020.

[39] A. K. Sharma and R. Sharma, "Comparative analysis of data protection laws and AI privacy risks in BRICS nations: A comprehensive examination," *Global Journal of Comparative Law*, vol. 13, no. 1, pp. 56–85, 2024.

[40] J. Babikian, "Securing rights: Legal frameworks for privacy and data protection in the digital era," *Law Research Journal*, vol. 1, no. 2, pp. 91–101, 2023.

[41] S. M and M. G. Sumithra, "Overview of spectrum sharing and dynamic spectrum allocation schemes in cognitive radio networks," *2022 8th International Conference on Advanced Computing and Communication Systems* (ICACCS), Coimbatore, India, 2022, pp. 934–937. doi: 10.1109/ICACCS54159.2022.9785048.

[42] O. Akanfe, D. Lawong, and H. R. Rao, "Blockchain technology and privacy regulation: Reviewing frictions and synthesizing opportunities," *International Journal of Information Management*, vol. 76, p. 102753, 2024.

[43] F. von der Dunk, "Legal aspects of satellite communications," in Handbook of Space Law, Edward Elgar Publishing, 2015, pp. 456–500.

[44] P. Bustamante, M. Gomez, P. Krishnamurthy, W. Lehr, I. Murtazashvili, A. Palida, and M. B. Weiss, "Unassigned spectrum: An institutional analysis of radio spectrum management," *SSRN Electronic Journal*, Available at SSRN 4528676, 2023.

[45] H. Jahankhani, S. Kendzierskyj and O. Hussien, "Approaches and Methods for Regulation of Security Risks in 5G and 6G," in *Wireless Networks: Cyber Security Threats and Countermeasures*, pp. 43–70, Cham: Springer International Publishing, 2023.

[46] W. K. Alsaedi, H. Ahmadi, Z. Khan, and D. Grace, "Spectrum options and allocations for 6G: A regulatory and standardization review," *IEEE Open Journal of the Communications Society*, vol. 4, pp. 1787–1812, 2023, doi: 10.1109/OJCOMS.2023.3301630.

[47] C. Koenig and A. Veidt, "Lifting a regulatory millstone around 5G investors' neck–5G network slicing versus EU-net neutrality?" *Telecommunications Policy*, pp. 102653, 2023.

[48] A. J. Gonzalez, J. Ordonez-Lucena, B. E. Helvik, G. Nencioni, M. Xie, D. R. Lopez, and P. Grønsund, "The isolation concept in the 5G network slicing," in *2020 European Conference on Networks and Communications (EuCNC)*, 2020, pp. 12–16.

[49] L. K. Branting, *Reasoning with Rules and Precedents: A Computational Model of Legal Analysis.* Springer Science & Business Media, 2013.

[50] J. Frankenreiter and M. A. Livermore, "Computational methods in legal analysis," *Annual Review of Law and Social Science*, vol. 16, pp. 39–57, 2020.

[51] R. Atat, L. Liu, H. Chen, J. Wu, H. Li, and Y. Yi, "Enabling cyber-physical communication in 5G cellular networks: Challenges, spatial spectrum sensing, and cyber-security," *IET Cyber-Physical Systems: Theory & Applications*, vol. 2, no. 1, pp. 49–54, 2017.

[52] M. Massaro, "Radio spectrum regulation as a matter of international affairs: Discussing the effectiveness of the European Union at world radiocommunication conferences," *Digital Policy, Regulation and Governance*, vol. 20, no. 5, pp. 373–398, 2018.

[53] European Commission, "General Data Protection Regulation (GDPR)," [Online]. Available: https://ec.europa.eu/commission/priorities/justice-and-fundamental-rights/data-protection/2019-reform-eu-data-protection-rules_en.

[54] S. L. Adams and D. M. Clark, "Privacy and security in wireless communication systems: Legal and ethical challenges," *Journal of Cybersecurity and Privacy*, vol. 2, no. 1, 2018.

[55] A. Winter, E. Ammenwerth, R. Haux, M. Marschollek, B. Steiner, and F. Jahn, "Technological perspective: Architecture, integration, and standards," in Winter, A., Ammenwerth, E., Haux, R., Marschollek, M., Steiner, B., Jahn, F. (Eds), *Health Information Systems: Technological and Management Perspectives*, Cham: Springer International Publishing, 2023, pp. 51–152. https://doi.org/10.1007/978-3-031-12310-8_3

[56] International Electrotechnical Commission, "IEC 62304: Medical device software – Software life cycle processes," 2019.

[57] D. Solomon, "Standards in Computer Science and Information Technology," *Teaching and Collecting Technical Standards,* pp. 109, 2023.

[58] International Organization for Standardization, "ISO/IEC 17025: General requirements for the competence of testing and calibration laboratories," 2020.

[59] C. E. Pereira, C. Diedrich and P. Neumann, "Communication protocols for automation," in *Springer Handbook of Automation*, pp. 535–560, 2023.

[60] E. Vernikos, "Investigating GDPR compliance in European telecommunication industries by using ISOIEC 27001: 2013 and ISOIEC 27701: 2019 Standards," Masters' Thesis. Department of Computer Science, Electrical and Space Engineering, Luleå University of Technology, 2023.

[61] A. Correia, P. Água, A. Frias, and M. Simões-Marques, "Security and privacy for interoperable organizations," *Human Factors and Systems Interaction*, vol. 84, p. 84, 2023.

[62] R. S. Pawar and D. R. Kalbande, "Optimization of quality of service using ECEBA protocol in wireless body area network," *International Journal of Information Technology*, vol. 15, no. 2, pp. 595–610, 2023.

[63] M. Burri, "Cross-border data flows and privacy in global trade law: Has trade trumped data protection?," *Oxford Review of Economic Policy*, vol. 39, no. 1, pp. 85–97, 2023.

[64] O. Gisca, M. Matinmikko-Blue, P. Ahokangas, S. Yrjolä and J. Gordon, "Regulatory challenges and implications of the European Electronic Communications Code (EECC) for local mobile communication network business," *Telecommunications Policy*, vol. 47, no. 10, pp. 102651, 2023.

[65] G. A. Karolyi, J. Sedunov and A. G. Taboada, "Cross-border bank flows and systemic risk," *Review of Finance*, vol. 27, no. 5, pp. 1563–1614, 2023.

[66] World Trade Organization (WTO), "Telecommunications services," [Online]. Available: www.wto.org/english/res_e/reser_e/ersd202005_e.htm

[67] European Commission, "International data flows," [Online]. Available: https://www.eeas.europa.eu/node/400005_fr?s=126#1168

[68] L. Mantl, "Law and Policy of Data from Space: Satellite Navigation and Remote Sensing," in *Space Law in a Networked World*, pp. 189–210, 2023.

[69] ITU Radiocommunication Sector (ITU-R), [Online]. Available: www.itu.int/en/ITU-R/Pages/default.aspx

[70] ITU Publications on Legal and Regulatory Aspects, [Online]. Available: //efaidnbmnnnibpcajpcglclefindmkaj/https://www.itu.int/dms_pub/itu-s/opb/gen/S-GEN-CAT.OL-2021-PDF-E.pdf

[71] E. Paasivirta, "The European union and the United Nations convention on the law of the sea," *Fordham Int'l LJ*, vol. 38, pp. 1045, 2015.

[72] S. Chowdhry, A. Sapir and A. Terzi, "The EU–Japan economic partnership agreement," *Bruegel Special Report*, September 2018.

[73] Y. Lu, X. Xu and L. Wang, "Smart manufacturing process and system automation–a critical review of the standards and envisioned scenarios," *Journal of Manufacturing Systems*, vol. 56, pp. 312–325, 2020.

[74] 3GPP, "5G Cybersecurity standards by 3GPP," [Online]. Available: www.3gpp.org/5g-security

[75] Y. Kamil, S. Lund, and M. S. Islam, "Information security objectives and the output legitimacy of ISO/IEC 27001: Stakeholders' perspective on expectations in private organizations in Sweden," *Information Systems and e-Business Management*, vol. 21, no. 3, pp. 699–722, 2023.

[76] International Standards Organization, "ISO/IEC 27001: Information security management system," 2020.

[77] Internet Engineering Task Force, "RFC 2616: Hypertext transfer protocol -- HTTP/1.1," IETF, 2020. www.ietf.org/rfc/rfc2616.txt

[78] K. N. Waltz, "Globalization and governance," *PS: Political Science & Politics*, vol. 32, no. 4, pp. 693–700, 1999.

[79] I. Abdikhakimov, "Jurisdiction over Transnational Quantum Networks," *International Journal of Law and Policy*, vol. 1, no. 8, 2023, pp 1–13.

[80] S. Hadzovic, S. Mrdovic, and M. Radonjic, "A path towards an Internet of Things and artificial intelligence regulatory framework," *IEEE Communications Magazine*, vol. 61, no. 7, pp. 90–96, 2023.

[81] Z. Rehman, "Beyond borders: International law and global governance in the digital age," *Journal of Accounting & Business Archive Review*, vol. 1, no. 1, pp. 1–12, 2023.

[82] J. L. Renaud, "The role of the International Telecommunication Union: Conflict, resolution and the industrialized countries," in Dyson, K., Humphreys, P. (Eds), *The Political Economy of Communications* (ist Edition), London: Routledge, 2023, pp. 33–57.

[83] Y. Guo and I. G. A. W. Puja, Eds., *Sustaining Peace in ASEAN and the Asia-Pacific: Preventive Diplomacy Measures*, vol. 8, World Scientific, 2022.

[84] A. Khan and M. A. H. S. Jiliani, "Expanding the boundaries of jurisprudence in the era of technological advancements," *IIUMLJ*, vol. 31, p. 393, 2023.

[85] L. L. Dhirani, N. Mukhtiar, B. S. Chowdhry, and T. Newe, "Ethical dilemmas and privacy issues in emerging technologies: A review," *Sensors*, vol. 23, no. 3, p. 1151, 2023.

[86] P. Sharma and S. Barua, "From data breach to data shield: The crucial role of big data analytics in modern cybersecurity strategies," *International Journal of Information and Cybersecurity*, vol. 7, no. 9, pp. 31–59, 2023.

[87] O. B. Ayoade, T. O. Oladele and A. L Imoize "Explainable artificial intelligence (XAI) in medical decision systems (MDSSs): Healthcare systems perspective," in *Explainable Artificial Intelligence in Medical Decision Support Systems*, IET, 2023.

[88] J. B. Awotunde, T. O. Oladele and A. L. Imoize "Explainable artificial intelligence (XAI) in medical decision support systems (MDSS): Applicability, prospects, legal implications, and challenges," in *Explainable Artificial Intelligence in Medical Decision Support Systems*, pp. 45, IET, 2023.

[89] A. L. Imoize and O. B. Ayoade. Eds., *Explainable Artificial Intelligence in Medical Decision Support Systems*, IET, 2023.

[90] A. T. Rufai and A. L. Imoize "XAI robot-assisted surgeries in future medical decision support systems," in *Explainable Artificial Intelligence in Medical Decision Support Systems*, pp. 167, IET, 2023.

[91] M. Usama "Conflict resolution in wireless sensor networks: A comprehensive review," *IEEE Access*, vol. 8, 2020.

[92] J. Ayling and A. Chapman, "Putting AI ethics to work: Are the tools fit for purpose?," *AI and Ethics*, vol. 2, no. 3, pp. 405–429, 2022.

[93] M. Anshari, M. Hamdan, N. Ahmad, E. Ali and H. Haidi "COVID-19, artificial intelligence, ethical challenges and policy implications," *Ai & Society*, vol. 38, no. 2, pp. 707–720, 2023.

[94] S. Gulyamov and S. Raimberdiyev, "Personal data protection as a tool to fight cyber corruption," *International Journal of Law and Policy*, vol. 1, no. 7, 2023.

[95] M. Kumar, A. Aijaz, O. Chattar, J. Shukla, and R. Mutharaju, "Opacity, transparency, and the ethics of affective computing," *IEEE Transactions on Affective Computing*, vol. 15, no. 1, pp. 4–17, 2023.

[96] C. Burr and D. Leslie, "Ethical assurance: A practical approach to the responsible design, development, and deployment of data-driven technologies," *AI and Ethics*, vol. 3, no. 1, pp. 73–98, 2023.

[97] C. Gomez, M. Unberath, and C. M. Huang, "Mitigating knowledge imbalance in AI-advised decision-making through collaborative user involvement," *International Journal of Human–Computer Studies*, vol. 172, p. 102977, 2023.

[98] N. K. Corrêa, J. W. Santos, C. Galvão, M. Pasetti, D. Schiavon, F. Naqvi, R. Hossain, and N. D. Oliveira, "Crossing the principle–practice gap in AI ethics with ethical problem-solving," *AI and Ethics*, pp. 1–18, 2024. https://doi.org/10.1007/s43681-024-00469-8

[99] M. T. Nguyen and M. Q. Tran, "Balancing security and privacy in the digital age: An in-depth analysis of legal and regulatory frameworks impacting cybersecurity practices," *International Journal of Intelligent Automation and Computing*, vol. 6, no. 5, pp. 1–12, 2023.

[100] S. Lucarini, M. Hossain and D. Garcia-Gonzalez, "Recent advances in hard-magnetic soft composites: Synthesis, characterisation, computational modelling, and applications," *Composite Structures*, vol. 279, pp. 114800, 2022.

[101] F. Ojeda, D. Mendez, A. Fajardo, and F. Ellinger, "On wireless sensor network models: A cross-layer systematic review," *Journal of Sensor and Actuator Networks*, vol. 12, no. 4, p. 50, 2023.

[102] S. Srinivasan, M. S. Vinmathi, S. N. Sivaraj, A. Karthikayen, C. Alakesan, and M. Preetha, "A novel approach integrating IoT and WSN with predictive modeling and optimization for enhancing efficiency and sustainability in smart cities," *Journal of Electrical Systems*, vol. 20, no. 4s, pp. 2228–2237, 2024.

[103] M. Fenwick, W. A. Kaal, and E. P. Vermeulen, "Regulation tomorrow: What happens when technology is faster than the law," *American University Business Law Review*, vol. 6, p. 561, 2016.

[104] M. Chessa, A. Van De Bruaene and K. Faroogi "Three-dimensional printing, holograms, computational modelling, and artificial intelligence for adult congenital heart disease care: An exciting future," *European Heart Journal*, vol. 43, no. 28, pp. 2672–2684, 2022.

[105] C. B. Collin, T. Gebhardt and M. Golebiewski "Computational models for clinical applications in personalized medicine—guidelines and recommendations for data integration and model validation," *Journal of Personalized Medicine*, vol. 12, no. 2, pp. 166, 2022.

[106] W. Shafik, "An overview of computational modeling and simulations in wireless communication systems," in Imoize, A.L., Montlouis, W., Obaidat, M.S., Popoola, S.I., Hammoudeh, M. (Eds), *Computational Modeling and Simulation of Advanced Wireless Communication Systems* (1st ed.), Boca Raton: CRC Press. pp. 8–40. https://doi.org/10.1201/9781003457428

[107] M. Crespo, "Fair, transparent and accountable algorithmic decision-making: What is the role of the human-in-the-loop? " iSChannel, vol. 17, no. 1, pp. 28–41, 2022.

[108] G. Lima, N. Grgić-Hlača, J. K. Jeong and M. Cha "The conflict between explainable and accountable decision-making algorithms," in *Proceedings of the 2022 ACM Conference on Fairness, Accountability, and Transparency*, pp. 2103–2113, 2022.

[109] S. Thakkar, Y. Upadhyay, and G. Thakur, "Analysis of compliance prediction models for radio frequency waves with measurement results," in *2014 International Conference on Circuits, Systems, Communication and Information Technology Applications (CSCITA)*, 2014, pp. 150–155.

[110] M. Mohammad Amini and M. Jesus "Artificial intelligence ethics and challenges in healthcare applications: A comprehensive review in the context of the European GDPR mandate," *Machine Learning and Knowledge Extraction*, vol. 5, no. 3, pp. 1023–1035, 2023.

[111] J. E. Cohen, "The regulatory state in the information age," *Theoretical Inquiries in Law*, vol. 17, no. 2, pp. 369–414, 2016.

[112] M. Ahmadalinejad and S. M. Hashemi, "A national model to supervise on virtual banking systems through the Bank 2.0 approach," *Advances in Computer Science: An International Journal*, vol. 4, no. 1, pp. 83–93, 2015.

[113] P. Clark, J. Kim and Y. Aphinyanaphongs, "Marketing and US food and drug administration clearance of artificial intelligence and machine learning enabled software in and as medical devices: A systematic review," *JAMA Network Open*, vol. 6, no. 7, pp. e2321792–e2321792, 2023.

[114] S. Kamaruddin and A. M. Mohammed "Compliance to GDPR Data Protection and Privacy in Artificial Intelligence Technology: Legal and Ethical Ramifications in Malaysia," in *2023 International Conference on Disruptive Technologies (ICDT)*, pp. 284–288, IEEE, May 2023.

[115] S. K. Khan, N. Shiwakoti, P. Stasinopoulos and M. Warren "Cybersecurity regulatory challenges for connected and automated vehicles–State-of-the-art and future directions," *Transport Policy*, vol. 143, pp. 58–71, 2023.

[116] Z. Deng and Z. Chen, "Balancing creative expression and societal well-being: A comprehensive regulatory framework for the Chinese video game industry," *Journal of the Knowledge Economy*, pp. 1–28, 2023.

[117] L. Kuang, S. Pobbathi, Y. Mansury and M. A. Shapiro "Predicting age and gender from network telemetry: Implications for privacy and impact on policy," *PloS one*, vol. 17, no. 7, pp. e0271714, 2022.

[118] A. L. Imoize, M. S. Obaidat, and H. H. Song, "Legal implications of federated learning integration in digital healthcare systems," in Imoize, A. L., Obaidat, M. S., Song, H. H. (Eds), *Federated Learning for Digital Healthcare Systems*, in Intelligent Data-Centric Systems, USA: Academic Press, 2024, pp. 355–385.

[119] M. Ositashvili, "The Compliance model as a catalyst for post-financial crisis efficiency in corporations," *ESI Preprints*, vol. 19, pp. 150–150, 2023.

[120] O. A. Cejas, M. I. Azeem, S. Abualhaija, and L. C. Briand, "NLP-based automated compliance checking of data processing agreements against GDPR," *IEEE Transactions on Software Engineering*, vol. 49, no. 9, pp. 4282–4303, 2023.

[121] I. Cheong, K. Xia, K. K. Feng, Q. Z. Chen, and A. X. Zhang, "(A) I am not a lawyer, but...: Engaging legal experts towards responsible LLM policies for legal advice," in *The 2024 ACM Conference on Fairness, Accountability, and Transparency*, June 2024, pp. 2454–2469.

[122] G. Pistilli, C. Muñoz Ferrandis and Y. Jernite "Stronger Together: On the Articulation of Ethical Charters, Legal Tools, and Technical Documentation in ML," in *Proceedings of the 2023 ACM Conference on Fairness, Accountability, and Transparency*, pp. 343–354, 2023.

[123] A. Lukács and S. Váradi, "GDPR-compliant AI-based automated decision-making in the world of work," *Computer Law & Security Review*, vol. 50, p. 105848, 2023.

[124] J. Wolff and N. Atallah, "Early GDPR penalties: Analysis of implementation and fines through May 2020," *Journal of Information Policy*, vol. 11, pp. 63–103, 2021.

[125] H. Du, J. Wang, D. Niyato, J. Kang, Z. Xiong, M. Guizani, and D. I. Kim, "Rethinking wireless communication security in semantic Internet of Things," *IEEE Wireless Communications*, vol. 30, no. 3, pp. 36–43, 2023.

[126] R. Alunge, "Breach of security vs personal data breach: Effect on EU data subject notification requirements," *International Data Privacy Law*, vol. 11, no. 2, pp. 163–181, 2021.

[127] D. Boakye, D. Sarpong, D. Meissner, and G. Ofosu, "How TalkTalk did the walk-walk: Strategic reputational repair in a cyber-attack," *Information Technology & People*, vol. 37, no. 4, pp. 1642–1673, 2024.

[128] S. Stevens, J. Dykstra, W. K. Everette and J. Chapman "Compliance Cautions: Investigating Security Issues Associated with US Digital-Security Standards," in NDSS, February 2020.

[129] S. D. Norton, "Suspicious activity reporting in the United Kingdom and the United States: Statutory obligations of auditors and optimal harvesting of information," *Journal of Money Laundering Control*, vol. 27, no. 3, pp. 432–444, 2023.

[130] A. N. Savin and C. E. Bagley, "On a strategic management approach to the new EU risk-based compliance regulations," *International In-House Counsel Journal*, vol. 16, no. 64, pp. 1–13, 2023.

[131] R. F. Syed, "Compliance with and enforcement mechanism of labour law: cost-benefits analysis from employers' perspective in Bangladesh," *Asian Journal of Business Ethics*, pp. 1–24, 2023.

[132] K. A. Bamberger, "Technologies of compliance: Risk and regulation in a digital age," *Texas Law Review*, vol. 88, p. 669, 2009.

[133] M. Alexandrova, "Risk factors in IT outsourcing partnerships: Vendors' perspective," *Global Business Review*, vol. 16, no. 5, pp. 747–759, 2015.

[134] L. Boffel, "The influence of artificial intelligence and emerging technologies on the regulation of insurance companies in the US: An exemplary analysis of California's rate making law," *Berkeley Business Law Journal*, vol. 20, p. 254, 2023.

[135] I. Cinemre, K. Mehmood, K. Kralevska, and T. Mahmoodi, "Direct-conflict resolution in intent-driven autonomous networks," in *European Wireless 2023; 28th European Wireless Conference*, VDE, October 2023, pp. 142–147.

[136] P. Tripathy, M. Shabaz, A. Zaidi, I. Keshta, U. Sharma, M. Soni, ... and D. P. Sharma, "Policy conflict detection approach for decision-making in intelligent industrial Internet of Things," *Computers and Electrical Engineering*, vol. 108, p. 108671, 2023.

[137] C. Adamczyk, "Challenges for conflict mitigation in O-RAN's RAN intelligent controllers," in *2023 International Conference on Software, Telecommunications and Computer Networks (SoftCOM)*, September 2023, pp. 1–6.

[138] D. Aragão, C. Rodrigues, D. Vieira, and M. F. de Castro, "A random access channel resources allocation approach to control machine-to-machine communication congestion over LTE-advanced networks," *International Journal of Communication Systems*, vol. 36, no. 10, p. e5493, 2023.

[139] A. Patil, S. Iyer, O. L. López, R. J. Pandya, K. Pai, A. Kalla, and R. Kallimani, "A comprehensive survey on spectrum sharing techniques for 5G/B5G intelligent wireless networks: Opportunities, challenges and future research directions," *Computer Networks*, vol. 253, p. 110697, 2024.

[140] B. Igried, A. Alsarhan, A. Sawalmeh, M. Anan, and I. Alkhawaldeh, "A novel game theoretic approach for market-driven dynamic spectrum access in cognitive radio networks," *Wireless Networks*, vol. 30, pp. 1–16, 2023.

[141] N. Alam, S. Mehfuz, H. Khan, and S. T. Siddiqu, "Interference limited area based resource allocation for device-to-device communication in an underlaying cellular network," *Ad Hoc & Sensor Wireless Networks*, vol. 56, no. 3/4, p. 253. DOI: 10.32908/ahswn.v56.9437 2023.

[142] M. Zheleva, C. R. Anderson, M. Aksoy, J. T. Johnson, H. Affinnih, & C. G. DePree, "Radio dynamic zones: Motivations, challenges, and opportunities to catalyze spectrum coexistence," *IEEE Communications Magazine*, vol. 61, no. 6, pp. 156–162, 2023.

[143] I. Bala and K. Ahuja, "Energy-efficient framework for throughput enhancement of cognitive radio network by exploiting transmission mode diversity," *Journal of Ambient Intelligence and Humanized Computing*, vol. 14, no. 3, pp. 2167–2184, 2023.

[144] D. Cavalcanti, C. Cordeiro, M. Smith and A. Regev, "WiFi TSN: Enabling deterministic wireless connectivity over 802.11," *IEEE Communications Standards Magazine*, vol. 6, no. 4, pp. 22–29, 2022.

[145] M. Beshley, N. Kryvinska, M. Seliuchenko and H. Beshley "End-to-End QoS 'smart queue' management algorithms and traffic prioritization mechanisms for narrow-band internet of things services in 4G/5G networks," *Sensors*, vol. 20, no. 8, pp. 2324, 2020.

[146] R. Figueiredo, H. Woesner, A. Kassler, and H. Karl, "Quality of service performance of multi-core broadband network gateways," in *2024 8th Network Traffic Measurement and Analysis Conference (TMA)*, May 2024, pp. 1–10.

[147] M. Davari and E. Bertino, "Access Control Model Extensions to Support Data Privacy Protection based on GDPR," in *2019 IEEE International Conference on Big Data (Big Data)*, pp. 4017–4024, IEEE, December 2019.

[148] X. Wang, J. Li, Z. Ning, Q. Song, L. Guo, S. Guo, and M. S. Obaidat, "Wireless powered mobile edge computing networks: A survey," *ACM Computing Surveys*, vol. 55, no. 13s, pp. 1–37, 2023.

[149] C. A. Ezeigweneme, C. N. Nwasike, O. O. Adekoya, P. W. Biu, and J. O. Gidiagba, "Wireless communication in electro-mechanical systems: Investigating the rise and implications of cordless interfaces for system enhancement," *Engineering Science & Technology Journal*, vol. 5, no. 1, pp. 21–42, 2024.

[150] G. Falco, B. Shneiderman, J. Badger and R. Carrier "Governing AI safety through independent audits," *Nature Machine Intelligence*, vol. 3, no. 7, pp. 566–571, 2021.

[151] I. D. Raji, S. C. Chock and D. Buolamwini, "Change from the outside: Towards credible third-party audits of AI systems," *Missing Links In Ai Governance*, vol. 5, 2023.

[152] J. Linåker and P. Runeson, "How to enable collaboration in open government data ecosystems: A public platform provider's perspective," *arXiv preprint arXiv:2208.00305*, 2022.

[153] D. Sziroczák and D. Rohács, "Automated conflict management framework development for autonomous aerial and ground vehicles," *Energies*, vol. 14, no. 24, pp. 8344, 2021.

[154] X. W. Chen and S. Y. Nof, "Automating Prognostics and Prevention of Errors, Conflicts, and Disruptions," in *Springer Handbook of Automation*, pp. 509–531, Springer International Publishing, 2023.

[155] L. E. Chatzieleftheriou, M. Gramaglia, M. Camelo, A. Garcia-Saavedra, E. Kosmatos, M. Gucciardo, ... and M. Fiore, "Orchestration procedures for the network intelligence stratum in 6G networks," in *2023 Joint European Conference on Networks and Communications & 6G Summit (EuCNC/6G Summit)*, June 2023, pp. 347–352.

[156] X. Deng, L. Wang, J. Gui, P. Jiang, X. Chen and F. Zeng "A review of 6G autonomous intelligent transportation systems: Mechanisms, applications and challenges," *Journal of Systems Architecture*, vol. 102929, 2023.

11 Government Policies and Economics of Computational Modelling in Wireless Networks

Emeka Ogbuju, Abubakar Aliyu, Agbotiname Lucky Imoize, Malik Adeiza Rufai, Ahmad Shehu Muhammad and Francisca Oladipo

11.1 INTRODUCTION

The economics of computational modelling in wireless networks and governmental policies play key roles in determining the current state of telecommunications [1]. These principles establish the guidelines for the operation of wireless networks and the incorporation of computational models [2]. The cost-effectiveness and overall influence of computational modelling on wireless communications are simultaneously determined by the economics of this integration [3]. In this investigation, we explore how government regulations in this area, such as spectrum allotment and net neutrality, affect the effective operation of wireless networks. We investigate the economic aspects, highlighting the potential advantages of computational modelling in the wireless communication industry, such as cost reduction, revenue generation, and market expansion. It is impossible to emphasise the importance of government regulations and economics in the field of computational modelling for wireless networks given the constantly changing telecommunications landscape [4]. These opposing forces not only influence the design and functionality of wireless networks but also have a big impact on how effective and useful computational modelling techniques are [5].

The guiding concepts that establish the parameters and requirements for wireless network operations are government policies [6]. These rules cover a wide range of important topics, including as net neutrality, regulation and standards, spectrum management, privacy, and data protection. They are made to protect user data, guarantee fair and equal access to wireless networks, encourage innovation, and keep networks operating efficiently [7]. Economic factors are equally important. Since computational modelling maximises the use of network resources and increases overall effectiveness, it has the potential to significantly reduce costs [8]. Through value-added services, data analytics, and individualised products, it also offers chances for revenue generation [9]. Computational modelling results in improved network performance and energy efficiency, which improves network efficiency and user experience [10]. These laws and economic forces play a bigger role in deciding how computational modelling influences the wireless communication landscape as the need for wireless communication keeps rising [11]. We will explore these influences on wireless networks in greater detail in this investigation, providing information on the advantages and disadvantages of integrating computational modelling methods.

Government policies have a big impact on how computational modelling is developed and used in wireless networks [12]. Spectrum management is a significant topic of policy. The management

DOI: 10.1201/9781003457428-12

and distribution of the radio frequency spectrum, which is necessary for wireless communications, is the responsibility of government organisations [13]. The effectiveness of wireless networks is directly impacted by policies on spectrum allocation, licencing, and usage [14]. When it comes to optimising spectrum use and making the best use of the available frequency bands, computational modelling is essential [15]. Another important component of governmental policies is regulation and standards [16]. These rules are in place to make sure wireless networks operate fairly and cooperatively [17]. They address topics including spectrum management, network architecture, data security, privacy, and quality of service [18]. The effective and seamless operation of computational modelling applications within the larger wireless network ecosystem depends on compliance with these standards [19].

Governments work to protect the core policy principle of net neutrality, It guarantees that Internet service providers, especially wireless carriers, handle all Internet traffic equally, including data produced by computational modelling programmes [20]. To keep the Internet free and impartial, discriminatory practices that favour particular types of data or services are forbidden [21]. Government policies prioritise concerns related to privacy and data protection. The privacy and security of user data in wireless networks, particularly those produced and processed by computational modelling applications, are protected by regulations [22]. According to these regulations, network operators must seek user consent, be transparent about their data handling procedures, and put strong security measures in place to safeguard personal data [23]. Government-set Quality of Service (QoS) requirements are essential for wireless services [24]. These requirements must be met by computational modelling software to guarantee dependable network performance and a superior user experience [25]. This ensures that computational models do not negatively impact user experience or network performance as a whole [26].

The promotion of a competitive environment among wireless network carriers depends heavily on competition policy [27]. Governments strive to suppress monopolistic behaviour and promote honest competition. Technologies for computational modelling must not promote anticompetitive behaviour or prevent new competitors from entering the market. Such regulations encourage creativity and give consumers access to better services and price options [23]. Government policies have a big impact on research funding as well, Research and development in computational modelling for wireless networks are supported by government investments [22]. Both the industry and the public profit from these funds' stimulation of technological development and encouragement of industry innovation [20]. The economics of computational modelling in wireless networks are multifaceted. One of the most significant benefits is Cost Reduction. For wireless networks, computational modelling can drastically save operating expenses [28].

Computational modelling reduces costs for both network operators and users by maximising resource allocation, forecasting network traffic patterns, and enhancing network efficiency. These cost reductions may result in lower service costs and more competitive market offerings [29]. Computational modelling may present prospects for money generation. With the help of data analytics, specialised offers, and premium services, advanced modelling can generate new revenue streams [30]. This translates to more revenue and improved client loyalty for network operators [31]. Service providers can deliver specialised goods and services, improving their potential for profit, by utilising the insights offered by computational models [32]. Computational modelling enhances the effectiveness of the network. A better user experience results from modeling-enhanced network speed and decreased congestion [33].

Customers are more satisfied and are more likely to stay customers when users have access to faster Internet speeds, lower latency, and dependable connections [34]. These developments increase the financial sustainability of wireless services [35]. Energy Efficiency is a crucial economic factor, in wireless networks, computational modelling can reduce energy usage, which is important for battery-powered devices like smartphones and Internet of Things sensors [36]. Network operators may save a lot of money and support environmental sustainability by employing computational models to reduce energy waste [37]. Computational modelling is increasingly being used, which promotes market expansion. Technology providers, software developers, and service businesses can benefit from this expansion, the demand for qualified experts in this industry is increasing, promoting job growth and economic development [38].

The monetization of data is one of the most exciting features. Computational modelling can be used to mine the massive amounts of data that wireless networks generate for insightful information [39]. These insights can be made money from in a variety of ways, including custom services, targeted advertising, and data-driven business models. As a result, network operators and data analytics firms now have additional revenue streams [40]. These economics and regulations support the sustainability of wireless networks over the long run [41]. They guarantee that networks can stay economically viable while adjusting to shifting user needs and technological improvements [39]. Global Connectivity is also a significant factor in the expansion of the economy [42], [43].

Computational modelling is essential for controlling the scope and complexity of networked devices and services across multiple locations as wireless networks spread internationally [38]. In the context of computational modelling for wireless communications, regulatory policies relate to the laws and regulations put in place by regulatory bodies and governments to control the use and usage of computational models in wireless networks [44]. These regulations cover a wide range of topics, such as network security, data privacy, spectrum management, and quality of service [45]. They make sure that computational modelling technologies are used in a way that is just, secure, and adheres to the law and ethical standards [46].

The opportunities and growth prospects associated with employing computational modelling to manage and optimise networks with a significant number of connected devices, such as those in the Internet of Things (IoT), are related to the market potentials for huge devices in wireless networks [47] [48]. For these devices, computational models can help with issues like resource allocation, network congestion, and data processing [49]. Large-scale device deployments are becoming more popular in sectors including smart cities, healthcare, and industrial IoT, which make up the potential market [50]. The possible benefits and drawbacks of implementing these technologies are evaluated as part of a cost-benefit analysis of newly developed computational models for dense wireless networks [51]. The possible return on investment as well as initial investment costs, ongoing operational costs, network performance improvements, and other considerations are all examined in this research [52]. It aids stakeholders in assessing the economic viability and alignment of adopting computational models in dense wireless environments with their strategic goals.

The possibility for businesses and investors to make use of these technologies is discussed in terms of commercialization and investment opportunities in the context of computational models in wireless communications [54]. This involves working with businesses to create computational modelling solutions, sponsoring research and development, and investing in start-ups [55]. The commercial potential covers a range of industries, including telecommunications, edge computing, the Internet of Things, and more, where computational models can improve network speed and efficiency [56]. With several benefits, computational modelling is crucial in the field of wireless networks [57]. It is essential for improving network efficiency, lowering congestion, and successfully managing the exponential growth in the number of connected devices, especially in the context of the Internet of Things (IoT) [58] [59]. These models play a key role in improving user experience by lowering latency, reducing packet loss, and providing a greater quality of service [60].

Computational modelling encourages in-the-moment judgement, enabling prompt and well-informed responses to network needs and problems [61]. By assisting in the detection of anomalies and potential security risks, it improves network security, protecting data and communications [62]. These models also improve energy efficiency, optimise energy use, lower operating costs, and have a smaller negative environmental impact [63]. As more effective network management results in lower operating costs, cost reduction is a significant advantage of computational modelling [64]. These approaches also offer prospects for revenue generation through value-added services, specialised advertising, and data monetization [65]. This combination of benefits guarantees that wireless networks continue to be affordable, versatile, and able to serve the constantly changing requirements of contemporary communication systems [66].

This chapter makes a distinctive contribution to the rapidly developing discussion on wireless networks by closely analysing how privacy, security, and government regulations interact with the financial aspects of computational models. Breaking down particular elements like network security, data encryption, access management, and authentication goes beyond the standard talks. Notably, it provides a comprehensive view of the impact of these security measures by shedding light on their economic implications. In addition, the chapter delves into new ground by examining spectrum security and emergency services security in the framework of computational models. This chapter offers a groundbreaking investigation of privacy, breaking down the dynamics of user consent, data security protocols, and the complex terrain of cross-border data transfer. This thorough and nuanced analysis closes important gaps in the literature by providing a new viewpoint that incorporates the complex interplay between security, privacy, legal requirements, and financial considerations in the rapidly evolving field of wireless computational modelling.

11.1.1 KEY CONTRIBUTIONS OF THE CHAPTER

The following are the main contributions made by the chapter Government Policies and the Economics of Computational Modelling in Wireless Networks;

1. The chapter offers a thorough description of the complex interplay between governmental regulations, computer modelling, and the financial implications of wireless networks. It draws attention to how important this intersection is to the current state of communications.
2. The main emphasis would be on how government policies will affect the economy. The chapter examines the effects of various regulations on the expenses and gains associated with applying computational models in wireless networks.
3. The chapter highlights topics that require more research, such as a comprehensive view of policy, global viewpoints, and interdisciplinary studies.
4. The chapter focuses on multidisciplinary cooperation, which emphasises the value of combining technical and policy skills to solve the difficulties and opportunities brought on by governmental policies.
5. Policymakers, network operators, and researchers are provided with policy proposals and consequences in the chapter conclusion.

The economics of computational modelling in wireless networks are examined in depth and with thorough knowledge in this chapter. It would be a useful tool for scholars, business people, and legislators navigating this dynamic and important sector.

11.1.2 CHAPTER ORGANISATION

There are eleven sections in this chapter. An Introduction (Section 11.1) gives a general summary of the subject at the outset. After that, Section 11.2 Related Work provides an overview of previous studies. The research methodology is described in Methodology (Section 11.3). Section 11.4 and Section 11.5 of the chapter, which explores the main topics, are titled Government Policies in Wireless Networks and Economics of Computational Modelling in Wireless Networks, respectively. The sections Cost-Benefit Analysis of Emerging Computational Models for Dense Wireless Networks (Section 11.7) and Market Potentials of Computational Models for Massive Devices in Wireless Networks (Section 11.6) examine possibilities and financial factors. Practical applications are examined in Commercialization and Investment Opportunities for Computational Models in

Wireless Communications (Section 11.8). Section 11.9. Lesson Learned summarises important takeaways, and Section 11.10 Recommendations offers helpful advice. Lastly, a Conclusion (Section 11.11) that highlights the major discoveries and contributions rounds off the chapter.

11.2 RELATED WORK

Table 11.1 provides a swift overview of key research documentation on government policy and the economics of computational modelling in wireless networks, highlighting their individual areas of focus and limitations.

TABLE 11.1
The Synopsis of Related Work

Reference	Description of Work Done	Limitation of the Work
[2]	Studied how government regulations affected the rollout of 5G networks and how to incorporate computational modelling to improve network performance and resource allocation.	Limited to a single case study, it does not offer a thorough analysis of global policies and their repercussions for the economy.
[67]	Investigated how government policies affect spectrum management and distribution in the context of wireless networks. They also discussed the financial advantages of effective spectrum use.	Failed to extensively explore the larger economic elements of computational modelling and instead concentrated on spectrum policies.
[68]	Examined how IoT connection can be optimised by computational modelling and government IoT network strategies. The study covered the financial advantages of IoT for different sectors.	Limited to IoT networks and does not thoroughly cover other wireless network characteristics.
[69]	Examined federal regulations and guidelines about data security and privacy in wireless communications. The study underlined both the financial consequences of data breaches and the need for computational models to ensure data safety.	Very little discussion of broader economic issues and a primary focus on data privacy and security regulations.
[70]	Examined how computational modelling in wireless networks will affect rural and underserved areas economically. The report emphasised how government programmes can help close the digital divide.	Focused on rural areas, making metropolitan or developed places less relevant. Additionally, it had a distinct geographical focus.
[71]	Analysed how governmental regulations affected how much frequency was set aside for emergency communications. The study covered the financial gains of providing trustworthy communication in emergencies.	Its application to other wireless network components is limited because it is primarily focused on emergency communications.
[72]	Investigated how government restrictions affect computational modelling for traffic management and their effects on wireless network neutrality. Discusses the financial aspects of net neutrality.	Confined to only one area of government policy and net neutrality, and not including a wider range of policies.
[15]	Following government regulations encouraging effective spectrum usage, we examined the financial advantages of computational modelling in spectrum sharing and dynamic spectrum access.	Concentrated on a single technique (dynamic spectrum access), which might not cover all aspects of wireless communication.

TABLE 11.1 (Continued)
The Synopsis of Related Work

Reference	Description of Work Done	Limitation of the Work
[20]	Investigated how computational models help to increase network coverage and government policies encouraging broadband deployment in rural areas. examined the financial effects on underdeveloped areas.	Concentrated primarily on rural broadband initiatives, and may not take into account the unique economic dynamics of metropolitan locations.
[17]	Examined how government regulations can aid in the creation of smart cities and the application of computer models to improve urban wireless infrastructure. Considering the financial benefits of smart city technology.	Restricted to smart city projects and might not explore more general wireless network regulations.
[73]	Examined the financial effects of government initiatives supporting the rollout of 5G and the incorporation of computer modelling in 5G network optimisation.	Primarily on 5G networks, possibly excluding other wireless technology generations.
[74]	Investigated how to optimise network infrastructure using computer models while taking economic sustainability into account, and regulatory rules governing environmental impact assessment for wireless network deployments.	Confined to environmental laws and may not touch on other policy issues.
[75]	Investigated the impact of government regulations on the cost and distribution of spectrum for wireless operators, stressing the financial effects of auction processes.	Consideration is primarily on spectrum auctions, and other wireless network rules might not be covered.
[76]	Discussed how computational models can improve emergency communication networks and examined government legislation connected to disaster response communications. Examined the financial advantages of effective emergency communications.	Restricted to emergency communications and maybe excluding normal wireless network rules.
[77]	Examined governmental rules about mobile number portability and the financial effects they would have on wireless companies. Discussed in this context how computational models can be used to optimise network operations.	Regulations specific to number portability, do not cover a wider range of policies.
[78]	Examined the financial advantages of using computer modelling to control interference in wireless networks and talked about how government laws can help to reduce interference problems.	They are focused on managing interference rather than encompassing a wider range of wireless policies.

These studies provide insightful information about the connection between governmental regulations, computer modelling, and wireless network economics. Each study has its limitations, though, so a thorough analysis ought to take into account a wider variety of economic and policy variables throughout the spectrum of wireless communication. Researchers can investigate these topics to learn more about how policy choices affect network economics and how computer modelling can help with associated problems and possibilities.

11.2.1 Gap Analyses

The research on government regulations and the economics of computational modelling in wireless networks is compiled in Table 11.1. Although this research provides insightful information, several gaps and restrictions can be found. The majority of the study currently conducted focuses on particular wireless network economic or governmental policy elements. Comprehensive studies that take into account a wider range of policies and economic issues are noticeably lacking. This discrepancy highlights the necessity of studies that provide a comprehensive understanding of how various government policies affect the overall economics of computational modelling in wireless networks. A few of the studies are regional or context-specific in their design. For instance, research on smart city projects or rural broadband policies may not be applicable everywhere. To close this gap, research is needed that considers the various regulatory regimes while looking at policy and economics in wireless networks globally.

Some research projects are noticeably specialised due to the rapid expansion of wireless technology, such as those that focus on IoT or 5G networks. Research that covers a wider spectrum of wireless generations and new technologies is lacking. It is necessary to take a more thorough look at the financial effects of computational modelling in various wireless situations. Although the studies discuss the technical and financial elements of computational modelling in wireless networks, interdisciplinary cooperation is frequently lacking. This analysis highlights the possibility for future research to provide a more thorough grasp of the topic by fusing economics, policy analysis, and technical skills. Several studies have touched on data privacy and security laws, but more research is needed in this crucial field. Further investigation is required to determine the economic ramifications and the function of computational modelling in assuring data protection as data breaches and privacy issues gain in popularity.

Numerous publications already published focus on the effects of regulations but may not go in-depth on the regulatory frameworks themselves. Studies that provide a thorough examination of the changing regulatory landscape and how it affects economic concerns are noticeably lacking. Comparative studies that evaluate the merits of various regulatory strategies in terms of their effects on the economy are a possibility. Policymakers and industry stakeholders can find best practises with the use of comparative research. More study is required to examine the economic implications of computational modelling in these cross-industry scenarios as wireless networks increasingly intersect with many industries, including healthcare, transportation, and energy. To close these gaps, multidisciplinary research projects are needed that cover a wider spectrum of wireless technologies, regulatory contexts, and policy considerations. The goal of the research should also be to provide a more comprehensive understanding of how economics and government policy interact when it comes to computational modelling in wireless networks. In a wireless environment that is always changing, such thorough insights help direct industrial practises, technology development, and legislative decisions.

11.3 REVIEW APPROACH

A mixed-methods research strategy is required for the thorough investigation of governmental policies and the economics of computational modelling in wireless networks. This approach integrates quantitative and qualitative research strategies to provide a comprehensive grasp of the complex topic matter. Government policies can be thoroughly explored by including qualitative research methods like interviews, questionnaires, and content analysis of policy texts. This aspect of the methodology reveals the subtleties of policy-making in the context of wireless networks, revealing the delicate elements of regulatory frameworks and their repercussions. The analysis is complemented by quantitative research tools that explore economic facts, trends, and statistical information. This data-driven methodology offers insights into potential cost-benefit scenarios, market trends, and the actual economic effects connected to certain policies.

A more comprehensive understanding of how governmental regulations influence the economics of computational modelling in wireless networks can be built by combining quantitative and qualitative methodologies. The mixed-methods approach uses case studies, cross-industry analysis, interdisciplinary collaboration, and longitudinal studies, all of which complement one another to address this complex subject. This approach promises to provide a profound, intelligent, and thorough analysis of the dynamic interaction between governmental regulations, computer modelling, and the business environment in wireless networks.

11.4 GOVERNMENT POLICIES IN WIRELESS NETWORKS

The use, implementation and management of wireless communication systems are governed by governmental policies [79], which are the rules, regulations, and directives that are established by governmental bodies [80]. These regulations serve several vital functions that guarantee the efficient and secure operation of wireless networks [81]. The following are some crucial facets of government wireless network policies;

a) *Spectrum Management*

   Government regulations control how radio frequency spectrum, a limited and priceless resource needed for wireless communication, is allotted and licensed [82]. These regulations establish who is permitted access to particular frequency bands, guaranteeing effective utilisation and avoiding service interference [83].

b) *Net Neutrality*

   All data carried over the Internet should be handled similarly by Internet service providers, including wireless carriers, according to net neutrality regulations [84]. They restrict actions that can jeopardise the open and impartial character of the Internet, such as blocking, throttling, or paid prioritization [85].

c) *Privacy and Data Protection*

   Rules for the protection of user data in wireless networks are established by government policies [85]. They demand that network operators acquire user consent, reveal their data processing policies, and put security measures in place to protect personal data [86].

d) *Quality of Service (QoS)*

   Minimum criteria for wireless services are established by QoS standards defined by government organisations [87]. Users are guaranteed a particular degree of service quality and dependability thanks to these standards [88].

e) *Emergency Services*

   Government regulations demand that cellular service providers enable public safety communication networks like 911 and other emergency services [89]. This guarantees that individuals may get assistance in times of need and that first responders have dependable communication equipment [90].

f) *Infrastructure Regulation*

   Tower placement, zoning rules, and environmental considerations are all covered by policies relating to the development of wireless network infrastructure, such as cell towers and tiny cells [91]. They try to strike a balance between the demand for effective network expansion and the interests of the neighbourhood and the environment [92].

g) *Consumer Protection*

   Access to emergency services, contract terms disclosure, and pricing transparency are all covered under wireless consumer protection policies [93]. Additionally, they offer procedures for handling customer grievances as well as problems with billing and service quality [94].

h) *Universal Service Fund*

To encourage access to telecommunications services, especially wireless, in underserved and rural areas, many governments establish universal service grants [95]. These grants support network construction to guarantee that everyone has access to crucial communication services [96].

i) *Competition Policy*

To avoid monopolistic practices, governments try to encourage competition among wireless network carriers [97]. A competitive market is aided by regulations on mergers, spectrum caps, and infrastructure sharing [98].

j) *Standardization and Interoperability*

To guarantee that wireless devices and networks can interoperate successfully, government agencies frequently take part in or support standards-setting groups [99]. This promotes a level playing field for various technologies and prevents vendor lock-in.

Government regulations for wireless networks are flexible and adaptable to changing user demands and technological advancements. They are essential for protecting consumer interests, promoting innovation, assuring resource efficiency, and preserving the overall integrity and security of wireless communication systems.

11.4.1 Regulation and Licensing

Government policies in the context of wireless networks include licensing and regulation as essential elements. This section briefly summarises the rules and requirements for wireless networks.

a) Regulation

Regulation in wireless networks involves the establishment of rules and guidelines by governmental authorities to ensure that wireless communication systems operate fairly, efficiently, and securely. These regulations cover various aspects of wireless networks, as depicted in Figure 11.1.

The regulatory landscape for wireless networks is depicted in Figure 11.1. The main elements of the legal frameworks controlling wireless communications are depicted in the diagram. Spectrum allocation, net neutrality, quality of service, privacy and data protection, and emergency services issues are some of its components. This illustration presents a thorough synopsis of the regulatory environment, highlighting the complex nature of laws that affect wireless networks.

1. *Spectrum Allocation*

Governments allocate and control the radio frequency spectrum, allocating particular frequency bands to different users including satellite services, mobile network operators, and broadcasters [14]. This distribution minimises interference and maximises spectrum use.

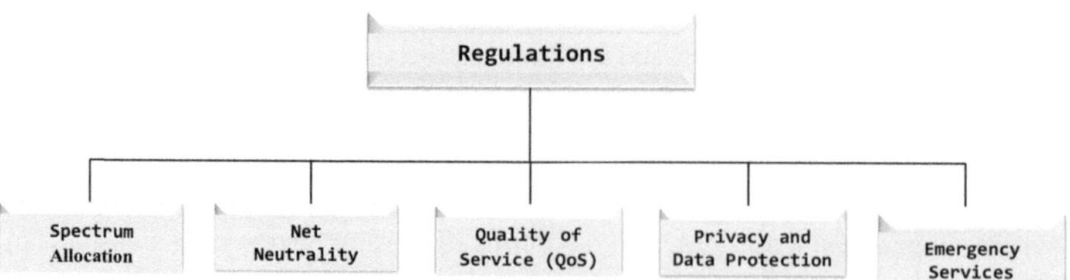

FIGURE 11.1 Wireless Networks Regulations.

2. *Net Neutrality*

   Internet service providers (ISPs), including cellular carriers, are required by law to treat all data equally, without favouring any particular programmes, services, or content [100]. This protects an impartial and inclusive Internet [101].

3. *Quality of Service (QoS)*

   Regulations frequently set minimum requirements for wireless services for quality of service standards. Parameters including data rates, latency, and dependability may be included in these standards [102].

4. *Privacy and Data Protection*

   Wireless network operators are required by laws about user privacy to obtain users informed consent before collecting and processing their data. Operators are also required to disclose data handling practices and put in place data protection measures [103].

5. *Emergency Services*

   According to regulations, wireless service providers must provide emergency services, enabling consumers to contact 911 and other crucial public safety communication channels [76].

b) Licensing

In wireless networks, licensing is the process of allowing and allocating rights to organisations that want to utilise particular frequencies in the radio spectrum for wireless communication. Important details about licensing are depicted in Figure 11.2.

Wireless network licencing is a complex field, as seen in Figure 11.2. The diagram illustration describes the different licensing-related elements inside the wireless communications framework. This covers the spectrum auctions, licence prices, terms, and duration as well as spectrum sharing. The goal of the figure is to give a clear visual representation of the essential components and difficulties of wireless network licencing.

   i. *Spectrum Auctions*

      Governments frequently hold spectrum auctions to give wireless operators access to particular frequency bands [104]. The highest bidder wins the license to utilise these frequency bands, which are up for grabs, Governments can make a substantial amount of money from these auctions.

   ii. *License Duration*

      Typically, wireless spectrum licenses are granted for predetermined periods, which can change according on the frequency range and national laws. When a license expires, holders are required to renew it [105].

   iii. *License Conditions*

      License holders are subject to several restrictions and conditions. These terms could deal with network build-out specifications, service coverage, and regulatory compliance [106].

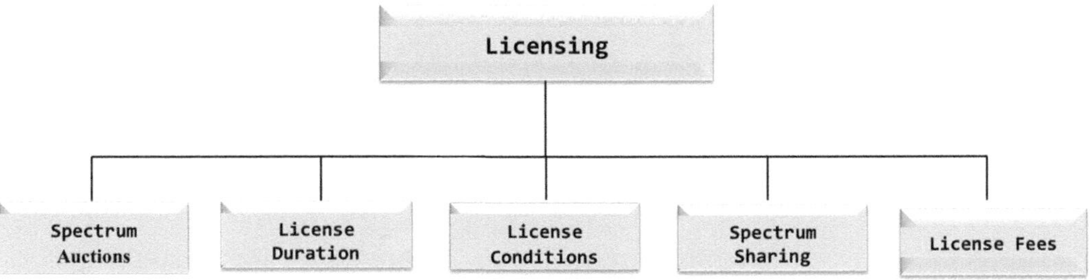

FIGURE 11.2   Wireless Networks Licensing.

iv. *Spectrum Sharing*

Governments occasionally approve spectrum sharing, enabling several organisations to use the same frequency channels with certain restrictions. This could boost competition and improve spectrum efficiency [107].

v. *License Fees*

It may be necessary for licensees to pay fees to obtain and maintain spectrum licenses. These levies can add significantly to government revenue, particularly when high-demand frequency bands are up for auction [106].

Government management and oversight of the radio frequency spectrum, a valuable and complicated resource, depends heavily on regulation and licensing. These safeguards guarantee that wireless networks function dependably, securely, and following moral and legal principles, all the while encouraging healthy competition and effective spectrum utilisation.

11.4.2 Security and Privacy

Government regulations about wireless networks must prioritise security and privacy, as shown in Figures 11.3 and 11.4, to protect user information, network infrastructure, and communications [81].

a) Security

The crucial combination of security, governmental regulations, and the financial aspects of computational modelling in wireless networks are examined in this section. It includes several aspects, including network security, which clarifies precautions against cyberattacks; data encryption techniques, which investigates ways to secure wireless data transmission; access control and authentication, which deals with methods to manage network access and confirm identities; spectrum security, which centres on procedures and policies for secure spectrum allocation; and emergency services security, which looks at procedures guaranteeing the stability of communication, as depicted in Figure 11.3.

The critical security component in the context of wireless computational models is depicted in Figure 11.3. The figure outlines the essential elements and factors to be taken into account when implementing security protocols using computational models in wireless networks. Network security, data encryption techniques, access control and authentication, spectrum security, and emergency

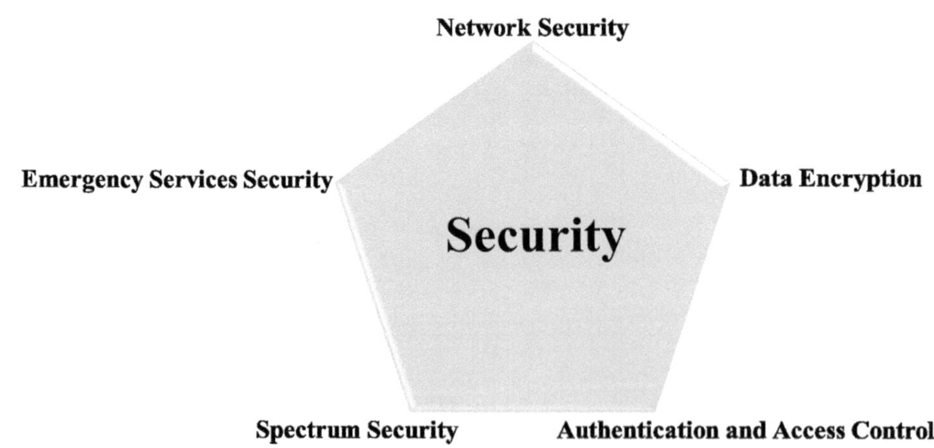

FIGURE 11.3   Security in Wireless Computational Model.

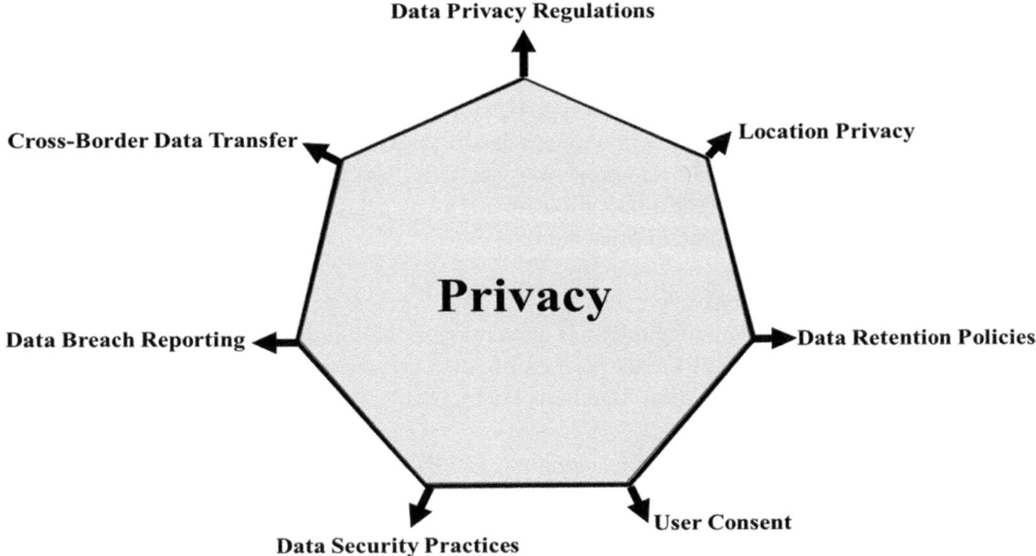

FIGURE 11.4 Privacy in Wireless Computational Model.

services security are all included. The purpose of this diagram depiction is to demonstrate how all-encompassing the security features built into wireless computational models are.

1. *Network Security*
   Wireless network providers are required by law to put safeguards in place to protect their infrastructure against intrusions, attacks, and unauthorised access. Protecting essential network components like base stations and data centres falls under this category [81].
2. *Data Encryption*
   In order to safeguard data carried via wireless networks from interception and eavesdropping, encryption standards are frequently controlled. To secure data in transit, government regulations may mandate the usage of particular encryption algorithms [108].
3. *Authentication and Access Control*
   Only those who are authorised are allowed access to the network, according to regulations. To stop unwanted access, strong access control measures and multi-factor authentication are frequently implemented [109].
4. *Spectrum Security*
   The security of the radio frequency spectrum is addressed by government rules, which seek to stop interference and illegal use that could impair wireless services [107].
5. *Emergency Services Security*
   The security of emergency services communications is governed by specific laws to guarantee their availability in times of need and their protection from hostile interference [110].

b) Privacy

A crucial issue of privacy in the context of computational modelling in wireless networks is the focus of this section. Covering important topics, the conversation includes user consent and how it explains the role that consumer agreement plays in the use of data; data security protocols and how they protect sensitive information; location privacy and how location-based services and user privacy are balanced; data retention rules and how they explore policies governing how long user data is stored; privacy laws and how they govern privacy in wireless communications; reporting of data breaches and the significance of open reporting mechanisms; as well as international data transfer as shown in Figure 11.4.

The crucial aspect of privacy in wireless computational models is depicted in Figure 11.4. The figure outlines important elements and factors to be taken into account when implementing computational models for privacy protection in wireless networks. This covers user consent, data security procedures, location privacy, data retention rules, privacy laws, reporting of data breaches, and cross-board data transmission. The illustration functions as a visual aid, illustrating the complex mechanisms incorporated to maintain privacy in the context of wireless computational models.

1. *Data Privacy Regulations*

   Governments set laws and guidelines governing how cellular service companies gather, store, and use user data [69]. User permission, data transparency, and protection against data breaches are mandated by these standards [111], [112].

2. *Location Privacy*

   For a variety of reasons, including emergency services, wireless devices frequently track user whereabouts [111]. Privacy laws guarantee that this information is treated carefully and not improperly [126].

3. *Data Retention Policies*

   Certain countries set data retention periods, mandating that wireless carriers keep customer data for a specific amount of time before securely discarding it [113].

4. *User Consent*

   Policies require cellular service providers to acquire consumers informed consent prior to gathering and utilising their data. This covers giving clear permission for location monitoring and other data-related uses [114].

5. *Data Security Practices*

   Wireless carriers may be subject to regulations that set data security standards and mandate the adoption of cybersecurity measures to safeguard sensitive user data [115].

6. *Data Breach Reporting*

   Government regulations in many areas mandate that wireless carriers notify affected users and regulatory bodies of data breaches as soon as possible. This promotes openness and prompt response in the event of a breach [116].

7. *Cross-Border Data Transfer*

   In the context of international communications and roaming, certain governments establish regulations for cross-border data transfer to protect user data when it leaves their borders [117].

In order to protect sensitive data and foster confidence among users of wireless networks, effective security and privacy standards are needed. These regulations are always changing to handle fresh privacy issues and security risks brought on by people using wireless technology more frequently in their daily lives.

11.4.3 Net Neutrality

A key idea pertaining to the equitable handling of all data on the Internet, especially wireless networks, is known as net neutrality [20]. The goal of government regulations about net neutrality is to guarantee that Internet service providers (ISPs), which comprise wireless carriers, handle all data sent across their networks in an impartial and non-discriminatory way [101].

Since mobile devices are becoming the major method of accessing the Internet, government laws regarding net neutrality are especially crucial in the context of wireless technology [72]. To promote an open and accessible digital environment, they make sure that mobile carriers handle data from websites, applications, and services equally and without discrimination. Maintaining the Internet innovative and democratic values requires these policies [100].

TABLE 11.2
Government Policies Regarding Net Neutrality

Reference	Policy	Significance
[118]	**Open Internet Access**	The goal of net neutrality regulations is to preserve a free and open Internet where users may access all types of information equally, including apps, services, and content. ISPs are prohibited from blocking, throttling, or prioritising Internet traffic according to its origin, content, or destination.
[20]	**Fair Competition**	Regulations pertaining to net neutrality promotes equitable competition in the Internet economy. ISPs are prevented from providing particular material or services preferential treatment, ensuring that even startups and smaller companies have an equal opportunity to compete.
[119]	**User Choice and Freedom**	Users now have more control over the services and content they access because of net neutrality. It stops ISPs from controlling users from access to particular websites or applications.
[120]	**Innovation**	Net neutrality encourages innovation by guaranteeing that new online services and applications can reach customers impartially. This encourages developers and business owners to produce and market new goods without encountering unnatural obstacles.
[101]	**Content Diversity**	By prohibiting ISPs from banning or impeding access to websites and services that might not be well-liked by some groups but yet serve the interests of others, it protects the diversity of information.
[20]	**Public Interest**	By keeping the Internet open, net neutrality laws promote the general welfare. They guarantee that the Internet continues to be a medium for innovation, free speech, and information sharing.
[20]	**Consumer Rights**	Maintaining net neutrality is crucial to defending consumer rights. Net neutrality guarantees that consumers who pay for an Internet connection receive the service they paid for, with equal access to all content that is permitted online.
[122]	**Data Privacy**	Data privacy is also related to net neutrality. In the absence of net neutrality, users may have privacy issues as ISPs may monitor and alter data transmission.
[123]	**Global Relevance**	The topic of net neutrality is worldwide. Although national policies differ, the idea of an open and impartial Internet is accepted globally.

11.4.4 Government Regulations on Equal Treatment of Internet Traffic

In order to guarantee that Internet service providers (ISPs) handle all data transferred over their networks in an impartial and non-discriminatory manner, government rules known as net neutrality legislation are put into place [20]. These rules aim to safeguard users and content producers rights while preserving an open and unrestricted online environment [72]. Table 11.3 outlined and describes the key government regulations on equal treatment of Internet traffic.

TABLE 11.3
Government Regulations on Equal Treatment of Internet Traffic

Reference	Regulation	Description
[122]	No Blocking	ISPs are generally not allowed by regulations to obstruct access to websites, content, apps, or lawful services. It should be possible for users to access any website or use any online service they choose.
[124]	No Throttling	In most cases, throttling intentionally reducing the speed of some Internet traffic is forbidden. This guarantees that all data, regardless of its source or substance, is transmitted at a constant speed.
[124]	No Paid Prioritization	In general, paid prioritisation is prohibited for ISPs. This implies that they are unable to slow down the traffic of content providers who do not pay additional fees while providing faster or better service to those who do.
[126]	Transparency	ISPs must frequently abide by regulations requiring them to be open and honest about their performance, network management procedures, and service agreement terms. This openness enables consumers to make wise decisions.
[127]	Content Neutrality	ISPs are guaranteed not to discriminate against data traffic based on its source, content, or destination thanks to net neutrality legislation. Every online service and piece of content should be equally accessible.
[128]	User Freedom	Users are free to access and use the Internet anyway they see fit thanks to these restrictions. Without interference from their ISP, they are free to select the websites they want to access, the apps they want to use, and the services they want to subscribe to.
[121]	Competition and Innovation	Fair competition and innovation are encouraged in the Internet marketplace by net neutrality. Fair competition between big and small content providers encourages innovation and gives users more options.
[129]	Consumer Rights	Rights of customers who pay for Internet access are safeguarded by regulations. Consumers should not be subjected to capricious restrictions on their Internet use; they expect to receive the service for which they paid.

The goal of government laws governing the equitable treatment of Internet traffic is to maintain the Internet status as a free medium for innovation, free speech, and information sharing. They are critical to maintaining the values of an open and impartial Internet, which benefits individual users as well as the larger digital ecosystem. The major objective of these laws, which differ from nation to nation, is to safeguard Internet users rights and preserve the Internet innovative and democratic character.

11.4.5 Implications of Net Neutrality on Computational Modelling in Wireless Networks

Significant changes to computational modelling in wireless networks are brought about by net neutrality [126]. It affects the deployment, accessibility, and integration of computational modelling technology inside these networks [72]. The main effects of net neutrality on computational modelling in wireless networks were defined and explained in Table 11.4.

TABLE 11.4
Implications of Net Neutrality on Computational Modelling in Wireless Networks

Reference	Implication	Description
[20]	**Equal Treatment of Data**	Regulations about net neutrality mandate that all Internet data be handled equally. This means that discriminatory practices cannot be applied to data created or processed by computational modelling tools. It guarantees that the performance and accessibility of these apps are equivalent to those of any other Internet service.
[72]	**Innovation and Accessibility**	An atmosphere of accessibility and innovation is promoted by net neutrality. Internet service providers (ISPs) are not a hindrance to the growth of computational modelling applications, which frequently depend on data-intensive procedures and real-time analysis. This inspires programmers to produce fresh modelling solutions.
[128]	**End-User Experience**	Net neutrality benefits users of computational modelling programmes by guaranteeing unimpeded data transmission. This is especially crucial for applications that need high bandwidth and low latency, such as augmented reality, autonomous systems, and the Internet of Things.
[122]	**Data Privacy**	Computational modelling data privacy may be indirectly impacted by net neutrality. Because equitable treatment of data prevents Internet service providers (ISPs) from tracking or manipulating traffic according to its content, it helps ensure the privacy of data relating to modelling.
[128]	**Competitive Environment**	Competition amongst suppliers of computational modelling is encouraged by an impartial Internet environment. Small and startup companies are equally able to provide modelling services without being at a competitive disadvantage to their larger rivals.
[130]	**Network Resource Allocation**	Net neutrality has an impact on how network resources are distributed. It is forbidden for ISPs to give particular data kinds or services priority. To provide a level playing field, computational modelling apps must share network resources with other online services in a fair manner.
[20]	**Regulatory Compliance**	Companies that offer computational modelling services have to abide by net neutrality laws to guarantee that their services are available to everyone. When developing and implementing these applications, compliance is a crucial factor.
[72]	**Access to Data Sources**	A lot of the time, computational modelling uses data from the Internet. Because of net neutrality, access to these data sources is guaranteed to be unrestricted and unreliant on agreements or collaborations between content providers and ISPs. This promotes a data environment that is both accessible and diverse.

In wireless networks, computational modelling applications are guaranteed fair play thanks to net neutrality. It protects their freedom to operate without discrimination or intervention from ISPs, encouraging creativity, user experience, and fair competition. The continuation of the expansion and advancement of computational modelling in wireless networks depends on the preservation of a neutral and open Internet environment, which is made possible by these regulations.

11.5 ECONOMICS OF COMPUTATIONAL MODELLING IN WIRELESS NETWORKS

The financial, operational, and strategic facets of applying modelling technology in the telecommunications industry are included in the economics of computer modelling in wireless networks [131]. Important factors to take into account while analysing the economics of computational modelling in wireless networks include;

a) *Cost Reduction*

In wireless network operations, computational modelling can result in significant cost reductions [131]. Modelling contributes to lower operating costs by maximising resource allocation, forecasting traffic patterns, and enhancing overall network efficiency. This includes financial savings on equipment expenditures, maintenance, and energy use [132].

b) *Revenue Generation*

Wireless network operators can generate income streams using sophisticated computer modelling. Offering premium services with improved quality, tailored options, and data analytics services that may be charged for are some examples of this. This service diversity can boost profitability [133].

c) *Network Efficiency*

Because computational modelling reduces congestion and enhances performance, it increases network efficiency. Better user experiences with quicker data rates, less latency, and more dependable connections result from this. Effective networks lower attrition and aid in client retention [134].

d) *Energy Efficiency*

Modelling can optimise wireless network energy consumption, which is important for battery-powered devices like smartphones and Internet of Things sensors. Not only does cutting energy use save operating expenses, but it also helps with sustainability initiatives [135].

e) *Market Growth*

The market is expanding as a result of wireless networks implementing computational modelling technology. This growth opens doors for software developers, technology suppliers, and service providers, promoting economic growth and the creation of jobs [136].

f) *Monetization of Data*

A huge volume of data is produced by wireless networks. This data can yield insightful information through computational modelling that can be used to generate revenue through tailored services, focused advertising, and data-driven company plans. This gives network operators and data analytics firms access to new revenue streams [65].

g) *Sustainability*

Wireless network long-term viability is supported by computational modelling economics [137]. These networks are economically viable because of their effective resource use, low energy consumption, and minimal operating expenses [68].

h) *Global Connectivity*

Computational modelling is essential for controlling the number and complexity of networked devices and services in different locations as wireless networks spread throughout the world [138]. There are economic ramifications to this worldwide connectivity, such as more trade and global business prospects [139].

i) *Investment in Research and Development*
The integration of computational modelling in wireless networks frequently necessitates significant R&D spending. These expenditures boost technological development and promote industry innovation, which is advantageous to the sector and customers [140].

j) *Job Creation*
The increasing use of computational modelling in wireless networks creates job opportunities in data science, network administration, and software development, among other domains. It helps the IT industry grow economically and create jobs [141].

The potential for both cost reduction and income generation is highlighted by the economics of computational modelling in wireless networks. Computational modelling tools are essential in determining the economics of the wireless communication industry since they optimise network operations, improve user experience, and open up new revenue streams.

11.5.1 ECONOMIC BENEFITS OF COMPUTATIONAL MODELLING IN NETWORK PLANNING AND OPTIMIZATION

Wireless network operators and the telecom sector as a whole benefit financially from computational modelling use in network planning and optimization [51]. Among the main financial benefits of computational modelling are;

I. *Cost Reduction*
Through the application of computational modelling, network resources can be used more effectively, negating the need for expensive infrastructure expansion. Network capacity and traffic control can be optimised by operators to prevent needless capital expenditures.

II. *Resource Efficiency*
Better resource allocation, from spectrum to hardware, is made possible by modelling, which lowers operating costs. One aspect of this is reducing energy usage, which is a major running cost for wireless networks [37].

III. *Improved Network Performance*
Computational modelling improves network performance by optimising traffic routing and network elements [135]. Increased user satisfaction and decreased churn rates are the results of faster Internet speeds, lower latency, and fewer dropped calls or data packets [37], [142].

IV. *Lower Maintenance Costs*
By identifying possible network problems early on, modelling helps avoid the need for expensive emergency maintenance and repairs. Predictive maintenance has the potential to drastically reduce operating costs [143].

V. *Spectrum Efficiency*
By optimising the use of the present spectrum, effective modelling helps to avoid or postpone the need to purchase new frequencies through pricey auctions. This results in time and money savings [98].

VI. *Infrastructure Planning*
To ensure that resources are distributed where they are most needed, network modelling helps with the planning of small cell and cell tower deployments. This reduces the amount of network infrastructure that is overbuilt and underutilized [144].

VII. *Revenue Generation*
Modelling enables improved network quality and dependability, which can increase customer happiness and draw in new users. Furthermore, modelling can spot the potential for specialised services and premium offers, which might bring in extra money [41].

VIII. *Competitive Advantage*

Operators who optimise their networks through computational modelling have a competitive advantage. They may be able to draw in more clients and hold on to their current clientele by providing higher-quality services and making more effective use of their resources [25].

IX. *Faster Deployment of Services*

Operators may roll out new services and technologies more quickly with the use of network planning and optimisation models. This may result in new offerings seeing a quicker return on investment [144].

X. *Scalability*

Wireless network scalability is supported via modelling. Operators don't need to upgrade their infrastructure much because they can effectively handle the increase in consumers and data traffic [83].

XI. *Data Monetization*

The information gathered and examined by modelling can be very beneficial. By providing insights to outside parties, operators can make money off of this data and establish new sources of income.

XII. *Regulatory Compliance*

Using computer modelling to optimise networks makes it frequently simpler and less expensive to meet regulatory criteria for network performance [37].

A more competitive and financially sustainable wireless network sector is a result of the significant economic advantages that computational modelling in network planning and optimisation offers. These advantages include cost reductions, resource efficiency, enhanced performance, and revenue generation.

11.5.2 Allocation of Budget and Resources for Modelling and Simulation

It takes careful planning and consideration of many criteria to allocate funds and resources for modelling and simulation (M&S) projects. The following are the main procedures and factors to take into account when allocating funds and resources for M&S efforts.

a) *Define Objectives and Scope*

Establish the goals and parameters of the M&S project in explicit terms first. Which particular issues or procedures will the simulation and modelling attempt to solve? Recognise the objectives to ascertain the required resources [146].

b) *Assess Resource Requirements*

Determine the technology, software, people resources (such as knowledgeable analysts and modellers), and data sources needed for the project. Assess the need for any new technology or tools [147].

c) *Cost Estimation*

Provide a thorough cost estimate for the undertaking. Take into account both the direct (such as hardware and software licences) and indirect (such as labour, maintenance, and training) costs. Make a budget that accounts for all project costs.

d) *Prioritize Resources*

Set a resource allocation priority list based on the most important requirements of the project. Distribute resources where they will most effectively contribute to the accomplishment of project goals [49].

e) *Resource Management*

Throughout the project, manage and allocate resources in an efficient manner. As project requirements change, this entails keeping an eye on resource use, making sure procurement is done on time, and optimising resource allocation [49].

f) *Human Resources*

Think about the knowledge and abilities required for modelling and simulation. Distribute human resources according to the availability, expertise, and skill sets needed [147]. This could entail outsourcing, hiring, or training [147].

g) *Technology and Infrastructure*

Verify that the required software and hardware resources are available. Invest in cutting-edge equipment and techniques to make modelling and simulation processes more efficient [148].

h) *Data Sources*

Determine and allot resources for gathering, preparing, and validating data. Reliable and pertinent data are essential for precise modelling and simulation [147].

i) *Risk Management*

Allocate supplemental funds for unanticipated difficulties or modifications to the project scope. Resources may be needed in ways not anticipated, thus having a backup plan is crucial [147].

j) *Monitoring and Reporting*

Establish reporting and monitoring systems to keep tabs on project development, resource usage, and cost management. Evaluate the alignment of resource allocation with project goals regularly and make appropriate adjustments [147].

k) *Timeframe*

Distribute resources in accordance with the project timeline. Make that resource availability corresponds with project deadlines and milestones [147].

l) *Funding Sources*

Ascertain the project financial sources. This could include funds from grants, outside funding sources, or organisational budget allocations [140].

m) *Stakeholder Communication*

Inform stakeholders on budget management and resource allocation. Effective communication and transparency contribute to increasing trust in the project success [147].

n) *Iterative Evaluation*

Analyse the efficiency of resource distribution regularly and make necessary adjustments. Since M&S projects are frequently iterative, be ready to reallocate resources when new information becomes available [150].

o) *Post-Project Assessment*

Conduct a detailed post-project review once the project is over to see whether its goals were met and how resources were allocated. Utilise this data to guide future decisions about the distribution of resources [147].

Project success depends on allocating resources for modelling and simulation in an efficient manner. Strategic planning, budgeting, hiring qualified staff, and effective management are all necessary to guarantee that resources are used to the fullest extent possible in order to meet project objectives.

11.5.3 COST SAVINGS THROUGH PREDICTIVE MODELLING AND PERFORMANCE EVALUATION

By using data-driven insights and making wise judgements, predictive modelling and performance evaluation can save costs. This is how these procedures can result in appreciable cost savings.

a) *Predictive Maintenance*

Predictive modelling, which is based on historical performance data, can estimate when maintenance is required in industries involving machinery and equipment [151]. Through early detection of possible malfunctions, this method avoids expensive downtime and lowers maintenance expenses.

b) *Energy Efficiency*

By predicting demand patterns, predictive modelling helps optimise energy use. This lowers energy expenses for businesses by enabling them to schedule energy use for off-peak hours when rates are lower [36].

c) *Inventory Management*

By using historical sales data to forecast inventory demands, predictive modelling helps avoid stockouts and overstocking. Effective inventory control lowers carrying costs and misses sales opportunities, which results in financial benefits [152].

d) *Supply Chain Optimization*

By optimising supply chain procedures and forecasting changes in demand, predictive models can lower storage and transportation expenses. This reduces the possibility of having too much inventory or not enough stock [153].

e) *Customer Churn Prevention*

Forecasting models can detect which clients are most likely to churn – that is, stop using a service or product. The expenses related to obtaining new clients can be reduced by putting focused retention tactics into place [154].

f) *Fraud Detection*

Real-time fraud detection and transaction identification are possible with predictive models. Organisations save money by stopping fraudulent transactions, which would otherwise be lost to false charges or chargebacks [155].

g) *Quality Control*

Quality control problems in manufacturing can be observed and predicted with predictive modelling. Early defect discovery lowers waste and rework expenses [156].

h) *Personnel Optimization*

By ensuring that the appropriate number of workers are scheduled at the appropriate times, predictive modelling can optimise workforce scheduling and lower labour and overtime expenditures.

i) *Asset Management*

Models for performance evaluation can track how well assets, such as cars or machines, are used and how efficiently. By optimising asset lifetime costs, this data-driven method assists organisations in determining whether to sell, replace, or repair assets.

j) *Process Optimization*

Business process bottlenecks and inefficiencies can be found through performance evaluation. Through process optimisation, organisations can reduce operating expenses and increase productivity.

k) *Marketing Efficiency*

By focusing on the most likely consumers, predictive models aid in the optimisation of marketing initiatives. As a result, less money is spent on audiences who are unlikely to convert, making better use of marketing funds.

l) *Resource Allocation*

Performance review sheds light on the use of resources, including money and staff. Through performance-based resource reallocation, organisations can optimise both efficiency and cost-effectiveness [29].

m) *Healthcare Cost Reduction*
Patients at high risk who could need expensive therapies can be identified using predictive modelling. Early intervention can save healthcare costs for high-risk patients by averting costly hospital stays.

n) *Customer Support Optimization*
Predictive models can predict the volume of customer support tickets and determine which issues need to be addressed right away. This makes it possible for businesses to deploy support resources more effectively, which lowers operating expenses and response times.

o) *Insurance Underwriting*
With the use of predictive models, insurers can more precisely assess risk, which improves pricing and lowers the possibility of expensive claims that exceed premiums.

Numerous uses of predictive modelling and performance evaluation can greatly save costs across a range of sectors. Organisations can save a significant amount of money by optimising processes, preventing problems, and allocating resources more effectively when they use data to make proactive decisions.

11.6 MARKET POTENTIALS OF COMPUTATIONAL MODELS FOR MASSIVE DEVICES IN WIRELESS NETWORKS

Computational models for large-scale device management in wireless networks have a significant and expanding market potential. Computational modelling is essential for tackling several issues as wireless networks develop and more devices are connected through the Internet of Things (IoT). The following are some salient features of the market potential for computational models in wireless network management of large numbers of devices.

a) *IoT Growth*
The increasing number of Internet of Things devices, ranging from industrial machinery to smart sensors, has led to a huge demand for effective device management. In Internet of Things networks, computational models are crucial for maximising device connectivity and data processing [47].

b) *Network Optimization*
In order to effectively manage the growing number of connected devices, wireless networks need to be optimized. Computational models help to ensure that devices receive the required bandwidth, manage network resources, and reduce congestion [102].

c) *Real-time Decision Making*
Real-time decision-making is made possible by computational models, and this is essential for controlling IoT devices. They can swiftly assess data and make decisions, such as changing network settings, routing information, or attending to requests unique to a particular device [157].

d) *Quality of Service (QoS)*
Sustaining a particular level of service quality is necessary when managing large devices. Computational models aid in guaranteeing that devices obtain the necessary quality of service, be it high bandwidth, low latency, or dependable connectivity [25].

e) *Security*
Security of devices is a major IoT concern. Computational models provide prompt reactions to safeguard the network and devices from cyberattacks by identifying anomalies and possible security concerns.

f) *Energy Efficiency*

Since most Internet of Things devices run on batteries, energy economy is vital. By optimising data transfer, computational models can lower linked devices energy consumption [10].

g) *Cost Reduction*

Cost reductions are the result of effective device management. Computational models assist in lowering operating costs associated with network administration and maintenance [132].

h) *Monetization Opportunities*

Computational models generate revenue prospects in addition to cost savings. Based on the insights obtained from device data, service providers can deliver value-added services, tailored advertising, and data analytics solutions.

i) *5G and Beyond*

Computational models are in more demand when 5G networks, which provide fast, low-latency communication, are deployed. These versions can efficiently handle more devices by utilising 5G capabilities [158].

j) *Diverse Applications*

Numerous industries, including smart cities, industrial automation, healthcare, agriculture, and more, have the potential to benefit from this industry. Computational models are flexible and can be applied to many industries.

k) *Global Reach*

Geographical restrictions do not affect the market for computational models in wireless networks. Given the widespread deployment of wireless networks and IoT, it has an international scope.

l) *Regulatory Compliance*

IoT and device management are becoming more and more important to governments and regulatory agencies. To ensure compliance with changing legislation, computational models are essential [91].

Computational models have enormous business potential for controlling large numbers of devices in wireless networks. Adoption and market growth will be fueled by the demand for effective, data-driven device management as the Internet of Things (IoT) ecosystem grows and wireless networks become more sophisticated. In this situation, companies and organisations that can leverage computational model capabilities stand to gain from increased network performance, lower costs, and new revenue streams.

11.7 COST-BENEFIT ANALYSIS OF EMERGING COMPUTATIONAL MODELS FOR DENSE WIRELESS NETWORKS

Evaluating the possible costs and benefits of putting these evolving computational models for dense wireless networks into practice is known as the cost-benefit analysis. Key cost and benefit analyses of a new computational model for dense wireless networks are presented and detailed in Tables 11.5 and 11.6.

Emerging computational models for dense wireless networks should be cost-benefit analysed with both short and long-term effects taken into account. Although there are upfront investment costs, they can be greatly outweighed in the long run by the advantages of increased revenue growth, customer happiness, and enhanced network efficiency. To fully utilise computational models in dense wireless networks, careful planning, strategic execution, and ongoing monitoring are essential.

TABLE 11.5

Cost Analysis of Emerging Computational Models for Dense Wireless Networks

Reference	Cost	Description
[17]	Infrastructure Upgrades	Adding base stations, tiny cells, or better network hardware may be necessary for the implementation of computational models in dense wireless networks. These changes could require a lot of money.
[15]	Hardware and Software	Purchasing the gear and software required for computational modelling might come with a hefty upfront price. This covers licensing for modelling software, data storage, and powerful servers.
[31]	Skilled Workforce	Skilled labour is required to create, implement, and maintain computational models. It can be expensive to hire data scientists, network engineers, and modelling specialists.
[70]	Data Acquisition	Data are necessary for computational models to function, and obtaining reliable data sources may incur costs. Data management systems, data feeds, and sensors are examples of this.
[145]	Training and Education	Employees can need training in order to use computational models properly. This covers instruction in network optimisation, model building, and data analysis.
[146]	Integration Costs	It could be necessary to make adjustments and alterations to the current network infrastructure to integrate computational models, which would cost extra.

TABLE 11.6

Benefit Analysis of Emerging Computational Models for Dense Wireless Networks

Reference	Benefit	Description
[151]	Network Optimization	Computational models aid in network performance optimisation, enhancing user experience and service quality. This can benefit cellular network operators by raising customer happiness and lowering attrition.
[28], [29]	Resource Efficiency	Computational models lower operating expenses by optimising resource allocation. This covers the effective utilisation of network and spectrum equipment, as well as energy efficiency.
[159]	Predictive Maintenance	Models can foresee equipment breakdowns and required repair, minimising unplanned downtime and related expenses.
[147]	Capacity Planning	Planning for capacity growth or densification, ensuring that network resources are used effectively, and preventing overinvestment in infrastructure can all be aided by computational models.
[150]	Data Monetization	Computational models have the ability to produce insights that can be sold. This includes providing data analytics services to outside parties or developing prospects for targeted advertising.
[154]	Competitive Advantage	Effective computational models give wireless network operators a competitive edge in terms of network performance, cost effectiveness, and creative service offers.
[154]	Customer Retention	Improved QoS and network optimisation can lower acquisition expenses by keeping current clients.
[53]	Revenue Growth	Operators can investigate new sources of income thanks to computational models. This includes value-added services, IoT connections, and premium services.
[51]	Sustainability	Long-term cost reductions may result from computational models facilitation of resource optimisation and energy efficiency, which are in line with environmental initiatives.
[136]	Regulatory Compliance	Regulatory fines and penalties may be avoided by adhering to network performance and quality of service criteria.

11.8 COMMERCIALIZATION AND INVESTMENT OPPORTUNITIES FOR COMPUTATIONAL MODELS IN WIRELESS COMMUNICATIONS

As the need for data-driven insights and optimisation in the telecommunications industry grows, there are several commercialization and investment prospects for computational models in wireless communications [104]. These are some significant chances in this field.

a) *Network Optimization Solutions*

Wireless carriers can benefit from network optimisation solutions provided by computational modelling companies [150]. Services that increase network efficiency, lower operating expenses, and raise service quality fall under this category [102].

b) *Predictive Maintenance Services*

By anticipating equipment failures and maintenance requirements, providers can help wireless network operators cut downtime and maintenance expenses.

c) *Data Analytics and Insights*

Companies can make money off of data by providing wireless carriers with analytics services. This entails trend analysis, data-driven suggestion-making, and network data analysis performance optimisation.

d) *IoT Management Solutions*

Solutions that effectively manage IoT devices are becoming more and more necessary as the Internet of Things grows. Computational models can be used by businesses to provide IoT device management services [159].

e) *Security Solutions*

When it comes to network security, computational models can be quite important. Offering cutting-edge security solutions to wireless carriers presents an opportunity for businesses that specialise in threat detection and prevention [8].

f) *Data Monetization*

Businesses can build systems that let cellular providers profit from their data by providing aggregated, anonymised, and insightful information to advertising and researchers, among other parties [65].

g) *Machine Learning as a Service (MLaaS)*

Companies can provide MLaaS platforms, which free wireless carriers from requiring internal knowledge to leverage the potential of computational modelling and machine learning.

h) *Consulting and Training*

Consulting services are becoming more and more necessary to assist wireless carriers in comprehending and successfully using computational models. To upskill the current workforce, there is also a demand for training and educational services.

i) *Hardware and Software Development*

Businesses that create specialised hardware and software for data processing and computational modelling in wireless networks have opportunities.

j) *Investment in Startups*

In the wireless communication industry, venture capital firms and angel investors might look into chances to invest in startups that concentrate on computational modelling, machine learning, and data analytics [140].

k) *Research and Development*

Research and development expenditures for computational modelling in wireless networks have the potential to yield novel, marketable technology and solutions [131].

l) *Global Expansion*

Companies that specialise in computational modelling can expand their services to wireless carriers all over the world and investigate global markets [138].

m) *Collaboration with Wireless Carriers*
   There are chances for enterprises to collaborate closely with wireless carriers to customise solutions to meet their unique requirements and challenges.

n) *Edge Computing Services*
   There are investment opportunities in developing edge computing solutions powered by computational models due to the proliferation of edge computing in wireless networks [147].

o) *Customized Solutions*
   delivering specialised solutions to meet certain issues that cellular carriers encounter, including special event coverage or rural network optimisation.

The requirement for effective network administration, data analysis, security, and optimisation in wireless communications is what is driving these commercialization and investment prospects. Companies and investors in this field are well-positioned to profit from the changing environment of computational modelling in wireless networks as technology advances and the need for wireless connectivity rises.

11.9 LESSON LEARNED

Among the important lessons provided in this chapter on Government Policies and the Economics of Computational Modelling in Wireless Networks are;

a) It was revealed to the readers that the implementation, administration, and enhancement of wireless networks are significantly influenced by government policies. Network operators, legislators, and industry stakeholders must comprehend this influence.

b) The economic dynamics of computational modelling in wireless networks were clarified in this chapter. It demonstrates how various policies might result in income opportunities, cost savings, and improvements in general economic efficiency.

c) The chapter would highlight the significance of combining qualitative and quantitative procedures to acquire a thorough grasp of the issue by giving a mixed-methods research methodology. This is a useful lesson for scholars who want to investigate intricate, diverse subjects.

d) The gap analysis in the chapter would help readers discover topics that require more investigation. For academics and politicians seeking to overcome gaps in current knowledge and policies, this lesson is essential.

e) The importance of multidisciplinary collaboration would draw attention to the need to combine technical and policy knowledge in the area of telecommunications. This lecture emphasises how crucial it is to work together to properly meet the opportunities and problems in wireless networks.

f) Policymakers can benefit from the policy recommendations from this chapter, which offer insightful advice on how to best implement measures to improve wireless networks financial performance. This course emphasises the possibility of making policy decisions based on evidence.

g) The significance of taking international considerations into account was made clear to readers. Wireless networks are a global issue, and business participants and policymakers must comprehend the effects of regulations in many countries.

The chapter teaches readers about the complex interactions that exist between the wireless network economy, computer modelling, and governmental regulations. These lessons offer insights to better effectively traverse the dynamic and growing world of telecommunications, and they would apply to a wide spectrum of readers, including scholars, policymakers, and industry personnel.

11.10 RECOMMENDATIONS

For different stakeholders in the wireless communications industry, this chapter Government Policies and the Economics of Computational Modelling in Wireless Networks provides several recommendations. These suggestions are meant to improve economic results, optimise policy frameworks, and encourage efficient use of computational modelling. Some possible suggestions are;

1. *For Policymakers and Regulators*
   a) Make sure that rules about computational modelling and wireless networks are precise, standardised, and updated often to stay up with new developments in technology. Industry participants are less uncertain when regulations are clear.
   b) Promote laws that facilitate the development of computational modelling and wireless technology innovation. While balancing regulatory concerns, policies should permit experimentation and the adoption of innovative technology.
   c) When creating policies, consult with representatives of the industry and technical specialists. Working together can produce well-informed laws that take into consideration the economic and technological aspects of a situation.
   d) Encourage international cooperation and policy alignment to remove obstacles to global connectivity and to enable cross-border wireless communication.
2. *For Wireless Network Operators*
   a) Maintain adherence to governmental directives while strategically utilising regulatory frameworks to enhance network performance. This could entail looking for ways to improve services and cut expenses within the framework of current policies.
   b) Prioritise data security and privacy while adhering to applicable policies. Think about making investments in effective data protection and security procedures.
   c) Align resource allocation and network performance with policy goals by using computational modelling. Utilise modelling to improve network operations efficiency and quality of service.
3. *For Researchers and Academics*
   a) Keep pointing out areas that need more research, particularly those about the financial effects of computational modelling in wireless networks. Fill in these gaps so that industry stakeholders and policymakers can benefit from insightful information.
   b) Examine the financial effects of computational modelling in relation to the different wireless network-dependent industries. Think about how this would affect the transportation, healthcare, and agricultural industries, among others.
4. *For Industry Professionals*
   a) To fully utilise modelling tools, collaborate with data scientists and professionals in computational modelling. Cooperation can lead to improved service delivery and more effective network management.
   b) Invest in computer modelling solutions that support the goals of network optimisation and policy objectives. Long-term financial gains from these expenditures may result from increased effectiveness and higher-quality services.
   c) To have an impact on the creation of policies, and interact with advocacy organisations and trade associations. Take part in regulatory framework negotiations to make sure they reflect the interests of the sector.
   d) Spend money on regular training for your staff so they are knowledgeable about the newest computational modelling approaches and how they are used in wireless networks.

These suggestions are intended to help researchers, industry experts, legislators, and wireless network operators navigate the intricate interactions that arise from the combination of economic factors, computer modelling, and governmental regulations in wireless networks. By putting these recommendations into practice, the wireless communications industry can be encouraged to innovate and become more efficient while also maximising policy outcomes and economic benefits.

11.11 CONCLUSION

To sum up, this chapter has offered a thorough examination of the complex interplay between political regulations and the financial aspects of computer modelling in wireless networks. We now know that government regulations have a significant impact on how wireless communications are developed, affecting everything from resource allocation and network development to optimisation. The costs and advantages of computer modelling are impacted by these policies, which have significant economic ramifications. It has been noted that the mixed-methods research technique, which combines qualitative and quantitative methodologies, is an effective way to get a thorough grasp of this complex topic. It gives us a comprehensive picture by enabling us to depict the complex interactions between economics, technology, and policy. The significance of multidisciplinary cooperation in bridging the knowledge gap between technical proficiency and policy analysis has also been emphasised in this chapter. It has demonstrated the necessity for regulations that are transparent, unbiased, and conducive to innovation to support the dynamic character of wireless networks. Our debate has included real-world applications, policy proposals, and a global perspective, providing useful insights and direction for researchers, industry experts, regulators, and wireless network operators. The lessons learned and suggestions made in this chapter will be helpful resources for navigating the intricate and dynamic world of government regulations and the economics of computational modelling in wireless networks as the telecommunications industry develops and grows. Achieving global connectivity and economic success through wireless networks can be ensured by stakeholders by heeding these insights and promoting cooperation, innovation, and policy alignment. This chapter paves the way for future research endeavours by outlining the need for novel economic models that are in line with emerging government policies in the constantly changing field of wireless networks, as well as a deeper investigation of adaptive security and privacy measures in response to evolving computational models.

REFERENCES

[1] Ezeigweneme, C. A., Umoh, A. A., Ilojianya, V. I., & Adegbite, A. O. (2024). Review of telecommunication regulation and policy: Comparative analysis USA and Africa. *Computer Science & IT Research Journal*, 5(1), 81–99.

[2] Gisca, O., Matinmikko-Blue, M., Ahokangas, P., Yrjölä, S., & Gordon, J. (2023). Regulatory challenges and implications of the European electronic communications code (EECC) for local mobile communication network business. *Telecommunications Policy*, 47(10), 102651.

[3] Myovella, G., Karacuka, M., & Haucap, J. (2020). Digitalization and economic growth: A comparative analysis of Sub-Saharan Africa and OECD economies. *Telecommunications Policy*, 44(2), 101856.

[4] Gupta, S. (2022). The interaction between technology, business environment, society, and regulation in ICT industries. *IIMB Management Review*, 34(2), 103–115.

[5] Liu, Y. F., Chang, T. H., Hong, M., Wu, Z., So, A. M. C., Jorswieck, E. A., & Yu, W. (2024). A survey of recent advances in optimization methods for wireless communications. *IEEE Journal on Selected Areas in Communications*.

[6] Oughton, E. J. (2023). Policy options for broadband infrastructure strategies: A simulation model for affordable universal broadband in Africa. *Telematics and Informatics*, 76, 101908.

[7] Chen, Y., Tian, X., Wang, Q., Jiang, J., Li, M., & Zhang, Q. (2020). SAFE: A general secure and fair auction framework for wireless markets with privacy preservation. *IEEE Transactions on Dependable and Secure Computing*, 19(3), 2038–2053.

[8] Akram, H. (2024). The economic and environmental impact of sustainable enterprise systems: Integrating cloud, web technology, attacks, AI, IoT, and security. *Journal of Information Technology and Informatics*, 3(1).

[9] Chen, Y., Kumara, E. K., & Sivakumar, V. (2021). Investigation of finance industry on risk awareness model and digital economic growth. *Annals of Operations Research, 45*(2), 1–22.

[10] Ale, L., Zhang, N., Fang, X., Chen, X., Wu, S., & Li, L. (2021). Delay-aware and energy-efficient computation offloading in mobile-edge computing using deep reinforcement learning. *IEEE Transactions on Cognitive Communications and Networking, 7*(3), 881–892.

[11] Okoro, Y. O., Oladeinde, M., Akindote, O. J., Adegbite, A. O., & Abrahams, T. O. (2023). Digital communication and us economic growth: A comprehensive exploration of technology's impact on economic advancement. *Computer Science & IT Research Journal, 4*(3), 351–367.

[12] Oughton, E. J. (2023). Policy options for broadband infrastructure strategies: A simulation model for affordable universal broadband in Africa. *Telematics and Informatics, 76*, 101908.

[13] Balachander, T., Ramana, K., Mohana, R. M., Srivastava, G., & Gadekallu, T. R. (2023). Cooperative spectrum sensing deployment for cognitive radio networks for Internet of Things 5G wireless communication. *Tsinghua Science and Technology, 29*(3), 698–720.

[14] Adams, C. J., & Garcia, L. M. (2022). Spectrum allocation policies and their ethical implications in wireless communications. *Ethics & Information Technology, 22*(4).

[15] Alsaedi, W. K., Ahmadi, H., Khan, Z., & Grace, D. (2023). Spectrum options and allocations for 6G: A regulatory and standardization review. *IEEE Open Journal of the Communications Society, 4*, 1787–1812, doi: 10.1109/OJCOMS.2023.3301630.

[16] Massaro, M. (2018). Radio spectrum regulation as a matter of international affairs: Discussing the effectiveness of the European Union at world radiocommunication conferences. *Digital Policy, Regulation and Governance, 20*(5), 373–398.

[17] Kumar, S. K. A., & Oughton, E. J. (2023). Infrastructure sharing strategies for wireless broadband. *IEEE Communications Magazine, 61*(7), 46–52.

[18] Parvini, M., Zarif, A. H., Nouruzi, A., Mokari, N., Javan, M. R., Abbasi, B., ... & Yanikomeroglu, H. (2023). Spectrum sharing schemes from 4G to 5G and beyond: Protocol flow, regulation, ecosystem, economic. *IEEE Open Journal of the Communications Society, 4*, 464–517.

[19] Daousis, S., Peladarinos, N., Cheimaras, V., Papageorgas, P., Piromalis, D. D., & Munteanu, R. A. (2024). Overview of protocols and standards for wireless sensor networks in critical infrastructures. *Future Internet, 16*(1), 33.

[20] Colangelo, G. (2024). Fair share of network costs and regulatory myopia: Learning from net neutrality mistakes. *Law, Innovation and Technology, 16*(1), 218–242.

[21] Strover, S., Riedl, M. J., & Dickey, S. (2021). Scoping new policy frameworks for local and community broadband networks. *Telecommunications Policy, 45*(10), 102171.

[22] Ylianttila, M., Kantola, R., Gurtov, A., Mucchi, L., Oppermann, I., Yan, Z., ... & Röning, J. (2020). 6G white paper: Research challenges for trust, security and privacy. *arXiv preprint arXiv:2004.11665*.

[23] Schade, F. (2023). Dark sides of data transparency: Organized immaturity after GDPR? *Business Ethics Quarterly, 33*(3), 473–501.

[24] Witkowski, D., Page, T., & Ong, L. (2023, November). Deployment. In *2023 IEEE Future Networks World Forum (FNWF)* (pp. 01–34), IEEE.

[25] Mazhar, T., Malik, M. A., Mohsan, S. A. H., Li, Y., Haq, I., Ghorashi, S., ... & Mostafa, S. M. (2023). Quality of service (QoS) performance analysis in a traffic engineering model for next-generation wireless sensor networks. *Symmetry, 15*(2), 513.

[26] Ahmed, A., Azizi, S., & Zeebaree, S. R. M. (2024). ECQ: An energy-efficient, cost-effective and Qos-aware method for dynamic service migration in mobile edge computing systems. *Wireless Personal Communications, 133*, 1–35.

[27] Tangirala, M. P. (2019). *Telecom Sector Regulation in India: An Institutional Perspective.* Taylor & Francis.

[28] Luong, N. C., Wang, P., Niyato, D., Liang, Y. C., Han, Z., & Hou, F. (2018). Applications of economic and pricing models for resource management in 5G wireless networks: A survey. *IEEE Communications Surveys & Tutorials, 21*(4), 3298–3339.

[29] Yang, S., Li, F., Trajanovski, S., Yahyapour, R., & Fu, X. (2020). Recent advances of resource allocation in network function virtualization. *IEEE Transactions on Parallel and Distributed Systems, 32*(2), 295–314.

[30] Demestichas, K., & Daskalakis, E. (2020). Information and communication technology solutions for the circular economy. *Sustainability, 12*(18), 7272.

[31] Awuku, E., Agyei, P. M., & Gonu, E. (2023). Service innovation practices and customer loyalty in the telecommunication industry. *PLOS One, 18*(3), e0282588.

[32] Mazumder, S., & Garg, S. (2021). Decoding digital transformational outsourcing: The role of service providers' capabilities. *International Journal of Information Management, 58*, 102295.

[33] Ochuba, N. A., Okafor, E. S., Akinrinola, O., Amoo, O. O., & Usman, F. O. (2024). Enhancing customer service in satellite telecommunications: A review of data-driven insights and methodologies for personalized service offerings. *International Journal of Management & Entrepreneurship Research, 6*(3), 582–593.

[34] Basalla, M., Schneider, J., Luksik, M., Jaakonmäki, R., & Vom Brocke, J. (2021). On latency of e-commerce platforms. *Journal of Organizational Computing and Electronic Commerce, 31*(1), 1–17.

[35] Ozturk, I., & Ullah, S. (2022). Does digital financial inclusion matter for economic growth and environmental sustainability in OBRI economies? An empirical analysis. *Resources, Conservation and Recycling, 185*, 106489.

[36] Ahmad, T., & Zhang, D. (2021). Using the Internet of Things in smart energy systems and networks. *Sustainable Cities and Society, 68*, 102783.

[37] Bharany, S., Sharma, S., Khalaf, O. I., Abdulsahib, G. M., Al Humaimeedy, A. S., Aldhyani, T. H., ... & Alkahtani, H. (2022). A systematic survey on energy-efficient techniques in sustainable cloud computing. *Sustainability, 14*(10), 6256.

[38] Khan, L. U., Yaqoob, I., Imran, M., Han, Z., & Hong, C. S. (2020). 6G wireless systems: A vision, architectural elements, and future directions. *IEEE Access, 8*, 147029–147044.

[39] Sunhare, P., Chowdhary, R. R., & Chattopadhyay, M. K. (2022). Internet of Things and data mining: An application oriented survey. *Journal of King Saud University – Computer and Information Sciences, 34*(6), 3569–3590.

[40] Fast, V., Schnurr, D., & Wohlfarth, M. (2023). Regulation of data-driven market power in the digital economy: Business value creation and competitive advantages from big data. *Journal of Information Technology, 38*(2), 202–229.

[41] Schüritz, R., Seebacher, S., & Dorner, R. (2017). Capturing value from data: Revenue models for data-driven services.

[42] Arvin, M. B., Pradhan, R. P., & Nair, M. (2021). Uncovering interlinks among ICT connectivity and penetration, trade openness, foreign direct investment, and economic growth: The case of the G-20 countries. *Telematics and Informatics, 60*, 101567.

[43] Pain, K., Van Hamme, G., Vinciguerra, S., & David, Q. (2016). Global networks, cities and economic performance: Observations from an analysis of cities in Europe and the USA. *Urban Studies, 53*(6), 1137–1161.

[44] Hadzovic, S., Mrdovic, S., & Radonjic, M. (2023). A path towards an Internet of Things and artificial intelligence regulatory framework. *IEEE Communications Magazine, 61*(7), 90–96.

[45] Olukoya, O. (2022). Assessing frameworks for eliciting privacy & security requirements from laws and regulations. *Computers & Security, 117*, 102697.

[46] Rao, S. P., Chen, H. Y., & Aura, T. (2023). Threat modeling framework for mobile communication systems. *Computers & Security, 125*, 103047.

[47] Shafique, K., Khawaja, B. A., Sabir, F., Qazi, S., & Mustaqim, M. (2020). Internet of Things (IoT) for next-generation smart systems: A review of current challenges, future trends and prospects for emerging 5G-IoT scenarios. *IEEE Access, 8*, 23022–23040.

[48] Nižetić, S., Šolić, P., Gonzalez-De, D. L. D. I., & Patrono, L. (2020). Internet of Things (IoT): Opportunities, issues and challenges towards a smart and sustainable future. *Journal of Cleaner Production, 274*, 122877.

[49] Goudarzi, S., Anisi, M. H., Ahmadi, H., & Musavian, L. (2020). Dynamic resource allocation model for distribution operations using SDN. *IEEE Internet of Things Journal, 8*(2), 976–988.

[50] Wang, W., Kumar, N., Chen, J., Gong, Z., Kong, X., Wei, W., & Gao, H. (2020). Realizing the potential of the Internet of Things for smart tourism with 5G and AI. *IEEE network, 34*(6), 295–301.

[51] Ali, A. A., & Uzmi, Z. A. (2019). Relays in cellular networks: A simplified cost–benefit analysis. In *Proceedings of the 15th ACM International Symposium on QoS and Security for Wireless and Mobile Networks* (pp. 61–69), November 2019.

[52] Zhong, C., Zhao, Z., Luo, C., Gursoy, M. C., Qiu, Q., Caicedo, C., Basti, F., & Solomon, A. A cost–benefit analysis to achieve command and control (C2) link connectivity for Beyond Visual Line of Sight (BVLOS) Operations. In *2020 Integrated Communications Navigation and Surveillance Conference (ICNS)* (pp. 2D1–1), September 2020, IEEE.

[53] Stahlberg, M. J., Sagnol, G., Ducke, B., & Klimm, M. Spatiotemporal reconstruction of ancient road networks through sequential cost– benefit analysis," *PNAS Nexus*, vol. 2, no. 2, pp. pgac313, 2023.

[54] Teece, D. J. (2018). Profiting from innovation in the digital economy: Enabling technologies, standards, and licensing models in the wireless world. *Research policy, 47*(8), 1367–1387.

[55] Ehrenhard, M., Wijnhoven, F., van den Broek, T., & Stagno, M. Z. (2017). Unlocking how start-ups create business value with mobile applications: Development of an App-enabled Business Innovation Cycle. *Technological forecasting and social change, 115*, 26–36.

[56] Ai, Y., Peng, M., & Zhang, K. (2018). Edge computing technologies for Internet of Things: a primer. *Digital Communications and Networks, 4*(2), 77–86.

[57] Canbaloğlu, G., Treur, J., & Wiewiora, A. (Eds.). (2023). *Computational modeling of multilevel organisational learning and its control using self-modeling network models* (Vol. 468). Springer Nature.

[58] Anitha, P., Vimala, H. S., & Shreyas, J. (2023). Comprehensive review on congestion detection, alleviation, and control for IoT networks. *Journal of Network and Computer Applications*, 103749.

[59] Zainaddin, D. A., Hanapi, Z. M., Othman, M., Ahmad Zukarnain, Z., & Abdullah, M. D. H. (2024). Recent trends and future directions of congestion management strategies for routing in IoT-based wireless sensor network: a thematic review. *Wireless Networks, 30*(3), 1939–1983.

[60] Figueiredo, R., Woesner, H., Kassler, A., & Karl, H. (2024, May). Quality of Service Performance of Multi-Core Broadband Network Gateways. In *2024 8th Network Traffic Measurement and Analysis Conference (TMA)* (pp. 1–10), IEEE.

[61] Swanson, H., Lawrence, L., Arnell, J., Jones, B., Sherin, B., & Wilensky, U. (2024). Computational models as tools for supporting for responsive teaching. *Behaviour & Information Technology*, 1–18.

[62] Lyu, M., Gharakheili, H. H., & Sivaraman, V. (2024). A survey on enterprise network security: Asset behavioral monitoring and distributed attack detection. *IEEE Access*.

[63] Srinivasan, S., Vinmathi, M. S., Sivaraj, S. N., Karthikayen, A., Alakesan, C., & Preetha, M. (2024). A Novel Approach Integrating IoT and WSN with Predictive Modeling and Optimization for Enhancing Efficiency and Sustainability in Smart Cities. *Journal of Electrical Systems, 20*(4s), 2228–2237.

[64] Li, H., Wang, S. X., Shang, F., Niu, K., & Song, R. (2024). Applications of large language models in cloud computing: An empirical study using real-world data. *International Journal of Innovative Research in Computer Science & Technology, 12*(4), 59–69.

[65] Shukla, S., Bisht, K., Tiwari, K., & Bashir, S. (2023). Data Monetization. In *Data Economy in the Digital Age* (pp. 37–62). Singapore: Springer Nature Singapore.

[66] Kumar, A., Sharma, S., Singh, A., Alwadain, A., Choi, B. J., Manual-Brenosa, J., ... & Goyal, N. (2021). Revolutionary strategies analysis and proposed system for future infrastructure in Internet of Things. *Sustainability, 14*(1), 71.

[67] M. S. Gupta, and K. Kumar, "Progression on spectrum sensing for cognitive radio networks: A survey, classification, challenges and future research issues," *Journal of Network and Computer Applications*, vol. 143, pp. 47–76, 2019.

[68] Ahmad, T., & Zhang, D. (2021). Using the Internet of Things in smart energy systems and networks. *Sustainable Cities and Society, 68*, 102783.

[69] Habibzadeh, H., Nussbaum, B. H., Anjomshoa, F., Kantarci, B., & Soyata, T. (2019). A survey on cybersecurity, data privacy, and policy issues in cyber-physical system deployments in smart cities. *Sustainable Cities and Society, 50*, 101660.

[70] Chen, X., Chen, K., Wang, M., & Li, R. (2023). Roles of wireless networks in bridging the rural smart infrastructural divide. *Infrastructures, 8*(11), 159.

[71] Damaševičius, R., Bacanin, N., & Misra, S. (2023). From sensors to safety: Internet of Emergency Services (IoES) for emergency response and disaster management. *Journal of Sensor and Actuator Networks, 12*(3), 41.

[72] Briglauer, W. (2024). Efficiency and effectiveness of net neutrality rules in the mobile sector: Relevant developments and state of the empirical literature.

[73] Patel, B., Yarlagadda, V. K., Dhameliya, N., Mullangi, K., & Vennapusa, S. C. R. (2022). Advancements in 5G technology: Enhancing connectivity and performance in communication engineering. *Engineering International, 10*(2), 117–130.

[74] Meddour, D. E., Rasheed, T., & Gourhant, Y. (2011). On the role of infrastructure sharing for mobile network operators in emerging markets. *Computer networks, 55*(7), 1576–1591.

[75] Abdulsalam, K. A., Babajide, O., Adebisi, J. A., Ndjuluwa, L., & Babatunde, O. (2024). Energy efficient cellular network model for spectrum sharing and management. *Adeleke University Journal of Engineering and Technology, 7*(1), 147–162.

[76] Debnath, S., Arif, W., Roy, S., Baishya, S., & Sen, D. (2022). A comprehensive survey of emergency communication network and management. *Wireless Personal Communications, 124*(2), 1375–1421.

[77] Alhassan, M. D., & Adam, I. O. (2022). The effect of mobile phone penetration on e-commerce diffusion: A global perspective using structural equation modelling. *International Journal of Business and Systems Research, 16*(5–6), 783–804.

[78] Pahlavan, K. (2022). Understanding of RF cloud interference measurement and modeling. *International Journal of Wireless Information Networks, 29*(3), 206–221.

[79] Howlett, M. (2009). Government communication as a policy tool: A framework for analysis. *Canadian Political Science Review, 3*(2), 23–37.

[80] Mitra, A. (2023, December). Adapting the regulation of spectrum and telecom networks to 5G technology – A cross country analysis. In *International Working Conference on Transfer and Diffusion of IT* (pp. 107–125), Cham: Springer Nature Switzerland.

[81] Nguyen, M. T., & Tran, M. Q. (2023). Balancing security and privacy in the digital age: An in-depth analysis of legal and regulatory frameworks impacting cybersecurity practices. *International Journal of Intelligent Automation and Computing, 6*(5), 1–12.

[82] Zhou, Y., Yu, L., Jiang, Z., Zhi, Z., Kang, J., & Han, Z. (2023). An improved spectrum trading design based on dynamic credit aggregate-signature blockchain. *IEEE Wireless Communications Letters, 12*(4), 625–629.

[83] Netalkar, P., Zahabee, A., Bastidas, C. E. C., Kadota, I., Stojadinovic, D., Zussman, G., & Raychaudhuri, D. (2023). Large-scale dynamic spectrum access with IEEE 1900.5.2 Spectrum Consumption Models. In *2023 IEEE Wireless Communications and Networking Conference (WCNC)* (pp. 1–6), March 2023, IEEE.

[84] Garrett, T., Setenareski, L. E., Peres, L. M., Bona, L. C., & Duarte Jr, E. P. (2022). A survey of network neutrality regulations worldwide. *Computer Law & Security Review, 44*, 105654.

[85] Stocker, V., Smaragdakis, G., & Lehr, W. (2020). The state of network neutrality regulation. *ACM SIGCOMM Computer Communication Review, 50*(1), 45–59.

[86] Hartono, J., Milano, A., Nugraha, X., & Felicia, S. A. (2023). Failing to protect personal data: Key aspects of electronic system operators' agreements. *Barelang Journal of Legal Studies, 1*(1), 31–55.

[87] Johnson, G. A., Shriver, S. K., & Goldberg, S. G. (2023). Privacy and market concentration: intended and unintended consequences of the GDPR. *Management Science, 69*(10), 5695–5721.

[88] Ariyoosu, D., & Akangbe, S. (2023). Consumer rights protection and its impact on service quality in the telecommunications sector. *Coventry Law Journal, 28*(1), 51–60.

[89] Lagen, S., Bojovic, B., Koutlia, K., Zhang, X., Wang, P., & Qu, Q. (2023). QoS management for XR traffic in 5G NR: A multi-layer system view & end-to-end evaluation. *IEEE Communications Magazine.*

[90] Spellman, F. R. (2023). *Emergency Services Sector Protection and Homeland Security.* Rowman & Littlefield.

[91] Takyi, S. A., Amponsah, O., Duamor, S. N., Azunre, G. A., & Ahadzie, D. K. (2022). An assessment of regulatory compliance and residents' perceptions of the siting of telecommunication masts: Insights from a rapidly urbanising Ghanaian city. *African Geographical Review, 41*(4), 411–432.

[92] Von Hirschhausen, C., Beckers, T., & Brenck, A. (2004). Infrastructure regulation and investment for the long-term – An introduction. *Utilities Policy, 12*(4), 203–210.

[93] Wang, S., Chen, Z., Xiao, Y., & Lin, C. (2021). Consumer privacy protection with the growth of AI-empowered online shopping based on the evolutionary game model. *Frontiers in Public Health, 9*, 705777.

[94] Dachyar, M., Zagloel, T. Y. M., & Saragih, L. R. (2020). Enterprise architecture breakthrough for telecommunications transformation: A reconciliation model to solve bankruptcy. *Heliyon, 6*(10).

[95] Humphreys, B. E. (2019). Demand for broadband in rural areas: Implications for universal access. Congressional Research Service Report, 46108.

[96] World Health Organization, & United Nations Children's Fund. (2022). *Global Report on Assistive Technology*. World Health Organization.

[97] Picard, R. G. (2022). Competition, monopoly, and antitrust issues. *The SAGE Handbook of the Digital Media Economy* (pp. 449–469).

[98] Calabrese, M., Roberson, D., Dombrowsky Jr, T. S., Russell, J., Gibson, M., & Sharkey, S. et al. (2023). Commerce Spectrum Management Advisory Committee (CSMAC).

[99] Drake-Brockman, J., Kripalani, M., Nordås, H. K., Lee-Makiyama, H., & Gopalakrishnan, B. N. (2023). The case for international digital standards for interoperability of trade in digital services.

[100] Briglauer, W., Cambini, C., Gugler, K., & Stocker, V. (2023). Net neutrality and high-speed broadband networks: Evidence from OECD countries. *European Journal of Law and Economics, 55*(3), 533–571.

[101] Alharbi, H. A., Elgorashi, T. E., & Elmirghani, J. M. (2020). Impact of the net neutrality repeal on communication networks. *IEEE Access, 8*, 59787–59800.

[102] Lafta, S. A., Abdulkareem, M. M., Ibrahim, R. K., Kareem, M. M., & Ali, A. H. (2021). Quality of service performances of video and voice transmission in universal mobile telecommunications system network based on OPNET. *Bulletin of Electrical Engineering and Informatics, 10*(6), 3202–3210.

[103] Petik, M. (2023). AI in 5G: High-speed dilution of the right to privacy and reducing the role of consent. *Intelligent and Autonomous: Transforming Values in the Face of Technology* (pp. 89–106). Brill.

[104] Matinmikko-Blue, M., Yrjölä, S., & Ahokangas, P. (2020). Spectrum management in the 6G era: The role of regulation and spectrum sharing. In *2020 2nd 6G Wireless Summit (6G SUMMIT)* (pp. 1–5), March 2020, IEEE.

[105] Saint, M., & Brown, T. X. (2019). A dynamic policy license for flexible spectrum management. *Telecommunications Policy, 43*(1), 23–37.

[106] Gomulkiewicz, R., Nguyen, X. T., & Conway, D. M. (2023). *Licensing Intellectual Property: Law and Application*. Aspen Publishing.

[107] Xiao, Y., Shi, S., Lou, W., Wang, C., Li, X., & Zhang, N. et al. (2023). BD-SAS: Enabling dynamic spectrum sharing in low-trust environment. *IEEE Transactions on Cognitive Communications and Networking, 8*(8), 6498–6518.

[108] Shen, X. S., Huang, C., Liu, D., Xue, L., Zhuang, W., Sun, R., & Ying, B. (2021). Data management for future wireless networks: Architecture, privacy preservation, and regulation. *IEEE Network, 35*(1), 8–15.

[109] Ahmed, I., Karvonen, H., Kumpuniemi, T., & Katz, M. (2020). Wireless communications for the hospital of the future: Requirements, challenges and solutions. *International Journal of Wireless Information Networks, 27*(1), 4–17.

[110] Ahmad, N. (2023). Data privacy issues and risks with sharing on social media: An inquiry. *Russian Law Journal, 11*(4).

[111] Nguyen, V. L., Hwang, R. H., Cheng, B. C., Lin, Y. D., & Duong, T. Q. (2023). Understanding privacy risks of high-accuracy radio positioning and sensing in wireless networks.

[112] Khan, Z., Lee, C., Girinathan, S., Mohammad Usman Fareed, M., Ferreira, S., Al Saeedi, F., … & Bamakhramah, A. (2023, March). Operational excellence through implementation of data governance policies and procedures covering data retention, disposal and archival: A collaborative case study between HQ and Group Company, Abu Dhabi, United Arab Emirates (UAE). In *SPE Gas & Oil Technology Showcase and Conference* (p. D021S021R003), SPE.

[113] Quach, S., Thaichon, P., Martin, K. D., Weaven, S., & Palmatier, R. W. (2022). Digital technologies: Tensions in privacy and data. *Journal of the Academy of Marketing Science, 50*(6), 1299–1323.

[114] Bandari, V. (2023). Enterprise data security measures: A comparative review of effectiveness and risks across different industries and organization types. *International Journal of Business Intelligence and Big Data Analytics, 6*(1), 1–11.

[115] Schlackl, F., Link, N., & Hoehle, H. (2022). Antecedents and consequences of data breaches: A systematic review. *Information & Management, 59*(4), 103638.

[116] Guamán, D. S., Del Alamo, J. M., & Caiza, J. C. (2021). GDPR compliance assessment for cross-border personal data transfers in android apps. *IEEE Access, 9*, 15961–15982.

[117] Garrett, T., Setenareski, L. E., Peres, L. M., Bona, L. C., & Duarte Jr, E. P. (2022). A survey of Network Neutrality regulations worldwide. *Computer Law & Security Review, 44*, 105654.

[118] Maillé, P., & Tuffin, B. (2022). *From Net Neutrality to ICT Neutrality* (pp. 1–179). Cham: Springer.

[119] Gadringer, S. (2020). Network neutrality in the European Union: A communications policy process analysis. *Internet Histories, 4*(2), 178–195.

[120] Øverby, H., & Audestad, J. A. (2020). Standards, regulations, and net neutrality in the digital economy, May 15, 2020.

[121] Comeig, I., Klaser, K., & Pinar, L. D. (2022). The paradox of (Inter) net neutrality: An experiment on ex-ante antitrust regulation☆. *Technological Forecasting and Social Change, 175*, 121405.

[122] Jitsuzumi, T. (2020). Net neutrality. *Telecommunications Policies of Japan* (pp. 201–220).

[123] Bauner, C., & Espin, A. (2023). Do subscribers of mobile networks care about Data Throttling? *Telecommunications Policy, 47*, 102665.

[124] Silva, H. B. G., & Ricardo, M. (2023). ISPs' discriminatory practices and their regulation in Andean America.

[125] Chung, W. Y., Nam, J., Ryong, K., & Lee, D. (2022). When, how, and what kind of information should Internet service providers disclose? A study on the transparency that users want. *Telematics and Informatics, 70*, 101799.

[126] Feng, J., Yu, Y., & Xu, T. (2023). Content regulation laws for Chinese ISPs: Legal responsibilities in free speech and filtering of harmful content. *Law and Economy, 2*(11), 53–59.

[127] Samuelson, P. (2020). Pushing back on stricter copyright ISP liability rules. *Michigan Technology Law Review, 27*, 299.

[128] Tully, S. (2014). A human right to access the Internet? problems and prospects. *Human Rights Law Review, 14*(2), 175–195.

[129] Elokda, E., Bolognani, S., Censi, A., Dörfler, F., & Frazzoli, E. (2024). A self-contained karma economy for the dynamic allocation of common resources. *Dynamic Games and Applications, 14*(3), 578–610.

[130] Ochuba, N. A., Amoo, O. O., Okafor, E. S., Usman, F. O., & Akinrinola, O. (2024). Conceptual development and financial analytics for strategic decision-making in telecommunications, focusing on assessing investment opportunities and managing risks in satellite projects. *International Journal of Management & Entrepreneurship Research, 6*(3), 594–607.

[131] Prasad, V., Haralayya, B., Jayashree, R. A., Prabavathy, A. K., & Selvarasu, S. (2023). Creating a cost effective network of operational resources for wireless networks. *AIP Conference Proceedings*, vol. 2523, no. 1, January 2023.

[132] Dai, M., Sun, G., Yu, H., & Niyato, D. (2023). Maximize the long-term average revenue of network slice provider via admission control among heterogeneous slices. *IEEE/ACM Transactions on Networking*.

[133] Mazhar, T., Malik, M. A., Mohsan, S. A. H., Li, Y., Haq, I., Ghorashi, S., ... & Mostafa, S. M. (2023). Quality of service (QoS) performance analysis in a traffic engineering model for next-generation wireless sensor networks. *Symmetry, 15*(2), 513.

[134] Fatima, Z., Rehman, A. U., Hussain, R., Karim, S., Shakir, M., Soomro, K. A., & Laghari, A. A. (2024). Mobile crowdsensing with energy efficiency to control road congestion in internet cloud of vehicles: A review. *Multimedia Tools and Applications, 83*(18), 53949–53974.

[135] Ahokangas, P., Atkova, I., Yrjölä, S., & Matinmikko-Blue, M. (2024). Business model theory and the becoming of new mobile communications technologies. In *Business Model Innovation: Game Changers and Contemporary Issues* (pp. 263–293). Cham: Springer International Publishing.

[136] Vohnout, R., Bukovsky, I., Chou, S. Y., Geyer, J., Budik, O., Sharma, R., ... & Wyckmans, A. (2023). Living lab long-term sustainability in hybrid access positive energy districts – A Prosumager smart fog computing perspective. *IEEE Internet of Things Journal, 10*(21), 18898–18908.

[137] Rikitianskaia, M. (2024). "The real ethernet": The transnational history of global Wi-Fi connectivity. *New Media & Society, 26*(6), 3568–3587.

[138] Xia, L., Baghaie, S., & Sajadi, S. M. (2024). The digital economy: Challenges and opportunities in the new era of technology and electronic communications. *Ain Shams Engineering Journal, 15*(2), 102411.

[139] Tula, O. A., Daraojimba, C., Eyo-Udo, N. L., Egbokhaebho, B. A., Ofonagoro, K. A., Ogunjobi, O. A., ... & Banso, A. A. (2023). Analyzing global evolution of materials research funding and its

influence on innovation landscape: A case study of us investment strategies. *Engineering Science & Technology Journal, 4*(3), 120–139.

[140] Alkholidi, A., Alsharabi, N. A., Hamam, H., & Alshammari, T. S. (2023, February). The 5G wireless technology and a significant economic growth and sustainable development. In *2023 International Conference on Smart Computing and Application (ICSCA)* (pp. 1–6), IEEE.

[141] Abdulkarem, M., Samsudin, K., Rokhani, F. Z., & Rasid, M. F. A. (2020). Wireless sensor network for structural health monitoring: A contemporary review of technologies, challenges, and future direction. *Structural Health Monitoring, 19*(3), 693–735.

[142] Logeshwaran, J., Shanmugasundaram, N., & Lloret, J. (2023). Energy-efficient resource allocation model for device-to-device communication in 5G wireless personal area networks. *International Journal of Communication Systems, 36*(13), e5524.

[143] Chaidir, J., & Haerofiatna, H. (2023). Network infrastructure development in Serang District. *International Journal of Management Technology, 10*(1), 11–19.

[144] Linde, L., Frishammar, J., & Parida, V. (2021). Revenue models for digital servitization: A value capture framework for designing, developing, and scaling digital services. *IEEE Transactions on Engineering Management, 70*(1), 82–97.

[145] Treaty, N. A. (2021). Guidelines for modelling and simulation (M&S) use risk identification, analysis, and mitigation.

[146] Šlapak, E., Gazda, J., Guo, W., Maksymyuk, T., & Dohler, M. (2021). Cost-effective resource allocation for multitier mobile edge computing in 5G mobile networks. *IEEE Access, 9*, 28658–28672.

[147] Cortes, L. A., Wong, M. K., Cortes-Morales, A., Wells, D., & Daly, J. (2021). Advance M&S in Acquisition T&E. (2021-11-19) (2023-06-21). www.mitre.org

[148] Mykoniatis, K., & Angelopoulou, A. (2020). A modeling framework for the application of multi-paradigm simulation methods. *Simulation, 96*(1), 55–73.

[149] Ostermeier, M., Henke, T., Hübner, A., & Wäscher, G. (2021). Multi-compartment vehicle routing problems: State-of-the-art, modeling framework and future directions. *European Journal of Operational Research, 292*(3), 799–817.

[150] Kathirgamanathan, A., De Rosa, M., Mangina, E., & Finn, D. P. (2021). Data-driven predictive control for unlocking building energy flexibility: A review. *Renewable and Sustainable Energy Reviews, 135*, 110120.

[151] Tadayonrad, Y., & Ndiaye, A. B. (2023). A new key performance indicator model for demand forecasting in inventory management considering supply chain reliability and seasonality. *Supply Chain Analytics, 3*, 100026.

[152] Yang, Y., & Peng, C. (2023). A prediction-based supply chain recovery strategy under disruption risks. *International Journal of Production Research, 61*(22), 7670–7684.

[153] Narina, P. (2023). Customer churn prediction tool using deep learning: A case of an ecommerce business operating in Kenya. Ph.D. dissertation, Strathmore University.

[154] Wei, Y. C., Lai, Y. X., & Wu, M. E. (2023). An evaluation of deep learning models for chargeback Fraud detection in online games. *Cluster Computing, 26*, (2), 927–943.

[155] Bintoro, B. K., Lutfiani, N., & Julianingsih, D. (2023). Analysis of the effect of service quality on company reputation on purchase decisions for professional recruitment services. *APTISI Transactions on Management (ATM), 7*(1), 35–41.

[156] Mohsin, M. A., Shahrouzi, S. N., & Perera, D. G. (2024). Composing efficient computational models for real-time processing on next-generation edge-computing platforms. *IEEE Access, 12*, 48–62.

[157] Silva, M. M. D., & Guerreiro, J. (2020). On the 5G and beyond. *Applied Sciences, 10*(20), 7091.

[158] Shamayleh, A., Awad, M., & Farhat, J. (2020). IoT based predictive maintenance management of medical equipment. *Journal of medical systems, 44*(4), 72.

[159] Banafaa, M., Shayea, I. Din, J., Azmi, M. H., Alashbi, A., Daradkeh, Y. I., & Alhammadi, A. (2023). 6G mobile communication technology: Requirements, targets, applications, challenges, advantages, and opportunities. *Alexandria Engineering Journal, 64*, 245–274.

12 An Enhanced Lightweight Cryptographic Algorithm Towards Securing Wireless Networks and Big Data

Joseph Bamidele Awotunde, Abidemi Emmanuel Adeniyi, Abdulrauf Olarenwaju Babatunde, Mukaila Olagunju, Agbotiname Lucky Imoize and Odunayo Dauda Olanloye

12.1 INTRODUCTION

The prevalence of installed wireless networks has made security one of the top worries in the telecommunications industry [1]. A major issue in the realm of guaranteeing wireless information sharing is the initial shared awareness between the source and the destination [2–3]. Wireless networks are regarded as one of the cutting-edge technologies that have garnered significant interest from both academia and industry because of their numerous applications, including industrial control, healthcare, and the military [4]. Sensor nodes, which are a small number of devices placed throughout a monitored region, are what make up wireless networks [5]. These nodes do not need fixed network connections for transmitting data or communicating wirelessly with one another. Nonetheless, the resource-constrained nature of sensor devices, including computation, energy, storage capacity, and bandwidth, is typically what distinguishes wireless networks [6]. In addition to their limited energy capacity, sensors placed in remote environments are thought to be challenging to recharge or to replace their batteries. As a result, this problem would negatively impact the network's lifespan.

A clustering approach was proposed to decrease a sensor node's energy consumption and hence increase the overall lifespan of a wireless network [7–9]. The collection of guidelines, safeguards, and offerings that guard a network against intrusions and unapproved access is known as network security. Many obstacles stand in the way of wireless network security, particularly for applications that demand extreme security, like emergency response, healthcare, and the military [10–11]. Although sensor devices are commonly used in unfriendly or even unsafe locations, they are more vulnerable to cyberattacks that compromise critical data and negatively impact network performance [12].

Moreover, by their very nature, wireless communications within wireless networks are unsafe. Hence, it is simple for an enemy to use a wireless device to eavesdrop on conversations between authorized nodes. Therefore, it is necessary to guarantee the bare minimum of security standards, including data integrity, confidentiality, and authentication. Designing a safe, effective, and lightweight method that takes into account the resource-constrained sensor nodes is also essential. Consequently, to secure the data, many encryption techniques are utilized in literature. The authors [13] suggested secure hashing algorithms and encryption methods to improve the effectiveness and power of data transfer in a cloud environment. A hash algorithm generates digital signatures to transfer patient data more securely and legitimately. A Lightweight Cryptography is designed to be efficient in terms of both computation and memory usage while providing adequate security for constrained environments such as wireless network, RFID tags, and other resource-constrained systems [14–17]. This is caused by a number of elements, such as the following:

i. Lightweight Cryptography offers low-cost fundamental security requirements in terms of energy usage, computational additional costs, and storage space. Lightweight Cryptography is therefore appropriate for resource-constrained devices, including wireless network nodes, thanks to this capability.

ii. In contrast to symmetric key cryptography, the key distribution in Lightweight Cryptography is simpler and more straightforward to administer.

iii. In contrast to conventional public-key infrastructure, a Lightweight Cryptography public key does not require a digital certificate since it is self-authenticated.

Lightweight cryptography focuses on providing security solutions optimized for resource-constrained environments, prioritizing efficiency in computation, memory, and energy usage. However, traditional cryptography often emphasizes stronger security guarantees with larger key sizes and more complex algorithms, suitable for general-purpose computing environments but potentially impractical for constrained devices due to higher resource requirements. In recent years, the application of wireless networks and Big Data has significantly increased and with that the need for proper security measures. This chapter presents a comprehensive look into the current state of security measures of wireless networks and Big Data, as well as the recommendations for improvements. The discussion will focus on using innovative computational models to ensure a secure wireless network and Big Data environment. Therefore, this chapter proposes a lightweight cryptographic algorithm towards securing wireless networks and big data.

12.1.1 Key Contributions of the Chapter

The following are the key contributions of the chapter:

i. The chapter emphasizes how the incorporation of computational models can improve the security and privacy of wireless networks, and big data. The utilization of computational models ensures the sacredness of wireless networks nodes and big data environments.

i. The contribution of this chapter, found in its investigation of how the use of lightweight cryptographic algorithm can be enabled to secure the wireless networks and big data, thus enhancing robust data access authorization in various systems.

ii. The proposed model is designed to execute efficiently on resource-constrained devices common in wireless networks and big data systems. The model was designed to optimize fewer resources like processing power and memory, ensuring minimal impact on system performance.

iii. The enhanced lightweight model facilitates quick encryption and decryption processes, vital for real-time data transmission and processing in wireless networks and big data applications. Their optimized design allows for swift cryptographic operations without compromising security.

12.1.2 Chapter Organization

The other sections are separated into five divisions: Section 12.2 discusses the general applications of wireless networks. Overview of the common security and privacy issues in wireless networks are discussed in Section 12.3. A case study of lightweight cryptographic for a secure wireless networks is presented in Section 12.4. Section 12.5 discusses the experimental results. The critical lesson learned from the chapter is discussed in Section 12.6, while Section 12.7 discusses the open issues and future research directions of the chapter. Finally, Section 12.8 concludes the chapter with results obtained from the experimental results.

12.2 THE GENERAL APPLICATIONS OF WIRELESS NETWORKS

Wireless networks have revolutionized communication and connectivity across various domains. Their applications are extensive and continually evolving. Wireless networks encompass a wide range of applications and variations, each tailored to specific needs and scenarios. Wireless networks have permeated almost every aspect of modern life due to their flexibility, convenience, and connectivity. The following are the various applications of wireless networks in different domains.

Mobile Communication: The primary application of wireless networks is in mobile communication, enabling voice calls, messaging, and data transfer through cellular networks (2G, 3G, 4G, and 5G) and technologies like Wi-Fi and Bluetooth [18–19]. Wireless networks play a pivotal role in mobile communication, providing connectivity and enabling various services on mobile devices. Wireless networks have fundamentally transformed mobile communication, providing the infrastructure that supports various applications and services essential in today's interconnected world. These networks, notably cellular technologies like 4G and 5G, are the backbone of mobile communication, enabling voice calls, messaging, and high-speed data transfer across smartphones and other mobile devices.

One of the primary applications of wireless networks in mobile communication is enabling high-speed Internet access on the go. Cellular networks, particularly 4G and the subsequent advancements in 5G technology, provide users with faster and more reliable connections, facilitating seamless access to the Internet, social media, video streaming, and other online services. These networks empower users to stay connected and access information from virtually anywhere, transforming how people communicate, work, and entertain themselves through mobile devices.

Moreover, wireless networks support various communication applications beyond voice calls and messaging. They enable video calling, conferencing, and multimedia messaging services, allowing users to engage in face to face communication, share content, and collaborate remotely. Additionally, wireless networks have facilitated the growth of mobile applications and services, providing a platform for a diverse range of apps that cater to productivity, entertainment, education, and numerous other purposes, further enhancing the capabilities and utility of mobile devices. These applications demonstrate how wireless networks have revolutionized mobile communication, providing users with seamless connectivity, access to information, and a wide array of services and functionalities on their mobile devices.

Internet of Things: Wireless networks play a fundamental role in the Internet of Things (IoT), enabling the interconnection of numerous devices and facilitating data exchange without the need for physical wired connections [20–21]. Wireless networks serve as the underlying infrastructure powering the vast and interconnected ecosystem of the IoT. This technology facilitates the seamless communication and data exchange among an extensive array of devices and sensors, enabling transformative applications across diverse industries [22–23].

One significant application of wireless networks in IoT is within smart homes and buildings. These networks enable the integration of smart devices like thermostats, security cameras, smart appliances, lighting systems, and voice assistants. Leveraging protocols such as Wi-Fi, Zigbee, Z-Wave, or Bluetooth, these devices communicate, allowing users to remotely control and automate various functions, enhancing convenience, energy efficiency, and security within residential and commercial spaces. Moreover, wireless networks play a pivotal role in Industrial IoT (IIoT) applications. In manufacturing plants and supply chains, sensors, actuators, and smart machinery are interconnected via wireless protocols. This connectivity enables real-time monitoring of equipment performance, predictive maintenance, process automation, and inventory management. Consequently, IIoT enhances operational efficiency, minimizes downtime, and streamlines production processes across various industries. The applications of wireless networks in IoT are extensive and diverse, impacting various sectors:

Smart Homes and Buildings: Wireless networks form the backbone of IoT ecosystems in smart homes. Smart heating and cooling systems, surveillance cameras, intelligent lighting controls, and digital appliances are examples of such gadgets, and voice assistants communicate through wireless protocols like Wi-Fi, Zigbee, Z-Wave, or Bluetooth. This connectivity enables users to remotely control and automate various functions, enhancing convenience, energy efficiency, and security.

Industrial IoT (IIoT): Wireless networks in IoT are integral to industries, facilitating the deployment of sensors, actuators, and smart devices in manufacturing plants, warehouses, and supply chains [24–25]. These networks monitor equipment performance, automate processes, optimize production lines, and enable predictive maintenance, leading to increased efficiency and reduced downtime.

Healthcare and Wearable Devices: Wireless connectivity is crucial in healthcare IoT, allowing Activity tracking devices, smartwatches, continual glucose monitoring, and therapeutic sensors are examples of wearable gadgets to real-time medical information is collected and transmitted to medical professionals or smartphone apps. Remote surveillance of patients, early illness identification, and individualized healthcare are all aided by these gadgets.

Smart Cities and Infrastructure: Wireless networks form the backbone of IoT applications in smart cities. Various sensors and connected devices, including smart streetlights, waste management systems, traffic monitoring systems, environmental sensors, and public safety devices, communicate via wireless networks. This connectivity aids in optimizing resource management, improving urban services, and enhancing overall city functionality.

Agriculture and Environmental Monitoring: Wireless networks enable IoT applications in precision agriculture, where sensors and drones equipped with environmental sensors gather information on soil moisture, climate, humidity, and crop health. This data helps farmers make choices based on data to maximize crop yields and avoid resource waste. Wireless connections are used by ecological tracking devices to measure contamination rates, weather conditions, and wildlife habitats for conservation purposes.

Logistics and Supply Chain Management: IoT devices powered by wireless networks track shipments, monitor inventory levels, and ensure goods' quality during transportation. Sensors, RFID tags, and GPS devices communicate through wireless protocols, providing real-time visibility and enabling efficient logistics operations.

Energy Management and Utilities: Wireless networks in IoT play a crucial role in smart grids and energy management systems. These networks facilitate the integration of smart meters, energy monitoring devices, and control systems, optimizing energy distribution, reducing energy consumption, and supporting the integration of renewable energy sources.

Overall, wireless networks are the cornerstone of IoT applications, enabling seamless connectivity, data transmission, and interaction between diverse devices and systems, thereby revolutionizing enhancing effectiveness, productivity, and taking decisions in numerous sectors processes.

Healthcare: Wireless networks have become integral to the modern healthcare system, revolutionizing patient care, medical device connectivity, and the overall efficiency of healthcare delivery. These networks facilitate various applications that significantly impact healthcare providers and patients alike. Monitoring patients remotely is one of the most common uses of wireless connections in healthcare. Wearable gadgets with sensors that gather and send real-time health data such as blood pressure, rhythm, glucose levels, and activity levels can be used by patients. These devices communicate via wireless connections to healthcare providers, allowing for continuous monitoring of patients' health conditions. Remote monitoring helps in early detection of abnormalities, enables proactive interventions, and reduces the requirement for regular in-person visits, which is especially advantageous for people suffering from chronic diseases or those who are elderly requiring post-operative care.

Moreover, wireless networks enable medical device connectivity within hospitals and healthcare facilities. Equipment such as patient monitors, infusion pumps, and electronic health records (EHRs) utilize wireless connectivity to seamlessly share patient data, giving utilization of reliable and current data for healthcare practitioners. This connectivity streamlines workflows, enhances communication between healthcare staff, and supports informed decision-making, ultimately improving patient care and safety [26].

Additionally, telemedicine and telehealth services heavily rely on wireless networks. Video teleconferencing, confidential communication, and remote access consultations, healthcare providers can connect with patients virtually, offering medical advice, diagnoses, and treatment recommendations. Wireless networks support the transmission of high-quality audio and video, enabling efficient communication between healthcare professionals and patients, particularly useful for rural or underserved areas where access to healthcare services may be limited. Wireless networks significantly improve healthcare systems by enabling remote patient monitoring, medical device connectivity, and telemedicine services. These applications enhance patient outcomes by facilitating proactive care and timely interventions, while also streamlining healthcare workflows for more efficient and effective delivery.

Entertainment and Media: Wireless networks have significantly transformed the landscape of entertainment and media, offering ubiquitous connectivity and accessibility to a plethora of content across various devices [27]. These networks underpin numerous applications that change the way we absorb and engage with media that entertains.

Streaming services have flourished due to the widespread availability of high-speed wireless networks. Platforms like Netflix, Amazon Prime Video, Hulu, and Disney+ leverage these networks to deliver on-demand movies, TV shows, documentaries, and original content to smartphones, smart TVs, tablets, and computers. The seamless transmission of high-definition video content over wireless connections has revolutionized entertainment consumption, allowing users to watch their favorite content anytime, anywhere.

Gaming is another area heavily influenced by wireless networks. Online gaming has thrived with the advent of high-speed Wi-Fi and cellular connections, enabling multiplayer experiences across consoles, PCs, and mobile devices. Games like Fortnite, PUBG Mobile, and Call of Duty: Mobile leverage wireless networks to facilitate real-time gaming experiences, connecting players worldwide and fostering a vibrant gaming community. Moreover, wireless networks have transformed how we access and enjoy music and other media content. Services like Spotify, Apple Music, and YouTube Music leverage these networks to offer vast libraries of music, podcasts, and other audio content to users on the go. Wireless connectivity allows seamless streaming and downloading of content to smartphones, ensuring users have access to their favorite tunes and podcasts wherever they are. In essence, wireless networks have fundamentally reshaped the entertainment and media industry by democratizing access to a wide array of content. From video streaming services and online gaming to music streaming platforms, these networks have enabled a new era of connectivity-driven entertainment experiences, empowering users to engage with diverse content across multiple devices at their convenience.

Education: Wireless networks have profoundly impacted the education sector by revolutionizing the way students access information, collaborate, and engage with educational resources. These networks play a crucial role in facilitating modern learning environments both within traditional classrooms and in remote or distance learning settings. In educational institutions, wireless networks, particularly Wi-Fi, provide ubiquitous connectivity, enabling students and educators to access a vast array of digital resources [28]. Students can use laptops, tablets, or smartphones to connect to the Internet, access e-books, educational websites, online research databases, and educational apps, fostering personalized and interactive learning experiences. Wireless networks also support collaborative tools and platforms that allow students to collaborate on projects, take part in virtual classes, and engage in conversations irrespective of their place of residence.

Business and Commerce: Wireless networks have become integral to businesses and commerce, offering flexibility, mobility, and enhanced connectivity in various operational aspects. These networks play a pivotal role in facilitating communication, transactions, and seamless operations within organizations, contributing significantly to their efficiency and productivity. One primary application of wireless networks in business is the establishment of robust and secure Wi-Fi infrastructure within office spaces. This infrastructure supports connectivity for computers, laptops, smartphones, printers, and other devices, allowing employees to access resources, collaborate, and communicate effectively. Wireless networks enable flexible work arrangements, allowing employees to work from various locations within the office premises, fostering collaboration and productivity.

Furthermore, wireless networks facilitate transactions and operations in commerce through mobile payment systems and wireless point-of-sale (POS) terminals. Retail businesses utilize wireless networks to process transactions, manage inventory, and provide customer service. Mobile payment solutions and contactless payments, powered by wireless connectivity such as NFC (Near Field Communication), enable seamless and secure transactions, offering convenience to both businesses and customers. Additionally, wireless networks support the integration of e-commerce platforms, enabling businesses to reach a broader audience and conduct online transactions securely. Wireless networks streamline operations, improve communication, and facilitate business transactions, offering flexibility, scalability, and mobility, enabling organizations to adapt to technological advancements and market changes.

Transportation: Wireless networks play a crucial role in revolutionizing the transportation sector, offering connectivity and advanced communication systems that enhance safety, efficiency, and convenience across various modes of transportation. These networks are integral to modern transportation systems, enabling real-time data transmission, navigation, and improved passenger experiences. One significant application of wireless networks in transportation is within the realm of intelligent transportation systems (ITS). These systems utilize wireless connectivity for Traffic management, vehicle-to-vehicle (V2V) interaction, and vehicle-to-infrastructure (V2I) interaction are all examples of V2V communication. Vehicles can connect with one another and with highway infrastructure via wireless connections, providing data regarding roadway conditions and road dangers, and optimizing traffic flow. This technology not only improves road safety by providing drivers with real-time warnings but also enhances traffic efficiency and reduces congestion.

Moreover, wireless networks play a pivotal role in enhancing passenger experiences in public transportation systems. Applications such as real-time passenger information systems, enabled by wireless connectivity, provide travelers with up-to-date information about schedules, routes, delays, and alternative options. Additionally, Wi-Fi connectivity in buses, trains, and airplanes allows passengers to access the Internet, work, or entertain themselves during their journeys, enhancing overall travel experiences. Wireless networks are crucial in transforming transportation by enabling smart, connected systems that improve safety, efficiency, and passenger satisfaction through real-time communication, intelligent choice-making and data sharing.

Agriculture: Wireless networks have revolutionized agriculture by enabling the deployment of precision farming techniques and sophisticated monitoring systems that optimize resource utilization and crop management. These networks facilitate the implementation of various smart agricultural solutions, allowing farmers making decisions based on data and increasing general efficiency.

One significant application of wireless networks in agriculture is through the utilization of wireless sensor networks (WSNs). These networks consist of interconnected sensors deployed across agricultural fields, collecting data on soil moisture, temperature, water content, crop development, and atmospheric variables in actual time. Through wireless connectivity, these sensors transmit data to a central system, supplying farmers with vital knowledge. Farmers will then be able to make educated judgments regarding irrigation schedules, crop health management, and the application of fertilizers or pesticides, as a consequence of which resource utilization is maximized, yields are increased, and the ecological impact is minimized.

Moreover, wireless networks facilitate the operation of drones and autonomous agricultural machinery in farming with pinpoint accuracy. Robots outfitted with cameras and sensors use wireless connectivity to capture aerial images and gather plant health information, pest infestations, and field conditions. This data, transmitted via wireless networks, helps farmers identify areas that need attention, enabling targeted interventions and efficient management practices. Additionally, autonomous machinery equipped with wireless communication capabilities enables precision planting, harvesting, and other agricultural operations, optimizing efficiency and reducing labor requirements. Wireless networks are revolutionizing agriculture by enabling precision farming, improving crop management, resource utilization, and productivity, and providing farmers with valuable insights and tools.

Military and Defense: Wireless networks play a critical role in military and defense applications, providing essential communication, surveillance, and connectivity solutions across various operational scenarios. These networks are vital components that enable secure and efficient data transmission, coordination, and strategic decision-making for defense forces globally.

One significant application of wireless networks in the military is secure communication systems. Military-grade wireless networks facilitate encrypted communication channels, safeguarding sensitive details privacy, reliability, and accessibility shared among troops, command centers, and other defense assets. These networks support real-time voice, data, and video transmissions, enabling seamless communication even in challenging and remote environments, contributing to operational efficiency and mission success.

Moreover, wireless networks are integral to military surveillance, reconnaissance, and battlefield management systems. Unmanned Aerial Vehicles (UAVs), ground-based sensors, and reconnaissance equipment utilize wireless connectivity to relay critical information about enemy positions, terrain analysis, and situational awareness back to military command centers. These networks enable the transmission of high-resolution imagery, sensor data, and real-time intelligence, empowering military decision-makers with valuable insights for strategic planning and tactical operations on the battlefield. Wireless networks are crucial in military and defense applications, providing communication infrastructure, data transmission, and operational enhancement, enhancing decision-making, and contributing to the success and safety of military missions.

Environmental Monitoring: Wireless networks play a significant function in environmental surveillance, allowing for real-time data collecting on various environmental factors and contributing to the assessment, protection, and conservation of ecosystems. These networks facilitate the deployment of sensors, remote monitoring devices, and data collection systems that aid in understanding and mitigating environmental issues.

One primary wireless connectivity network applications in tracking environmental conditions is through the use of wireless sensor networks (WSNs). These networks consist of interconnected

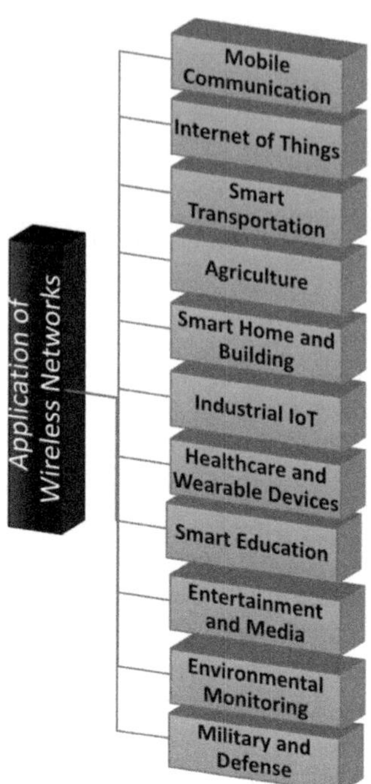

FIGURE 12.1   The general applications of wireless networks.

sensors deployed in remote or challenging environmental conditions to collect data on atmospheric water purity, soil moisture, and climate, humidity, and pollution levels. Through wireless connectivity, these sensors transmit collected data to central databases or monitoring stations, enabling environmental scientists and researchers to analyze the data, detect anomalies, and make informed decisions regarding environmental conservation and management.

Moreover, wireless networks enable the deployment of remote monitoring systems in natural reserves, wildlife habitats, and ecological zones. Cameras, acoustic sensors, and GPS tracking devices, equipped with wireless communication capabilities, assist in monitoring wildlife behavior, biodiversity, and ecological changes. This data, transmitted through wireless networks, aids in wildlife conservation efforts, ecosystem preservation, and early detection of environmental changes or threats, allowing for timely intervention and protection of sensitive ecological areas. Wireless networks are crucial for environmental monitoring, enabling data collection, transmission, and analysis, supporting conservation efforts, and contributing to sustainable natural resource management (see Figure 12.1).

12.3   OVERVIEW OF THE COMMON SECURITY AND PRIVACY ISSUES IN WIRELESS NETWORKS

These days, security is increasingly taking center stage in the debate over wireless network research. For example, mobile devices can cooperate with one another on wireless connections with limited bandwidth and organize themselves independently in an ad hoc smartphone network. In a mobile area network, mobile nodes can function as both network hosts and network routers; a network host is used for sending and receiving packets, while a network router handles packet routing [29]. The smartphone ad hoc network's flexible architecture is constantly evolving. Numerous scholars have put forth several distributed algorithms that ascertain network organization, routing, and link scheduling. On the other hand, several security-related issues arise from the special characteristics of wireless networks. Prior to focusing only on the major security flaws in the wireless network that are the main cause for concern, the following briefly illustrates the general issues:

a. *Insecure Wireless Environment:* The fact that processing is done everywhere and has widespread access is one of the main problems with utilizing wireless networks [30]. Multiple kinds of proximity services are commonly utilized with the diverse range of services provided in a wireless environment, requiring users to rely on many identities to access these services. This occurrence makes the wireless environment extremely vulnerable in terms of security, particularly when it comes to the networking equipment they use. Validating and approving users primarily present in a wireless network is challenging due to similar features that apply to them.

b. *Absence of Central Points:* Entry and exit points, such as routers and gateways, are not conceptualized in wireless networks [31]. The absence of such a consolidated location makes network attack detection extremely difficult, particularly in the context of a wireless network environment that is highly dynamic and predominantly scaled [32].

c. *Constrained Resources:* The mobile nodes that are part of the wireless network system are often battery-operated, tiny devices with little processing power. One of the main causes of a node's selfish behavior is its limited power supply, especially when it senses that its power supply has dropped below a certain point [33]. Furthermore, a wireless network system channel capacity for data transport is severely constrained.

d. *Node Mobility:* The term "mobile" in networking not only denotes sophisticated tendencies but also the related complexities. Although nodes in wireless networks are movable, they can travel in any direction at any moment, leave the network on their own, and join it; thus, the network topology is dynamic, since it is subject to rapid change. Node mobility causes frequent disruptions in the link; similarly, to handle highly dynamic topologies, conventional Intrusion Detection System (IDS) methodologies become inadequate [34–36].

e. *Scalability:* In conventional wired networks, the amount of bandwidth is essentially set before implementation and is rarely altered while in operation. This is not the case, however, with wireless networks, whose magnitude is believed to be uncertain or to be constantly changing over time. Therefore, it is difficult to anticipate the status (variety of connections and network structure), future state, of the network, therefore associated protocols and services need to be appropriately suited to its constantly changing scale.

f. *Nature of Unpredictability:* Anywhere on Earth can employ a mobile ad hoc networking system as long as there are several nodes with routing capabilities. Because of this, rendering this scenario in real-time applications will need that there be numerous nodes where it is nearly impossible to, and occasionally challenging, to determine which node is authentic [37]. The network fluctuating scale, randomly migrating nodes, and dynamically altering linkages all add to the unpredictable asset.

g. *Interference:* Signals from other devices, physical obstructions, or competing networks can interfere with wireless transmissions, leading to degraded performance or connectivity issues. Interference in wireless networks refers to the disruption or degradation of the wireless signal caused by various external or internal factors. This interference can significantly impact the performance, reliability, and speed of wireless communications.

h. *Security Concerns:* Wireless connections are vulnerable to security risks such as illegal access and data eavesdropping, and hacking. Encryption methods and secure configurations are essential to prevent breaches. Security concerns in wireless networks are paramount due to the inherent nature of wireless transmissions, which can be stolen more readily than connected ones. Unidentified individuals can intercept wireless communications, exposing them to listen in on critical information being carried across the network. Predators could try to obtain illicit access to a wireless connection by leveraging safety protocol flaws or by using brute-force attacks to crack weak passwords.

i. *Signal Range and Coverage:* The coverage area of a wireless network is limited by the signal range of the access points or routers. Dead zones or weak signal areas can occur, impacting connectivity. Signal range and coverage in wireless networks refer to the distance over which a wireless access point or router can effectively transmit its signal and provide connectivity to devices. The transmission power of the access point or router determines the signal strength and consequently affects the coverage area. Higher transmission power typically results in a larger coverage area but can also lead to interference in densely populated areas. Different frequency bands, such as 2.4 GHz and 5 GHz, have varying ranges. For instance, 2.4 GHz signals can travel farther distances but may be more susceptible to interference, while 5 GHz signals have shorter ranges but offer higher data transfer speeds and less interference. Physical obstacles like walls, buildings, electronic devices, or environmental factors can obstruct or weaken the wireless signal, reducing the effective coverage area and signal strength.

j. *Bandwidth Limitations:* High user demand can strain the available bandwidth of a wireless network, causing slower speeds and congestion, especially in densely populated areas. Bandwidth limitations in wireless networks refer to the maximum data transfer rate or capacity that the network can handle at any given time. Different wireless standards (e.g., 802.11n, 802.11ac, 802.11ax) offer varying maximum theoretical speeds. Older standards generally have lower maximum speeds compared to newer ones. In densely populated areas or locations with multiple wireless networks operating in the same frequency band, channel congestion can occur. This congestion leads to interference and reduced available bandwidth for each network, impacting performance. Interference from other devices operating in the same frequency range, physical obstacles, or environmental factors can degrade the quality of the wireless signal. This interference can reduce the effective bandwidth available for data transmission. As devices move farther away from the access point or router, the signal strength weakens. Reduced signal strength can lower the achievable data rates, affecting the available bandwidth for connected devices.

k. *Reliability and Stability:* Reliability and stability in wireless networks refer to the consistency, dependability, and robustness of the network's performance over time. Wireless connections may suffer from instability due to environmental factors, device interference, or technical issues, leading to dropped connections or inconsistent performance. Adequate signal strength and coverage throughout the desired area are essential for reliable connectivity. Weak signals or dead zones can lead to dropped connections or intermittent connectivity issues. Interference from other devices, neighboring networks, or physical obstacles can disrupt wireless signals, causing instability. Additionally, network congestion due to numerous connected devices can impact stability by reducing available bandwidth. Implementing QoS mechanisms allows for the prioritization of certain types of traffic (such as voice or video data) over the network, ensuring a consistent user experience for critical applications.

l. *Device Compatibility:* Compatibility issues between different devices or network protocols can create complications in establishing connections or achieving optimal performance. Device compatibility in wireless networks refers to the capacity of numerous devices to connect, interact, and share information function effectively within the network environment. Compatibility issues can arise due to differences in hardware, software, or protocols among devices. Different devices may support different wireless standards (e.g., 802.11n, 802.11ac, 802.11ax) or protocols, leading to compatibility issues when trying to connect to the network. Older devices might not support newer standards or features, limiting their compatibility. Devices operating on different frequency bands (2.4 GHz or 5 GHz) might encounter compatibility issues, especially if one device operates only on a specific frequency band that others do not support. Outdated drivers or firmware on devices can lead to compatibility problems with the network, affecting connectivity and performance. Some devices may have hardware limitations that prevent them from supporting certain network features, leading to compatibility issues when attempting to utilize those features.

However, because wireless networks have a dispersed topology, it is far more difficult to detect intruders, so that hackers can take advantage of the weak points in wireless networks through malevolent actions. The recent advancements in wireless networks, which provide quick network deployment and interaction without predetermined setup, are also relevant for a variety of IoT-based application fields in smart communities [38–40]. There have been numerous attacks documented at nearly every tier of the wireless network – for example, network, MAC or data link, application, transport, and physical. These networks are vulnerable to a number of assaults that damage mobile nodes' trust and reputation [41–43]. The cost of the infiltration and subsequent attacks, which seriously impair communication and have a negative effect on application performance overall, eavesdropping, huge intrusion, and data loss, falls on the dependable mobile nodes.

When considering a wireless network system on a big scale, like in Internet of Things metropolitan settings, the cost of an attack scenario is substantially higher. Reviewing relevant publications revealed that there are much more routing protocol-based solutions than other security-based mechanisms used in wireless networks. However, it should be acknowledged that if routing-based techniques are the only ones prioritized, then hostile conduct in large-scale wireless network applications may go unnoticed. This is due to the fact that while automating the routing strategy, a number of features, like node behavior and the dynamics of methods on various nodes are difficult to solve. Table 12.1 shows the security and privacy commonly associated with wireless networks.

TABLE 12.1
The Wireless Connection Safety and Confidentiality Concerns

Issue	Description	Mitigation
Eavesdropping	Unauthorized interception of wireless transmissions, leading to the exposure of sensitive data. This in wireless networks is to the unauthorized interception of wireless transmissions, allowing a third party to monitor, capture, or access data being transmitted over the network. This activity is a significant security and privacy concern as it can lead to the exposure of sensitive information, including personal data, passwords, financial details, or confidential business communications. Wireless transmissions, such as Wi-Fi signals, travel through the air and can be intercepted by individuals using specialized equipment or software. Attackers eavesdrop on these transmissions to gather information without the knowledge or consent of the parties involved in the communication. They may exploit weaknesses in the encryption protocols, capture unencrypted data, or use sophisticated techniques to decrypt encrypted data.	Encryption, Virtual Private Networks (VPNs), Secure Authentication, Regular Security Updates, Monitoring and Intrusion Detection,

(continued)

TABLE 12.1 (Continued)
The Wireless Connection Safety and Confidentiality Concerns

Issue	Description	Mitigation
Rogue Access Points	Unsecured points of entry are created by hackers to impersonate genuine systems, enabling data transmission interception or manipulation. This issue represents a significant security threat in wireless networks. These unauthorized access points are set up by individuals or attackers without permission from the network administrator or organization. Rogue APs can lead to several security and operational issues such as Unauthorized Access, Data Interception, Man-in-the-Middle Attacks, and Network Performance Issues	Wireless Intrusion Detection Systems (WIDS), Regular Scanning and Auditing, Secure Configurations and Policies, and Geolocation and Monitoring Tools
Denial-of-Service (DoS) Attacks	Deliberate attempts to disrupt or overload a network's resources, causing legitimate users to be unable to access services. The attacks in wireless networks aim to disrupt the normal functioning of network services by overwhelming or disabling them. These attacks target the availability of the network, rendering it inaccessible to legitimate users. DoS attacks pose a significant threat to wireless networks' availability. Implementing a combination of preventive and reactive measures can help mitigate the impact of such attacks and ensure network resilience. Regular monitoring and updating security protocols are crucial in mitigating the risks associated with DoS attacks.	Intrusion Discovery and Preventative Systems (IDPS), Traffic Sorting, and Network Separation are all examples of network security measures, Rate Limiting, Jamming Detection and Prevention, and Behavioral Analysis
Man-in-the-Middle (MitM) Attacks	Interception and modification of two-way communication, permitting hackers to listen in or change data. This attacks in wireless networks involve a Without the parties' expertise, an outsider intercepts and ultimately modifies their conversation. These assaults can occur in various forms within wireless environments, exploiting vulnerabilities in the network to intercept, read, or modify data. Implementing various mitigations preventive measures may considerably minimize the dangers of MitM attacks in wireless connections while improving general network security. To counteract emerging threats, regular upgrades and preventive security measures are required in wireless environments	Encryption, Secure Authentication, Digital Certificates, ARP Spoofing Prevention, Continuous Monitoring, and Security Awareness
Wi-Fi Phishing	Wi-Fi phishing, also known as "evil twin" attacks or Wi-Fi network spoofing, is a type of cyber-attack that involves the creation of fraudulent wireless networks that appear legitimate. Attackers set up fake access points (APs) that mimic genuine Wi-Fi networks, tricking users into connecting to them. False Wi-Fi networks created to trick users into connecting, enabling attackers to steal login credentials or sensitive information.	Verify Network Authenticity, Avoid Unsecured Networks, Use Virtual Private Networks (VPNs), Security Awareness Training, and Two-Factor Authentication (2FA)
Insider Threats	Insider threats in wireless networks refers to the safety hazards created by persons within a corporation who granted access to sensitive information the network infrastructure, data, or systems, and deliberately or inadvertently compromise security. These threats can be particularly challenging to detect and mitigate because insiders already have some level of access and familiarity with the network. Malicious activities initiated by authorized users within the network, posing a significant security risk.	Access Control and Monitoring, Employee Training and Awareness, Encryption and Segmentation, Regular Security Audits and Reviews, Behavioral Analytics, and Incident Response Plan.

TABLE 12.1 (Continued)
The Wireless Connection Safety and Confidentiality Concerns

Issue	Description	Mitigation
Lack of Encryption	Failure to implement strong encryption methods, leading to data transmissions being vulnerable to interception and exploitation. The lack of encryption in wireless networks poses significant security risks as it leaves data transmitted over the network vulnerable to interception and unauthorized access. Without encryption, data such as login details, passwords, and financial data, and sensitive business data can be easily captured by malicious actors, leading to various security threats. wireless networks	Enable Encryption Protocols, Regularly Update Security Settings, Implement Virtual Private Networks (VPNs), Segregate Network Traffic Strong Authentication Mechanisms, and Regular Security Audits.
Weak Authentication	Insufficient authentication mechanisms allowing unauthorized users to gain access to the network resources. Weak authentication mechanisms in wireless networks pose a significant security risk as they can enable unauthorized access to network resources, compromise sensitive data, and leave the network vulnerable to various attacks. Weak authentication can take several forms, including default passwords, easily guessable credentials, or outdated authentication protocols, and it can lead to several security threats. By implementing strong authentication measures and staying vigilant against potential threats, organizations can significantly reduce the risk of unauthorized access and enhance the security posture of their wireless networks.	Use Strong Authentication Protocols, Implement Complex Password Policies, Multi-Factor Authentication (MFA), Regularly Update and Patch Devices, and Monitor and Audit Authentication Logs.
Lack of Patching/ Updates	Failure to regularly update firmware, software, or security protocols, leaving systems vulnerable to known exploits and vulnerabilities. The lack of patching or updates in wireless networks can create significant security vulnerabilities and expose these networks to various risks and potential threats. Failure to regularly update devices, access points, routers, and other components within the wireless infrastructure can lead to several security issues. By prioritizing regular updates, implementing a robust patch management strategy, and staying proactive in addressing vulnerabilities, risks linked with enterprises can be considerably reduced unpatched wireless networks and enhance overall network security.	Regular Patch Management, Vendor Support and Notifications, Network Segmentation, Vulnerability Assessments and Penetration Testing, Implement Change Management, and Backup and Recovery Plans
Physical Security	Vulnerabilities arising from physical access to devices or infrastructure components, leading to potential data breaches. Physical security plays a crucial role in protecting wireless networks, as the physical infrastructure and devices themselves can be vulnerable to various threats. Businesses may build a more secure environment by integrating strong physical safety precautions with thorough cybersecurity policies resilient and secure wireless network environment, safeguarding against potential physical threats and unauthorized access to network infrastructure.	Physical Security Audits, Remote Site Security, Disposal and Decommissioning, Asset Inventory and Tracking, Employee Training on Physical Security, Visitor and Guest Access Control, Physical Cable Security, Protection from Environmental Factors, Surveillance and Monitoring, Secure Placement of Access Points, and Access Control to Network Equipment.

12.4　A CASE STUDY OF LIGHTWEIGHT CRYPTOGRAPHIC FOR A SECURE WIRELESS NETWORK

The proposed lightweight cryptographic use the Elliptic Curve Cryptography to secure wireless networks. Elliptic Curve Cryptography (ECC) is based on the mathematics of elliptic curves over finite fields. The mathematical model for ECC involves the algebraic structures associated with elliptic curves.

An elliptic curve is expressed by the formula $y^2 = x^2 + ax + b$, where a and b are fixed, and the curvature has certain mathematical properties. The set of points (x, y). An additive set is made up of points that fulfill the curve formula and a point at infinite.

Given two points P and Q on the elliptic curve, the sum $P + Q$ is another point on the curve. The addition operation is defined geometrically and has algebraic expressions involving the coordinates of the points. The addition operation is designed such that the resulting structure forms a group, and it is associative, commutative, and has an identity element (the point at infinity).

Scalar multiplication involves repeatedly adding a point to itself a certain number of times. Given a point P and a scalar k, the result of k. P is obtained by adding P to itself k times. Scalar multiplication is a fundamental operation in ECC and serves as the basis for key generation, encryption, and digital signatures.

Key Pair Generation: In ECC, a user's public and private keys are generated as points on the elliptic curve. The secret key is a selected at random scalars, while the public key is the outcome of the secret key's scalar combination of a base point (generator point).

Diffie-Hellman Key Exchange: The Elliptic Curve Diffie-Hellman (ECDH) key exchange involves each party selecting a private key, generating a public key, and then exchanging public keys. The shared secret is derived through scalar multiplication of the other party's public key with one's own private key.

Digital Signatures: ECC is employed for electronic signatures in which a secret key is utilized for signing a message and the associated public key is used to validate the signature. The mathematical procedures associated with signing and confirming are based on elliptic curve features. ECC safety is based on the complexity of the elliptic curve discrete logarithm issue, which entails locating the secret key with the public key and the curve's base point. The mathematical model for ECC involves the algebraic structures of elliptic curves, the geometric operations on these curves, and the application of scalar multiplication in various cryptographic protocols.

12.4.1　ENCRYPTION PROCESS

The mapping procedure guarantees that the resulting point is on the elliptic curve. The SSK, an elliptic curve point, is enlarged using the points obtained by the mapping procedure. Algorithm 1 describes the encryption procedure.

Algorithm 1: Encryption Process ECCA
Module ECCencryption(SSK, mappedpoints)
Output: encryptedpoint

1. START
2. INITIALISE array of encrypredpoint
3. FOR i = 1 to LENGTH(mappedpoints)
        encryptedpoint[i] = mappedpoint[i] + SSK
4. STOP

ECC is frequently used for transferring keys and electronic signatures, and encryption. When it comes to encryption, ECC is often used in combination with symmetric encryption algorithms rather than being a standalone encryption method. One common scenario is using ECC for transferring keys, followed by symmetrical encryption utilizing the transferred key. It is worth noting that ECC itself is not a symmetric encryption algorithm. Instead, It is employed with symmetrical encryption methods to provide a safe interaction route. The combination of ECC for key exchange and a symmetric encryption algorithm for data encryption offers an excellent blend of safety and effectiveness. The specific symmetric encryption algorithm and mode (e.g., AES in CBC mode) can vary depending on the application and security requirements. The use of ECC for key exchange, combined with symmetric encryption, is a common approach in protocols like TLS (Transport Layer Security) for securing communication over the wireless networks.

12.4.2 Decryption Process

To acquire the decrypted point, the encoded region and SSK are sent into the decoding procedure, which it then subtracts from the encrypted point. The steps of the decryption algorithm are described in Algorithm 2.

Algorithm 2: Decryption ECCA scheme
Module ECCdecryption(SSK, encryptedpoint)
Output: decryptedpoint

1. START
2. INITIALISE array decryptedpoint of decrypted points
3. FOR i = 1 to LENGTH(encryptedpoint)
        decryptedpoint[i] = encryptedpoint[i] – SSK
4. STOP

The shared secret obtained from the ECDH process is used as an input to A Key Derivation Function (KDF) and is used to generate one or more symmetrical keys. The resulting symmetrical key is then applied for symmetric decryption using a symmetric encryption algorithm (e.g., AES). The ciphertext (encrypted data) is broken down into segments, and each block can be decoded with the symmetrical key. When the symmetrical decryption is completed, the original plaintext data is obtained.

ECC is primarily utilized to trade keys, and derived symmetrical key from this process is then used for symmetric encryption and decryption. The symmetric decryption process itself depends on the specific symmetric encryption algorithm and mode used. It is worth noting that ECC-based key exchange is commonly employed in secure communication protocols like TLS, where it facilitates the creation of a safe connection between two parties. The symmetric key derived from the key exchange is then used for efficient and secure data decryption using symmetric encryption algorithms.

12.4.3 Performance Parameters

To secure the connectivity of wireless networks nodes to extend the connection's lifespan, the suggested plan leverages four essential performance metrics in the evaluation:

Execution speed of ECC algorithm: The performance of our ECC technique employing the AES cipher is evaluated using a variety of cryptographic parameters [44], and it is contrasted with other lightweight encryption algorithms and traditional methods.

Round numbers and block size: the influence of varied quantity of cycles and sizes of blocks on system longevity is investigated.

Nodes power consumption and residual power: numerous simulations timeframes are used to examine the influence of the ECC dynamic framework on wireless node utilization of power and leftover energy as compared to TEA.

Network lifetime: to investigate the influence of the ECC dynamic approach on network longevity, several beginning energy values are explored.

12.5 THE EXPERIMENTAL RESULTS

To give certain wireless network nodes time to reach 1, the experiment begins at 20 seconds. In the first case, the mapped ECC cipher was used to assess how the time taken to execute is affected by the cycle frequency and block size. 128 bytes with round sizes that vary from 2 to 32 and block sizes varying from 4 to 128 bits are used to calculate the encryption processing time. Figure 12.3 shows that the encryption time grows exponentially as the round quantity and block dimensions get larger. When r was equal to 2 and b was 4 bits in size, the encoding duration was 0.67 milliseconds. When r was equivalent to 32 and b was equivalent to 128 bits, this increased to 5.12 milliseconds. The time spent by the repeated routines, which grows as the number of blocks and rounds increases, is what is causing this increase in execution time. Therefore, it may be successfully regulated by the computing the key issues with limited in resources devices are cost, battery life, storage use, and network latency by using variable cryptographic settings.

The enhanced ECC cipher is evaluated for its performance efficiency in comparison to the AES encryption technique and the TEA lightweight cipher. The average of the estimated processing times obtained by encoding data of varying sizes with block counts that vary from 4 to 128 and cycles varying from 4 to 32 is used here. The outcomes are contrasted with encryption timings that were attained following the use of the TEA and AES techniques for encrypting data. AES is a 128-bit block cipher with ten to twelve cycles with key lengths of 128 bits, 192 bits, and 256 bits. The TEA

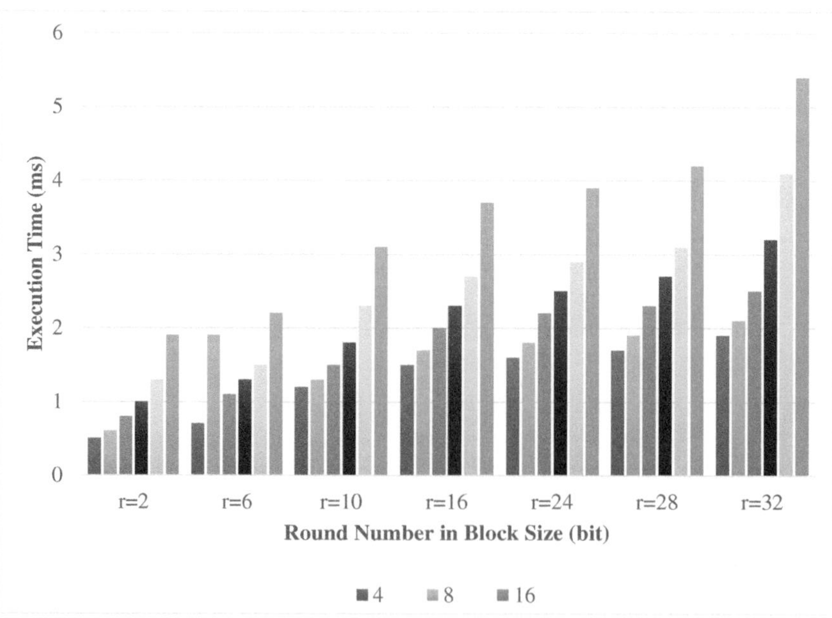

FIGURE 12.2   The encrypted data of 128byte data in milliseconds in execution time utilizing various block dimensions and round digits.

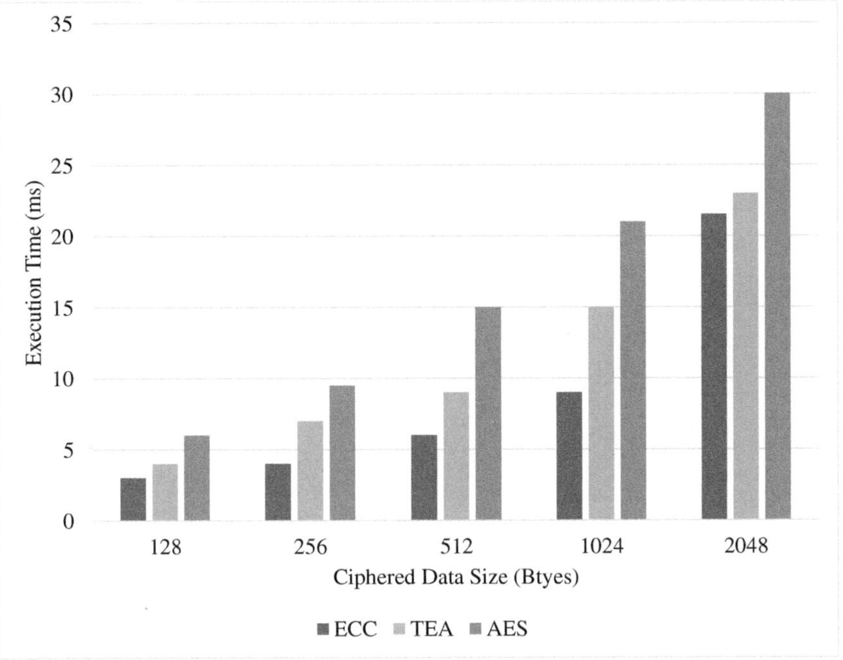

FIGURE 12.3 The encryption times in milliseconds for ECC, TEA and AES-256 using data of various sizes.

is a tiny cipher with an entire block size of 64 bits and a key size of 128 bits. It employs a Feistel structure with 64 rounds. Encoding times are compared for the TEA, AES-256, and ECC ciphers is shown in Figure 12.3. Based on the results, we can see that the ECC cipher works better than TEA and AES-256, which both encrypt data using constant parameters. The ECC cipher has an average encryption throughput of 91.23 byte/ms, surpassing AES-256 and TEA ciphers' 62.36 and 40.78 byte/ms respectively.

We examine the impact in the subsequent circumstance, the flexibility of the total number of cycles and the quantity of blocks on the network's durability. To investigate the effect of cycle amount and block dimension on network lifetime, we set the total amount of cycles to 10 and the block size to 8 bits. Figure 12.4 depicts the impact of altering the number of cycles and block size on network lifespan with regard to network bandwidth and packet dimensions.

As shown in figure 4(a), the network lifetime decreases from 11.8s to 3s with increasing encryption complexity, affecting power consumption and illustrating a decrease in the round number from 10 to 40. The increase in encryption complexity leads to a decrease in network lifetime, which in turn reduces power consumption. In Figure 4(b), the network lifespan drops from 19.7s at 4 bits block size to 2.5s at 64 bits block size. Increased block size reduces network lifespan by a factor of two, owing to increased energy use throughout encoding and key administration activities.

The third step of the investigation looks at the effect of flexible cryptography parameter choice on network node electrical usage and leftover energy. Therefore, we can contrast the outcomes of the suggested scheme, which allows for discretionary choice of encrypting cycle frequency and block dimension with those of the identical scheme, which employs a fixed number of rounds and block size at various simulation timeframes. The AES and TEA model has a set number of sixteen rounds and a block size of 32 bits [45]. In the meantime, the block size and number of rounds in the suggested elastic ECC work vary from 4 to 32 bits. $\lambda=12, \lambda1=8, and\ \lambda2=4$ are the range from which the block size is chosen. The selection of round number and block size values is based on ω value, as shown in Figure 12.5, which analyzes network power consumption. Consider the residual

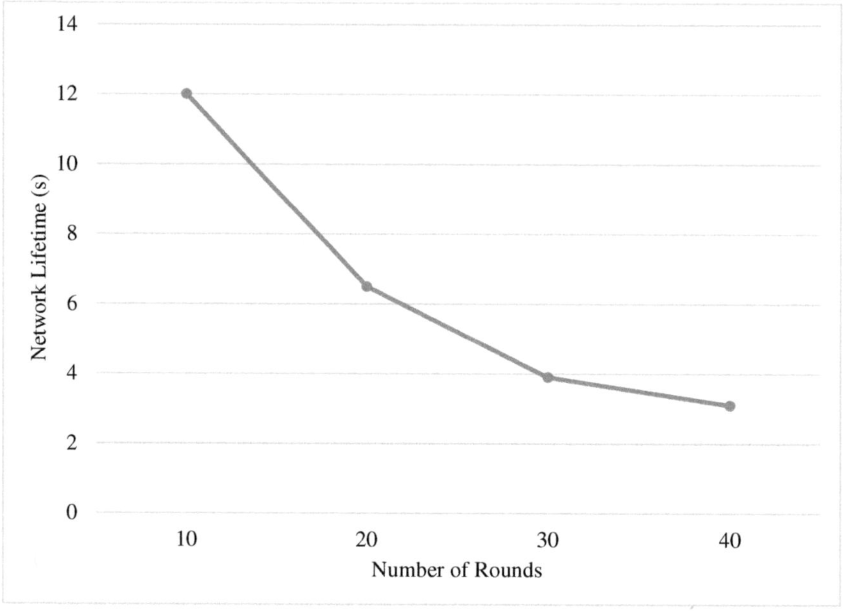

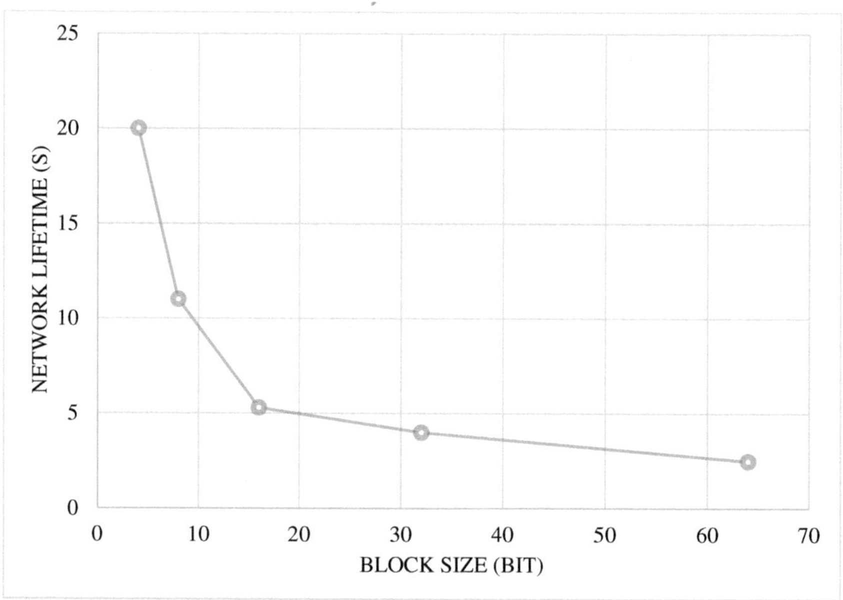

FIGURE 12.4 the impact of changing (i) the amount of cycles and (ii) the block size on network lifespan.

as the simulation time increases from 10 to 40 seconds for all wireless network nodes in relation to the used packet size.

The power consumption increases with the increase in simulation time, as depicted in Figure 12.5. This increase is the result of a positive correlation between the amount of power used and the lengthening of use. The residual power decreases in a energy and time usage have a beneficial link, as depicted in Figure 12.6. The proposed technique, which dynamically selects round and block sizes, significantly reduces over all simulation timeframes, this reduces network electrical

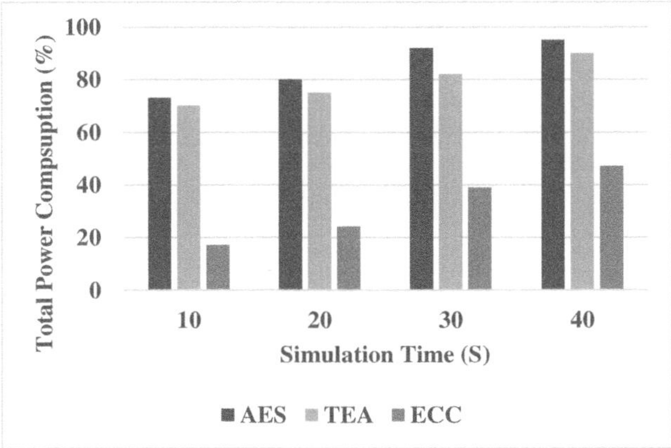

FIGURE 12.5 Simulation time based on network power consumption.

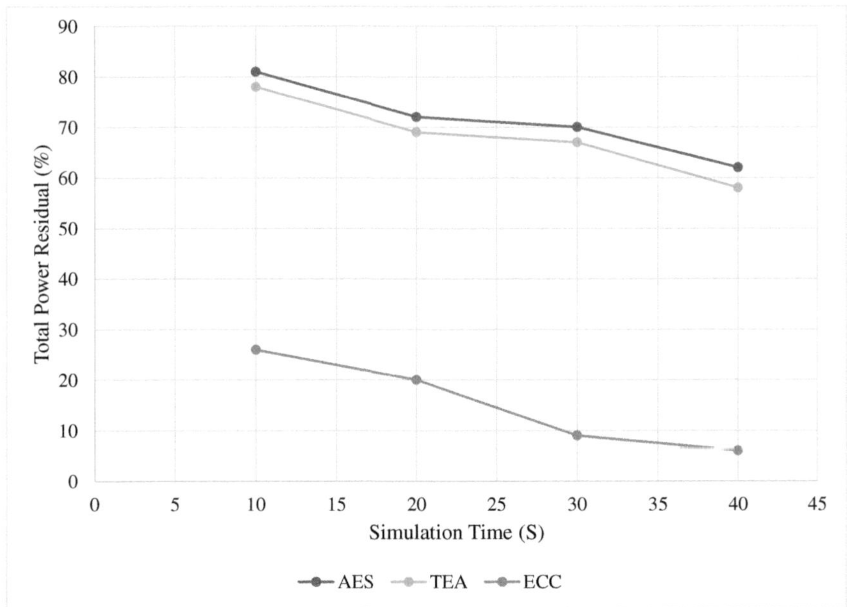

FIGURE 12.6 Simulation time based on network power residual.

consumption by 66% and boosts leftover energy by 79%. The proposed model determines the quantity of cycles and block dimensions for each node depending on leftover energy and the distance to its present position.

The outcome demonstrates that when the initial power of the network nodes grows, ECC, TEA, and AES network lifetimes increases. The results, however, clearly show that our flexible ECC scheme has significantly extended the network lifetime by around 89%, 97%, and 93% in comparison to AES and TEA, respectively. It does this by dynamically selecting as the node sends data, the amount of cycles and the dimension of blocks. This illustrates how such a technique (ECC) can be used to preserve data security while allowing wireless networks to operate for as long as possible in difficult environments.

12.6 CRITICAL LESSONS LEARNED

Implementing an enhanced lightweight cryptographic algorithm for securing wireless networks involves several critical lessons learned, which are essential for achieving robust security without compromising the performance and efficiency of the network. Here are key lessons:

1. *Performance Optimization:* Efficient Algorithms: Developing cryptographic algorithms that strike a balance between security and performance, ensuring encryption and decryption processes are streamlined. Implementing key generation processes that are optimized for minimal computational overhead.
2. *Low Memory Footprint*: Designing cryptographic algorithms with minimal memory requirements, using compact data structures to store keys and intermediate results. Optimizing cryptographic operations to minimize memory consumption, especially in resource-constrained devices.
3. *Energy Efficiency:* Integrating power-aware design principles to minimize energy consumption during cryptographic operations. Implementing mechanisms for dynamically adjusting cryptographic strength based on the available energy resources.
4. *Resistance against Attacks*: Incorporating countermeasures against power evaluation and timing assaults are examples of side-channel assaults. Putting in place strong key management processes, such as protected key preservation, movement, and dissemination.
5. *Scalability*: Developing scalable key management strategies suitable for large-scale wireless networks. Creating algorithms that can adapt to changes in network size and topology without compromising security.
6. *Interoperability*: Ensuring compatibility with cryptographic standards to facilitate interoperability with various devices and network protocols. Providing implementations that work seamlessly across different devices, operating systems, and network configurations.
7. *Usability and User Acceptance*: Designing key management processes that are intuitive and user-friendly. Ensuring that cryptographic operations have low latency and do not disrupt the overall user experience of network applications.
8. *Continuous Monitoring and Adaptation*: Conducting periodic security inspections to discover and correct possible flaws. Implementing mechanisms that can adapt to evolving threats and security requirements over time.
9. *Integration with Emerging Technologies*: Ensuring that the cryptographic algorithm is compatible with and can secure emerging Internet of Things (IoT) and 5G wireless networks are examples of such technology.
10. *Ongoing Research and Development*: Remaining vigilant about new developments in cryptographic attacks and countermeasures. Engaging in ongoing research and development to enhance the algorithm's security, efficiency, and adaptability.

These lessons are crucial for the successful implementation and deployment of lightweight cryptographic algorithms in wireless networks, providing a foundation for achieving strong security while meeting the unique challenges of resource-constrained environments.

12.7 OPEN ISSUES AND FUTURE RESEARCH DIRECTIONS

Here are some open issues and potential future research directions in the context of lightweight cryptographic algorithms for securing wireless networks and big data. It should be noted that cryptographic is a continuous science, and new breakthroughs may have emerged since then.

12.7.1 Open Issues

- *Security-Performance Trade-offs:* Striking a balance between lightweight designs and maintaining a high level of security remains a challenge. Further research is needed to explore innovative approaches that optimize cryptographic performance without compromising security.
- *Quantum-Resistant Algorithms*: With the potential advent of quantum computing, there's a need for lightweight cryptographic algorithms that are resistant to quantum attacks. Research should focus on developing quantum-resistant primitives suitable for resource-constrained devices.
- *Post-Quantum Cryptography in Wireless Networks*: Investigating the impact of post-quantum cryptographic algorithms on the performance of wireless networks and developing lightweight solutions that can withstand quantum threats.
- *Efficient Key Management*: Lightweight key management schemes need to be designed and evaluated for wireless networks and big data environments, ensuring secure key distribution, storage, and update mechanisms without overwhelming computational and communication resources.
- *Side-Channel Resistance*: Continued efforts are required to address side-channel attacks in lightweight cryptographic algorithms, especially in scenarios where power consumption and timing information can be exploited by adversaries.
- *Dynamic Network Environments*: Adapting lightweight cryptographic algorithms to dynamic changes in wireless network topologies, such as node additions, departures, or changes in communication patterns.
- *Scalability to Big Data*: Extending lightweight cryptographic algorithms to handle the unique challenges posed by big data, including efficient processing of large datasets and scalable key management in distributed big data environments.

12.7.2 Future Research Directions

- *Homomorphic Encryption for Big Data:* Exploring the integration of homomorphic encryption techniques with lightweight algorithms to enable secure computation on encrypted big data without decryption, opening avenues for privacy-preserving analytics.
- *Machine Learning and Cryptography Integration:* Investigating the intersection of machine learning and cryptography for enhancing security solutions in wireless networks and big data, such as using machine learning for anomaly detection in encrypted data.
- *Blockchain and Lightweight Cryptography:* Studying the synergy between lightweight cryptographic algorithms and blockchain technology to provide secure and efficient solutions for decentralized and distributed applications, including secure data sharing.
- *Secure Multi-Party Computation (SMPC):* Advancing research on lightweight SMPC protocols for secure computation over distributed data, allowing multiple parties to jointly analyze data without revealing sensitive information.
- *Energy-Efficient Cryptography in IoT:* Developing lightweight cryptographic solutions tailored specifically for IoT devices, with a focus on minimizing energy consumption while ensuring strong security.
- *Standardization and Interoperability:* Working towards standardization of lightweight cryptographic algorithms to promote interoperability and facilitate their adoption across diverse wireless networks and big data platforms.
  - *Privacy-Preserving Data Aggregation:* Researching techniques for secure and in wireless connections, effective aggregating of encrypted information and big data scenarios to enable meaningful analysis while preserving privacy.

- *Robustness Against Advanced Attacks:* Investigating the resilience of lightweight cryptographic algorithms against emerging and advanced attacks, including machine learning-based attacks, to ensure long-term security.

Continued partnership between scholars and practitioners in the business, and standardization bodies is essential to address these open issues and propel the development of effective lightweight cryptographic solutions for securing wireless networks and big data.

12.8   CONCLUSION

In recent years, the application of wireless networks and Big Data has significantly increased. The use of wireless networks and big data in everyday life has made them vulnerable to various cyber-attacks, data breaches, and privacy issues. It is therefore essential to explore technologies that can effectively secure these networks and data. This chapter presents a comprehensive look into the current state of security measures of wireless networks and big data, as well as the recommendations for improvements. The discussion focused on using innovative computational models and lightweight cryptographic techniques to ensure a secure wireless network and Big Data environment. The first section of the chapter outlined the crucial aspects of wireless networks and Big Data that necessitate security measures. This included topics such as authentication, encryption, and authorization. Attention was given to issues that may arise depending on the device or environment, as well as any security concerns with the underlying applications. The second section move on to discussed the current security practices for wireless networks and Big Data. Considerations was given to existing security protocols, infrastructure design, and risk management. This section then addresses the approaches for improving the security system, for example, using methods such as artificial intelligence, cryptography, and machine learning. The third section of the chapter focused on a present case study using lightweight cryptographic technique for securing wireless networks and Big Data. A comparison is made between these methods and conventional security practices. The lesson learned, open issues and future research trends were discussed. Finally, the chapter provided a conclusion on the state of security practices for wireless networks, as well as recommendations for the future.

REFERENCES

[1] Awotunde, J.B., Imoize, A.L., Adeniyi, E.A., ...Jimoh, R.G., & Falola, P.B. (2023). Artificial intelligence-enabled security Systems for 6G wireless networks: Algorithms, strategies, and applications. *Security and Privacy Schemes for Dense 6G Wireless Communication Networks*, 2023, pp. 61–87.

[2] Ren, H., Xu, Z., Liang, W., Xia, Q., Zhou, P., Rana, O. F., ... & Wu, G. (2020). Efficient algorithms for delay-aware NFV-enabled multicasting in mobile edge clouds with resource sharing. *IEEE Transactions on Parallel and Distributed Systems*, *31*(9), 2050–2066.

[3] Oladipupo, E. T., Abikoye, O. C., Imoize, A. L., Awotunde, J. B., Chang, T. Y., Lee, C. C., & Do, D. T. (2023). An efficient authenticated elliptic curve cryptography scheme for multicore wireless sensor networks. *IEEE Access*, *11*, 1306–1323.

[4] Mezrag, F., Bitam, S., & Mellouk, A. (2022). An efficient and lightweight identity-based scheme for secure communication in clustered wireless sensor networks. *Journal of Network and Computer Applications*, *200*, 103282.

[5] Jimoh, R. G., Imoize, A. L., Awotunde, J. B., Ojo, S., Akanbi, M. B., Bamigbaye, J. A., & Faruk, N. (2022, November). An enhanced deep neural network enabled with cuckoo search algorithm for intrusion detection in wide area networks. In *2022 5th Information Technology for Education and Development (ITED)* (pp. 1–5). IEEE.

[6] Ganesan, D., Cerpa, A., Ye, W., Yu, Y., Zhao, J., & Estrin, D. (2004). Networking issues in wireless sensor networks. *Journal of Parallel and Distributed Computing, 64*(7), 799–814.

[7] Fanian, F., & Rafsanjani, M. K. (2019). Cluster-based routing protocols in wireless sensor networks: A survey based on methodology. *Journal of Network and Computer Applications, 142*, 111–142.

[8] Yousefpoor, M. S., Yousefpoor, E., Barati, H., Barati, A., Movaghar, A., & Hosseinzadeh, M. (2021). Secure data aggregation methods and countermeasures against various attacks in wireless sensor networks: A comprehensive review. *Journal of Network and Computer Applications, 190*, 103118.

[9] Mezrag, F., Bitam, S., & Mellouk, A. (2017, December). Secure routing in cluster-based wireless sensor networks. In *GLOBECOM 2017-2017 IEEE Global Communications Conference* (pp. 1–6). IEEE.

[10] Benayache, A., Bilami, A., Barkat, S., Lorenz, P., & Taleb, H. (2019). MsM: A microservice middleware for smart WSN-based IoT application. *Journal of Network and Computer Applications, 144*, 138–154.

[11] Jain, U., & Hussain, M. (2020). Securing wireless sensors in military applications through resilient authentication mechanism. *Procedia Computer Science, 171*, 719–728.

[12] Jiang, J., Han, G., Wang, H., & Guizani, M. (2019). A survey on location privacy protection in wireless sensor networks. *Journal of Network and Computer Applications, 125*, 93–114.

[13] Abiodun, M. K., Imoize, A. L., Awotunde, J. B., Lee, C. C., Adeniyi, A. E., Chioma, U., & Li10, C. T. (2023). Analysis of a double-stage encryption scheme using hybrid cryptography to enhance data security in cloud computing systems. *Journal of Library and Information Studies, 21*(2), 1–26.

[14] Sogani, A., & Jain, A. (2019). Energy aware and fast authentication scheme using identity based encryption in wireless sensor networks. *Cluster Computing, 22*, 10637–10648.

[15] Kim, J. Y., Hu, W., Sarkar, D., & Jha, S. (2019). Long-term secure management of large scale Internet of Things applications. *Journal of network and computer applications, 138*, 15–26.

[16] Saeed, M. E. S., Liu, Q. Y., Tian, G., Gao, B., & Li, F. (2019). AKAIoTs: Authenticated key agreement for Internet of Things. *Wireless Networks, 25*, 3081–3101.

[17] Mishra, S., Yaduvanshi, R., Dubey, K., & Rajpoot, P. (2021). ESS-IBAA: Efficient, short, and secure ID-based authentication algorithm for wireless sensor network. *International Journal of Communication Systems, 34*(8), e4764.

[18] Kiruthiga Devi, M., & Padma Priya, M. (2023). Evolution of next generation networks and its contribution towards industry 5.0. *Resource Management in Advanced Wireless Networks*, 45–80.

[19] Liu, H. (2024). Network and communication protocols in cyber-physical systems. In *A Practical Guide on Security and Privacy in Cyber-Physical Systems: Foundations, Applications and Limitations* (pp. 25–88), World Scientific Publisher, https://doi.org/10.1142/9789811273551_0002.

[20] Awotunde, J. B., Gaber, T., Prasad, L. N., Folorunso, S. O., & Lalitha, V. L. (2023). Privacy and security enhancement of smart cities using hybrid deep learning-enabled blockchain. *Scalable Computing: Practice and Experience, 24*(3), 561–584.

[21] Imoize, A. L., Balas, V. E., Solanki, V. K., Lee, C. C., & Obaidat, M. S. (eds.) (2023). *Handbook of Security and Privacy of AI-Enabled Healthcare Systems and Internet of Medical Things.* CRC press.

[22] Awotunde, J. B., Misra, S., Ajagbe, S. A., Ayo, F. E., & Gurjar, R. (2023, May). An IoT machine learning model-based real-time diagnostic and monitoring system. In *Machine Intelligence Techniques for Data Analysis and Signal Processing: Proceedings of the 4th International Conference MISP 2022, Volume 1* (pp. 789–799). Singapore: Springer Nature Singapore.

[23] Gaber, T., Awotunde, J. B., Folorunso, S. O., Ajagbe, S. A., & Eldesouky, E. (2023). Industrial Internet of Things Intrusion detection method using machine learning and optimization techniques. *Wireless Communications and Mobile Computing, 2023*, 1–15.

[24] Awotunde, J. B., Chakraborty, C., & Adeniyi, A. E. (2021). Intrusion detection in industrial internet of things network-based on deep learning model with rule-based feature selection. *Wireless communications and mobile computing, 2021*, 1–17.

[25] Awotunde, J. B., Folorunso, S. O., Imoize, A. L., Odunuga, J. O., Lee, C. C., Li, C. T., & Do, D. T. (2023). An ensemble tree-based model for intrusion detection in industrial internet of things networks. *Applied Sciences, 13*(4), 2479.

[26] Ajagbe, S. A., Adigun, M. O., Awotunde, J. B., Oladosu, J. B., & Oguns, Y. J. (2023). Internet of Things enabled convolutional neural networks: Applications, techniques, challenges, and prospects. *IoT-enabled Convolutional Neural Networks: Techniques and Applications,* 27–63.

[27] Okpok, M., & Kihei, B. (2023). Challenges and opportunities for multimedia transmission in vehicular Ad Hoc networks: A comprehensive review. *Electronics, 12*(20), 4310.

[28] Singh, S. K., Mishra, A. K., & Singh, R. K. (2024). Deep Learning-Based 5G networks and IoVs: Advances, meta-data analysis, and future direction. *International Journal of Intelligent Systems and Applications in Engineering, 12*(4s), 762–770.

[29] Khan, B. U. I., Olanrewaju, R. F., Anwar, F., & Shah, A. (2014). Manifestation and mitigation of node misbehaviour in adhoc networks. *Wulfenia Journal, 21*(3), 462–470.

[30] Rong, C., Zhao, G., Yan, L., Cayirci, E., & Cheng, H. (2013). Wireless network security. In *Computer and Information Security Handbook* (pp. 301–316). Morgan Kaufmann.

[31] Islam Khan, B. U., Olanrewaju, R. F., Mir, R. N., Baba, A., & Adebayo, B. W. (2015). Strategic profiling for behaviour visualization of malicious node in manets using game theory. *Journal of Theoretical & Applied Information Technology, 77*(1), 25–43.

[32] Gaber, T., Awotunde, J. B., Torky, M., Ajagbe, S. A., Hammoudeh, M., & Li, W. (2023). Metaverse-IDS: Deep learning-based intrusion detection system for Metaverse-IoT networks. *Internet of Things, 24,* 100977.

[33] Awotunde, J. B., Ayo, F. E., Panigrahi, R., Garg, A., Bhoi, A. K., & Barsocchi, P. (2023). A multi-level random forest model-based intrusion detection using fuzzy inference system for Internet of Things Networks. *International Journal of Computational Intelligence Systems, 16*(1), 31.

[34] Ayo, F. E., Awotunde, J. B., Folorunso, S. O., Adigun, M. O., & Ajagbe, S. A. (2023). A genomic rule-based KNN model for fast flux botnet detection. *Egyptian Informatics Journal, 24*(2), 313–325.

[35] Aluvala, S., & Rajasekhar, K. (2024). An Overview of Constraints and Diverse Attacks in Mobile Ad-hoc Networks. *International Journal of Intelligent Systems and Applications in Engineering, 12*(1), 419–423.

[36] Reka, R., Karthick, R., Ram, R. S., & Singh, G. (2024). Multi head self-attention gated graph convolutional network based multi-attack intrusion detection in MANET. *Computers & Security, 136,* 103526.

[37] Adeniyi, A. E., Jimoh, R. G., & Awotunde, J. A review on elliptic curve cryptography algorithm for Internet of Things: Categorization, application areas, and security. *Application Areas, and Security.* http://dx.doi.org/10.2139/ssrn.4683742

[38] Awotunde, J. B., Sur, S. N., Jimoh, R. G., Aremu, D. R., Do, D. T., & Lee, B. M. (2023). FL_GIoT: Federated learning enabled edge-based Green Internet of things system: A comprehensive survey. *IEEE Access, 11,* 136150–136165.

[39] Rath, M., & Panigrahi, C. R. (2016, March). Prioritization of security measures at the junction of MANET and IoT. In *Proceedings of the Second International Conference on Information and Communication Technology for Competitive Strategies.* ICTCS '16: Second International Conference on Information and Communication Technology for Competitive Strategies, Udaipur India (pp. 1–5).

[40] Bellavista, P., Cardone, G., Corradi, A., & Foschini, L. (2013). Convergence of MANET and WSN in IoT urban scenarios. *IEEE Sensors Journal, 13*(10), 3558–3567.

[41] Awotunde, J. B., Imoize, A. L., Jimoh, R. G., Adeniyi, E. A., Abdulraheem, M., Oladipo, I. D., & Falola, P. B. (2024). AIoMT enabling real-time monitoring of healthcare systems: Security and privacy considerations. *Handbook of Security and Privacy of AI-Enabled Healthcare Systems and Internet of Medical Things,* 97–133.

[42] Abdulraheem, M., Adeniyi, E. A., Awotunde, J. B., Imoize, A. L., Jimoh, R. G., Oladipo, I. D., & Falola, P. B. (2024). Artificial Intelligence of Medical Things for Medical Information Systems Privacy and Security. In *Handbook of Security and Privacy of AI-Enabled Healthcare Systems and Internet of Medical Things* (pp. 63–96). CRC Press.

[43] Khan, B. U. I., Olanrewaju, R. F., Anwar, F., Najeeb, A. R., & Yaacob, M. (2018). A survey on MANETs: Architecture, evolution, applications, security issues and solutions. *Indonesian Journal of Electrical Engineering and Computer Science, 12*(2), 832–842.

[44] Hafsa, A., Sghaier, A., Malek, J., & Machhout, M. (2021). Image encryption method based on improved ECC and modified AES algorithm. *Multimedia Tools and Applications, 80*, 19769–19801.

[45] Khashan, O. A., Ahmad, R., & Khafajah, N. M. (2021). An automated lightweight encryption scheme for secure and energy-efficient communication in wireless sensor networks. *Ad Hoc Networks, 115*, 102448. https://doi.org/10.1016/j.adhoc.2021.102448

13 Computational Models Enabling Smart Teaching and Learning in Wireless Communication Systems

Abidemi Emmanuel Adeniyi, Rasheed Gbenga Jimoh, Joseph Bamidele Awotunde, Mukaila Olagunju, Deborah Olufemi Ninan, Odunayo Dauda Olanloye, Halleluyah Oluwatobi Aworinde and Abdulrauf Olarenwaju Babatunde

13.1 INTRODUCTION

Education is on the verge of a technological revolution, fueled by the pervasive presence of wireless communication technology [1, 2]. This research digs into the historical, social, and pedagogical contexts in which these devices emerge in the classroom, analyzing their potential to change learning experiences and transcend educational gaps. The pervasive presence of wireless communication technology is driving a fundamental revolution in the educational landscape [3]. These technologies, which include mobile devices, high-speed Internet, and cutting-edge platforms, are more than just tools; they are catalysts for reinventing learning experiences, broadening access to information, and empowering learners in unprecedented ways [4]. To really grasp the significance of wireless communication technologies in education, we must examine their historical background, social ramifications, educational uses, and current problems.

The voyage of education with wireless communication began with radio transmissions in the early twentieth century, democratizing information access and pioneering remote learning efforts [5]. Later, the mobile revolution ushered in a paradigm shift, allowing for anytime, anywhere learning via phones and smartphones, opening the door for collaborative information exchange and individualized learning paths [6, 7]. The digital gap, on the other hand, remains an urgent problem, underlining the need for fair infrastructure and digital literacy measures to guarantee that technology benefits all learners [8]. Wireless communication devices have enormous promise for bridging educational gaps and promoting social good [9, 10]. Virtual classrooms and cross-cultural collaboration platforms promote globalization of education by improving learning settings and cultivating global citizenship [11]. Furthermore, these technologies appeal to a wide range of learners by providing tailored learning paths and removing geographical barriers for students with impairments. Mobile learning supports self-directed learning and increases educational possibilities beyond the restrictions of traditional brick-and-mortar schools in low-resource settings.

Wireless communication devices are reshaping the educational environment by encouraging active learning and student involvement. Inquiry-based learning is supported by mobile devices and interactive platforms, which encourage problem-solving abilities and critical thinking [12]. Adaptive learning software adapts material and routes to individual requirements and learning

DOI: 10.1201/9781003457428-14

styles, improving learning outcomes and enabling individualized, self-paced learning [13]. Virtual classrooms and communication technologies improve collaborative learning by facilitating peer-to-peer engagement, project-based learning activities, and global knowledge exchange [14]. Immersive experiences are brought to life by educational virtual reality (VR) and augmented reality (AR) technology, enhancing engagement and knowledge in a wide range of disciplines, ranging from anatomy to historical events. Mobile learning, made possible by smartphones and tablets, allows for on-the-go access to educational resources and technologies, encouraging self-directed learning and flexibility [15]. Blended learning provides customization and accommodates a variety of learning styles by smoothly mixing online and offline components. Social learning platforms foster knowledge sharing, peer support, and global cooperation among learners by using online communities and interactive technologies [16]. To make learning interesting and effective, educational games and gamified learning experiences use intrinsic motivation and active involvement [17].

While wireless communication technology has clear potential in education, managing this shift demands careful assessment of difficulties and future possibilities. Cybersecurity and data privacy continue to be top priorities, necessitating strong digital citizenship practices and ethical technology integration [18]. Teacher education and professional development are critical to ensuring that educators have the skills and knowledge to successfully use technology in their classrooms. To close the digital divide, infrastructural limitations, financial concerns, and cultural obstacles to technology access must all be addressed. To enhance justice and inclusion in learning experiences, addressing possible biases in algorithms and educational technologies necessitates ethical considerations and intentional decisions.

Wireless communication devices are more than just tools; they are catalysts for an educational revolution. These technologies hold enormous promise for building a future where education is engaging, accessible, and allows learners to attain their full potential by supporting individualized learning, global cooperation, and equal access to information [1]. However, realizing this promise necessitates resolving current issues and adopting appropriate technology integration. Education must embrace the capabilities of wireless communication networks while maintaining ethical considerations, diversity, and an emphasis on human-centered learning experiences as we move forward. We can unleash the actual potential of technology to transform education and enable learners to succeed in the twenty first century and beyond with careful preparation and teamwork.

Learning is becoming more interactive, readily available and linked with the technology changes influencing our society by incorporating computer modeling and simulation into education. This contemporary method has profoundly transformed the educational landscape away from rote memorization and passive learning and toward a more active, engaging, and successful paradigm. Computational modeling and simulation have improved education in a variety of ways over traditional techniques.

1. Students can better understand complicated and abstract subjects through visualization and engagement. In simulations, they may observe the direct impacts of altering variables, which improves their learning and recall.
2. Computational models may be tailored to the pace and manner of the student, allowing for a more customized education. Students can repeat, adapt, or expand on simulations to meet their own requirements, which is frequently not achievable in a typical learning setting.
3. Students that use simulations must make judgments, solve issues, and consider the consequences and procedures critically. This active learning strategy is more engaging and successful than traditional teaching's passive learning styles.
4. Anyone, at any time may access modeling and simulations thanks to the Internet. This democratizes education and opens doors to a broader audience. It also enables scalability in education, allowing it to reach more pupils than traditional approaches.

13.1.1 KEY CONTRIBUTIONS OF THE CHAPTER

The following are the significant contributions of this chapter:

i. The contributions of this chapter include: provision of a useful framework, identifying possible benefits and obstacles, and encouraging continued study and development toward a future in which technology enables educators and learners to create individualized, engaging, and powerful learning experiences.

ii. This study lays the groundwork for future research into the convergence of wireless communication technology and computational models in the context of education. It underlines the unique potential and problems that this confluence presents, opening the path for more study and development.

iii. The chapter presents a complete framework for integrating computational models into wireless communication networks in order to improve teaching and learning. This framework describes crucial components, including data gathering, model creation, feedback systems, and tailored learning routes.

iv. The study underlines the endless possibilities of computational models to customize learning experiences through adaptive learning platforms, real-time feedback, and targeted interventions. This individualized approach addresses individual student requirements and learning styles, encouraging deeper engagement and better learning results.

v. The study demonstrates how wireless communication networks may create active learning settings by utilizing mobile devices and interactive platforms. Educational games, simulations, and collaborative activities are easily accessible and usable, allowing students to participate actively in their learning journey.

vi. The research looks at how computational models might be utilized to give students continuous assessment and feedback. This real-time data may be used to create tailored learning paths, pinpoint areas of difficulty, and measure progress, so empowering both students and educators.

13.1.2 CHAPTER ORGANIZATION

The remainder of the chapter is divided into ten sections; Section 13.2, a discussion on the role of computation models in improving teaching and learning. The fundamentals of wireless communication is discussed in Section 13.3. Wireless Education Systems' Digital Landscape: Challenges and Opportunities is discussed in Section 13.4. Section 13.5 discusses the application Scenarios of Computational Modeling for Teaching and Learning. Section 13.6 discusses the simulation models for wireless education systems. Machine Learning Models for Personalized Learning is discussed in Section 13.7. The computational models enabling reforms in wireless communication systems are discussed in Section 13.8. The lesson learned from the chapter and future trends is discussed in Section 13.9, while Section 13.10 concludes and provides key findings of the chapter.

13.2 THE ROLE OF COMPUTATIONAL MODELS IN IMPROVING TEACHING AND LEARNING: A DATA-PEDAGOGY SYMPHONY

The educational environment is changing as a result of the incorporation of computational models into the fundamental fabric of teaching and learning. These models, which are powered by data-driven algorithms and machine learning, provide a symphony of options for personalizing, optimizing, and revolutionizing the learning process. Let us look at how computational models may help improve teaching and learning in a variety of ways.

13.2.1 Personalization at its Finest

Visualize a classroom where each student receives a curriculum that is personalized to their own requirements and learning styles. Adaptive learning systems that use advanced algorithms to assess student performance data and modify the learning route in real-time make this ambition a reality [19]. Adaptive systems propose learning resources, modify difficulty levels, and provide focused feedback to ensure that each learner develops at their ideal rate. This tailored approach encourages greater involvement, autonomy, and ownership of the learning experience.

13.2.2 Identifying and Addressing Learning Gaps

Computational models may be used as diagnostic tools, identifying areas where students struggle and forecasting prospective problems. Predictive analytics detect pupils who are at danger of falling behind, allowing teachers to engage proactively with tailored help and resources [20]. For eliminating learning gaps and reducing frustration and disengagement, early detection and intervention are critical. Furthermore, educational chatbots powered by natural language processing may provide individualized feedback and answer student queries in real time, functioning as virtual tutors and providing continuous assistance [21].

13.2.3 Changing the Teacher's Role

Computational models are not here to replace instructors; rather, they are great allies. Models free up crucial instructor time for more effective interactions with students by automating chores like grading and offering individualized feedback. Teachers may use their time to encouraging discussions, promoting inquiry-based learning, and developing relationships with their pupils, harnessing the distinctive human touch that technology cannot mimic [22]. Teachers may now become mentors and facilitators, leading students on their individualized learning journeys, thanks to a transformation in the educational environment.

13.2.4 Unlocking the Power of Data

The genuine magic of computational models resides in their capacity to harness the power of data. Educational statistics, which include student performance, learning activities, and feedback, can give important insights into student requirements and learning patterns. Models may discover patterns, forecast outcomes, and enhance evidence-based decision-making in education policy and practice by evaluating these massive datasets. This data-driven approach encourages continual development, allowing educators to modify their teaching approaches and adapt to their students' changing requirements.

13.2.5 Challenges and the Road Ahead

While the possibility of computational models is apparent, managing this shift necessitates careful consideration of difficulties. Ethical issues about data privacy, algorithmic bias, and the risk of overreliance on technology must be addressed by rigorous rules and responsible implementation [23]. Furthermore, providing fair access to technology and digital literacy training for educators and students is critical to preventing the digital gap from spreading further.

Computational models are more than just tools; they are the driving force behind a new age of customized, data-driven, and student-centered learning. We can unlock a symphony of possibilities for improving teaching and learning by embracing their promise while reducing difficulties and emphasizing ethical issues [24]. As we move forward, let us utilize technology to empower educators and students to build a learning landscape in which every learner may grow and realize their full potential, rather than to replace the human touch of education.

13.3 FUNDAMENTALS OF WIRELESS COMMUNICATION

The unseen symphony of wireless communication pervades our modern environment. These technologies have altered the way we interact, learn, and work, from the pocket-sized symphony of smartphones to the towering antennas conversing with satellites. In this chapter, we look at the fundamentals of wireless communication networks, decoding the magic underlying every text message, phone conversation, and streaming video.

13.3.1 THE SIGNAL ORCHESTRA

Wireless communication begins with the creation of signals, which are electrical or electromagnetic waves that carry information. Consider dipping a tuning fork into a bowl of water; the ripples that form represent the signal as it travels through the air. To carry information, these signals can be modulated or shaped. In amplitude modulation (AM), the signal's intensity (amplitude) fluctuates according to the information being conveyed, like a dimmer switch transmitting Morse code. Figure 13.1 depicts the Frequency Modulated (FM) wave.

FM, on the other hand, changes the frequency (pitch) of the transmission dependent on the information, much like a musical tune delivering a message. AM and FM are both often used in radio broadcasting, with FM providing higher noise immunity due to its shifting frequency.

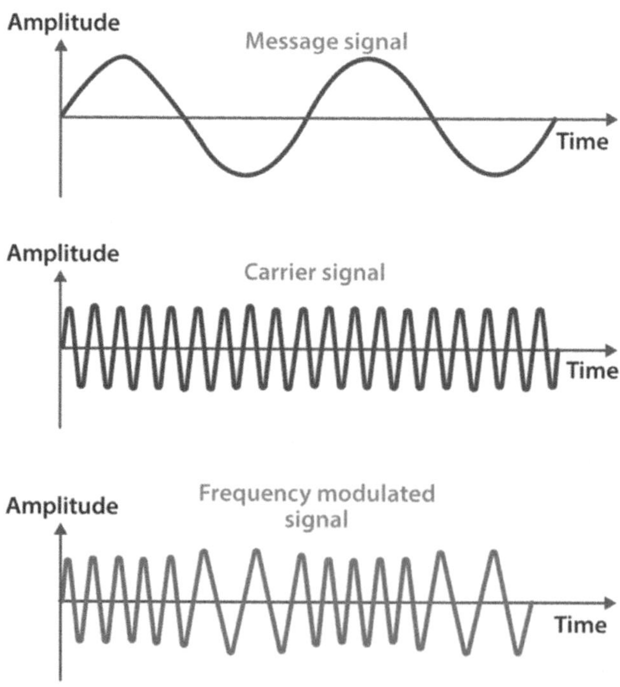

FIGURE 13.1   Graphical representation of Frequency Modulated Wave.

13.3.2 The Electromagnetic Spectrum

These modulated signals then travel over the electromagnetic spectrum, which is a large ocean of invisible waves spanning from radio waves to gamma rays. Different frequencies in this spectrum have different qualities and are employed for different things. Figure 13.2 shows the low-frequency radio waves, for example, penetrate buildings effectively and are suitable for AM radio transmissions, but high-frequency microwaves transport vast quantities of data and power WiFi and cellular networks [25].

Figure 13.2 displays the electromagnetic spectrum, with various properties across the range of frequencies and wavelengths.

13.3.3 Modulation and Demodulation

Signals are sometimes broadcast at significantly higher frequencies than the information itself to avoid interference over long distances. This is known as carrier modulation. The signal is demodulated at the receiving end, recovering the original information from the carrier wave [26]. Consider it like peeling an onion: the modulated signal is the outside layer, and the information is the core.

13.3.4 Transmitters and Receivers

Wireless communication is based on the coordinated movement of transmitters and receivers. Transmitters, such as radio towers or smartphone antennas, turn electrical impulses into electromagnetic waves and broadcast them into the atmosphere. Receivers, such as radios or smartphones, collect these waves and turn them back into electrical signals, allowing the information they contain to be extracted.

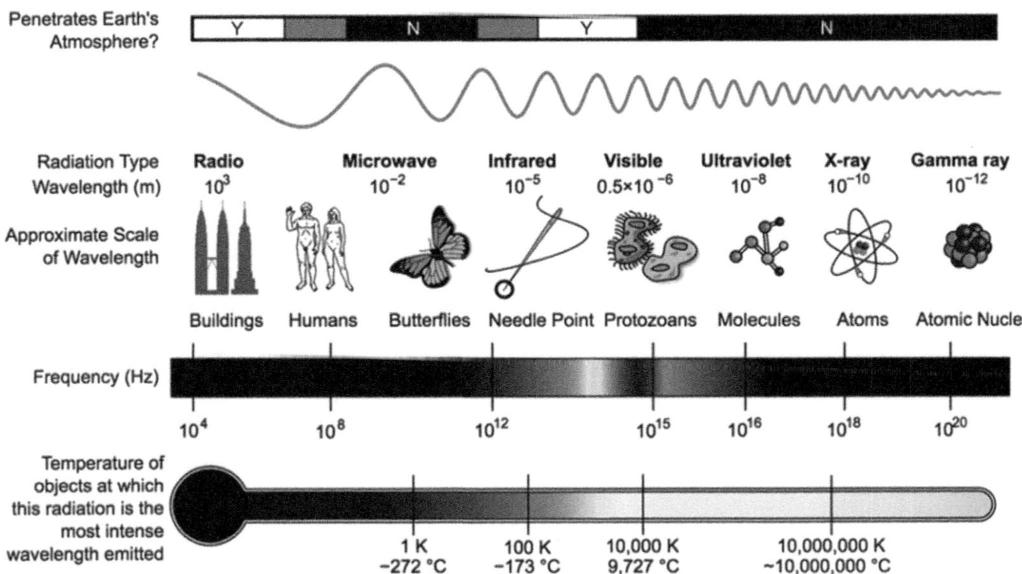

FIGURE 13.2 Electromagnetic spectrum.

13.3.5 NOISE AND INTERFERENCE

A signal's path is not always smooth. The air is packed with noise, random electromagnetic waves that can interfere with the signal and distort the information it contains, much as a noisy restaurant might make communication difficult. Furthermore, numerous transmissions with the same frequency might overlap and interfere with one other, causing the data to be scrambled.

Various strategies are used to overcome these obstacles. Error correction coding increases information redundancy, allowing receivers to discover and rectify mistakes produced by noise. To limit interference, frequency reuse assigns multiple frequencies to neighboring emitters, similar to how musicians play alternate songs to avoid clashing notes.

13.3.6 BEYOND THE FUNDAMENTALS

Wireless communication is an ever-changing industry. Multiple-input multiple-output (MIMO) innovations use numerous antennas to boost signal quality and data level, while cellular networks change to 5G and beyond, announcing higher speeds and less latency for exciting virtual reality adventures and instantaneous interaction with autonomous vehicles. Understanding the fundamentals of wireless communication networks provides a glimpse into this fascinating world, where invisible waves transport the voices, movies, and data that create our lives. The symphony of signals will only get more complicated and powerful as technology advances, linking us in ways we can only begin to comprehend.

13.3.7 WIRELESS COMMUNICATION IN EDUCATION: REVOLUTIONIZING LEARNING IN THE DIGITAL AGE

The pervasive presence of wireless communication technology is driving a fundamental revolution in the educational landscape. These technologies, which include mobile devices, high-speed Internet, and cutting-edge platforms, are more than just tools; they are catalysts for reinventing learning experiences, broadening access to information, and empowering learners in unprecedented ways. To really grasp the significance of wireless communication technologies in education, we must examine their historical background, social ramifications, educational uses, and current problems [27].

13.3.8 BRIDGING THE DIGITAL DIVIDE: ACCESS TO THE PAST, PRESENT, AND FUTURE

The voyage of education with wireless communication began with radio transmissions in the early twentieth century, democratizing information access and pioneering remote-learning efforts [28]. Later, the mobile revolution ushered in a paradigm shift, allowing for anytime, anywhere learning via phones and smartphones, opening the door for collaborative information exchange and individualized learning paths [28]. The digital gap, on the other hand, remains an urgent problem, underlining the need for fair infrastructure and digital literacy measures to guarantee that technology benefits all learners.

13.3.9 SOCIAL AND ECONOMIC BENEFITS INCLUDE A GLOBAL OUTLOOK AND INCLUSIVE LEARNING

Wireless communication devices have enormous promise for bridging educational gaps and promoting social good. Virtual classrooms and cross-cultural collaboration platforms promote globalization of education by improving learning settings and cultivating global citizenship [29]. Furthermore, these technologies appeal to a wide range of learners by providing tailored learning paths and removing geographical barriers for students with impairments. Mobile learning supports self-directed learning and increases educational possibilities beyond the restrictions of traditional brick-and-mortar schools in low-resource settings [30].

13.3.10 Implications for Education: Active Learning, Personalization, and Immersive Experiences

Wireless communication devices are reshaping the educational environment by encouraging active learning and student involvement. Inquiry-based learning is supported by mobile devices and interactive platforms, which encourage problem-solving abilities and critical thinking [31]. Adaptive learning software adapts material and routes to individual requirements and learning styles, improving learning outcomes and enabling individualized, self-paced learning. Virtual classrooms and communication technologies improve collaborative learning by facilitating peer-to-peer engagement, project-based learning activities, and global knowledge exchange. Immersive experiences are brought to life by educational virtual reality (VR) and augmented reality (AR) technology, enhancing engagement and knowledge in a wide range of disciplines ranging from anatomy to historical events [32].

13.3.11 Mobile Learning, Blended Learning, and Gamification: Examples of Specific Technologies and Use Cases

Mobile learning, made possible by smartphones and tablets, allows for on-the-go access to educational resources and technologies, encouraging self-directed learning and flexibility [33–35]. Blended learning provides customisation and accommodates a variety of learning styles by smoothly mixing online and offline components. Social learning platforms foster knowledge sharing, peer support, and global cooperation among learners by using online communities and interactive technologies. To make learning interesting and effective, educational games and gamified learning experiences use intrinsic motivation and active involvement [36].

13.3.12 Cybersecurity, Teacher Training, and Ethical Considerations: Challenges and Future Directions

While wireless communication technology have clear potential in education, managing this shift demands careful assessment of difficulties and future possibilities. Cybersecurity and data privacy continue to be top priorities, necessitating strong digital citizenship practices and ethical technology integration. Teacher education and professional development are critical to ensuring that educators have the skills and knowledge to successfully use technology in their classrooms. To enhance justice and inclusion in learning experiences, addressing possible biases in algorithms and educational technologies necessitates ethical considerations and intentional decisions.

13.3.13 A Symphony of Educational Opportunities for the Future

Wireless communication devices are catalysts for an educational revolution. These technologies hold enormous promise for building a future where education is engaging, accessible, and allows learners to attain their full potential by supporting individualized learning, global cooperation, and equal access to information [37]. However, realizing this promise necessitates resolving current issues and adopting appropriate technology integration. Education must embrace the capabilities of wireless communication networks while maintaining ethical considerations, diversity, and an emphasis on human-centered learning experiences as we move forward. We can unleash the entire symphony of possibilities given by technology and reshape education for the twenty-first century and beyond with careful planning and teamwork.

13.4 WIRELESS EDUCATION SYSTEMS' DIGITAL LANDSCAPE: CHALLENGES AND OPPORTUNITIES

Wireless communication technologies have emerged as revolutionary tools in today's fast-expanding educational scene, transforming the way we teach, study, and access knowledge. While new technologies offer enormous promise for improving educational experiences and broadening access to learning opportunities, they also present a slew of obstacles that must be carefully considered and mitigated in advance.

13.4.1 CHALLENGES

a. Wireless education systems invariably manage a large quantity of sensitive student data, prompting worries about cybersecurity and data privacy. To secure student information from illegal access or misuse, strong security measures, data encryption techniques, and rigorous data protection legislation must be implemented.

b. Teacher Training and Digital Literacy: To effectively integrate wireless devices in the classroom, thorough teacher training and digital literacy efforts are required. Educators must be prepared to use these technologies successfully for individualized education, blended learning techniques, and collaborative activities.

c. The increased acceptance and use of wireless technology in education risk aggravating the current digital divide, expanding the difference between those who have access to these resources and those who do not. It is critical to address infrastructural gaps, financial concerns, and cultural obstacles to technology access in order to ensure fair educational opportunities for all students.

d. Educational technology, such as adaptive learning platforms and automated assessment systems, frequently rely on algorithms that might perpetuate biases and disparities. It is critical to carefully evaluate these algorithms in order to avoid prejudice and promote fair and inclusive learning experiences.

e. While wireless technology has the potential to improve student involvement, there is a danger of abuse or misuse, which can distract students from the learning process and impair their capacity to focus and absorb information properly. It is critical to carefully choose and moderate instructional information, as well as strike a balance between technology and conventional teaching approaches.

13.4.2 OPPORTUNITY

a. Wireless technologies provide tailored learning paths, adapting training to specific student needs, learning styles, and pace of advancement. Adaptive learning software can optimize learning results by adjusting material, tests, and suggestions based on real-time data.

b. Wireless communication devices allow global cooperation and virtual learning experiences by breaking down geographical barriers and connecting students with peers and professionals from across the world. This encourages global citizenship and a diversified learning environment.

c. Flexibility and Anytime, Anywhere Learning: Wireless technologies enable anytime, anywhere learning by allowing students to access educational resources and participate in virtual classrooms or online conversations from the comfort of their own homes or other remote locations. This adaptability improves accessibility and accommodates a variety of learning methods.

d. Immersive Learning Experiences and Gamification: Using educational technologies such as virtual reality (VR) and augmented reality (AR), immersive learning experiences may be created, increasing engagement and giving interactive simulations and representations of ideas and processes. Through game-like themes and prizes, gamification approaches may encourage and engage pupils.

e. Accessibility for Diverse Learners and Learners with impairments: Wireless technologies, including text-to-speech capabilities, sign language translation, and adaptive content formats, can provide accessible educational resources and assistance for learners with impairments. This fosters inclusive learning and lowers participation obstacles.

Wireless communication technology provides several chances to modify and improve education. However, resolving the aforementioned problems is critical to ensuring their responsible and fair implementation. We can harness these technologies to provide inclusive, engaging, and individualized learning experiences for all children by emphasizing cybersecurity, teacher training, bridging the digital gap, minimizing algorithmic bias, and supporting balanced technology usage. We must seek to harness the power of wireless education systems while maintaining the values of fairness, accessibility, and ethical technology integration as we traverse the ever-changing digital world.

13.5 APPLICATION SCENARIOS OF COMPUTATIONAL MODELING FOR TEACHING AND LEARNING

The educational environment is experiencing tremendous change, fueled by the unstoppable march of technology. Computational modeling is at the vanguard of this transformation, transforming teaching and learning by promoting individualized paths, predictive insights, and immersive experiences that were long reserved for science fiction. In this investigation, we will delve into the essence of computational modeling, reveal its varied uses, and negotiate the hurdles and ethical issues that come with its incorporation into the classroom.

Computational modeling in education, at its most basic, relates to the use of algorithms, data analytic tools, and statistical models to simulate and evaluate educational processes. These models make use of computers' immense computing capacity to comprehend student behavior, forecast learning outcomes, tailor learning courses, and generate interactive learning experiences. Consider a complicated network of interrelated variables that indicate cognitive skills, learning styles, degrees of engagement, and external effects. These factors are analyzed by computational models, which find trends and generate predictions about how students will interact with and respond to instructional content.

13.5.1 EXPLOITING THE POTENTIAL: APPLICATION CASE STUDIES

Computational modeling applications in education are as varied as the learning types they cater to.

1. *Individual Journeys in Personalized Learning*. Consider a learning environment in which individual students follow a route suited to their own requirements and speed. This is customized learning's vision, and computational models are making it a reality. Adaptive learning platforms driven by advanced algorithms assess real-time student performance data, offering best paths, altering difficulty levels, and delivering focused feedback [38]. Consider students who are having difficulty understanding fractions; the model detects this gap and guides them to interactive, video-based lectures and gamified practice sessions, closing the knowledge gap without frustration. As a result, students who become active participants in their own

educational journeys experience deeper engagement, higher learning results, and a sense of autonomy.

2. *Proactive Intervention for At-Risk Students*. Consider a teacher who is equipped with the capacity to identify children who are at risk of falling behind before they fall behind. This is a possibility thanks to predictive analytics models based on historical data and current performance analysis. These approaches identify students who may be suffering by assessing attendance patterns, engagement levels, and knowledge gaps, allowing instructors to give targeted intervention before problems escalate. Consider a model highlighting a student who repeatedly misses deadlines and has difficulty understanding. With this knowledge, the instructor may provide individualized support, set up check-ins, and prescribe study tactics, so avoiding possible failure and building the student's confidence.

3. *Intelligent Feedback and Evaluation: Beyond the Red Line*. Consider receiving feedback that extends beyond a red X on a piece of paper. Natural language processing-powered intelligent feedback systems may examine student opportunities for progress [39]. This eliminates the subjectivity of traditional grading and gives students meaningful recommendations on how to improve their work and comprehension [40]. Furthermore, automated evaluation technologies can save teachers time and allow them to focus on targeted advice and mentorship.

4. *Beyond the Textbook: Immersive Learning and Gamification*. Learning is not limited to a textbook or a whiteboard. Powered by computational models, virtual reality (VR) and augmented reality (AR) technologies may build immersive learning experiences that take students to the heart of history, examine the complexities of the human body, or explore the furthest reaches of the cosmos [38]. Consider students using a VR headset to explore ancient Rome or doing virtual dissections in an AR setting. These interactive experiences increase engagement, strengthen knowledge, and promote a passion of learning that goes beyond traditional approaches. Another use of computational models is gamification, which adds a layer of fun and motivation to the learning process. Learning becomes an exciting challenge by introducing game-like features such as points, badges, and leaderboards, pushing students to push their bounds and strive for mastery.

5. *Collaborative Learning and Global Communities: Education not a one-man show*. Computational models can help students from different countries and cultures connect through virtual classrooms and online communities [41]. Consider pupils collaborating on a science project with classmates from another nation, exchanging ideas, exchanging data, and presenting their findings to a worldwide audience. This promotes intercultural awareness, helps students develop critical thinking and communication skills, and prepares them for a future where collaboration crosses borders.

13.5.2 TYPES OF COMPUTATIONAL MODELS (E.G., SIMULATION, MACHINE LEARNING, ANALYTICAL)

Computational models play an important part in the symphony of scientific discovery and technological growth, generating songs of insight from our world's data-rich symphonies. Understanding these models is more than simply a pedagogical exercise; it is necessary for navigating the increasingly data-driven landscapes of research, analysis, and decision-making. This chapter goes into the three primary types of computational models: simulation, machine learning, and analytical, revealing their distinct contributions and showing the many ways they organize understanding across domains.

1. Simulation: Predicting and Exploring by Mimicking Reality

Consider entering a virtual world where scientific experiments take place, economic forces collide, and historical events are replayed. This is the domain of simulation models, which are digital reconstructions of real-world systems that imitate their behavior and reaction to different inputs. Simulations provide essential insights and chances for research, from climate change models that anticipate future weather patterns to flight simulators that educate pilots for every imaginable eventuality. The complex algorithms that regulate the interactions between simulated things are at the heart of simulation models. To anticipate temperature changes and probable weather extremes, a climate model can include equations representing ocean currents, atmospheric physics, and greenhouse gas concentrations. Simulations' strength rests in their capacity to test theories, detect possible dangers, and improve systems in a safe, controlled setting before going into the complexity of the real world.

2. Machine Learning: Pattern Recognition and Prediction

Machine learning models, as opposed to simulations, excel in uncovering hidden patterns and correlations within large datasets. Consider sending mountains of medical information to a computer and seeing it uncover small links between symptoms and illnesses, paving the path for individualized care. This is the beauty of machine learning: algorithms learn from previous data to generate accurate future predictions. There are several types of machine learning models, each suited to a certain job. For example, supervised learning learns from labeled datasets with paired inputs and outputs. Based on previously identified samples, a computer evaluating email spam trends may learn to identify future spam messages. Unsupervised learning, on the other hand, identifies patterns in unlabeled data, such as grouping consumer groups based on their online buying histories without the use of pre-defined categories.

Machine learning has several uses, ranging from improving corporate choices to projecting market trends, customizing content suggestions, and even identifying people in images. It is important to remember, however, that machine learning models are only as good as the data they are trained on. Biases and limits in data can result in skewed predictions, emphasizing the need of ethical concerns and proper data processing.

3. Analytical Models: Uncovering Connections and Improving Solutions

Analytical models focus on understanding the underlying links and mechanisms that drive complex systems, whereas simulations and machine learning excel at prediction and pattern identification [42]. Consider examining the complex network of variables that influence traffic flow in a city. An analytical model may employ mathematical equations and optimization algorithms to determine ideal traffic signal timings and road infrastructure upgrades, with the goal of reducing congestion and improving traffic flow. Analytical models are frequently mathematical equations or statistical models that depict the connections between key variables. Scientists and engineers can get significant insights into the behavior of the system and develop optimal solutions to specific issues by modifying these equations or studying their outputs. Analyzing a financial market model, for example, may identify optimal investment strategies or forecast impending economic swings.

It is critical to understand that these three types of models are not mutually exclusive; they frequently collaborate to create complete and powerful insights. Simulations may produce data for machine learning algorithms, which can subsequently inform analytical models, resulting in a synergistic discovery interaction. A simulation model of a protein's function, for example, may yield data that a machine-learning algorithm might use to predict its interactions with possible medications, eventually leading to the creation of innovative therapies.

13.6 SIMULATION MODELS FOR WIRELESS EDUCATION SYSTEMS

Traditional approaches form the basis of the educational symphony, but wireless technologies are emerging as dynamic instruments, weaving a tapestry of individualized and engaging learning experiences. Simulation models for wireless education systems are at the center of this revolution, a powerful force impacting the future of learning by recreating and evaluating the intricate interactions inside these dynamic settings.

Consider a conductor who methodically tunes their orchestra, altering settings, and anticipating potential problems throughout a performance. Simulation models serve a similar purpose in that they enable educators and researchers to examine and optimize wireless education systems before deploying them in real-world classrooms. These models replicate the behavior of various network components, user interaction patterns, and environmental conditions, revealing possible bottlenecks, coverage flaws, and optimal resource allocation.

13.6.1 UNCOVERING THE SYMPHONY'S OPPORTUNITIES: SIMULATION MODEL APPROACHES

Network Design and Optimization: Creating a strong wireless education system is similar to writing a complex piece of music. Simulation models can aid in the optimization of network architecture, the prediction of signal coverage in a variety of classroom contexts, and the identification of possible interference sources [42]. This provides continuous connectivity, less irritating dropouts, and clear hearing of all instruments in the instructional orchestra.

Resource Allocation and Content Delivery: Consider an orchestra that is unable to perform owing to a lack of sheet music. Simulation models may monitor bandwidth utilization, estimate peak demand during online lectures or video streaming, and optimize resource allocation to ensure seamless delivery of instructional content. This promotes a steady information flow, avoiding frustrating pauses and enabling the instructional melody to flow freely.

13.6.2 CASE STUDIES DEMONSTRATING THE USE OF SIMULATION MODELS IN TEACHING AND LEARNING

Instead of dull textbooks and interminable lectures, how about modeling a historical event, constructing a metropolis, or even piloting a spaceship? That is the power of simulation models in education: they allow students to actively experiment, make errors, and learn from the outcomes in a secure, virtual setting. Let us look at some real-world examples of how these approaches are changing teaching and learning:

Case 1: Creating a Future Sustainable City
> Consider middle-school students working together to develop a prosperous, eco-friendly metropolis using a city-building simulation. This is exactly what New York City students did. They used SimCityEdu to address problems like pollution, renewable energy, and traffic congestion, while also learning about urban planning, resource management, and sustainability. As students debated green areas, public transit networks, and energy sources, they were immersed in the intricate realm of urban planning.

Case 2: Flying Through History During World War II
> Forget dull chronology and textbook reports; students in California donned virtual pilot helmets and flew over World War II skies using a flight simulator. This interactive experience brought the Battle of Britain to life, allowing students to fly Spitfires, manage radar systems, and engage in aerial combat. This direct experience brought history to life, allowing us a better grasp of tactics, technology, and the human cost of conflict.

Case 3: Disease Diagnosis Like a Medical Detective
> Medical school may be demanding, but what if there was a method to practice illness diagnosis before entering a real hospital? Enter virtual reality-based medical simulations in which students put on virtual reality headsets and assume the virtual role of medical investigators.

Students refine their diagnostic abilities, learn about diverse diseases, and gain confidence in a risk-free environment by examining virtual patients, assessing symptoms, and making treatment options.

Case 4: Establishing a Business Empire

Have you ever wanted to be a CEO? Business students in Australia were able to fulfill their dreams thanks to the use of a business simulation model. They were separated into virtual businesses that competed in a simulated market. They made price decisions, launched marketing efforts, and navigated economic situations directly, gaining firsthand experience with the joy and challenges of running a business.

Case 5: Empowering Empathy with Virtual Reality

VR empathy simulations propel social and emotional development forward. Consider yourself in the shoes of someone experiencing homelessness, prejudice, or mental health issues. This is what students in Canada experienced through VR simulations, which cultivated understanding, compassion, and a greater awareness for different points of view. The impact was powerful, inspiring kids to become activists for social change and to tear down prejudiced barriers.

These are just a handful of the ways simulation models are transforming education. These models unleash learning potential, encourage critical thinking, and prepare students for future problems and opportunities by providing dynamic, engaging, and often emotionally significant experiences. So, abandon rote memorizing in favor of the excitement and power of learning by doing – the future of education has arrived, and it is simulated!

13.7 MACHINE LEARNING MODELS FOR PERSONALIZED LEARNING

The educational environment is experiencing a profound transformation, fueled by the inexorable march of technological progress. Machine learning (ML) is at the vanguard of this change, a powerful force giving a vision of customized learning: a paradigm in which students embark on individual journeys tailored to their own requirements, pace, and learning styles [44]. This chapter looks into the delicate melody of machine learning models in customized learning, examining their promise, negotiating their limitations, and eventually discovering the harmonic ensemble of technology and pedagogy set to alter education.

13.7.1 UNDERSTANDING THE MAESTRO: THE POWER OF MACHINE LEARNING MODELS

Consider a learning environment in which curriculum and teaching adjust dynamically to individual variances. This is the promise of machine learning models, which are algorithms trained on massive datasets and can assess student performance, foresee obstacles, and offer ideal learning routes in real time. Let us look at some of the instruments that these models add to the educational orchestra:

1. *Adaptive Learning Platforms*: Think of an adaptive learning platform as one that smoothly adapts difficulty levels, recommends appropriate resources, and delivers customized feedback depending on each student's progress. This is the magic of adaptive learning, powered by machine learning models that identify knowledge gaps, tailor material delivery, and keep students engaged on their best learning trajectory [45]. Consider how you struggle with fractions; the model identifies this and recommends interactive video lectures and gamified practice sessions to fill the gap without causing irritation.

2. *Proactive Support for Struggling Pupils*: Predictive Insights Consider a teacher who has a crystal ball and can predict which kids will fall behind before they do. This is made possible by predictive analytics models based on historical data and current performance assessments.

These approaches identify kids who may be suffering by assessing attendance patterns, engagement levels, and knowledge gaps, allowing instructors to give targeted intervention before problems escalate. This proactive strategy eliminates academic failure and builds student confidence, guaranteeing that no kid is left out of the educational symphony.

3. *Beyond the Red Line*: Intelligent Feedback and Evaluation: Consider receiving feedback that extends beyond a red X on a piece of paper. Intelligent feedback systems based on natural language processing may examine student writings, code, and even spoken comments, delivering precise, tailored feedback that identifies strengths, flaws, and opportunities for progress. This eliminates the subjective nature of traditional grading and delivers actionable information rather than simply a decision. Consider a writer who is suffering with sentence structure; the system identifies cumbersome phrases and proposes alternative constructs, allowing them to improve their work and get a better understanding. This tailored feedback becomes an important part of the learning symphony.

13.7.2   Supervised and Unsupervised Learning Approaches in Educational Contexts

1. The Conductor's Baton in Supervised Learning.

Consider a teaching assistant wielding a magical wand and directing pupils toward certain learning objectives. This is the essence of supervised learning, in which algorithms are trained on labeled data sets, with each data point corresponding to a desired outcome. Consider evaluating student writings: labeled data contains essays with grades as well as full rubrics. The computer learns to discover patterns linked with high and poor grades by methodically researching these pairs, eventually predicting performance on unseen writings. This enables instructors to identify students' strengths and shortcomings, provide focused comments, and provide treatments suited to their specific needs.

2. Supervised Learning Excels at Tasks such as these:

Predictive analytics: identifying students who are at danger of falling behind based on trends of attendance, engagement, and achievement [46].

Adaptive learning is the dynamic adjustment of difficulty levels, the recommendation of resources, and the provision of tailored feedback based on real-time.

Analysis of essays, code, and written comments to offer extensive feedback and individualized score breakdowns.

3. This "Conductor's Baton" Strategy Has Limitations, However.

Data Dependency: The accuracy and quantity of labeled data determine the quality of forecasts and recommendations. Biased or inadequate data might result in biased or erroneous results.

Interpretability: Because black-box algorithms might be difficult to grasp, explaining how predictions are formed or identifying potential biases can be difficult.

Overfitting: Excessively customized models may fail to generalize to new data, limiting their usefulness outside the training data set.

4. The Symphony's Unsupervised Learning: The Hidden Chorus.

Unsupervised learning goes into the inner workings of the orchestra, revealing hidden patterns and correlations within the data itself, whereas supervised learning follows the conductor's baton. Consider studying student involvement records without labels. Unsupervised algorithms uncover hidden clusters, exposing student groupings with similar learning methods, engagement patterns, or knowledge gaps. This enables educators to personalize instructional tactics, develop targeted interventions, and promote collaborative learning environments for students who share traits.

13.7.3 Unsupervised Learning Excels in various Challenges as Listed Here:

1. Student segmentation is the classification of students based on their learning habits, engagement patterns, and preferred learning methods.
2. Curriculum customization is identifying common knowledge gaps and areas of difficulty in order to change curriculum and resources dynamically.
3. Identifying hidden trends: Discovering unanticipated links between variables such as learning styles, socioeconomic backgrounds, and academic outcomes.

However, the "symphony's hidden chorus" is not without its difficulties:

1. Interpretation: Deciphering the meaning behind data clusters and patterns may be a difficult undertaking that demands meticulous investigation and experience.
2. Transforming abstract patterns into tangible instructional solutions can be difficult, necessitating collaboration between data scientists and educators.
3. Uncertainty: Unlike supervised learning, which uses labeled data sets, unsupervised learning is fraught with uncertainty, necessitating rigorous validation and interpretation of outcomes.

13.7.4 The Harmonious Ensemble: Combining Methods for Educational Insights

The ultimate promise of ML in education rests in the synergistic combination of supervised and unsupervised learning, much as a symphony thrives on the interaction of varied instruments. Labeled data from supervised learning may assist unsupervised algorithms, adding context and structure to their data exploration. Unsupervised learning, on the other hand, can reveal hidden patterns and trends that help to influence the construction of more specific supervised learning models. This partnership yields deeper insights, resulting in the following:

1. Student profiles that are more robust and nuanced: Combining supervised evaluations with unsupervised cluster analysis results in a full picture of each student's strengths, weaknesses, and learning styles.
2. Human-interpretation predictive models: Unsupervised learning can reveal previously unknown elements impacting student results, spurring further inquiry and refining supervised models with new data points.
3. Learning environments that are dynamic and adaptable: Educators may modify curriculum, resources, and teaching practices to changing student needs and trends by regularly assessing data using both supervised and unsupervised approaches.

13.8 COMPUTATIONAL MODELS ENABLING REFORMS IN WIRELESS COMMUNICATION SYSTEM

Wireless communication is a dynamic playground where data speeds rocket, towers sprout like digital trees, and connection threads itself into the fabric of our lives. A powerful force driving the future of communication is at the center of this transformation: Computational models. These complex software constructions serve as digital architects, assessing, forecasting, and improving every aspect of wireless networks, ushering in a new era of efficiency, resilience, and accessibility.

Consider conductors who methodically tune their orchestras, analyze acoustics, forecasts to avoid possible problems, and optimizes instrument placement. Computational models provide a similar job, orchestrating the intricate symphony of wireless networks by fulfilling critical roles:

1. Network Design and Optimization: Computational models swing their digital wands before a single tower graces the skyline. They examine the landscape, forecast signal coverage, optimize tower placement, and locate probable interference sources. Consider modeling network performance in a variety of metropolitan environments, identifying coverage gaps in rural locations, and providing continuous connectivity even in congested city centers. This thorough planning ensures that the network is deployed efficiently, that dropouts are minimized, and that everyone has equal access.

2. Resource Allocation and Spectrum Management: Consider the wireless spectrum to be a thriving marketplace with precious capacity for sale. Market analysts are computational models that estimate peak demand and optimize resource allocation, ensuring optimal spectrum usage. They evaluate data traffic trends, predict spikes during online events like video streaming, and dynamically assign resources to ensure continuous data flow. This enables seamless communication, avoids aggravating network congestion, and optimizes the value of every valuable megabyte.

3. Network Security and Threat Detection: Malicious actors lurk in the digital environment, aiming to disrupt communication and exploit weaknesses. Computational models serve as watchful sentinels, continually assessing network activity, detecting irregularities, and forecasting future assaults. They scan data streams for suspected patterns, detect suspicious device behavior, and sound alerts before dangers emerge. This proactive defense improves network security, protects user privacy, and maintains communication channel integrity.

4. Service Quality (QoS) and User Experience: Every user instrument deserves to be heard clearly in the symphony of communication. By prioritizing data flows, computational models assure equal performance by ensuring flawless streaming for entertainment, crystal-clear conversations for business meetings, and uninterrupted online learning sessions. They assess user requirements, prioritize bandwidth allocation, and dynamically modify latency to offer a seamless and gratifying experience for everybody, independent of data use or network traffic.

5. Emerging Technologies and Long-Term Planning: Wireless communication's future is a kaleidoscope of innovation, with technologies such as 5G, the Internet of Things (IoT), and low-power wide-area networks (LPWAN) taking center stage. Computational models are the visionary architects, laying the path for these developments by evaluating network performance under various situations, improving resource allocation for new devices, and anticipating possible issues. This strategy to future-proofing guarantees that networks stay ahead of the curve by smoothly integrating new technology and adjusting to changing user requirements.

13.9 LESSON LEARNED AND FUTURE TRENDS

Computational models are not panaceas, but rather strong tools that should be utilized alongside traditional instructional techniques with caution and ethical responsibility. We can build a dynamic and inclusive educational environment where all students may thrive and attain their full potentials by harnessing their potential while addressing the problems.

13.9.1 OPPORTUNITIES

Computational models may adjust learning experiences to individual requirements, learning styles, and speed, increasing deeper engagement and greater outcomes.

Predictive Insights: These models can identify children who are at danger of falling behind, allowing for preemptive interventions and help before problems worsen.

Intelligent Feedback and Assessment: They can give thorough, tailored feedback in addition to grades, allowing students to fine-tune their work and find areas for development.

Immersive Learning: Model-powered virtual and augmented reality technologies may generate compelling and interactive learning experiences that bring abstract concepts to life.

Global cooperation: They have the potential to bridge geographical divides by linking students and educators from all around the world, boosting intercultural understanding and cooperation.

13.9.2 CHALLENGES

Data Privacy and Security: Protecting student data and ethical concerns are critical in order to prevent violating privacy and ensuring responsible technology use.

Addressing biases in training data and models is critical to preventing unfair and inequitable learning experiences for specific groups of students.

Human connection and conventional educational approaches must remain fundamental to education, complementing rather than replacing technology.

Accessibility and equity: To ensure that all students have access to and benefit from these technologies, the digital divide must be bridged as well as socioeconomic imbalances addressed.

Continuous Learning and Improvement: Models must adapt to changing problems and educational trends, which necessitates continuing training and development.

13.9.3 ADDITIONAL LESSONS

Collaboration is critical for building and implementing effective smart teaching and learning models.

Balance between Automation and Human Intervention: Technology should strengthen teachers rather than replace them. Finding a happy medium is essential for effective and responsible teaching.

Ethics and openness: Maintaining trust and preventing misuse requires ethical concerns and openness in data utilization and model algorithms.

Focus on Learning Outcomes: At the end of the day, the goal of these approaches is to improve learning outcomes and student achievement. The emphasis should continue to be on student learning and growth.

13.9.4 KEY FINDINGS FROM THE CHAPTER

The following are the key findings:

1. Computational models have enormous potential for individualized learning, forecasting demands, and providing immersive experiences.
2. Future trends indicate that customization, immersive learning, and global cooperation will become more prevalent.
3. Data privacy, algorithmic prejudice, and the digital divide are all issues that must be addressed.
4. Collaboration, ongoing learning, ethical concerns, and a focus on learning outcomes are all critical components of responsible implementation.
5. Using technology to its full potential may help to create a dynamic, individualized, and inclusive learning environment for all.

13.9.5 FUTURE TRENDS

The potential for teaching in wireless communication networks are endless. We can use the power of computational modeling to build a dynamic, customized, and inclusive learning environment where every student may thrive and realize their full potential by addressing the hurdles and emphasizing ethical issues. To fully realize the promise of computational modeling in wireless communication networks for smart teaching and learning, we must do the following:

Encourage interdisciplinary cooperation: Interdisciplinary collaboration among educators, data scientists, engineers, and policymakers is critical for building and implementing responsible and successful models.

Continuous learning and adaptation: Education and models must change in tandem. To remain relevant and responsive to changing requirements, educational techniques must adapt to technological changes, and models must be regularly updated.

Transparency, accountability, and the ethical use of models and data are vital for maintaining confidence and ensuring fair learning opportunities for all.

Concentrate on learning outcomes: The ultimate purpose of these strategies is to improve student learning and development. Technology should be used to enhance rather than replace the power of teaching and learning.

13.10 CONCLUSION

The foreseeable future of education shimmers with the melody of possibilities as we stand at the intersection of technology and pedagogy. Computational models, integrated into the fabric of wireless communication networks, provide a powerful conductor's baton, poised to command a symphony of intelligent teaching and learning. This vision of individualized learning, predictive insights, and immersive experiences offers enormous potential, but there are murmurs of conflicting problems. Consider an orchestra in which individual students learn at their own pace, aided by adaptive routes shaped by data-driven algorithms. Consider early warnings identifying problematic students and preventing their notes from fading in the midst of the symphony. Consider virtual environments that convey pupils to the heart of history, inspiring their desire to learn. This is the power of computational models: to weave a tapestry of individual progress and global collaboration.

However, as with any strong instrument, these models must be handled with caution. Ensuring data privacy and security necessitates strong encryption and ethical concerns, mistrust destabilize the balance. Combating algorithmic bias requires ongoing monitoring in order to ensure that every student has an equal voice in the orchestra. Bridging the digital gap is critical, or else certain instruments would go quiet. It is critical to strike a balance between human engagement and AI instruction, because even the most virtuoso conductor cannot replace the human touch. Finally, ethical innovation and sustainable development must govern every technological advance, ensuring that this symphony does not come at the expense of the environment.

Navigating these problems necessitates a collaborative mindset. Educators, data scientists, technologists, and politicians must work together to create a unified score for the future of learning. The models will be refined via ongoing research and development, ensuring that they progress in tandem with educational best practices. Transparency and ethical concerns must be woven into every algorithm in order to create trust and avoid dissonance. Above all, the emphasis must remain on learning outcomes, with technology serving as a supplement to pedagogy rather than a replacement for human connection and critical thinking.

The promise of computational models in wireless communication networks for smart teaching and learning is a dynamic composition waiting to be produced, not a pre-recorded track. We can build a symphony of education in which all students discover their own voices, add to the melody, and flourish within the harmonic orchestra of learning by embracing technology with open minds, confronting difficulties with a collaborative attitude, and emphasizing student achievement.

REFERENCES

[1] A. Collins, & R. Halverson. (2018). *Rethinking Education in the Age of Technology: The Digital Revolution and Schooling in America*. Teachers College Press. pp. 1–11.

[2] R. McHaney. (2023). *The New Digital Shoreline: How Web 2.0 and Millennials are Revolutionizing Higher Education*. Taylor & Francis. pp. 80–88.

[3] D. M. West. (2014). *Going Mobile: How Wireless Technology is Reshaping Our Lives*. Brookings Institution Press.

[4] A. A. Emmanuel, A. E. Adedoyin, O. Mukaila, & O. O. Roseline. (2020). Application of Smartphone Qrcode Scanner as a Means of Authenticating Student Identity Card. *International Journal of Engineering Research and Technology*, *13*(1), 48–53.

[5] A. Koutropoulos, & A. Girelli. (2015). The Intersection of Learning, Globalization and Technology: Historical Perspectives and Future Outlooks. *The Sage Handbook of Research in International Education*, Sage Publications Ltd, 246–261.

[6] W. H. Dutton, G. Law, D. Groselj, F. Hangler, G. Vidan, L. Cheng. ... & B. Wang. (2014). Mobile Communication Today and Tomorrow. *A Quello Policy Research Paper, Quello Center, Michigan State University*. 12(4), 1–74.

[7] K. M. Abiodun, E. A. Adeniyi, D. R. Aremu, J. B. Awotunde, & E. Ogbuji. (2021, November). Predicting Students Performance in Examination using Supervised Data Mining Techniques. In *International Conference on Informatics and Intelligent Applications*. Cham: Springer International Publishing. pp. 63–77.

[8] C. H. Ei, & C. Soon. (2021). *Towards a Unified Framework for Digital Literacy in Singapore*. IPS Working Papers vol. 39.no. 4.

[9] J. Wu, S. Guo, H. Huang, W. Liu, & Y. Xiang. (2018). Information and Communications Technologies for Sustainable Development Goals: State-of-the-Art, Needs and Perspectives. *IEEE Communications Surveys & Tutorials*, *20*(3), 2389–2406.

[10] M. Adebiyi, F. Oladeji, D. Ajiboye, R. O. Ogundokun, & A. E. Adeniyi. (2019). RFID-Based Human Tracking System in Tertiary Institution. *Journal of Engineering and Applied Sciences*, *14*(7), 2345–2351.

[11] M. Kopish, & W. Marques. (2020). Leveraging Technology to Promote Global Citizenship in Teacher Education in the United States and Brazil. *Research in Social Sciences and Technology*, *5*(1), 45–69.

[12] V. M. Cardullo, N. S. Wilson, & V. I. Zygouris-Coe. (2018). Enhanced Student Engagement Through Active Learning and Emerging Technologies. *Student Engagement and Participation: Concepts, Methodologies, Tools, and Applications*, IGI Global Publisher, 399–417.

[13] A. Christodoulou, & C. Angeli. (2022, June). Adaptive Learning Techniques for a Personalized Educational Software in Developing Teachers' Technological Pedagogical Content Knowledge. *Frontiers in Education*, *7*, 789397.

[14] A. R. Carvalho, & C. Santos. (2022). Developing Peer Mentors' Collaborative and Metacognitive Skills with a Technology-Enhanced Peer Learning Program. *Computers and Education Open*, *3*, 100070.

[15] H. G. Seyhan. (2022). Examining the Effect of Using Mobile Technologies in Chemistry Laboratory on Self-Directed Learning Readiness: An Action Research. *Education Quarterly Reviews*, *5*(2), 313–325.

[16] G. J. Ajamu, J. B. Awotunde, T. B. Jimoh, E. A. Adeniyi, K. M. Abiodun, I. D. Oladipo, & M. Abdulraheem. (2024). Online Teaching Sustainability and Strategies during the COVID-19 Epidemic. In *The Role of Sustainability and Artificial Intelligence in Education Improvement*. Chapman and Hall/CRC, United Kingdom. pp. 106–132.

[17] M. K. Abiodun, J. B. Awotunde, E. A. Adeniyi, R. O. Ogundokun, & S. Misra. (2022). Analytics of IoT-Based System for Monitoring Students' Progress in Educational Environment. In *New Frontiers in Cloud Computing and Internet of Things*. Cham: Springer International Publishing. pp. 323–342.

[18] H. A. Althibyani, & A. M. Al-Zahrani. (2023). Investigating the Effect of Students' Knowledge, Beliefs, and Digital Citizenship Skills on the Prevention of Cybercrime. *Sustainability*, *15*(15), 11512.

[19] T. Kabudi, I. Pappas, & D. H. Olsen. (2021). AI-Enabled Adaptive Learning Systems: A Systematic Mapping of the Literature. *Computers and Education: Artificial Intelligence*, *2*, 100017.

[20] A. J. Kehinde, A. E. Adeniyi, R. O. Ogundokun, H. Gupta, & S. Misra. (2022). Prediction of Students' Performance with Artificial Neural Network using Demographic Traits. In *Recent Innovations in Computing: Proceedings of ICRIC 2021*. Singapore: Springer Singapore. *Volume 2*. pp. 613–624.

[21] D. Ramandanis, & S. Xinogalos. (2023). Investigating the Support Provided by Chatbots to Educational Institutions and Their Students: A Systematic Literature Review. *Multimodal Technologies and Interaction*, 7(11), 103.

[22] S. K. Chu, R. B. Reynolds, N. J. Tavares, M. Notari, & C. W. Lee. (2021). *21st Century Skills Development Through Inquiry-Based Learning from Theory to Practice*. Springer International Publishing.

[23] R. Schwartz, A. Vassilev, K. Greene, L. Perine, A. Burt, & P. Hall. (2022). Towards a Standard for Identifying and Managing Bias in Artificial Intelligence. *NIST Special Publication*, 1270(10), 6028.

[24] A. Ishaque, & S. Rehman. (2023). Psychological Landscape: A Symphony of Disciplines to Unravel the Complexity of Human Society. *Social Sciences Spectrum*, 2(1), 30–40.

[25] Robertson, I., Somjit, N., & Chongcheawchamnan, M. (2016). *Microwave and Millimetre-Wave Design for Wireless Communications*. John Wiley & Sons.

[26] D. Zibar, R. Sambaraju, A. Caballero, J. Herrera, U. Westergren, A. Walber, ... & I. T. Monroy. (2011). High-Capacity Wireless Signal Generation and Demodulation in 75-to 110-GHz Band Employing All-Optical OFDM. *IEEE Photonics Technology Letters*, 23(12), 810–812.

[27] A. Collins, & R. Halverson, R. (2018). *Rethinking Education in the Age of Technology: The Digital Revolution and Schooling in America*. Teachers College Press.

[28] K. D. Good. (2020). *Bring the World to the Child: Technologies of Global Citizenship in American Education*. MIT Press.

[29] L. N. Duffy, G. A. Stone, J. Townsend, & J. Cathey. (2022). Rethinking Curriculum Internationalization: Virtual Exchange as a Means to Attaining Global Competencies, Developing Critical Thinking, and Experiencing Transformative Learning. *SCHOLE: A Journal of Leisure Studies and Recreation Education*, 37(1-2), 11–25.

[30] R. O. Ogundokun, J. B. Awotunde, J. B. S. Misra, T. Segun-Owolabi, E. A. Adeniyi, & V. Jaglan. (2021, February). An Android Based Language Translator Application. *Journal of Physics: Conference Series*, 1767(1), 012032.

[31] C. M. Plump, & J. LaRosa. (2017). Using Kahoot! in the Classroom to Create Engagement and Active Learning: A Game-Based Technology Solution for eLearning Novices. *Management Teaching Review*, 2(2), 151–158.

[32] N. Pellas, S. Mystakidis, & I. Kazanidis. (2021). Immersive Virtual Reality in K-12 and Higher Education: A Systematic Review of the Last Decade Scientific Literature. *Virtual Reality*, 25(3), 835–861.

[33] M. I. Soclo, O. D. Ninan, and K. C. Olufokunbi. (2022). Mobile Game-Based Learning System for a Local Language, *International Journal of Mobile Learning and Organisation*, 16(1), 52–73.

[34] O. D. Ninan, A. R. Iyanda, and A. E. Akinde. (2019). A Mobile Game-Based Learning System for Diacritic Insertion, *International Journal of Smart Technology and Learning*, 1(4), 344–360.

[35] K. O. Jeong. (2022). Facilitating Sustainable Self-Directed Learning Experience with the Use of Mobile-Assisted Language Learning. *Sustainability*, 14(5), 2894.

[36] A. Topîrceanu. (2017). Gamified Learning: A Role-Playing Approach to Increase Student In-Class Motivation. *Procedia Computer Science*, 112, 41–50.

[37] Z. T. Zhu, M. H. Yu, & P. Riezebos. (2016). A Research Framework of Smart Education. *Smart Learning Environments*, 3, 1–17.

[38] T. Kabudi, I. Pappas, & D. H. Olsen. (2021). AI-Enabled Adaptive Learning Systems: A Systematic Mapping of the Literature. *Computers and Education: Artificial Intelligence*, 2, 100017

[39] L. B. Nilson, & C. J. Stanny. (2015). *Specifications Grading: Restoring rigor, Motivating Students, and Saving Faculty Time*. Routledge.

[40] D. Demszky, D. Yang, D. S. Yeager, C. J. Bryan, M. Clapper, S. Chandhok, S., ... & J. W. Pennebaker. (2023). Using Large Language Models in Psychology. *Nature Reviews Psychology*, 2(11), 688–701.

[41] C. A. Lenkaitis, & B. Loranc. (2021). Facilitating Global Citizenship Development in Lingua Franca Virtual Exchanges. *Language Teaching Research*, 25(5), 711–728.

[42] C. Janiesch, P. Zschech, & K. Heinrich. (2021). Machine Learning and Deep Learning. *Electronic Markets*, 31(3), 685–695.

[43] T. W. Janssen, J. K. Grammer, M. G. Bleichner, C. Bulgarelli, I. Davidesco, S. Dikker, ... & N. van Atteveldt. (2021). Opportunities and Limitations of Mobile Neuroimaging Technologies in Educational Neuroscience. *Mind, Brain, and Education*, 15(4), 354–370.

[44] E. A. Adeniyi, S. A. Ajagbe, O. A. Oki, A. O. Adebayo, & O. Olasupo. (2023). Application of Machine Learning Algorithm in Cloud-to-edge Computing: Analysis and Limitations. *2023 IEEE AFRICON*, Nairobi, Kenya, 1–6. doi: 10.1109/AFRICON55910.2023.10293346

[45] S. Sedkaoui, & M. Khelfaoui. (2019). Understand, Develop and Enhance the Learning Process with Big Data. *Information Discovery and Delivery, 47*(1), 2–16.

[46] A. J. Kehinde, A. E. Adeniyi, R. O. Ogundokun, H. Gupta, & S. Misra. (2022). Prediction of Students' Performance with Artificial Neural Network using Demographic Traits. In *Recent Innovations in Computing: Proceedings of ICRIC 2021, Volume 2*. Singapore: Springer Singapore. pp. 613–624.

14

14 Stochastic Geometry and Federated Learning in Computational Modeling of Communication Systems

Samuel Ibukun Olotu

14.1 INTRODUCTION

A computational model is a set of mathematical equations that provide a description of a physical system. The physical system, usually interpreted as real-life problems, is simplified and represented as mathematical problems. An important system that affects our daily lives is the communication system. It is an important system that enables communicating users to talk to each other and also share information and resources from any location around the world. It is reported that 86% of the world population uses smartphones to stay connected to the globe over different sophisticated communication networks. The description of the architecture, functionality and node movements of communication networks can assist network administrators and designers get better understanding on their functionalities and operations. One modeling approach that has been applied to communication systems and networks to describing their operations and the distribution of users properly is stochastic geometry. Stochastic geometry (SG) uses probability theory to examine random phenomena on the plane or dimensions. It is successfully applied in various domains such as astronomy, biology and material science and so forth. It is also used in the modeling of various types of communication networks [1]. Stochastic geometry offers a mathematical modeling technique for wireless networks to include the operational characteristics and behaviour. Stochastic geometry-based systems achieves efficient analysis by capturing the spatial randomness inherent in wireless networks. Other conditions of uncertainties accounted for in stochastic geometry modeling include fading, shadowing, and power control. The analysis is able to yield a closed-form expressions that defines the behaviour and understanding about the operations of the system that may often be difficult to obtain from only computational simulations [2].

Federated learning (FL) utilizes distributed computing and machine learning techniques to allow for train data locally while avoiding data from being uploaded directly to the cloud server. The technique is developed by Google to predict the text input of users while keeping data on devices [3]. The global model is acquired by each local server, trained with local dataset and transmitted back to the server. The learning technique offers users an opportunity to benefit from the shared model that is trained from a rich data without centrally storing it. Due to the absence of centralized training in FL, the privacy of users is effectively protected [4]. Some properties of problems suited for federated learning are as follows: (1) When it is more beneficial to train the model on data acquired from a mobile device rather than from a remote data center. (2) When there is need for privacy of data kept locally on the devices as compared with those placed in the central data center mainly during the model training. With a supervised approach, the label on the data is simply acquired or obtained from user interaction.

DOI: 10.1201/9781003457428-15

Some existing review works on modeling and analysis of communication networks using stochastic geometry and federated learning are usually either limited to specific communication network type or an application domain. The aim of this chapter is to present a literature review of recent works on computational modeling across several communication systems. The review covers research works on modeling and analysis of communication systems using stochastic geometry and federated learning.

14.1.1 KEY CONTRIBUTIONS

The significant contributions of this chapter are as follows:

i. Discussions on key applications of stochastic geometry and federated learning in communication networks.
ii. A literature review of recent research works on modeling and analysis of communication networks using stochastic geometry and federated learning, and
iii. A literature review of recent surveys on computational modeling and analysis of communication systems and models based on stochastic geometry and federated learning.

14.1.2 ORGANIZATION OF THE CHAPTER

Section 14.2 presents a literature review of recent research works on computational modeling in stochastic geometry and federated learning. Section 14.3 presents stochastic geometry modeling techniques and its application in modeling wireless ad hoc, cognitive radio and cellular networks. In Section 14.4 the federated and machine learning models are discussed. Section 14.5 discusses research trends, lessons learned and future directions based on the study. The chapter is concluded in section 14.6 with some future directions.

14.2 RELATED WORKS

This section presents a review of recent literatures on stochastic geometry and federated learning in computational modeling of communication systems. The selected research work also covers the performance analysis and model of different communication systems.

14.2.1 STOCHASTIC GEOMETRY

Ouamri [5] presents a downlink coverage probability for mmWave small cell network (SCN) according to signal-to-interference parameter (SINR) with the assumption of directional beamforming. Ouamri et al. [6] present an expression for the analysis of the coverage probability of UAV-assisted cellular networks with imperfect beam alignment using stochastic geometry techniques. The analysis reveals that at low UAV altitude, the beam alignment errors reduce the coverage performance. Nguyen et al. [7] defined a closed-form expression to analyse the coverage probability in a typical Voronoi cell of a wireless communication network. The work analyzed the effects of geographic node deployment on the coverage probability and model aggregation rate in a small-cell cellular network. Sun and Ding [8] utilized stochastic geometry to implement the coverage probability analysis for low earth orbit (LEO) satellite networks. The contributions of work include user fairness and transmission reliability based on the characterization of LEO satellite networks.

Ouamri et al. [9] present an analysis of coverage probability and ergodic capacity for UAV-assisted mm-Wave cellular network. The Matern cluster process (MCP) is used to model the user location. The cluster size and UAV density are used to compute the association probability, while SINR is used to compute the coverage probability. The result reveals that the association probability depends on the altitude of the UAV. Xu et al. [10] derives the coverage probability for users in Space-Air-Ground-Sear integrated networks (SAGSINs) with focus on wireless communication over a body of water such as the Atlantic Ocean and the Pacific Ocean. Wang and Chun [11] present the spatial distribution of millimeter (mmWave) nodes and the cluster nodes in Terahertz (THz) band. The proposed derives provides insights on the spatial dependence between THz and mmWave tier and the effect of clustering on the network performance based on association probability, distance distribution and convergence probability.

Lin and Yu [12] analyse the reconfigurable intelligent surface (RIS)-assisted ultra-dense network (UDN) system. The work utilizes the ternary Campbell's theorem and compute the ternary probability generating functional (PGFL) of stochastic geometry. The work analyzes the coverage probability, signal statistical characteristics, area energy efficiency (AEE), area spectral efficiency (ASE) and energy coverage efficiency (ECE) of the system.

Cheng et al. [13] analyzed the large-scale heterogeneous cellular network (HCN) assisted by Intelligent reflection surface (IRS). The work derives the coverage probability and the network spatial throughput of the downlink IRS using stochastic geometry. Zhao et al. [14] analysed a RIS-assisted cellular network according to the stochastic geometry model by providing the close-form expression of coverage probability and user association probability. Zhang et al. [15] derive the expression for UE detection probability to model grant-free random access (CF-mMIMO) system. The distribution of the access points (APs) and user equipments (UEs) are according to Poisson point process (PPP).

Schmidt et al. [16] propose stochastic geometry model for interference prediction for wireless networks. The approach performs the prediction using autoregressive moving average (ARMA) model based on previous interference observations. Qiu et al. [17] introduce the interference analysis of massive non-geostationary orbit (NGSO) networks. The work models the satellites locations as random distributed point process using stochastic geometry and analyses the interference encountered by ground users in massive NGSO.

Liu et al. [18] utilize stochastic geometry to analyse the violation probability of QoS in Ultra-Reliability and Low-Latency Communications (URLLC). The work builds a Random Edge Graph Neural Network (REGNN) and reduces the QoS violation probability to outperform the model-based exhaustive search method. Yoon et al. [19] derive a criterion for evaluating LTE spectrum resources in terms of edge quality of service (QoS). The work considers a cooperative schema for joint resource utilization of associated and base stations for resource allocation to edge user. The approach uses ergodic spectral efficiency at the edge user to devise a QoS evaluation.

Ayoubi and Spagnolini [20] derived a model for the link blockage and antenna configuration and device density in 5G dense wireless network. The work analyze the service the probability reveals that an optimal configuration is achieved based on the height of the AP and the density of the device. The authors derive tractable expressions for the characteristic function of aggregate interference power for the homogeneous distribution of UEs using the SG framework placed at an arbitrary height. The model proposed is used for link blockages, antenna configuration and device density network parameters.

Sun et al. (2023) [21] analysed the outage performance of downlink non-orthogonal multiple access (N-NOMA) using stochastic geometry. A model for the random locations of communication nodes in coordinated multi-point (CoMP) system is derived using stochastic geometry. The result of the work shows that the outage probabilities of N-NOMA outperform the OMA based CoMP scheme

Zhu et al. [22] use stochastic geometry to analyze the performance of large intelligent surface (LIS)-based cellular networks. The model consists of many users, 1-hop reflection and transmitters from LIS. The analysis is performed by considering 2-steps user association system deployment constraints in mmWave networks. The result shows that LIS enhances the area spectrum efficient (ASE) and average rate.

Shafique et al. [23] analyse the downlink performance of effect of intelligent reflecting surface (IRS)-assisted transmissions in cellular networks. The work used generalized gamma random scale to model the desired signal power. The Laplace transform of the aggregate interference is used to compute the ergodic capacity, coverage probability and energy efficiency.

Gontier et al. [24] analysed the downlink performance of Homogeneous and Inhomogeneous Cellular Networks using stochastic geometry in terms of electromagnetic field (EMF) exposure. The results of the analysis show that the optimal parameters values are able to maximize the coverage and minimize the EMF exposure.

Maeng et al. [25] derived closed-form expression for the performance of co-existing UAV radar and communication networks using stochastic geometry. The work analyses the performance of successful ranging probability (SRP) and transmission capacity (TC) in spectrum overlay multiple access (SOMA) and time-division multiple access (TDMA). The result shows that SOMA outperforms TDMA on both SRC and TCP.

Qin et al. [26] studied the uplink transmission of UAV wireless networks with inversion power control and proposes a stochastic geometry framework that models the locations of users and UAVs. The result of the work show that the altitude of UAV impacts the system reliability more than the transmit power at low value of SINR threshold.

Fadoul (2020) [27] derive a model to capture the downlink multi-antenna heterogeneous cellular network. The analysis is performed in terms of coverage probability and rate coverage achieved by a typical user. Chae et al. [28] applied stochastic geometry to analyse the downlink performance of low Earth Orbit (LEO) satellite communication and derives the expressions for coverage probability and the ergodic rate as a function of the beamwidth. The work reveals the effect of satellite density and altitude on the optimal beamwidth. Kaj and Morozova [29] analyses the operation of a slotted time Poisson-Voronoi type stochastic model of a wireless cellular communication system. The focus of the work is on the downlink traffic between the base station and user equipment to determine the possibility of the link data rate input and estimate the loss and delay.

Zhang et al. [30] investigates the packet-level performance of unmanned aerial vehicle (UAV)-assisted cellular networks using stochastic geometry. The analysis shows that by optimizing the network the throughput and energy efficiency can be improved.

Stegehuis and Weedage [31] derived an analytic expression for link strength in wireless networks. The work estimates the impact of the co-location on resource sharing by deriving the optimal range with which base stations ensure users are connected based on the amount of co-located base stations.

Ma et al. [32] analyzed the secrecy capacity and ergodic capacity performances of a downlink wireless network system using stochastic geometry. The result shows that an optimal amount of density of APs and the number of antennas per AP achieves a best secrecy transmission rate performance. Chamkhia et al. [33] derive secrecy outage probability (SOP) and average Ergodic capacity (AEC) expression for physical-layer security (PLS)-based IoT system. The physical-layer security (PLS) is proposed to provide a secured transmission, low power consumption and low computational complexity.

Feng et al. [34] developed a secured D2D communication of cellular networks that leverages covert D2D communication to hide its presence from adversaries that detect D2D communications based on the received signal powers. The system carries out the system-level security for the D2D cellular network being threatened and model the spatial configuration of the network using stochastic geometry.

Kalamkar et al. [35] presents modeling and analysis of beam management in dense 5G cellular network using stochastic geometry. The work maintains a balance between inter-site distance (ISDs), path loss, interference, and signaling overhead due to beam management in order to improve the area spectral efficiency (ASE).

14.2.2 Federated Learning

Yang and Quek [36] derive the convergence rate of FL in wireless network by taking into considering the effects of scheduling schemes, inter-cell interference and resource allocation. The convergence rate presented based on three practical scheduling policy which include, round robin (RR), random scheduling (RS) and proportional fair (PF). The result of the work shows that the iteration time of FL with PF is smaller compared to that of RS and RR when the network operates under high SINR threshold. When the SINR threshold increases, however, RR is more.

Shi et al. [37] present a joint bandwidth allocation and scheduling problem to capture the convergence rate performance of FL. The problem is solved by decoupling it into two sub-problems; bandwidth allocation and device scheduling. In the former, the derived optimal solution advices to allocate more bandwidth to device with worst channel conditions or weaker computation capabilities. The later sub-problem, is solved by revealing the trade-off between the number of rounds required to attain a certain model and latency per round.

Chen et al. [38] analyses the performance and the convergence rate for the FL algorithms with a view to minimizing the convergence time and training loss of the FL model. The work proposes a probabilistic user selection scheme that connects the base stations to users with local FL models that impact the global FL model with high probabilities. The aim is to optimize the user selection and resource block (RB) allocation. Wan et al. [39] present the convergence rate analysis of FL training with joint consideration of computing training and communication. The work considered a joint cost of the delay and power to discover the trade-off between FL convergence and system cost. The work proposes the theoretical principle for hyper-parameters and scheduling algorithms for better interactions between clients in wireless networks.

Zhao et al. (2022) [40] proposed a technique to mitigate the impact of divergence due to non-independent and identically distributed (non-IID) data in wireless network using federated learning technique. The work presents a sharing scheme to harmonize the distribution divergence. The evaluation result shows that the proposed work can improve the performance of federated learning with non-IID in wireless networks.

Lee et al. [41] propose a technique that performs the detection and aggregation of the local gradients using a Bayesian federated learning with the aim to minimize the mean-squared error (MSE). The work shows that the proposed scalable-BFL (SBFL) outperforms the conventional stochastic gradient descent algorithm. Chen et al. [42] performs a joint model aggregation, spectrum allocation and routing over multi-hop wireless networks as an optimization problem. The work transforms the non-convex mixed-integer linear programming (MILP) into a coarse-grained fixing procedure to solve the problem efficiently.

Nguyen et al. [43] presents an extension of FedAvg federated learning algorithm to mitigate the problem of heterogeneity of UE data and resource allocation in wireless Internet-of-Things (IoT) networks. The proposed technique executes stochastic gradient descent on sample sets of UEs with replacement on every global round. The results show that a small number of the UEs that participated in the training process every round still guarantees convergence. Salehi and Hossain [44] propose a FL algorithm that is suitable for resource-constrained and unreliable cellular wireless networks. The work allocates weights to each device based on the scheduling policy and transmission success probability. The work uses the sub-optimal scheduling policy to improve the convergence rate.

Xu et al. [45] presents an adaptive hierarchical FL system for resource control and edge aggregation control with the aim of reducing the latency and loss during training. The evaluation of the result show that a higher learning performance can be achieved with low latency in the training. Chen et al. [46] presents a communication-efficient FL that mitigate the scheduling and resource allocation problem. The work uses the Lagrange multiplier method to achieve efficient resource allocation during the FL training. The proposed technique achieves an improved learning performance and communication efficiency of FL over wireless IoT networks compared to a uniform resource allocation approach.

Dinh et al. [47] proposes a technique that solve the data privacy problem for heterogeneous user equipment using a federated learning algorithm. The work aims to solve the data privacy problem of the heterogeneity across user data and the UE characteristics. The proposed FL algorithm solve the resource allocation problem by considering the trade-offs between computation, communication latencies and energy consumption.

Wang et al. [48] presents a asynchronous learning technique that adapts to learning tasks, heterogeneity of users and communication environment. The work implements a centralized fusion algorithm on base station to efficiently utilize the information from local models and eliminate bias convergence. The impact of each user equipment's local dataset is enforced to be proportional to the sample share.

Yang et al. [49] derive a federated learning model for time and energy consumption using the convergence rate. The authors formulate an optimization problem for the joint communication and learning process. From the result, it is shown that the proposed technique achieves lower energy consumption to 59% when compared with conventional FL methods.

Amiri and Gunduz [50] present a distributed stochastic gradient descent technique that enable the selection of devices for transmission based on the states of the channel. The proposed analog compressed analog DSGD (CA-DSGD) scheme obtains the gradient estimates from the accumulation of errors from the last iterations and creates a spar vector of low-dimension for bandwidth reduction. The results show that when device datasets are not IID the gap between the digital and analog schemes increases.

14.2.3 RELATED REVIEWS

This subsection discusses recent related reviews on the application of stochastic geometry and federated learning in computational modeling of communication systems.

Stochastic Geometry

Amodu et al. [51] presents a survey on the analysis of D2D cellular communication, that is based on relay, using the stochastic geometry tools. The work introduces the D2D communication modes, studies the spectrum sharing models and presents an overview on the performance metrics for evaluating the relay-assisted D2D networks. Hmamouche et al. [52] presents a survey of statistical tools, stochastic geometry system model preferences and emerging research in modeling and analysis of 5G/B5G and cognitive wireless networks using stochastic geometry. The work also covers researches on the evaluation of performance metrics in the molecular communication, cooperation of stochastic geometry with machine learning, and Grothendieck proposes mathematical concept in SG. Salem et al. [53] presents a survey of contributing technologies to ultra-dense network (UDN). The technologies considered include mmWave, massive MIMO, D2D, IoT, C-RAN, Multi-RATE and hybrid beamforming technologies. The work also presents the coverage performance analysis of dense network based on stochastic geometry. Lu et al. [54] presents a survey of research on the performance analysis of large-scale wireless network using stochastic geometry with consideration for the signal-to-interference ratio. The work aims to provide the stochastic geometry analysis that captures the spatial-temporal interference correlation. Okegbile et al. [55] presents a survey and classification on stochastic geometry (SG)-based interference models in cognitive radio network (CRN). The work reviews literatures on the application of stochastic geometry to control and manage interference in cognitive radio network (CRN). The survey shows that existing techniques do not discuss the impact of the distribution of users and traffic dynamics on interference models in CRN.

Federated Learning

Bouzinis et al. [56] presents a review of the application of wireless federated learning (WFL) in 6G cellular networks. The work also discussed the challenges of in the engagement of WFL and wireless communication networks. Naser, et al. [57] presents review in the literature on the application of FL in visible light communication (VLC) networks. The work also discusses the design and promising solutions of FL based VLC systems. Shome et al. [58] presents a review of works on FL over wireless communications. The aim of the work is to excavate the bidirectional relationship between FL and wireless communication. The work discusses the role of smart radio environments for the performance improvement of the FL. A survey on existing works on FL communications that maintains the secure and smart properties is presented in [59]. The work focuses on the communication challenges and constraints in an FL environment which are bandwidth and limited computation.

Gadekallu et al. [60] presents a review on the applications of FL services in big data. The services include acquisition, storage, analytics, and privacy preservation. The potential applications include smart city, smart healthcare, smart transportation, smart gird and social media. Sirohi et al. [61] presents a survey of research done on the application areas of FL in space, air, ground, and under water. The authors also present various deployments of the FL in various applications in different domains and propose privacy and security measures in them, including smart cities, smart buildings, transportation, smart healthcare, Internet/Web, Internet-of-Things (IoT), UAVs, Internet-of-under water devices and so forth.

Du et al. [62] conducts a study on the application of FL in vehicular IoT. The work further discussed the challenges in the application of FL in wireless IoT environments. Khan et al. [63] presents evolutions of federated learning towards federated learning-powered IoT applications. The work presents a review of federated learning over IoT networks. [64] provides a survey of emergence of FL applications in IoT networks.

The potential FL-enabled IoT services explored includes, data offloading and caching, IoT data sharing, attack detection, mobile crowd-sensing, IoT privacy and security and localization. Various integrations of FL-IoT applications surveyed includes smart transportation, smart healthcare Unmanned Aerial Vehicles (UAVs), smart industry and smart cities. Parimala et al. [65] surveys the approaches and techniques involved in the integration of FL and Industrial IoT (IIoT) based on privacy, resource and data management. The work discussed the advantages of the extension based on communication cost, computation power and energy consumption.

Lim et al. [66] discusses the challenges of the application of FL in a mobile edge network involving devices with varying constraints. Among the challenges are resource allocation, communication cost, data security and privacy. A review of approaches to mitigate these challenges is also presented. Abreha et al. [67] provides a survey of literature on the implementation of FL in edge computing (EC) environments. The work also discusses the challenges in the implementation of FL at the edge, which include a communication and computation efficiency, heterogeneity, privacy and security, client selection and resource allocation, and service pricing. The work further presents an overview of existing developments and solutions to the challenges in the EC paradigm. Brecko et al. provides an overview of the methods used in FL with focus on edge devices with limited computational resources. The paper also presents FL frameworks that provide communication between clients and servers. The discussion on the architecture, communication, design and analysis of FL is also presented in the work.

Liu et al. [68] presents a survey of existing functional architecture, distributed training and techniques for data manipulation of federated learning systems.

The review on the application of federated learning in software-defined networks (SDN) is presented by [69]. Three major issues with the application of FL over SDN are given as (i) incentives mechanisms to encourage the participation of clients in FL training and more contributions towards global model aggregation; (ii) increase in security and privacy strategies for a risk-free communication; (iii) high performance global model aggregation mechanism

14.3 STOCHASTIC GEOMETRY MODELING

This section discusses some important concepts that are used in the modeling of communication systems using stochastic geometry modeling such as signal interference, geometric point processes and communication networks.

14.3.1 SIGNAL INTERFERENCE

Some factors that affect the efficiency of the operations in a wireless network include the shared wireless medium, the transmit power decay, and the network geometry. The signal power decays with increase in the distance of the receiver from the transmitter and the receiver [70] as described in equation (14.1).

$$P_r(y) = P_t(x) A h_{xy} \|x - y\|^{-\eta} \tag{14.1}$$

Where $x \in \mathbb{R}^d$ represent the spatial locations of the transmitter while $y \in \mathbb{R}^d$ refers to that of the receiver. $P_t(x)$ is the transmission power, h_{xy} represents the power gain between locations x and y. $x - y$ represents the Euclidean norm, A and η is a propagation constant and the path-loss respectively. The network geometry affects the SINR at the receivers in the network as defined in equation (14.2).

$$SINR(y) = \frac{P_i(x_0) A h_{xoy} \|x_0 - y\|^{-\eta}}{W + \sum_{x \in \mathcal{I}} P_t(x) A h_{xy} \|x_0 - y\|^{-\eta}} \tag{14.2}$$

Where y and x_0 represent the transmitter and receiver locations respectively.

14.3.2 GEOMETRIC POINT PROCESSES

A point process (PP) is used to model events at random intervals based on the time axis or the space axis, $\mathbb{X}$. The space $\mathbb{X}$ is either a space of compact subsets of $\mathbb{R}^d$, or the space of flats of a certain dimension in $\mathbb{R}^d$ [71]. The work in [72] defines PP as a random point in a measure space of wireless networks, given in equation (14.3).

$$\Phi = \{X_i, i \in \mathbb{N}\} \tag{14.3}$$

PP consists of a distribution of points in $\mathbb{R}^d$ based on some probabilistic law expressed using finite-dimensional distribution [73] as defined in equation (14.4).

$$Pr\big(\Phi(B_1) = n_1, \ldots, \Phi(B_k) = n_k\big) \tag{14.4}$$

Where the notation $\Phi(B_k) = n_i$ represents n points, Φ is some PP, B is a subset of space $\mathbb{R}^d$. PP represents a finite point measure obtained locally and only assumes integer values [74] as given in equation (14.5).

$$\Phi = \sum_i \delta_{X_i} \tag{14.5}$$

Where Φ and X_i represent point process and the random variables with Euclidean space values ($\mathbb{R}^d$) respectively.

Poisson point process (PPP) is a common PPs modeling technique applied in wireless networks. It gives an analytically tractable description of networks based on the statistical properties where nodes are independently distributed [75].

The location of the base station in real-life deployments are neither regular (hexagonal) nor entirely random (PPP model) and it is usually located in clustered form with deployments in a region with more concentration of people [75].

The Poisson cluster process consists of Poisson process Φ with density λ_p and replacing individual points $x \in \Phi$ by some cluster Z_x which are finite point processes (8). The Poisson process obtained by the superposition of all cluster is given in equation (14.6).

$$Y = \bigcup_{x \in \Phi} Z_x \tag{14.6}$$

14.3.3 Communication Networks

Stochastic geometry has been utilized to model ad hoc and sensor networks and properly model with support for their intrinsic random spatial nature. Some applications of PPP include the analysis of ad hoc networks, cognitive radio networks and cellular networks.

14.3.3.1 Ad Hoc Networks

Wireless ad hoc network refers to a local area network that is formed spontaneously and allows users with wireless devices to be connected without a dedicated central server. The transmitting nodes yield a stationary Poisson point process on the domain with each transmitting node having a unique associated receiver. Assume a wireless ad hoc network, the containing transmitters that are distributed spatially will follow the Poisson process point process (PPP) defined by equation (14.7).

$$\Phi_{Tx} = \{x_i \mid i \in \mathbb{N}\} \subset \mathbb{R}^2 \tag{14.7}$$

Where λ represent density.

14.3.3.2 Cognitive Radio Networks (CRN)

CRN enables users to share the wireless spectrum without the need to acquire spectrum licenses. Two categories of users in CRN are primary and secondary user which are licensed and unlicensed users respectively. The condition for the secondary users is to ensure that data of primary users is protected [76]. Existing cognitive radio networks (CRNs) environment research use stochastic geometry in analyzing the interference, coverage probability and outage probability of the system [77].

(i) Interference

The interference problem occurs in CRN when the primary and secondary users share the limited frequency channels with the following types:

1. The interference that occurs at the primary receiver from signal transmission from the primary transmitter.
2. The interference that occurs at the primary receivers from the cognitive transmitters.
3. The interference at the cognitive receiver from transmission from cognitive transmitters.

4. Interference that occurs at the cognitive receivers from transmissions from the primary transmitters.

It is therefore necessary to model the distributed users (primary and secondary) with appropriate point process technique in CRN to achieve the desired interference control and management.

The distribution of primary and secondary users is therefore assumed to follow HPPP distribution and an independent PPP distribution respectively.

The signal-to-noise plus interference ratio (SINR) is an important parameter that is used to characterize the interference in the network. Considering a CRN in which the primary transmitters follow the distribution HPPP given as Φ_p with an intensity defined as λ_p. Let the secondary transmitters follow a distribution of independent PPP Φ_s with an of intensity λ_s. Equation (14.8) defines the SINR experienced at the primary receiver y_k^p [55].

$$SINR_{y_k^p} = \frac{P_p h_{X_{pp}} \left\| X_{pp} \right\|^{-a}}{I_{pp} + I_{sp} + \sigma^2} \tag{14.8}$$

Where $\sigma^2, \left\| X_{pp} \right\|, h_{X_{pp}}, I_{pp}$ represent the noise signal power, Euclidean distance between an active primary transmitter, random channel gain in the presence of fading and shadowing between any active primary transmitter-primary receiver pair, interference power from both active primary transmitters and active secondary transmitters respectively.

(ii) Coverage Probability
The probability that the SINR at receiver is higher than the pre-defined threshold is the coverage probability. The coverage probability is given by equation (14.9)

$$P_{co}^p = P\left(SINR_{y_k^p} > \theta_p \right) \tag{14.9}$$

Where $SINR_{y_k^p}$ and θ_p represent the SINR at the receiver and the threshold respectively.

14.3.3.3 Cellular Networks

There was an initial assumption that cellular networks are spatially deployed according to a hexagonal grid but studies show that the cellular network deviates from the hexagonal grid structure and follows an irregular topology that randomly changes from one geographical location to another [78]. The cellular network consists of a directional mobile terminal that moves randomly along a straight line with a given speed as shown in Figure 14.1. The line of the mobile terminal movement is considered along the x-axis and pass through the origin. It is assumed that each base station serves one mobile terminal at a time and the mobile terminal achieves this by associating itself with the nearest base station. The associated result in base station cells shows a form of Poisson-Voronoi tessellation.

(i) PPP-based Modeling
Consider a downlink (single-tier) cellular network. The base stations are modeled as various points of a homogeneous PPP of a given density. The mobile terminals are also modeled as homogeneous PPP of a given density that are independent. Each base station has some orthogonal Resource Blocks (RBs) for serving the mobile terminals and avoid interference within the cells. Each base station

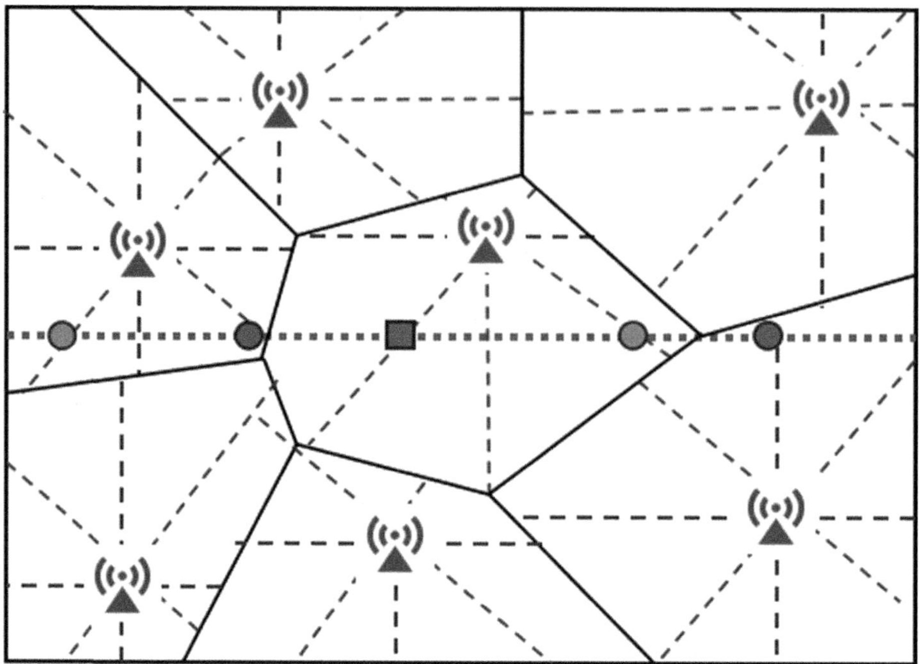

FIGURE 14.1   Poisson cellular network with mobile terminals.

transmits with constant power in each resource block. Some transmit power are budget for each base station and some other used by each resource block.

(ii) Link State Modeling

Consider an arbitrary link distance from a base station to a mobile terminal. Each link can be in different states due to some environmental-dependent blockages. If the set of states is denoted by S, the probability of being in state, s, is defined by equation (14.10).

$$\sum_{s \in S} p_s(r) = 1 \tag{14.10}$$

Where r is the link between a base station and mobile terminal, S represent the set of link states, and $p_s(.)$ is the probability that a link is in state $s \in S$.

(iii) Channel Modeling

The distribution probability of path-loss, shadowing and fast-fading experienced at the channels of a cellular network depends on the link states. The links in the same state are independent and identically distributed (IID) [79]. Considering a generic link of length in state $s \in S$. The path-loss [80] is defined by equation (14.11).

$$l(r_n) = \mathcal{K} r_n^{\beta} \tag{14.11}$$

Where r_n denotes the length of the link. $\mathcal{K}$ and β are the path-loss constant and the path-loss slope, respectively.

The shadowing follows a log-normal distribution with mean equal to μ_s (in dB) standard deviation equal to σ_s (in dB) [79] as given in equation (14.12).

$$f_{X_s}(x) = \frac{10}{Ln(10)} \frac{1}{\sqrt{2\pi}\sigma_s x} \, exp\left(\frac{\left(10\log_{10} x - \mu_s\right)^2}{2\pi_s^2}\right) \tag{14.12}$$

Where s denotes the link state, μ_s represent the mean and σ_s is standard deviation of the value

The power gain due to small-scale fading, g_s, follows a gamma distribution with fading parameter m_s and mean Ω_s. The probability distribution function is given by equation (14.13).

$$f_{g_s}(x) = \frac{m_s^{m_s} x^{m_s - 1}}{\Omega_s^{m_s} \Gamma(m_s)} \, exp\left(-\frac{m_s x}{\Omega_s}\right) \tag{14.13}$$

where $\Gamma(.)$ represent the gamma function.

(iv) Cell Association Criterion

A cell association criterion is based on the highest average received power. The serving base station, BS_0 is obtained by equation (14.14) while the intended link is given in equation (14.15).

$$BS_0 = argmax_{BS_n \in \Psi_{BS}} \left\{1 / L_n\right\} \tag{14.14}$$

$$L_n = min_{r_n \in \Psi_{BS}} \left\{L_n\right\} \tag{14.15}$$

14.4   FEDERATED LEARNING MODELING

A federated learning (FL) system consists of a central server and a group of distributed clients. Figure 14.2 shows a FL with a cellular network that is made up of a base station for users. The connection between the server and clients is via wireless cellular channel. Every user maintains a local dataset with which they use to train the local models. All the users and base station implement the federated learning algorithm in a cooperative manner over wireless networks. The federated learning model trained by the datasets of individual user is referred to the local FL model while the model generated by the base station from collective FL models from all users is called the global FL model.

Let vector w represents parameters related to the global FL model. The loss function $f\left(w, x_{kl}, y_{kl}\right)$ is introduced to captures the FL performance over input vector x_{kl} and output y_{kl}. The loss function for learning tasks based on linear regression and logistic regression are given by equation (14.16) and equation (14.17) respectively.

$$f_j\left(w, x_{kl}, y_{kl}\right) = \frac{1}{2}\left(x_{kl}^T - y_{kl}\right)^2 \tag{14.16}$$

$$f_j\left(w, x_{kl}, y_{kl}\right) = -log\left(1 + exp\left(-y_{kl} x_{kl}^T w\right)\right) \tag{14.17}$$

The FL system consists of the data set of clients j is given by equation (14.18) [39].

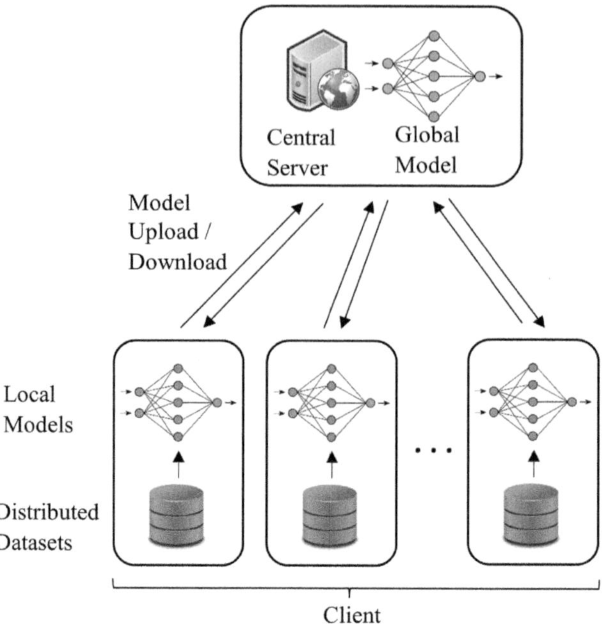

FIGURE 14.2 Federated learning architecture.

$$S_j = \left\{ \left(x_1^{(j)}, y_1^{(j)} \right), \left(x_2^{(j)}, y_2^{(j)} \right) \ldots, \left(x_{D_j}^{(j)}, y_{D_j}^{(j)} \right) \right\}$$

(14.18)

Where D_j is the data samples. The models are first separately trained in distributed manner by the clients then aggregated centrally. The loss function is given by equation (14.19).

$$f_j(w) = \frac{1}{D_j} \sum_{i=1}^{D} 1\left(w, \left(x_i^{(j)}, y_i^{(j)} \right) \right)$$

(14.19)

Where D_k denote the dataset of user j, $f_j(w)$ is the loss function, w is the model weight. The global loss function is given by equation (14.20).

$$f(w) = \sum_{j=1}^{n} q_j f(w)$$

(14.20)

Where $Q = \left(q_1, q_1, \ldots, q_N \right)$ denote the weights of data set S_1.

14.4.1 ENABLING TECHNOLOGIES

Al-Quraan et al. [81] present a review on the technologies and application of federated learning in wireless networks. The technologies discussed include multi-access edge computing (MEC), network slicing and blockchain. The applications include Internet-of-Vehicular Networks, unmanned aerial vehicle (UAV) networks, cellular networks, IoT networks and reconfigurable intelligence surface (RIS)-assisted networks.

A result of advancements in the applications of Internet-of-Things (IoT) has yielded a massive range of communication-ready end devices. The operations of these devices have resulted in the generation of huge amount of data at the edge of wireless networks. Due to the limited computation power of the device, it is usually necessary to transmit the generated data to a cloud server for necessary computation, access and storage. The limitation of the cloud computing on resource-limited real-time applications is due to an increase in overhead, occurring in the network according to energy and spectrum resources, latency and compromised security. This issue has brought about the introduction of MEC paradigm, which enables application to run closer to the customer where they are being used, the edge, rather than on the centralized cloud to reduce the network congestion, latency, improve the network operations and enhance customer experience. The application of MEC and smart end devices provides a suitable environment for employing FL algorithms. MEC enables FL algorithms to optimize latency for real-time applications with solutions for data aggregation, analytics, and computation within user proximity. An edge server has the capacity to act as an FL server while the edge devices can served as FL clients [81].

FL is a decentralized learning algorithm with two benefits, which are load balancing and privacy-preserving. The limitation of FL, however, still include the inability to keep records of participants' training contribution along with reliance of a central server prone to a sing point of failure. The decentralized data of blockchain is managed by distributed nodes and is able to play an important role in FL. The blockchain serves as a global ledger of transactions for the cryptocurrency Bitcoin. Its operation relies on the storage of multiple transactions in blocks and their encryption with the use of hash signature to obtain a chain of encrypted transactions in order. This feature ensures security in blockchain in that an agreement from all connected nodes is required to alter the content of any block. The blockchain maintains a level of transparency by ensuring that all incoming and outgoing transactions are observed by all nodes on the blockchain.

Network slicing is an enabler of B5G wireless network that tasks the physical structure of the network to create several logical and independent networks known as slices. The slices are made up of end-to-end network that fulfils several application requirements. The process enables the application of millimeter wave (mmWave) and terahertz communications (THz) in B5G/6G network to improve that capacity for devices operating in small coverage areas and the development of various Internet-of-Everything (IoE) networks. Network slicing provides each network segments with isolated and dedicated slice for a particular service. A challenge of this technique is the development of intelligent technique to efficiently management the network slices. This issue makes FL a promising enabler for network slicing [82].

14.4.2 Aggregation Algorithm

This is the method of combining local models updates from all participating clients of the training round to form the global model. The most popular aggregation algorithms are FedAvg algorithm, Scaffold, Adaptive federation optimization, FedBoost, FedProx, FedMA and the secure aggregation algorithms [83].

The FedAvg algorithm [84] is developed by Google according to stochastic gradient descent (SGD) optimization algorithm which is an efficient type of client-server architecture. Gradient descent (GD) is a commonly used algorithm in machine learning that aims to find the minimum of a differentiable function. As the dataset gets bigger the computational complexity of GD increases, which makes it unsuitable for FL systems as a result of slow convergence rate. Stochastic GD (SGD) implements gradient computation over a subset of data to significantly improve the converge rate. The training process of FedAvg algorithm is implemented by sharing the global model parameters with a group of clients selected randomly from a pool of clients. The clients then perform SGD on the local dataset a number of times in order to train the global model. The technique is repeated many times and developed to support unbalanced and non-IID data distribution. Some of the limitation of the FedAvg algorithm is the convergence problem that includes the lack of adaptive learning rate and client drifting [85, 86].

Karimireddy et al. [86] presents an algorithm that computes the direction of updates on the service model and of each client. It calculates a measure of the client drifting based on the difference. It mitigates the challenge of client drifting by implementing the local update with variance reduction technique. Some other challenges the algorithm mitigates include client heterogeneity and reduction of the communication rounds in model convergence.

The adaptive federation optimization [84] introduces adaptability in server optimization. The adaptive learning rates allow knowledge to be incorporated from previous iterations to keep the server optimization more informed. The framework used for the client optimizer minimizes loss by using local data over multiple training epochs. The global model is updated when the server performs gradient-based optimization on the average of the model updates of client. The FedAvg algorithm use SGD as client and server optimizers, incorporates the adaptive learning rates in server optimization and maintains the client storage or communication costs. It is compatible with cross-device FL but fails to remove the effect of client heterogeneity.

The FedBoost [87] is ensemble learning-based technique that trains an ensemble of pretrained-based predictors through federated learning. It uses gradient compression and model compression approach to reduce the cost of both server-client and client-server communication. It also includes computational speedups, convergence guarantees, privacy, and the optimality of the solution for density estimation.

The FedProx algorithm [88] addresses two types of heterogeneity challenges, system and statistical heterogeneity. System heterogeneity specifies the significant variable characteristics of the participating system or device in federated learning. The statistical heterogeneity specifies non-IID data within the network. Various amount of work, such as running different local epochs, can be performed locally based on the available resources. The aggregation accepts partial solutions from devices due to resource constraints. The local solver objective of the device is used to control the effect of the amount of the variable to local updates by the introduction of a proximal term.

The FedMA algorithm [89] enable federated learning to be introduced in modern network architectures for deep learning. The algorithm constructs a shared global model at the server by performing a layer-wise matching and averaging according to similarity of features across the channels of convolutional layers and fully connected layer neurons.

The secure aggregation algorithm [90] uses the concept of the secure multiparty computation algorithm. It ensures that the privacy of the client model is maintained by only sharing the learnable parameters derived from aggregation while not sharing private information of the mutually distrustful parties. The algorithm works as long as 2/3 of the clients engage in the aggregation making it fault-tolerant up to 1/3rd of users.

14.4.3 Wireless Network Applications

Research to extend FL is continually conducted to enhance, promote and determine better usage of the distributed learning algorithm. One such advancement is the adoption of FL in wireless networks. This subsection presents some main applications of FL in wireless networks according to [83].

A common wireless network is the cellular as shown in Figure 14.3 that is composed with a base station and mobile devices connected within each cell. The development of 5G has introduced the emergence of many commercial services with enhanced features provided by the cellular technology. One of such is the homogeneous cellular networks, a low frequency wireless network that is made up of macrocells with wide coverage. A major concern with this network type is with heterogeneous devices with different computational and communication capabilities. Heterogeneous network (HetNet) is composed of different cell types and used to expand the coverage of wireless networks. A benefit for implementing federated learning in HetNets is for resource allocation. The

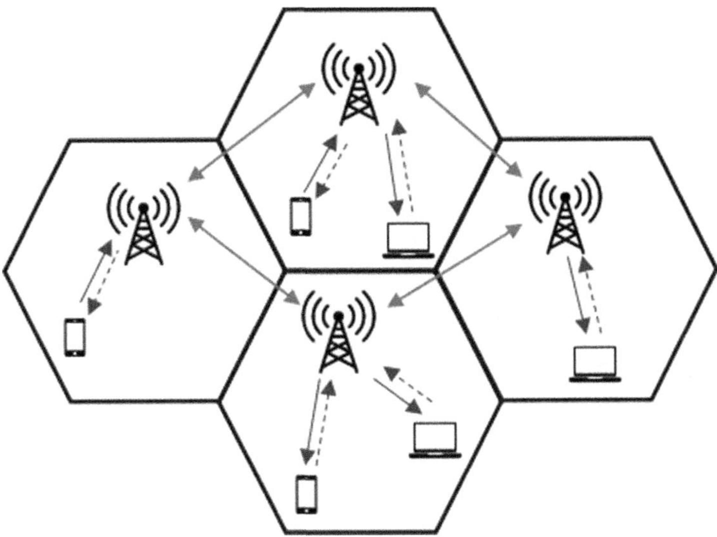

FIGURE 14.3   Cellular wireless network.

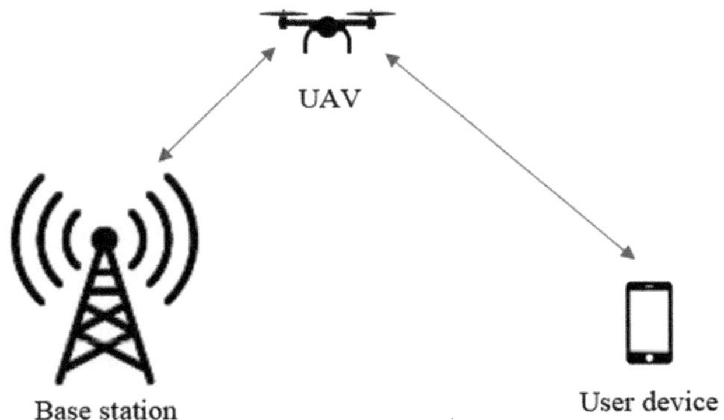

FIGURE 14.4   Unmanned Aerial Vehicle network.

high dynamicity of mmWave systems enhances the performance of massive MIMO systems by estimating the channel state information.

The Internet-of-Vehicular Networks emerged to enable intelligent transportation system (ITSs) by combining two key concepts, namely, vehicle networking and intelligence [91]. The aim of Internet-of-Vehicular Networks paradigm is to achieve smart information interaction between a vehicles and network entities. Intelligent transport system (ITS) aims to achieve communication reliability and latency without which it results in unpleasant consequence to human lives. The heterogeneous data assists in improving the accuracy of the model by considering scenarios of the network experienced by each vehicle. Some limitations of FL in server centralization are the server overhead from the exchange of large amount of updates between the server and the participating clients. The pressure on the central server challenge can be mitigated by using distributed ledgers of blockchain technique while maintaining the global and private models at each vehicle and server.

The Unmanned Aerial Vehicle (UAV) network utilizes flying vehicles with features such as low cost, mobility flexibility, and ease of deployment, enabling them to participate in many tasks considered hard to perform. Figure 14.4 shows a typical setup of UAV communication consisting of flying UAV with communication links with a base station and user device. Some application of AI algorithms in UAV technology includes AI-empowered UAV. The combination between AI and UAV networks is used to explore more complicated tasks. Other application combines UAV and Ad-hoc in a manner to create a local network and enable UAVs to jointly perform tasks [92].

Reconfigurable Intelligent Surface (RIS) enables the emergence applications such as mmWave and THz applications with concerns for attributes such as vulnerability, signal blockage and shadowing effects. RIS is composed of several reflective elements that can be artificially engineered to control the wireless signals properties and functionalities, including wave splitting, reflection, absorption, and so forth.

Internet-of-Things (IoT) has the computing capabilities to connect a wide range of objects and things to the networks. The technology aims to enhance customer services and application. AI techniques such as machine learning and deep learn are recently adopted in IoT to enable intelligent IoT systems due its ability to explore IoT data and extract insights for realizing different smart applications such as human activity classification, vehicular traffic control, weather prediction.

14.5 RESEARCH TRENDS, LESSONS LEARNED AND FUTURE DIRECTIONS

This subsection discusses the research trends of recent works in stochastic geometry and federated learning in the modeling of communication systems. The lessons learned and future directions on the open challenges are also presented.

14.5.1 RESEARCH TRENDS

A summary of the reviewed works on the stochastic geometry-based modeling and federated learning modeling of communications networks is shown in Figure 14.1 and Figure 14.2 respectively.

The works on stochastic geometry considered cellular networks, satellite communication networks or a combination of both. The cellular communication systems covered include large intelligent surface (LIS)-based cellular networks, heterogeneous cellular network, 5G cellular networks, mmWave small cell network (SCN), LTE, sensor networks, WiFi network, Ultra-reliability and low-latency communications, Intelligent reflecting surface (IRS)-assisted cellular, small-cell cellular network, Terahertz (THz) band network, grant-free random access (CF-mMIMO) system, coordinated multi-point (CoMP) system, urban cellular networks, NOMA/PDM-based IoT system, device-to-device (D2D communication. The satellite communications systems include low Earth orbit (LEO) satellite networks, space-air-ground-sear integrated network (SAGSINs) and non-geostationary orbit networks (NGSO). Some of the works consider networks with some assistance such as intelligent reflecting surface (IRS)-assisted cellular, UAV-assisted cellular networks, and reconfigurable intelligent surface (RIS)-assisted ultra dense network (UDN).

The objectives of the reviewed works include area spectrum efficient, coverage probability, rate coverage, inter-site distance (ISDs), path loss, signal-to-interference-plus-noise (SINR), interference prediction, QoS violation probability, link blockage, antenna configuration, device density, energy efficiency, ergodic capacity, beam alignment, model aggregation, ergodic capacity, association probability, spatial dependence, bias ratio, average energy efficiency (AEE), area spectral efficiency (ASE), energy coverage efficiency (ECE), network spatial throughput, association probability, joint resource utilization, cooperative allocations, outage probability, EMF exposure, busy-link probability, loss probability, throughput, energy efficiency, link strength, sharing gain, secrecy capacity, secrecy outage probability (SOP), average ergodic capacity (AEC), link reliability. The satellite network objective includes user fairness, transmission reliability, network characterization, coverage probability, association probability, interference analysis, ergodic rate. Works that consider a combination of cellular and satellite communication have the objectives that include successful ranging probability (SRP) and transmission capacity (TC), The satellite objectives include user fairness, transmission, network characterization, coverage probability, association probability, interference analysis, and ergodic rate.

The review on recent works on federated learning addresses the optimization problems such as convergence rate, joint model aggregation, spectrum allocation and routing over multi-hop wireless networks, User selection and resource block allocation, detection and aggregation of local gradients, problems scheduling, resource allocation, transmission success probability, resource control and edge aggregation control problems, bandwidth allocation, device scheduling, resource allocation, data privacy and bias convergence. Also, the techniques used in the reviewed works include Bayesian federated learning, Mixed integer linear programming (MILP), centralized fusion algorithm, Asynchronous learning. Furthermore, the metrics considered in the analysis include latency and loss during training, Computation, communication latencies and energy consumption trade-off, time and energy consumption using convergence rate. The communication models on federated learning include random wireless networks, cellular wireless networks, heterogeneous wireless networks, multi-hop wireless networks, wireless Internet-of-Things (IoT), cellular wireless networks. The objectives of cellular wireless networks include selecting devices for transmission based on the states of the channel, package loss rate, user selection, resource scheduling, heterogeneity across user data, UE characteristics, improve convergence rate using sub-optimal scheduling policy, maximize the convergence rate of the FL training based on time and not iterations, asynchronous FL framework that adapts to the heterogeneity of users, communication environments and learning tasks, minimize the total energy consumption for location computation and wireless transmission and optimize accuracy performance and cost balance. Table 14.1 and Table 14.2 show the summary of related papers on stochastic geometry and federated learning, respectively, over communication systems.

14.5.2 Lessons Learned

Some of the lesson learned about federated learning over wireless network include the following:

1. Optimizing the wireless resource allocation and user selection can minimize the loss function in FL.
2. The convergence speed can be improved by solving the wireless resource allocation, sub-optimal scheduling policy and based on time and not iterations.
3. The total computation and transmission energy of network can be minimized by optimizing the learning and communication problem.
4. The communication cost, resource allocation, data privacy and data security are consistent demand in federated learning.
5. Optimizing resource allocation minimizes energy consumption and completion time.

The study of the stochastic geometry modeling reveals the following:

1. LISs can enhance the average rate and area spectral efficiency (ASE) of mmWave networks but the reflected interference and phase-shift energy consumption will limit the performance of LIS-assisted networks when the density of the base station is higher than that of the LIS density.
2. The coverage probability of a downlink model reduces when the power control increases in the uplink.
3. The density of small base stations affects the coverage and rate probability under certain assumptions.
4. The prediction of the interference in wireless networks can improve medium access, scheduling, and radio resource allocation during interference management.

TABLE 14.1
Summary of Related Papers on Stochastic Geometry Over Communication Systems

Authors	Year	Networks	Technology	Objectives
Zhu et al [22]	2020	Large intelligent surface (LIS)-based cellular networks	Cellular	Area spectrum efficient (ASE) and average rate
Fadoul [27]	2020	Multi-antenna heterogeneous cellular network	Cellular	Coverage probability and rate coverage
Kalamkar et al. [35]	2020	Dense 5G cellular networks	Cellular	Inter-site distance (ISDs), path loss, interference, and signaling overhead due to beam management, area spectral efficiency (ASE)
Ouamri [9]	2021	mmWave small cell network (SCN)	Cellular	SINR
Schmidt et al. [16]	2021	LTE, sensor networks, and coexistence of Wi-Fi and LTE	Cellular	Interference prediction
Liu et al. [18]	2022	Ultra-Reliability and Low-Latency Communications (URLLC), Random Edge Graph Neural Network (REGNN)	Cellular	QoS violation probability
Ayoubi and Spagnolini [20]	2022	Dense 5G cellular networks	Cellular	Aggregate interference, link blockage, antenna configuration, device density
Shafique et al. [23]	2022	Intelligent reflecting surface (IRS)-assisted cellular networks	Cellular	Energy efficiency, ergodic capacity, download performance
Ouamri et al. [6]	2023	UAV-assisted cellular networks	Cellular	beam alignment
Nguyen et al. [7]	2023	Small-cell cellular network	Cellular	model aggregation
Sun and Ding [8]	2023	Low earth orbit (LEO) satellite networks	Satellite	user fairness, transmission reliability, network characterization
Ouamri et al. [6]	2023	UAV-assisted mm-Wave cellular network	Cellular	Coverage probability, ergodic capacity and association probability
Xu et al. [10]	2023	Space-Air-Ground-Sea integrated networks (SAGSINs)	Satellite	Coverage probability and association probability
Wang and Chun [11]	2023	Millimeter (mmWave) and Terahertz (THz) band network	Cellular	Association probability, Spatial dependence, bias ratio
Lin and Yu [12]	2023	Reconfigurable intelligent surface (RIS)-assisted ultra-dense network (UDN)	Cellular	Area energy efficiency (AEE), area spectral efficiency (ASE) and energy coverage efficiency (ECE)
Cheng et al. [13]	2023	Large-scale Heterogeneous Cellular Networks	Cellular	Coverage probability, network spatial throughput
Zhao et al. [14]	2023	Reconfigurable intelligent surfaces (RIS)-assisted cellular network	Cellular	Coverage performance, association probability
Zhang et al. [15]	2023	Grant-free random access (CF-mMIMO) system	Cellular	Coverage probability, user association probability, SINR, GFRA
Qiu et al. [17]	2023	Non-geostationary orbit (NGSO) networks	Satellite	Interference analysis

Yoon et al. [19]	2023	LTE network	Cellular	Joint resource utilization and cooperative allocations
Sun et al. [21]	2023	Coordinated multi-point (CoMP) system	Cellular	Outage probability
Gontier et al. [24]	2023	Urban Cellular Networks	Cellular	EMF exposure, coverage probability, download performance
Maeng et al. [25]	2023	Cellular-connected unmanned aerial vehicles (UAVs)	Cellular/Satellite	Successful ranging probability (SRP) and transmission capacity (TC)
Qin et al. [26]	2023	UAV-assisted cellular networks	Cellular/Satellite	Signal-to-interference-plus-noise (SINR)
Chae et al. [28]	2023	Low Earth Orbit (LEO) satellite communication	Satellite	Coverage probability and the ergodic rate as a function of the beam width
Kaj and Morozova [29]	2023	Cellular Networks	Cellular	Coverage probability, busy-link probability, throughput, loss probability and delay
Zhang et al. [15]	2023	Unmanned aerial vehicle (UAV)-assisted heterogeneous cellular networks	Cellular	Throughput and energy efficiency
Stegehuis and Weedage [31]	2023	Cellular Network	Cellular	Coverage probability, link strength, sharing gain
Ma et al. [32]	2023	Cellular Networks	Cellular	Secrecy capacity and ergodic capacity
Chamkhia et al. [33]	2023	NOMA/PDM-based IoT system	Cellular	Secrecy Outage Probability (SOP), and the Average Ergodic Capacity (AEC)
Feng et al. [34]	2023	Device-to-device (D2D) communication	Cellular	SINR, Link reliability

TABLE 14.2
Summary of Related Papers on Federated Learning over Communication Systems

Authors	Year	Communication model	Objectives
Yang and Quek [36]	2020	Random wireless networks	Random scheduling (RS), round robin (RR), and proportional fair (PF) scheduling schemes, inter-cell interference and resource allocation
Shi et al. [37]	2020	Cellular wireless networks	Maximize the convergence rate of the FL training based on time and not iterations
Lee et al. [41]	2020	Heterogeneous wireless networks	Minimize mean-squared error (MSE)
Amiri and Gunduz [50]	2020	Cellular wireless networks	Select devices for transmission based on the states of the channel
Chen et al. [93]	2021	Cellular wireless networks	Package loss rate, user selection and resource scheduling
Nguyen et al. [47]	2021	Wireless Internet-of-Things (IoT) networks	Minimize energy consumption or completion time
Salehi and Hossain [44]	2021	Cellular wireless networks	Improve convergence rate using Sub-optimal scheduling policy
Dinh et al. [94]	2021	Cellular wireless networks	Heterogeneity across users' data and the UEs' characteristics
Yang et al. [49]	2021	Cellular wireless networks	Minimize the total energy consumption for local computation and wireless trans- mission
Zhao et al. [40]	2022	Cellular wireless networks	Optimize accuracy performance and cost balance
Chen et al [42]	2022	Multihop Wireless Networks	Model aggregation, spectrum allocation and routing
Xu et al. [45]	2022	Wireless networks	Achieve higher learning performance with lower training latency
Chen et al. [46]	2022	Wireless Internet-of-Things (IoT) networks	Client scheduling subproblem and a resource allocation
Wang et al. [48]	2022	Cellular wireless networks	Asynchronous FL framework that adapts to the heterogeneity of users, communication environments and learning tasks,

5. The altitude of the deployed UAV affects the coverage performance, association probability and system reliability.
6. The IRSs can significantly enhance the network performance of HCN while not changing the asymptotic behaviour of the conventional HCN.

14.5.3 Future Directions

The following are some future directions on the computational modeling of communication system using stochastic geometry:

1. The distributed antenna systems (DAS) can be further investigated and extended with different types of multi-AP coordination at the base stations and the user devices.
2. The analysis of the coverage based on 3D directional beamforming.
3. The modulation scheme for THz communication with consideration for clustering of THz nodes.

The federated learning also shows the following future directions:

1. Implementation of over-the-air computation and multi-task learning when sharing wireless medium through the mobile devices
2. Personalized FL models to adapt to newly collected data.
3. Development of prototypes/testbeds to validate the proposed FL performance presented in this work.
4. Investigation on the heterogeneity of the system based on the computation capabilities and mobility methods of the mobile devices.

14.6 CONCLUSION

This chapter presents a review of computational modeling and analysis of communication systems and networks based on stochastic geometry and federated learning. Stochastic geometry provides a consolidated mathematical technique to model communication systems and networks, understand their behaviour, and characterise their operation. A stochastic geometry-based analysis is able to capture the spatial randomness inherent in wireless networks. Federated learning (FL) is a distributed machine learning scheme that allows data to be trained locally and avoids the upload of data directly to the cloud server. The learning technique gives users an opportunity to benefit from the shared model, which is trained from rich data without centrally storing it. Some concepts discussed include the signal-to-interference metric, geometric point processes and the communication networks that can be modeled using stochastic geometry. The discussion on federated learning modeling covers the enabling technologies, aggregation algorithms and the applications in wireless networks. Most of the existing reviewed on modeling and analysis of communication system using stochastic geometry and federated learning only cover specific communication networks and application domains. The aim of the chapter is to review recent works on computational modeling and analysis of available communication systems and application domains using stochastic geometry and federated learning. The work discusses the concepts, applications and review of research on modeling and analysis of communication systems using stochastic geometry and federated learning. The research trends and lesson learned from the review are also discussed. Some future directions of the work from the survey are then presented in the chapter.

REFERENCES

[1] M. Haenggi, S. Member, J. G. Andrews, and S. Member, "Stochastic Geometry and Random Graphs for the Analysis and Design of Wireless Networks," *IEEE Journal on Selected Areas in Communications*, vol. 27, no. 7, pp. 1029–1046, 2009.

[2] H. ElSawy, A. Sultan-Salem, M. S. Alouini, and M. Z. Win, "Modeling and Analysis of Cellular Networks Using Stochastic Geometry: A Tutorial," *IEEE Commun. Surv. Tutorials*, vol. 19, no. 1, pp. 167–203, 2017, doi: 10.1109/COMST.2016.2624939.

[3] H. Brendan McMahan, E. Moore, D. Ramage, S. Hampson, and B. Agüera y Arcas, "Communication-efficient learning of deep networks from decentralized data," in Proceedings of the 20th International Conference on Artificial Intelligence and Statistics, AISTATS 2017, 2017.

[4] J. Feng, L. Liu, Q. Pei, and K. Li, "Min-Max Cost Optimization for Efficient Hierarchical Federated Learning in Wireless Edge Networks," *IEEE Trans. Parallel Distrib. Syst.*, vol. 33, no. 11, pp. 2687–2700, 2022, doi: 10.1109/TPDS.2021.3131654.

[5] M. A. Ouamri, "Stochastic Geometry Modeling and Analysis of Downlink Coverage and Rate in Small Cell Network," *Telecommun. Syst.*, vol. 77, no. 4, pp. 767–779, 2021, doi: 10.1007/s11235-021-00770-5.

[6] M. A. Ouamri, R. Alkanhel, C. Gueguen, M. A. Alohali, and S. S. M. Ghoneim, "Modeling and Analysis of UAV-Assisted Mobile Network with Imperfect Beam Alignment," *Comput. Mater. Contin.*, vol. 74, no. 1, pp. 453–467, 2023, doi: 10.32604/cmc.2023.031450.

[7] K. A. Nguyen, Q. A. Nguyen, and J. P. Hong, "Federated Learning in Small-Cell Networks: Stochastic Geometry-Based Analysis on the Required Base Station Density," *Sensors*, vol. 23, no. 16, pp. 1–13, 2023, doi: 10.3390/s23167184.

[8] Y. Sun and Z. Ding, "A Fine Grained Stochastic Geometry Based Analysis on LEO Satellite Communication Systems," *IEEE Netw. Lett.*, vol. PP, pp. 1–1, 2023, doi: 10.1109/lnet.2023.3269818.

[9] M. A. Ouamri, D. Singh, M. A. Muthanna, A. Bounceur, and X. Li, "Performance Analysis of UAV Multiple Antenna-Assisted Small Cell Network with Clustered Users," *Wirel. Networks*, vol. 29, no. 4, pp. 1859–1872, 2023, doi: 10.1007/s11276-023-03240-9.

[10] J. Xu, M. A. Kishk, and M. S. Alouini, "Space-Air-Ground-Sea Integrated Networks: Modeling and Coverage Analysis," *IEEE Trans. Wirel. Commun.*, vol. 22, no. 9, pp. 6298–6313, 2023, doi: 10.1109/TWC.2023.3241341.

[11] C. Wang and Y. J. Chun, "Stochastic Geometric Analysis of the Terahertz (THz)-mmWave Hybrid Network With Spatial Dependence," *IEEE Access*, vol. 11, no. February, pp. 25063–25076, 2023, doi: 10.1109/ACCESS.2023.3253790.

[12] H. Lin and Q. Yu, "Ternary Stochastic Geometry Theory for Performance Analysis of RIS-Assisted UDN," *arXiv Prepr. arXiv2307.08200*, pp. 1–29, 2023, [Online]. Available: http://arxiv.org/abs/2307.08200

[13] H. Cheng, H. Sun, L. Zhang, J. Li, X. Wangg, and T. Q. S. Quek, "Understanding the Gain of Deploying IRSs in Large-Scale Heterogeneous Cellular Networks," *ICC 2023-IEEE Int. Conf. Commun.*, pp. 5787–5792, 2023, [Online]. Available: http://arxiv.org/abs/2302.03186

[14] D. Zhao, G. Wang, G. Chen, and S. Jia, "Coverage Analysis for Reconfigurable Intelligent Surface-Assisted Cellular Networks over Nakagami-m Fading Channels," *IEEE Commun. Lett.*, vol. 27, no. 3, pp. 1035–1039, 2023, doi: 10.1109/LCOMM.2023.3234913.

[15] Q. Zhang, J. Zhang, and S. Jin, "Grant-Free Random Access in Cell-Free Massive MIMO Systems with UE Detection Thresholds: A Stochastic Geometry Approach," *IEEE Trans. Veh. Technol.*, vol. 72, no. 6, pp. 8117–8121, 2023, doi: 10.1109/TVT.2023.3237870.

[16] J. F. Schmidt, U. Schilcher, M. K. Atiq, and C. Bettstetter, "Interference Prediction in Wireless Networks: Stochastic Geometry Meets Recursive Filtering," *IEEE Trans. Veh. Technol.*, vol. 70, no. 3, pp. 2783–2793, 2021, doi: 10.1109/TVT.2021.3059032.

[17] Z. Qiu, W. Wang, J. Geng, and Y. Liu, "Interference Analysis of Multi-tier NGSO Based on Stochastic Geometry," *IEEE Wirel. Commun. Netw. Conf. WCNC*, vol. 2023-March, pp. 1–6, 2023, doi: 10.1109/WCNC55385.2023.10119066.

[18] Y. Liu, C. She, Y. Zhong, W. Hardjawana, F.-C. Zheng, and B. Vucetic, "Interference-Limited Ultra-Reliable and Low-Latency Communications: Graph Neural Networks or Stochastic Geometry?," *arXiv Prepr. arXiv2207.06918*, pp. 1–30, 2022, [Online]. Available: http://arxiv.org/abs/2207.06918

[19] J. Yoon, Y. Kim, H. Kwon, and S. Park, "Stochastic Geometry-Based Evaluation of Edge User QoS with Cooperative Allocation," *IEEE Wirel. Commun. Lett.*, vol. 12, no. 9, pp. 1553–1557, 2023, doi: 10.1109/LWC.2023.3282588.

[20] R. Aghazadeh Ayoubi and U. Spagnolini, "Performance of Dense Wireless Networks in 5G and beyond Using Stochastic Geometry," *Mathematics*, vol. 10, no. 7, pp. 1–30, 2022, doi: 10.3390/math10071156.

[21] Y. Sun, Z. Ding, X. Dai, M. Zhou, and Z. Ding, "Stochastic Geometry Based Modeling and Analysis on Network NOMA in Downlink CoMP Systems," *IEEE Trans. Veh. Technol.*, vol. PP, pp. 1–7, 2023, doi: 10.1109/TVT.2023.3301627.

[22] Y. Zhu, G. Zheng, and K. K. Wong, "Stochastic Geometry Analysis of Large Intelligent Surface-Assisted Millimeter Wave Networks," *IEEE J. Sel. Areas Commun.*, vol. 38, no. 8, pp. 1749–1762, 2020, doi: 10.1109/JSAC.2020.3000806.

[23] T. Shafique, H. Tabassum, and E. Hossain, "Stochastic Geometry Analysis of IRS-Assisted Downlink Cellular Networks," *IEEE Trans. Commun.*, vol. 70, no. 2, pp. 1442–1456, 2022, doi: 10.1109/TCOMM.2022.3141097.

[24] Q. Gontier *et al.*, "Joint Metrics for EMF Exposure and Coverage in Real-World Homogeneous and Inhomogeneous Cellular Networks," *Jt. Metrics EMF Expo. Cover. Real-World Homog. Inhomogeneous Cell. Networks*, pp. 1–30, 2023, [Online]. Available: http://arxiv.org/abs/2302.03559

[25] S. J. Maeng, J. Park, and I. Guvenc, "Analysis of UAV Radar and Communication Network Coexistence with Different Multiple Access Protocols," *IEEE Trans. Commun.*, pp. 1–15, 2023, doi: 10.1109/TCOMM.2023.3305509.

[26] Y. Qin, M. A. Kishk, and M. S. Alouini, "On the Uplink SINR Meta Distribution of UAV-Assisted Wireless Networks," *IEEE Wirel. Commun. Lett.*, vol. 12, no. 4, pp. 684–688, 2023, doi: 10.1109/LWC.2023.3239264.

[27] M. M. Fadoul, "Rate and Coverage Analysis in Multi-Tier Heterogeneous Network Using Stochastic Geometry Approach," *Ad Hoc Networks*, vol. 98, p. 102038, 2020, doi: 10.1016/j.adhoc.2019.102038.

[28] S. H. Chae, H. Lim, H. Lee, and B. C. Jung, "Performance Analysis of Dense Low Earth Orbit Satellite Communication Networks with Stochastic Geometry," *J. Commun. Networks*, vol. 25, no. 2, pp. 208–221, 2023, doi: 10.23919/jcn.2023.000009.

[29] I. Kaj and T. Morozova, "Retransmission Performance in a Stochastic Geometric Cellular Network Model," *arXiv Prepr. arXiv2306.16200*, pp. 1–10, 2023, [Online]. Available: http://arxiv.org/abs/2306.16200

[30] Y. Zhang *et al.*, "Packet-Level Throughput Analysis and Energy Efficiency Optimization for UAV-Assisted IAB Heterogeneous Cellular Networks," *IEEE Trans. Veh. Technol.*, vol. 72, no. 7, pp. 9511–9526, 2023, doi: 10.1109/TVT.2023.3252093.

[31] C. Stegehuis and L. Weedage, "Resource Sharing in Wireless Networks with Co-Location," *arXiv Prepr. arXiv2304.11881*, pp. 1–18, 2023, [Online]. Available: http://arxiv.org/abs/2304.11881

[32] X. Ma, X. Lei, X. Zhou, and X. Tang, "Secrecy Performance Evaluation of Scalable Cell-Free Massive MIMO Systems: A Stochastic Geometry Approach," *IEEE Trans. Inf. Forensics Secur.*, vol. 18, pp. 2826–2841, 2023, doi: 10.1109/TIFS.2023.3268443.

[33] H. Chamkhia, A. Erbad, A. Mohamed, A. R. Hussein, A. Al-Ali, and M. Guizani, "Stochastic Geometry-based Physical Layer Security Performance Analysis of a Hybrid NOMA-PDM based IoT System," *IEEE Internet Things J.*, vol. PP, p. 1, 2023, doi: 10.1109/JIOT.2023.3292262.

[34] S. Feng, X. Lu, K. Zhu, D. Niyato, and P. Wang, "Covert D2D Communication Underlaying Cellular Network: A System-Level Security Perspective," *arXiv Prepr. arXiv2302.01745*, pp. 1–29, 2023, [Online]. Available: http://arxiv.org/abs/2302.01745

[35] S. S. Kalamkar, F. M. Abinader, F. Baccelli, A. S. M. Fani, and L. G. U. Garcia, "Stochastic Geometry-Based Modeling and Analysis of Beam Management in 5G," *IEEE Glob. Commun. Conf.*, pp. 1–6, 2020, doi: 10.1109/GLOBECOM42002.2020.9348187.

[36] H. H. Yang, Z. Liu, T. Q. S. Quek, and H. V. Poor, "Scheduling Policies for Federated Learning in Wireless Networks," *IEEE Trans. Commun.*, vol. 68, no. 1, pp. 317–333, 2020, doi: 10.1109/TCOMM.2019.2944169.

[37] W. Shi, S. Zhou, and Z. Niu, "Device Scheduling with Fast Convergence for Wireless Federated Learning," *IEEE Int. Conf. Commun.*, vol. 2020-June, 2020, doi: 10.1109/ICC40277.2020.9149138.

[38] M. Chen, Z. Yang, W. Saad, C. Yin, H. V. Poor, and S. Cui, "A Joint Learning and Communications Framework for Federated Learning over Wireless Networks," *IEEE Trans. Wirel. Commun.*, vol. 20, no. 1, pp. 269–283, 2021, doi: 10.1109/TWC.2020.3024629.

[39] S. Wan, J. Lu, P. Fan, Y. Shao, C. Peng, and K. B. Letaief, "Convergence Analysis and System Design for Federated Learning over Wireless Networks," *IEEE J. Sel. Areas Commun.*, vol. 39, no. 12, pp. 3622–3639, 2021, doi: 10.1109/JSAC.2021.3118351.

[40] Z. Zhao *et al.*, "Federated Learning with Non-IID Data in Wireless Networks," *IEEE Trans. Wirel. Commun.*, vol. 21, no. 3, pp. 1927–1942, 2022, doi: 10.1109/TWC.2021.3108197.

[41] S. Lee, C. Park, S.-N. Hong, Y. C. Eldar, and N. Lee, "Bayesian Federated Learning over Wireless Networks," *arXiv Prepr. arXiv2012.15486*, pp. 1–30, 2020, [Online]. Available: http://arxiv.org/abs/2012.15486

[42] X. Chen, G. Zhu, Y. Deng, and Y. Fang, "Federated Learning Over Multihop Wireless Networks with In-Network Aggregation," *IEEE Trans. Wirel. Commun.*, vol. 21, no. 6, pp. 4622–4634, 2022, doi: 10.1109/TWC.2022.3168538.

[43] V. D. Nguyen, S. K. Sharma, T. X. Vu, S. Chatzinotas, and B. Ottersten, "Efficient Federated Learning Algorithm for Resource Allocation in Wireless IoT Networks," *IEEE Internet Things J.*, vol. 8, no. 5, pp. 3394–3409, 2021, doi: 10.1109/JIOT.2020.3022534.

[44] M. Salehi and E. Hossain, "Federated Learning in Unreliable and Resource-Constrained Cellular Wireless Networks," *IEEE Trans. Commun.*, vol. 69, no. 8, pp. 5136–5151, 2021, doi: 10.1109/TCOMM.2021.3081746.

[45] B. Xu, W. Xia, W. Wen, P. Liu, H. Zhao, and H. Zhu, "Adaptive Hierarchical Federated Learning over Wireless Networks," *IEEE Trans. Veh. Technol.*, vol. 71, no. 2, pp. 2070–2083, 2022, doi: 10.1109/TVT.2021.3135541.

[46] H. Chen, S. Huang, D. Zhang, M. Xiao, M. Skoglund, and H. V. Poor, "Federated Learning over Wireless IoT Networks With Optimized Communication and Resources," *IEEE Internet Things J.*, vol. 9, no. 17, pp. 16592–16605, 2022, doi: 10.1109/JIOT.2022.3151193.

[47] C. T. DInh *et al.*, "Federated Learning over Wireless Networks: Convergence Analysis and Resource Allocation," *IEEE/ACM Trans. Netw.*, vol. 29, no. 1, pp. 398–409, 2021, doi: 10.1109/TNET.2020.3035770.

[48] Z. Wang *et al.*, "Asynchronous Federated Learning over Wireless Communication Networks," *IEEE Trans. Wirel. Commun.*, vol. 21, no. 9, pp. 6961–6978, 2022, doi: 10.1109/TWC.2022.3153495.

[49] Z. Yang, M. Chen, W. Saad, C. S. Hong, and M. Shikh-Bahaei, "Energy Efficient Federated Learning over Wireless Communication Networks," *IEEE Trans. Wirel. Commun.*, vol. 20, no. 3, pp. 1935–1949, 2021, doi: 10.1109/TWC.2020.3037554.

[50] M. M. Amiri and D. Gündüz, "Federated Learning over Wireless Fading Channels," *IEEE Trans. Wirel. Commun.*, vol. 19, no. 5, pp. 3546–3557, 2020, doi: 10.1109/TWC.2020.2974748.

[51] O. A. Amodu, M. Othman, N. O. R. K. Noordin, and I. Ahmad, "Relay-Assisted D2D Underlay Cellular Network Analysis using Stochastic Geometry: Overview and Future Directions," *IEEE Access*, vol. 7, pp. 1–30, 2019, doi: 10.1109/ACCESS.2019.2933915.

[52] Y. Hmamouche *et al.*, "New Trends in Stochastic Geometry for Wireless Networks: A Tutorial and Survey," *Proc. IEEE*, vol. 109, no. 7, pp. 1200–1252, 2021, doi: 10.1109/JPROC.2021.3061778.

[53] A. A. Salem, S. El-Rabaie, and M. Shokair, "Survey on Ultra-Dense Networks (UDNs) and Applied Stochastic Geometry," *Wirel. Pers. Commun.*, vol. 119, pp. 2345–2404, 2021.

[54] X. Lu, M. Salehi, M. Haenggi, E. Hossain, and H. Jiang, "Stochastic Geometry Analysis of Spatial-Temporal Performance in Wireless Networks: A Tutorial," *IEEE Commun. Surv. Tutorials*, vol. 23, no. 4, pp. 2753–2801, 2021, doi: 10.1109/COMST.2021.3104581.

[55] S. D. Okegbile, B. T. Maharaj, and A. S. Alfa, "Stochastic Geometry Approach Towards Interference Management and Control in Cognitive Radio Network: A Survey," *Comput. Commun.*, vol. 166, pp. 174–195, 2021, doi: 10.1016/j.comcom.2020.12.011.

[56] P. S. Bouzinis, S. Member, P. D. Diamantoulakis, S. Member, and K. George, "Wireless Federated Learning (WFL) for 6G Networks – Part I: Research Challenges and Future Trends," *IEEE Commun. Lett.*, vol. 26, no. 1, pp. 3–7, 2021.

[57] S. Naser, S. Member, L. Bariah, S. Member, and S. Muhaidat, "Towards Federated Learning-Enabled Visible Light Communication in 6G Systems," *IEEE Wirel. Commun.*, vol. 29, no. 1, pp. 48–56, 2021.

[58] D. Shome, O. Waqar, and W. U. Khan, "Federated Learning and Next Generation Wireless Communications: A Survey on Bidirectional Relationship," *Trans. Emerg. Telecommun. Technol.*, vol. 33, no. 7, pp. 1–19, 2022.

[59] S. Pouriyeh, O. Shahid, R. M. Parizi, Q. Z. Sheng, and G. Srivastava, "Secure Smart Communication Efficiency in Federated Learning: Achievements and Challenges," *Appl. Sci.*, vol. 12, no. 18, p. 8980, 2022.

[60] T. R. Gadekallu, Q. Pham, and T. Huynh-the, "Federated Learning for Big Data: A Survey on Opportunities, Applications, and Future Directions," *arXiv Prepr. arXiv2110.04160*, pp. 1–25.

[61] D. Sirohi, N. Kumar, P. Singh, and R. Sudeep, *Federated Learning for 6G - Enabled Secure Communication Systems: A Comprehensive Survey*, vol. 56, no. 10. Springer Netherlands, 2023. doi: 10.1007/s10462-023-10417-3.

[62] Z. Du, C. Wu, S. Member, and T. Y. Member, "Federated Learning for Vehicular Internet of Things: Recent Advances and Open Issues," *IEEE Open J. Comput. Soc.*, vol. 1, pp. 45–61, 2020, doi: 10.1109/OJCS.2020.2992630.

[63] L. U. Khan, W. Saad, Z. Han, and E. Hossain, "Federated Learning for Internet of Things: Recent Advances, Taxonomy, and Open Challenges," *IEEE Commun. Surv. Tutorials*, vol. 23, no. 3, pp. 1759–1799, 2021.

[64] D. C. Nguyen, M. Ding, P. N. Pathirana, A. Seneviratne, J. Li, and H. Vincent Poor, "Federated Learning for Internet of Things: A Comprehensive Survey," *IEEE Communications Surveys and Tutorials*, vol. 23, no. 3. pp. 1622–1658, 2021. doi: 10.1109/COMST.2021.3075439.

[65] M. Parimala, S. P. R. M, Q. Pham, and K. Dev, "Fusion of Federated Learning and Industrial Internet of Things: A Survey," *arXiv Prepr. arXiv2101.00798.*, pp. 1–24, 2021.

[66] W. Y. B. Lim, N. C. Luong, D. T. Hoang, Y. Jiao, and Y. Liang, "Federated Learning in Mobile Edge Networks: A Comprehensive Survey," *IEEE Commun. Surv. Tutorials*, vol. 22, no. 3, pp. 2031–2063, 2020.

[67] H. G. Abreha, "Federated Learning in Edge Computing: A Systematic Survey," *Sensors*, vol. 22, no. 2, pp. 1–45, 2022.

[68] J. Liu, J. Huang, Y. Zhou, S. Ji, H. Xiong, and D. Dou, "From Distributed Machine Learning to Federated Learning: A Survey," *Knowl. Inf. Syst.*, vol. 64, no. 4, pp. 885–917, 2022.

[69] X. Ma, L. Liao, Z. Li, R. X. Lai, and M. Zhang, "Applying Federated Learning in Software-Defined Networks: A Survey," *Symmetry (Basel).*, vol. 14, no. 2, pp. 1–27, 2022.

[70] H. Elsawy, E. Hossain, and M. Haenggi, "Stochastic Geometry for Modeling, Analysis, and Design of Multi-Tier and Cognitive Cellular Wireless Networks: A Survey," *IEEE Commun. Surv. Tutorials*, vol. 15, no. 3, pp. 996–1019, 2013, doi: 10.1109/SURV.2013.052213.00000.

[71] D. Hug and M. Reitzner, "Introduction to Stochastic Geometry," *Bocconi Springer Ser.*, vol. 7, pp. 145–184, 2016, doi: 10.1007/978-3-319-05233-5_5.

[72] J. G. Andrews, A. K. Gupta, and H. S. Dhillon, "A Primer on Cellular Network Analysis Using Stochastic Geometry," *arXiv Prepr. arXiv1604.03183*, pp. 1–46, 2016.

[73] A. Tukmanov, Z. Ding, S. Boussakta, and A. Jamalipour, "On the Impact of Network Geometric Models on Multicell Cooperative Communication Systems," *IEEE Wirel. Commun.*, vol. 20, no. 1, pp. 75–81, 2013.

[74] M. Haenggi, J. G. Andrews, F. Baccelli, O. Dousse, and M. Franceschetti, "Stochastic Geometry and Random Graphs for the Analysis and Design of Wireless Networks," *IEEE J. Sel. Areas Commun.*, vol. 27, no. 7, pp. 1029–1046, 2009, doi: 10.1109/JSAC.2009.090902.

[75] C.-H. Lee, C.-Y. Shih, and Y.-S. Chen, "Stochastic Geometry Based Models for Modeling Cellular Networks in Urban Areas," *Wirel. networks*, vol. 19, pp. 1063–1072, 2013, doi: 10.1007/s11276-012-0518-0.

[76] Y. C. Liang, K. C. Chen, G. Y. Li, and P. Mahonen, "Cognitive Radio Networking and Communications: An Overview," *IEEE Trans. Veh. Technol.*, vol. 60, no. 7, pp. 3386–3407, 2011.

[77] S. R. Sabuj and M. Hamamura, "Energy Efficiency Analysis of Cognitive Radio Network using Stochastic Geometry," *2015 IEEE Conf. Stand. Commun. Networking, CSCN 2015*, pp. 245–251, 2016, doi: 10.1109/CSCN.2015.7390452.

[78] S. S. Kalamkar, F. Baccelli, F. M. Abinader, A. S. M. Fani, and L. G. U. Garcia, "Beam Management in 5G: A Stochastic Geometry Analysis," *IEEE Trans. Wirel. Commun.*, vol. 21, no. 4, pp. 2275–2290, 2022, doi: 10.1109/TWC.2021.3110785.

[79] M. Di Renzo, W. Lu, and P. Guan, "The Intensity Matching Approach: A Tractable Stochastic Geometry Approximation to System-Level Analysis of Cellular Networks," *IEEE Trans. Wirel. Commun.*, vol. 15, no. 9, pp. 5963–5983, 2016, doi: 10.1109/TWC.2016.2574852.

[80] J. Liu, M. Sheng, L. Liu, and J. Li, "Effect of Densification on Cellular Network Performance with Bounded Pathloss Model," *IEEE Commun. Lett.*, vol. 21, no. 2, pp. 346–349, 2017, doi: 10.1109/LCOMM.2016.2615298.

[81] M. Al-quraan, G. S. Member, L. Mohjazi, S. Member, S. Muhaidat, and S. Member, "Edge-Native Intelligence for 6G Communications Driven by Federated Learning: A Survey of Trends and Challenges," *IEEE Trans. Emerg. Top. Comput. Intell.*, pp. 1–24, 2023.

[82] X. Foukas, G. Patounas, A. Elmokashfi, and M. K. Marina, "Network Slicing in 5G: Survey and Challenges," *IEEE Commun. Mag.*, vol. 55, no. 5, pp. 94–100, 2017, doi: 10.1109/MCOM.2017.1600951.

[83] K. M. J. Rahman, G. S. Member, and F. Ahmed, "Challenges, Applications and Design Aspects of Federated Learning: A Survey," *IEEE Access*, vol. 9, pp. 124682–124700, 2021, doi: 10.1109/ACCESS.2021.3111118.

[84] S. J. Reddi et al., "Adaptive Federated Optimization," in *International Conference on Learning Representations*, 2021, pp. 1–38.

[85] M. Moshawrab, M. Adda, A. Bouzouane, H. Ibrahim, & A. Raad. "Reviewing Federated Learning Aggregation Algorithms; Strategies, Contributions, Limitations and Future Perspectives," *Electronics*, vol. 12, no. 10, p. 2287, 2023.

[86] S. P. Karimireddy, S. Kale, M. Mohri, S. J. Reddi, S. U. Stich, and A. T. Suresh, "SCAFFOLD: Stochastic Controlled Averaging for Federated Learning," in *International conference on machine learning*, 2020, pp. 5132–5143.

[87] J. Hamer, M. Mohri, and A. T. Suresh, "FedBoost: Communication-Efficient Algorithms for Federated Learning," *Int. Conf. Mach. Learn.*, pp. 3931–3941, 2020.

[88] T. Li, A. K. Sahu, M. Zaheer, M. Sanjabi, A. Talwalkar, and V. Smith, "Federated Optimization in Heterogeneous Networks," in *Machine learning and systems*, 2020, pp. 429–450. [Online]. Available: http://arxiv.org/abs/1812.06127

[89] H. Wang, M. Yurochkin, Y. Sun, D. Papailiopoulos, and Y. Khazaeni, "Federated Learning With Matched Averaging," *8th Int. Conf. Learn. Represent. ICLR 2020*, pp. 1–16, 2020.

[90] K. Bonawitz *et al.*, "Practical Secure Aggregation for Privacy-Preserving Machine Learning," *Proc. ACM Conf. Comput. Commun. Secur.*, pp. 1175–1191, 2017, doi: 10.1145/3133956.3133982.

[91] A. Hammoud, H. Sami, A. Mourad, H. Otrok, R. Mizouni, and J. Bentahar, "AI, Blockchain, and Vehicular Edge Computing for Smart and Secure IoV: Challenges and Directions," *IEEE Internet Things Mag.*, vol. 3, no. 2, pp. 68–73, 2020.

[92] Z. Yu, Y. Gong, S. Gong, and Y. Guo, "Joint Task Offloading and Resource Allocation in UAV-enabled Mobile Edge Computing," *IEEE Internet Things J.*, vol. 7, no. 4, pp. 3147–3159, 2020.

[93] M. Chen, H. V. Poor, W. Saad, and S. Cui, "Wireless Communications for Collaborative Federated Learning," *IEEE Commun. Mag.*, vol. 58, no. 12, pp. 48–54, 2020, doi: 10.1109/MCOM.001.2000397.

[94] C. T. Dinh *et al.*, "Federated Learning over Wireless Networks: Convergence Analysis and Resource Allocation," *IEEE/ACM Trans. Netw.*, vol. 29, no. 1, pp. 398–409, 2021, doi: 10.1109/TNET.2020.3035770.

15 Wireless Network Encryption
Stream Ciphers, Computational Modeling, and Security Analysis

Oleksandr Kuznetsov, Emanuele Frontoni, Natalia Kryvinska, Vladyslav Chevardin and Oleksii Smirnov

15.1 INTRODUCTION

In the contemporary era of digital communication, the significance of cryptographic security in advanced wireless communication systems cannot be overstated [1], [2]. As these systems become increasingly integral to our daily lives, spanning from personal communication to critical infrastructure, the need for robust and efficient cryptographic solutions becomes paramount [3], [4]. This introduction sets the stage for a comprehensive exploration of high-speed stream ciphers, their design, simulation, and the pivotal role they play in securing wireless communication systems [5].

Wireless communication has undergone a remarkable evolution, transitioning from basic telephonic services to complex networks that support a myriad of applications [6], [7]. This evolution has been accompanied by an exponential increase in the volume of data transmitted wirelessly, raising significant concerns about data integrity and confidentiality [8], [9]. The open nature of wireless communication channels makes them inherently vulnerable to eavesdropping and other forms of cyberattacks, necessitating the implementation of strong cryptographic measures [2].

In this context, stream ciphers emerge as a critical component of wireless network security [10]. Known for their efficiency in encrypting data streams, stream ciphers operate by generating a seemingly random sequence of bits (the keystream) which is then combined with the plaintext to produce ciphertext [11], [12]. The security of these ciphers hinges on the randomness and unpredictability of the keystream [12]; any pattern or predictability could be exploited by attackers to breach the system.

The challenges in ensuring cryptographic security in wireless networks are multifaceted [4], [8]. They include the need for algorithms that are not only secure but also efficient enough to operate in environments with limited computational resources. Additionally, the rapid advancement in technology and the emergence of new threats, such as quantum computing, pose ongoing challenges to existing cryptographic paradigms.

Addressing these challenges requires more than theoretical cryptographic knowledge; it demands practical insights that can only be gained through computational modeling and simulation. These tools allow for the thorough testing and evaluation of cryptographic algorithms under various scenarios, providing invaluable insights into their real-world performance and resilience.

This chapter aims to delve deeply into the world of high-speed stream ciphers, exploring their fundamental principles, design methodologies, and the challenges unique to wireless communication systems. It discusses the importance of computational modeling and simulation in evaluating these ciphers and highlights the role of statistical testing in ensuring their effectiveness. The chapter also looks ahead, considering emerging trends and potential advancements in the field of cryptographic security for wireless systems.

DOI: 10.1201/9781003457428-16

In conclusion, the exploration of high-speed stream ciphers in advanced wireless communication systems is not just a technical endeavor but a necessary pursuit to safeguard the integrity and confidentiality of wireless communications in an increasingly connected world.

15.1.1 KEY CONTRIBUTIONS OF THE CHAPTER

i. Comprehensive Analysis of Cryptographic Security: This chapter provides an in-depth exploration of cryptographic security challenges in advanced wireless communication systems, emphasizing the critical role of high-speed stream ciphers.

ii. Integration of Computational Modeling and Simulation: A key contribution is the detailed discussion on the use of computational modeling and simulation techniques, particularly focusing on their application in the evaluation of stream ciphers using NIST STS and DIEHARD methodologies.

iii. Insights into Statistical Testing: The chapter offers valuable insights into the importance and methodology of statistical testing for stream ciphers, highlighting its significance in ensuring cryptographic security in wireless networks.

iv. Practical and Theoretical Balance: Balancing theoretical knowledge with practical applications, the chapter bridges the gap between cryptographic theory and its real-world implications in wireless communication systems.

15.1.2 CHAPTER ORGANIZATION

The chapter is organized to provide a thorough understanding of cryptographic security in wireless communication systems. It begins with a review of related work in the field, setting the stage for a deeper exploration of the challenges and solutions in cryptographic security. The chapter then delves into the essence of cryptographic security in wireless networks, highlighting the importance of stream ciphers and the challenges they face. A significant focus is placed on computational modeling as a crucial tool in cryptographic security, discussing how it aids in the evaluation and development of cryptographic algorithms. The chapter further explores the methodologies of NIST STS and DIEHARD for assessing the statistical security of stream ciphers, providing a detailed analysis of these testing suites and their application in wireless communication systems. The results of computational experiments using both NIST STS and DIEHARD are presented, offering empirical insights into the performance of various cryptographic ciphers. This is followed by a comprehensive discussion of the results, where the findings are analyzed and contextualized within the broader scope of cryptographic security in wireless networks. Overall, the chapter provides a holistic view of the current state and future directions of cryptographic security in advanced wireless communication systems, blending theoretical perspectives with practical insights.

15.2 RELATED WORK

The realm of wireless network security is rapidly evolving, with researchers continuously exploring innovative cryptographic techniques and security protocols. This literature review examines recent scholarly contributions, highlighting their significance in advancing the field of wireless network security and identifying gaps that our research aims to fill.

Deng et al. (2023) [13]: This study introduces a certificateless anonymous signcryption scheme for healthcare wireless sensor networks, enhancing data privacy and sensor anonymity. It represents a significant step in securing sensitive health data transmitted over public channels.

Gonzalez-Arango et al. (2021) [14]: This paper evaluates the performance of symmetric cryptographic algorithms in resource-constrained hardware for Wireless Sensor Networks (WSNs), addressing the critical balance between security, performance, and energy consumption in WSNs.

Halak et al. (2022) [15]: The authors present a comparative analysis of the energy costs of asymmetric versus symmetric encryption-based security applications, highlighting the energy efficiency of symmetric systems and their resilience to quantum computing attacks.

Halbouni et al. (2023) [16]: This systematic literature review on Wireless Security Protocol WPA3 discusses its improvements over previous protocols, focusing on enhanced encryption methods and key sharing, thus contributing to the advancement of wireless network security.

Harn et al. (2021) [17]: The paper proposes a lightweight aggregated data encryption method for WSNs, addressing the challenge of encrypting small bits of data with short-length keys, thereby enhancing the efficiency of data encryption in WSNs.

Helmy et al. (2023) [18]: This research introduces a hybrid encryption framework for robust 3D image communication over wireless channels, combining various encryption techniques with the Rubik's cube method, showcasing an innovative approach to secure image transmission.

Hrovatin et al. (2023) [19]: The authors describe a privacy-preserving protocol for WSNs that enables nodes to compute functions without disclosing private inputs, using Onion Routing for traffic distribution, thus enhancing data and query privacy in WSNs.

Hua and Maksud (2023) [20]: This study explores continuous encryption functions for network security, focusing on physical layer encryption and biometric template security, offering insights into the development of efficient and secure encryption functions.

Khashan et al. (2021) [21]: The paper presents FlexCrypt, an automated lightweight cryptographic scheme for WSNs, addressing key management, authentication, and power resource management, contributing to the security and efficiency of WSNs.

Li et al. (2022) [22]: This research proposes an efficient secure data transmission and node authentication scheme for wireless sensing networks, combining compressed sensing, lightweight hashing, and homomorphic encryption to enhance data confidentiality.

Park and Hong (2023) [23]: The authors introduce an encryption device based on wave-chaos for enhanced physical security in wireless communication, demonstrating a novel approach to secure wireless transmission using wave-chaotic cavities.

Ramadevi et al. (2023) [24]: This paper offers a security strategy for WSNs using Improved Elliptic Key Cryptography (IEKC), focusing on encryption, decryption, and node authentication, thereby enhancing the security of WSNs against various attacks.

Urooj et al. (2023) [25]: The study discusses cryptographic data security for reliable WSNs, proposing a novel algorithm combining AES and ECC cryptography with clustering, aiming to improve energy efficiency and network security.

Xu et al. (2023) [26]: This article explores the potential of physical layer security (PLS) in UAV networks, providing a comprehensive overview of various communication channels and advanced PLS techniques, contributing to the enhancement of information confidentiality in UAV systems.

The reviewed literature underscores the diverse approaches and methodologies employed in enhancing wireless network security. However, there remains a gap in the comprehensive analysis and simulation of high-speed stream ciphers, specifically in the context of advanced wireless communication systems. Our research aims to bridge this gap by providing an in-depth exploration of stream ciphers, their computational modeling, and statistical testing, thereby contributing to the advancement of cryptographic security in wireless networks.

15.3 CRYPTOGRAPHIC SECURITY CHALLENGES IN ADVANCED WIRELESS COMMUNICATION SYSTEMS

In the rapidly evolving landscape of Advanced Wireless Communication Systems, cryptographic security stands as a critical pillar [27], [28]. The increasing complexity and sophistication of cyber threats pose significant challenges to the integrity and confidentiality of wireless communications. This section delves into the multifaceted problem of cryptographic security in these systems, with a particular focus on the pivotal role of computational modeling and statistical testing of stream ciphers.

15.3.1 THE ESSENCE OF CRYPTOGRAPHIC SECURITY IN WIRELESS NETWORKS

Cryptographic security in wireless networks is a domain marked by its complexity and criticality. At its core, it involves the protection of data transmitted over airwaves, inherently more vulnerable to interception and unauthorized access compared to wired networks [2]. The essence of cryptographic security here lies not only in encoding the data but also in ensuring that the encryption mechanisms are resilient, adaptable, and robust against an array of sophisticated cyber threats [4].

Wireless networks, characterized by their dynamic nature and diverse range of applications, from personal communications to critical infrastructure, face unique security challenges. The open-air transmission medium of these networks makes them susceptible to eavesdropping, interception, and more sophisticated forms of cyberattacks, like man-in-the-middle and replay attacks [1]. As such, the cryptographic algorithms employed must be capable of providing strong encryption while being efficient enough to operate within the constraints of wireless environments, such as limited bandwidth and power resources.

Stream ciphers play a pivotal role in this landscape. Their design for processing data streams bit by bit or byte by byte makes them particularly suitable for the high-speed requirements of wireless communications [10], [29]. However, the strength of these ciphers lies in their ability to generate keystreams that are as random as possible. Any predictability in the keystream can lead to vulnerabilities, making the cipher susceptible to various forms of cryptanalysis.

The challenge, therefore, is twofold: ensuring the absolute randomness of the keystream and maintaining the efficiency of the encryption process. This is where the balance between computational complexity and operational efficiency becomes crucial. Stream ciphers must be lightweight enough to not overburden the network resources while being complex enough to withstand cryptographic attacks.

The evolving nature of cyber threats poses another significant challenge. As attackers develop more advanced techniques, the cryptographic algorithms must also evolve. This continuous cycle of threat and response necessitates a proactive approach to cryptographic security. It involves not only the implementation of current best practices but also the anticipation of future vulnerabilities and the development of new, more secure algorithms.

In essence, cryptographic security in wireless networks is about creating a secure, trustworthy channel for data transmission in an environment that is inherently insecure. It requires a careful balance of operational efficiency and cryptographic robustness, a deep understanding of the evolving threat landscape, and a proactive approach to security. Stream ciphers, with their efficiency and potential for high cryptographic strength, stand at the forefront of this effort, playing a crucial role in safeguarding wireless communications against the myriad of cyber threats.

15.3.2 COMPUTATIONAL MODELING: A CRUCIAL TOOL IN CRYPTOGRAPHIC SECURITY

In the intricate world of cryptographic security for wireless networks, computational modeling emerges as an indispensable tool. It serves as a bridge between theoretical cryptographic concepts and their practical applications in real-world scenarios. Computational modeling allows researchers and practitioners to simulate the behavior of cryptographic algorithms under a variety of conditions, providing invaluable insights into their performance, resilience, and potential vulnerabilities [30].

One of the primary advantages of computational modeling is its ability to replicate complex real-world environments [31]. This includes simulating various attack scenarios, network conditions, and usage patterns. For instance, modeling can demonstrate how a stream cipher would perform under different levels of network traffic or in the presence of sophisticated cyberattacks. Such simulations are crucial in understanding the practical implications of theoretical security measures and in identifying potential weaknesses that might not be apparent in a purely theoretical analysis.

Computational models are particularly effective in evaluating the performance of stream ciphers [5]. By simulating the generation of keystreams and the encryption process, these models can assess the randomness of the output, the speed of encryption, and the overall efficiency of the cipher. This is vital in wireless networks where the balance between security and performance is critical. For example, a cipher that offers strong encryption but is too resource-intensive may not be suitable for use in mobile devices with limited battery life.

Moreover, computational modeling plays a pivotal role in the development of new cryptographic algorithms. Through simulations, developers can test and refine their algorithms, ensuring they meet the required security standards before deployment. This iterative process of modeling, testing, and refinement is essential in creating ciphers that are not only secure but also optimized for the specific needs of wireless networks.

The challenges posed by wireless networks, such as limited bandwidth, variable connection quality, and the need for low-power consumption, make computational modeling even more crucial. Models can be used to simulate these constraints, allowing for the development of cryptographic solutions that are tailored to the unique requirements of wireless environments. This ensures that security measures do not impede the functionality or efficiency of the network.

Figure 15.1 provides a visual representation of the crucial aspects of computational modeling in cryptographic security, particularly in the context of wireless networks. It highlights the importance of bridging theory and practice, simulating real-world scenarios, evaluating cipher performance, aiding in cipher development, addressing wireless network challenges, and enhancing overall security through a proactive approach.

In conclusion, computational modeling is a critical tool in the field of cryptographic security for wireless networks. It bridges the gap between theory and practice, enabling the thorough

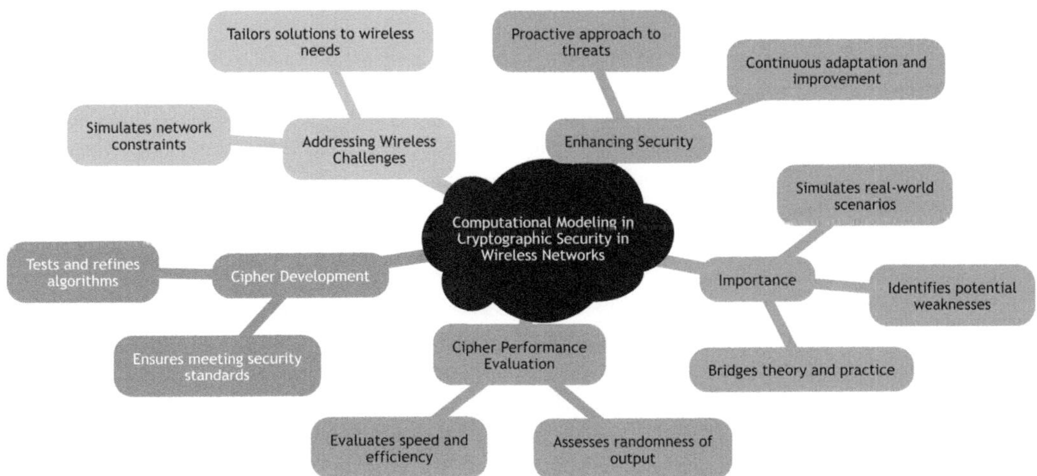

FIGURE 15.1 Computational Modeling in Cryptographic Security in Wireless Networks.

evaluation and refinement of cryptographic algorithms. By simulating real-world scenarios and network conditions, computational modeling helps ensure that the cryptographic solutions deployed in wireless networks are robust, efficient, and capable of withstanding the challenges of an ever-evolving digital landscape.

15.3.3 Statistical Testing of Stream Ciphers: Ensuring Cryptographic Security in Wireless Networks

In the domain of wireless network security, the statistical testing of stream ciphers is a critical process that ensures the robustness and effectiveness of cryptographic protocols [32], [33]. This subsection delves into the importance and methodology of statistical testing, focusing on its role in maintaining and enhancing the cryptographic security of wireless networks.

Stream ciphers, integral to the security framework of wireless networks, rely on generating keystreams that are indistinguishable from random sequences [10], [29]. The randomness of these keystreams is paramount; any predictability can lead to vulnerabilities, making the network susceptible to various cryptographic attacks [34], [35]. Statistical testing of these ciphers, therefore, becomes an essential step in validating their randomness and, by extension, their security.

Methodology of statistical testing involves subjecting the output of stream ciphers to a series of statistical tests, such as those offered by the NIST STS (National Institute of Standards and Technology Statistical Test Suite) [35], [36] and the DIEHARD suite [37], [38]. These tests are designed to rigorously evaluate the randomness of the cipher outputs, checking for uniformity, independence, and absence of predictable patterns. The tests range from simple frequency analysis to more complex evaluations like the Maurer's Universal Statistical Test, each probing different aspects of randomness.

In wireless networks, where data is transmitted over airwaves and is inherently more exposed to interception, the need for statistically sound cryptographic measures is even more pronounced [10], [29]. Stream ciphers in these networks encrypt data on-the-fly, often with constraints on computational resources and power [2]. The statistical testing of these ciphers ensures that despite these constraints, the security of the data is not compromised.

One of the challenges in statistical testing is the dynamic nature of wireless networks. As new technologies and protocols emerge, the ciphers must be re-evaluated to ensure their continued effectiveness. Additionally, the balance between computational efficiency and cryptographic strength is a constant consideration. Stream ciphers must be lightweight enough for efficient operation in wireless environments but robust enough to pass stringent statistical tests.

15.4 COMPUTATIONAL MODELING AND SIMULATION: NIST STS AND DIEHARD IN STREAM CIPHER TESTING

In the sphere of advanced wireless communication systems, where data integrity and confidentiality are paramount, the reliability of cryptographic algorithms is non-negotiable. Computational modeling and simulations play a pivotal role in this context. They allow researchers and practitioners to virtually assess the performance of ciphers under various conditions and scenarios, which would be impractical, time-consuming, or even impossible to replicate in real-world settings. This computational approach not only saves resources but also provides a deeper understanding of how these algorithms would perform in actual wireless communication environments. Moreover, the importance of computational calculations in this process cannot be overstated. They enable the precise analysis of statistical properties and the generation of large datasets required for rigorous testing. This computational rigor ensures that the evaluation of stream ciphers is thorough and accurate, providing confidence in their deployment in wireless systems.

This section delves into the significance of these methodologies, emphasizing the role of computational modeling and simulations in the realm of cryptographic research.

15.4.1 Methodology of NIST STS for Assessing Statistical Security of Stream Ciphers

The National Institute of Standards and Technology (NIST) has developed a Statistical Test Suite (STS) for evaluating the randomness of binary sequences produced by cryptographic random number generators, which is crucial in assessing the statistical security of stream ciphers [35], [36]. This suite is a comprehensive collection of tests designed to detect non-randomness in binary sequences, a key indicator of potential weaknesses in cryptographic systems.

NIST STS is designed to analyze the randomness of binary sequences generated by cryptographic algorithms, including stream ciphers. Randomness is a fundamental property that ensures unpredictability and uniformity in cryptographic applications.

The suite consists of multiple statistical tests, each targeting specific types of non-random behavior or patterns that could indicate weaknesses in the cipher's output [35], [36].

✓ Frequency (Monobit) Test:
  - Objective: To determine if the number of ones and zeros in a sequence are approximately the same as would be expected for a truly random sequence.
  - Methodology: The test calculates the proportion of ones in the sequence and compares it to the expected distribution.
✓ Frequency Test within a Block:
  - Objective: Similar to the monobit test, but the sequence is divided into blocks, and the test is performed on each block.
  - Methodology: The proportion of ones in each block is compared to the expected distribution.
✓ Runs Test:
  - Objective: To analyze the occurrence of runs of ones or zeros in the sequence (a run is an uninterrupted sequence of identical bits).
  - Methodology: The test checks if the number of runs of various lengths matches what would be expected for a random sequence.
✓ Longest Run of Ones in a Block Test:
  - Objective: To determine whether the length of the longest run of ones within the tested blocks is consistent with the behavior of a random sequence.
  - Methodology: The sequence is divided into blocks, and the longest run of ones in each block is determined and evaluated.
✓ Binary Matrix Rank Test:
  - Objective: To check for linear dependencies among fixed-length substrings of the original sequence.
  - Methodology: The test involves creating matrices from the subsequences and determining their rank.
✓ Discrete Fourier Transform (Spectral) Test:
  - Objective: To identify periodic features in the tested sequence that would indicate a deviation from randomness.
  - Methodology: The test applies a Discrete Fourier Transform to the sequence and analyzes the peak heights in the spectrum. It detects periodic patterns that would be unlikely to appear in a random sequence.
✓ Non-Overlapping Template Matching Test:
  - Objective: To detect the presence of pre-defined templates (patterns) within the tested sequence.

- • Methodology: The test counts the number of occurrences of each template and compares this to the expected number for a random sequence.
- ✓ Overlapping Template Matching Test:
  - • Objective: Similar to the non-overlapping template test, but allows for overlapping of the templates.
  - • Methodology: The sequence is scanned for occurrences of a specific template, and the number of matches is compared against the expected distribution for a random sequence.
- ✓ Maurer's "Universal Statistical" Test:
  - • Objective: To evaluate the randomness of the sequence by analyzing the distances between identical patterns.
  - • Methodology: The test calculates the average distance between identical patterns in the sequence and compares it to the expected value for a random sequence.
- ✓ Serial Test:
  - • Objective: To check for the frequency of all possible overlapping m-bit patterns across the entire sequence.
  - • Methodology: The frequency of each pattern is compared to the expected frequency for a random sequence.
- ✓ Approximate Entropy Test:
  - • Objective: To measure the regularity and unpredictability of overlapping subsequences.
  - • Methodology: The test compares the frequency of overlapping blocks of two consecutive lengths against the expected result for a random sequence.
- ✓ Cumulative Sums (Cusum) Test:
  - • Objective: To determine whether the cumulative sum of the partial sequences occurring in the tested sequence is too large or too small relative to the expected behavior of a random sequence.
  - • Methodology: The test calculates the cumulative sums of the sequence and analyzes the extremes to which these sums wander.
- ✓ Random Excursions Test and Random Excursions Variant Test:
  - • Objective: To detect deviations from randomness in the number of visits to a particular state in a random walk.
  - • Methodology: The sequence is treated as a series of steps in a random walk, and the number of visits to each state is compared to what would be expected for a random sequence.

The NIST STS is typically run as a suite where multiple tests are applied to the binary sequence generated by the stream cipher. The results are then analyzed collectively to assess the overall randomness.

The outcome of each test is a p-value, which indicates the probability that a perfect random number generator would have produced a sequence less random than the sequence being tested. A sequence is usually considered random if all or most of the p-values are within acceptable limits.

The NIST Statistical Test Suite provides a comprehensive and rigorous framework for evaluating the randomness and, by extension, the statistical security of stream ciphers [35], [36]. By applying a diverse array of tests, each targeting different aspects of randomness, the suite offers a nuanced view of a cipher's performance, highlighting potential vulnerabilities and strengths. This methodology is crucial in the cryptographic validation process, ensuring that stream ciphers deployed in advanced wireless communication systems are robust and secure.

15.4.2 The DIEHARD Statistical Test Suite for Stream Cipher Evaluation

In the realm of cryptographic analysis, the DIEHARD suite of statistical tests plays a crucial role in assessing the randomness and, consequently, the security of stream ciphers [37], [38]. Developed by George Marsaglia, DIEHARD is a comprehensive collection of statistical tests designed to evaluate the randomness of binary sequences generated by cryptographic algorithms. This section provides an in-depth overview of the DIEHARD test suite, elucidating its methodology, significance, and application in the field of stream ciphering.

The DIEHARD suite is specifically designed to detect non-randomness in binary sequences, a critical aspect in evaluating the strength and reliability of stream ciphers. It is renowned for its rigorous and diverse set of tests, each targeting different potential weaknesses in random number generators. Developed in the late 1990s, DIEHARD has been a benchmark in the field of cryptography for testing the randomness of binary sequences. Its comprehensive approach has set a high standard for randomness testing in cryptographic applications [37], [38].

- ✓ Birthday Spacings Test:
  - Objective: To assess the uniform distribution of spacings between certain events, akin to the "birthday problem" in probability theory.
  - Methodology: The test involves choosing random points in a large interval and checking the spacings between these points for uniformity.
- ✓ Overlapping Permutations Test:
  - Objective: To analyze the frequency of all possible overlapping 5-permutations of consecutive elements in the sequence.
  - Methodology: The test counts the occurrences of each permutation and compares them to the expected distribution for a random sequence.
- ✓ Ranks of Matrices Test:
  - Objective: To evaluate the rank of disjoint sub-matrices of the entire sequence.
  - Methodology: The test involves forming matrices from the sequence bits and determining the ranks, which are then compared to the theoretical distribution of ranks for random matrices.
- ✓ Monkey Tests:
  - Objective: Based on the infinite monkey theorem, these tests check for sequences that a random typing monkey might produce.
  - Methodology: The tests involve analyzing the frequency of overlapping m-bit patterns and comparing them to the expected distribution for a random sequence.
- ✓ Count the 1's Test:
  - Objective: To examine the number of 1's in specific bytes of the sequence.
  - Methodology: The test counts the 1's in either a stream of bytes or specific bytes within a stream and assesses the counts against expected values for random sequences.
- ✓ Parking Lot Test:
  - Objective: To simulate the process of parking a car in a parking lot and assess the randomness based on the distribution of the cars.
  - Methodology: The test involves randomly placing unit squares within a larger square and examining the distribution and overlaps, akin to parking cars in a lot.
- ✓ Minimum Distance Test:
  - Objective: To evaluate the minimum distance between randomly placed points.
  - Methodology: Points are placed in a two-dimensional space, and the distances between each pair are calculated to assess the randomness based on their minimum distances.
- ✓ Random Spheres Test:
  - Objective: To analyze the distribution of a certain number of points randomly placed in a space.
  - Methodology: The test involves placing points in a three-dimensional space and examining the distribution and overlap of spheres centered at these points.
- ✓ The Squeeze Test:
- ✓ Objective: To test the ability of the sequence to withstand "squeezing" – reducing a large range of integers to a smaller one.

✓ Methodology: The test repeatedly reduces a large integer by dividing it by integers generated from the sequence and counts the number of reductions needed to reach zero.

✓ Overlapping Sums Test:
  - Objective: To assess the sums of overlapping segments of the sequence.
  - Methodology: The test calculates the sums of specific lengths of overlapping segments and analyzes their distribution.

✓ Runs Test:
  - Objective: To evaluate the occurrence and length of runs of ones within the sequence.
  - Methodology: The test examines the sequence for runs of ones of various lengths and compares the occurrences to expected values for a random sequence.

✓ The Craps Test:
  - Objective: Based on the casino game Craps, this test assesses the number of wins and the number of throws per game.
  - Methodology: The test simulates playing a large number of Craps games and analyzes the win/loss ratio and the number of throws per game against expected values for random sequences.

The DIEHARD suite is typically executed as a comprehensive battery of tests, where each test is applied to the binary sequence generated by the stream cipher [37], [38].

The outcome of each test in the DIEHARD suite is usually a p-value, which indicates the probability that a perfect random number generator would have produced a sequence less random than the sequence being tested. A sequence is generally considered random if the p-values fall within acceptable ranges, indicating no significant deviations from randomness.

15.4.3 COMPUTATIONAL MODELING AND TESTING OF CRYPTOGRAPHIC CIPHERS IN WIRELESS COMMUNICATION SYSTEMS

In the pursuit of robust security for Wireless Communication Systems, a diverse array of cryptographic ciphers is subjected to rigorous computational modeling and testing. This section delves into the specifics of various ciphers, including both block and stream types, highlighting their relevance and application in the context of wireless security. The focus is on understanding the unique attributes and security parameters of each cipher, with an emphasis on their suitability for high-speed, secure wireless communications. For block ciphers, particular attention is paid to their potential for fast stream cipher modes of operation.

1. AES-128 (Advanced Encryption Standard – 128 bit) [39], [40]
   - Description: AES-128 is a widely recognized block cipher, known for its balance of speed and security. It employs a 128-bit key, making it robust against brute-force attacks while maintaining efficiency.
   - Parameters: Utilizes a block size of 128 bits and a key size of 128 bits. It operates on a 10-round process for encryption and decryption.
   - Significance: In wireless systems, AES-128 can be deployed in stream modes like CTR (Counter Mode) for high-speed data encryption, ensuring both security and performance.
2. AES-256 (Advanced Encryption Standard – 256 bit) [39], [40]
   - Description: An extension of AES, AES-256 provides enhanced security through a longer key size. It is designed to counter threats from future advancements in computational power.
   - Parameters: Features a block size of 128 bits and an extended key size of 256 bits, with a 14-round encryption process.
   - Significance: AES-256 is particularly relevant for systems requiring heightened security. Its application in stream modes makes it suitable for securing high-volume wireless communications without compromising speed.

3. CryptMT [41], [42], [43]
   - Description: CryptMT is a stream cipher derived from the Mersenne Twister algorithm, known for its high-quality randomness and speed.
   - Parameters: It generates keystreams using the Mersenne Twister pseudorandom number generator, ensuring a high degree of unpredictability in its output.
   - Significance: CryptMT's efficiency and strong randomness properties make it a suitable candidate for wireless systems where both speed and security are paramount. Its robustness against common cryptographic attacks adds an extra layer of security in wireless communications.

4. DECIM [44]
   - Description: DECIM is a hardware-oriented stream cipher designed primarily for resource-constrained environments. It is part of the eSTREAM portfolio and is known for its low power consumption and minimal hardware footprint.
   - Parameters: DECIM operates with an 80-bit key and an initialization vector (IV) of variable length. It employs an irregular clocking mechanism and a non-linear feedback shift register (NLFSR) for keystream generation.
   - Significance: In wireless communication systems, particularly in IoT and sensor networks, DECIM's low-resource requirement makes it an ideal choice. Its design caters to environments where power efficiency and compactness are crucial, without compromising the security integrity of the data transmission.

5. Enocoro [45], [46], [47]
   - Description: Enocoro is a lightweight stream cipher known for its simplicity and efficiency in software implementations. It is designed to provide a high level of security with a small code size and low memory usage.
   - Parameters: Enocoro comes in two variants: Enocoro-80 with an 80-bit key and Enocoro-128v2 with a 128-bit key. It features a simple structure with a small internal state, making it suitable for quick data processing.
   - Significance: For wireless systems that require fast and efficient encryption with minimal computational overhead, such as RFID systems and smart cards, Enocoro offers an optimal solution. Its compactness and efficiency make it a viable option for securing data in systems with stringent resource limitations.

6. Grain [48], [49]
   - Description: Grain is a stream cipher that is part of the eSTREAM portfolio, designed for low-resource hardware environments. It is known for its simplicity, small size, and high efficiency.
   - Parameters: Grain operates with a 80-bit key and a 64-bit IV. Its structure consists of a linear feedback shift register (LFSR) and a non-linear feedback shift register (NFSR), combined with a non-linear output function.
   - Significance: Grain's design makes it particularly suitable for applications where hardware resources are limited, such as in embedded systems and wireless sensor networks. Its ability to provide secure encryption with minimal hardware requirements makes it a valuable tool in the realm of wireless communications, especially in applications where power and space are at a premium.

7. HC-128 [50], [51]
   - Description: HC-128 is a stream cipher known for its high performance and strong security profile. It is part of the eSTREAM portfolio and is designed for software implementations with an emphasis on speed and security.
   - Parameters: HC-128 operates with a 128-bit key and a 128-bit initialization vector (IV). It utilizes a unique design consisting of two secret internal 512-bit state tables, updated using non-linear functions.

- Significance: In wireless communication systems, HC-128 is particularly valued for its balance between speed and cryptographic strength. It is well-suited for applications requiring fast data processing without compromising security, such as in high-speed wireless networks and VPNs.

8. HC-256 [50], [51]
   - Description: HC-256 is an extended version of HC-128, offering enhanced security features. It is designed for environments where higher security levels are required without a significant compromise on performance.
   - Parameters: This cipher uses a 256-bit key and a 256-bit IV. Similar to HC-128, it employs two 1024-bit internal state tables, which are updated using more complex non-linear functions compared to HC-128.
   - Significance: HC-256 is ideal for wireless systems where enhanced security is paramount. Its robust design makes it resistant to various cryptographic attacks, making it a reliable choice for securing sensitive data in wireless communication networks.

9. KCipher-2 [52], [53], [54]
   - Description: KCipher-2 is a stream cipher that has gained recognition for its reliability and versatility. Originally developed in Japan, it has been used in various commercial applications, including wireless networks.
   - Parameters: It operates with a key size of 128 bits and an IV size of 128 bits. KCipher-2 is known for its unique algorithm that combines linear and non-linear components to produce a secure keystream.
   - Significance: KCipher-2's adaptability and proven security track record make it a suitable choice for diverse wireless communication applications. Its efficiency and security balance ensure that it can be deployed in various scenarios, from mobile communications to wireless sensor networks, providing robust data protection.

10. Mickey2 [55]
    - Description: Mickey2 is a stream cipher that is part of the eSTREAM portfolio, known for its compact design and suitability for hardware implementations. It is designed to provide a high level of security with a relatively simple structure.
    - Parameters: Mickey2 operates with an 80-bit key and a 40-bit initialization vector (IV). Its mechanism is based on irregular clocking of two linear feedback shift registers (LFSRs) and a non-linear combining function.
    - Significance: In wireless communication systems, Mickey2's small footprint and efficient design make it particularly suitable for environments where hardware resources are limited, such as in embedded systems and RFID technology. Its security features and low power consumption are advantageous for securing data in resource-constrained wireless devices.

11. MUGI [56], [57], [58]
    - Description: MUGI is a stream cipher developed in Japan, recognized for its high security and suitability for both hardware and software implementations. It was designed to meet the requirements of fast and secure data encryption in various applications.
    - Parameters: MUGI uses a 128-bit key and a 128-bit initialization vector. Its architecture is based on a large internal state and a non-linear transformation function, providing a strong resistance against cryptographic attacks.
    - Significance: MUGI's versatility and robust security profile make it a strong candidate for use in diverse wireless communication systems, including mobile and broadband networks. Its ability to provide secure encryption at high speeds aligns well with the demands of modern wireless communication technologies.

12. RC4 [59], [60], [61]
    - Description: RC4 is one of the most widely used stream ciphers, known for its simplicity and speed in software implementations. Despite some vulnerabilities identified over the years, it remains popular in various applications.
    - Parameters: RC4 operates with a variable key size, typically ranging from 40 to 256 bits. It generates a long keystream by permuting an array of bytes and combining it with plaintext using an XOR operation.
    - Significance: RC4 has been extensively used in wireless protocols such as WEP and WPA for securing Wi-Fi networks. Its ease of implementation and speed make it suitable for real-time data encryption in wireless environments. However, due to known vulnerabilities, its use in new systems is generally discouraged, and it is being replaced by more secure alternatives.

13. Salsa20 [62], [63]
    - Description: Salsa20 is a stream cipher known for its high speed and security, designed by Daniel J. Bernstein. It stands out for its simplicity and efficiency, particularly in software implementations.
    - Parameters: Salsa20 operates with either a 128-bit or 256-bit key and a 64-bit IV. The cipher uses a unique 'hash-like' computation to generate its keystream, consisting of 20 rounds of simple, fast operations.
    - Significance: In wireless communication systems, Salsa20 is valued for its ability to provide secure encryption rapidly, making it suitable for high-throughput environments. Its design ensures a high level of security, making it a reliable choice for encrypting data in various wireless applications, including VPNs and secure messaging.

14. Snow 2 [64], [65], [66]:
    - Description: Snow 2 is a stream cipher that is part of the eSTREAM portfolio, an evolution of the earlier Snow cipher. It is designed for high performance in both hardware and software implementations.
    - Parameters: Snow 2 uses a 128-bit or 256-bit key and a 128-bit IV. The cipher's structure includes a linear feedback shift register (LFSR) and a finite state machine (FSM), which work together to produce the keystream.
    - Significance: Snow 2's efficiency and robust security make it well-suited for use in wireless communication systems where high data rates and strong encryption are required. Its adaptability to different platforms makes it versatile for various wireless security applications.

15. Sosemanuk [67], [68]:
    - Description: Sosemanuk is another stream cipher from the eSTREAM portfolio, combining elements from the Snow 2 and Serpent ciphers. It is recognized for its strong security and high throughput.
    - Parameters: Sosemanuk operates with a 128-bit to 256-bit key and a 128-bit IV. The cipher employs an LFSR for bit generation and a non-linear transformation inspired by Serpent for enhanced security.
    - Significance: In wireless systems, Sosemanuk offers an excellent balance between speed and security. Its ability to provide fast encryption with a high level of security makes it suitable for various wireless applications, including secure voice and video communications.

16. Strumok-256 [69], [70]:
    - Description: Strumok-256 is a stream cipher that offers a high level of security with a focus on resistance to various cryptographic attacks. It is designed to be efficient in both hardware and software implementations.

- Parameters: This cipher operates with a 256-bit key, providing a robust level of security. It employs complex algorithms for keystream generation, ensuring unpredictability and resistance to known forms of cryptanalysis.
- Significance: In wireless communication systems, Strumok-256 is particularly suitable for applications requiring a higher degree of security, such as secure financial transactions or confidential communications. Its strong security profile makes it a reliable choice for protecting sensitive data in wireless networks.

17. Strumok-512 [69], [70]:
    - Description: Strumok-512 is an advanced version of Strumok-256, offering enhanced security features. It is designed for scenarios where extremely high security is paramount.
    - Parameters: The cipher uses a 512-bit key, placing it among the most secure stream ciphers in terms of key size. Like its counterpart, it utilizes sophisticated algorithms for generating its keystream, providing an additional layer of security.
    - Significance: For wireless systems where the utmost security is required, such as in military or government communications, Strumok-512 offers an optimal solution. Its extended key size and advanced cryptographic mechanisms provide a very high level of security, suitable for protecting highly sensitive information.

18. Trivium [71], [72], [73]:
    - Description: Trivium is a stream cipher known for its simplicity, high speed, and security. It is part of the eSTREAM portfolio and is designed to be particularly efficient in hardware implementations.
    - Parameters: Trivium operates with an 80-bit key and an 80-bit IV. Its design is based on a unique combination of three shift registers with non-linear feedback, producing a highly secure keystream.
    - Significance: Trivium's efficiency and security make it well-suited for a wide range of wireless applications, from IoT devices to high-speed network communications. Its minimalistic design allows for fast data encryption and decryption, making it a practical choice for systems where both speed and security are essential.

In the dynamic realm of wireless communication, the cryptographic landscape is rich and varied, encompassing a range of ciphers each tailored to specific security needs and operational environments. This comprehensive exploration crystallizes the unique advantages of each cipher, highlighting their roles in ensuring secure and efficient wireless communications.

- AES Variants (AES-128 and AES-256): These block ciphers are renowned for their balance of speed and impenetrable security. AES-128 offers a swift encryption process ideal for general applications, while AES-256 provides an enhanced security level for more sensitive data. Their adaptability to stream modes like CTR enhances their applicability in high-speed wireless networks.
- CryptMT: This stream cipher stands out for its exceptional randomness quality and high-speed operation, making it ideal for applications requiring rapid data encryption without compromising security.
- DECIM, Enocoro, and Grain: These ciphers are specifically designed for resource-constrained environments. DECIM's low-resource requirement makes it ideal for IoT and sensor networks. Enocoro offers simplicity and efficiency, perfect for RFID systems and smart cards. Grain provides a balance of simplicity and efficiency, suitable for embedded systems and wireless sensor networks.
- HC-128 and HC-256: These ciphers are notable for their high-speed processing capabilities and robust security mechanisms, making them well-suited for environments where both performance and security are paramount.

- KCipher-2: This cipher is distinguished by its unique algorithmic structure, offering versatility and reliability for a wide range of wireless applications.
- Mickey2, MUGI, and RC4: Mickey2 is optimized for hardware efficiency, ideal for constrained environments. MUGI's robustness makes it suitable for a broad spectrum of applications. RC4, historically significant for its speed, has been pivotal in wireless security, though its usage is now more limited.
- Salsa20, Snow 2, and Sosemanuk: Salsa20 is celebrated for its speed and efficiency, Snow 2 for its high performance, and Sosemanuk for its balance of speed and security, demonstrating the diverse capabilities required for wireless communication security.
- Strumok-256, Strumok-512, and Trivium: Strumok-256 and Strumok-512 offer high and ultra-high security levels, ideal for highly sensitive data protection. Trivium provides a perfect blend of simplicity, speed, and security, suitable for diverse wireless environments.

Each cipher in this review brings unique strengths to the cryptographic domain, addressing the evolving and diverse needs of wireless communication systems. Their implementation underscores a commitment to maintaining high standards of security, ensuring data integrity and confidentiality across various wireless applications and platforms.

15.5   RESULTS OF THE COMPUTATIONAL EXPERIMENT

In the realm of cryptographic research, particularly in the context of stream ciphers for wireless communication systems, the statistical evaluation of cryptographic properties holds paramount importance. This section delves into the experimental investigations of modern stream ciphers using the National Institute of Standards and Technology's Statistical Test Suite (NIST STS), a benchmark for assessing the randomness and cryptographic robustness of algorithms. The analysis presented here is crucial for validating the security and effectiveness of these ciphers in real-world applications.

15.5.1   COMPUTATIONAL MODELING AND EXPERIMENTAL INVESTIGATIONS USING NIST STS

The experimental research conducted as per the NIST STS methodology focused on a range of contemporary stream ciphers, including AES, CryptMT, DECIM, Enocoro, Grain, HC, KCipher, Mickey2, MUGI, Rabbit, RC4, Salsa20, Snow 2, Sosemanuk, Strumok, and Trivium. For each algorithm, 100 sequences were generated using a random key K and a random initialization vector IV. Each sequence, with a length of 10^6 bits, was considered an independent test. The mean number of passed statistical tests and the standard deviations were calculated across all 100 sequences, resulting in a sample size of 10^8 bits for each algorithm. The results of these tests are summarized in Table 15.1.

The Table 15.1 presents the statistical testing results of various stream ciphers, providing key metrics such as the mean number of passed tests (M099, M096) and the standard deviations (D099, D096, S099, S096) for different confidence levels. The minimum number of passed tests (Min096) is also reported, offering a comprehensive view of each algorithm's performance under rigorous statistical scrutiny.

The results from the NIST STS reveal that all the examined stream ciphers exhibit high cryptographic standards. The number of successfully passed tests ranged from 130 to 134 for the criterion $P_j \geq 0,99$ and 186 to 187 for $P_j \geq 0,96$, except for CryptMT, which scored 158. These high scores, achieved with significant reliability, demonstrate the robustness and reliability of these ciphers in producing random, unpredictable outputs – a key requirement for secure cryptographic systems. The minimum values of passed tests varied between 179 and 186, indicating a consistent level of performance across different algorithms. Notably, KCipher exhibited the highest number of passed tests, underscoring its exceptional statistical strength.

TABLE 15.1
Results of NIST STS for Modern Stream Ciphers

Encryption algorithm	M099	D099	S099	M096	D096	S096	MIN
AES-128	127.07	20.456	4.438	186.63	0.3191	0.554	185
CryptMT	130.89	52.988	7.279	158.56	5.284	2.299	181
DECIM	132.44	19.358	4.399	186.44	0.9136	0.956	185
Enocoro	132.92	51.22	7.157	187.17	0.79	0.89	185
Grain	132.36	57.32	7.571	186.92	1.414	1.185	182
HC-256	133.75	36.44	6.04	186.66	1.93	1.381	182
KCipher	131.29	11.061	3.3258	186.71	0.489	0.699	186
Mickey 2	133.53	61.65	7.85	186.6	2.302	1.51	179
MUGI	132.23	53.721	7.329	186.5	0.978	0.989	185
Rabbit	132.65	16.87	4.017	187.22	0.451	0.657	185
RC4	133.7	67.01	8.186	186.3	1.61	1.269	184
Salsa20	134.16	28.055	5.27	187.001	1.01	0.99	183
Snow2.0	132.78	23.93	4.89	186.79	0.43	0.656	183
Sosemanuk	131.73	49.36	6.991	186.8	2.240	1.49	184
Strumok-256	130.01	23.6	4.86	186.45	1.4555	1.206	184
Strumok-512	132.83	56.516	7.518	186.90	0.802	0.896	185
Trivium	130.24	99.683	9.935	187.15	1.49	1.214	182

The Figure 15.2 provides a comprehensive view of the performance of various stream ciphers under rigorous statistical scrutiny. It highlights the mean number of tests passed at different confidence levels (99% and 96%) and the minimum number of tests passed. The results demonstrate the robustness and reliability of these ciphers in producing random, unpredictable outputs, which is crucial for secure cryptographic systems. Notably, some ciphers like KCipher and Salsa20 show exceptional statistical strength, while CryptMT has a notably lower score in the M096 metric.

The comprehensive statistical analysis using the NIST STS methodologically affirms the cryptographic efficacy of the examined stream ciphers. These results are integral to our understanding of the security landscape of modern cryptographic algorithms, particularly in the context of wireless communication systems. The high performance of these ciphers in statistical tests reinforces their suitability for securing wireless communications, where the integrity and confidentiality of data are of utmost importance. This analysis not only validates the current cryptographic standards but also sets a benchmark for future developments in stream cipher technologies.

15.5.2　Computational Modeling and Experimental Investigations Using DIEHARD

The DIEHARD Test Suite stands as a rigorous benchmark for evaluating the statistical properties of cryptographic algorithms, particularly stream ciphers. This section presents an in-depth analysis of various modern stream ciphers using the DIEHARD suite, focusing on their randomness and statistical robustness. The results from these tests are crucial in assessing the reliability and security of these ciphers in wireless communication systems.

The DIEHARD tests are designed to assess the randomness of sequences generated by cryptographic algorithms. A key characteristic of truly random sequences is a uniform distribution of probabilities across the unit interval. Deviations from this uniformity indicate the presence of ordered patterns in the sequence, potentially compromising its cryptographic integrity.

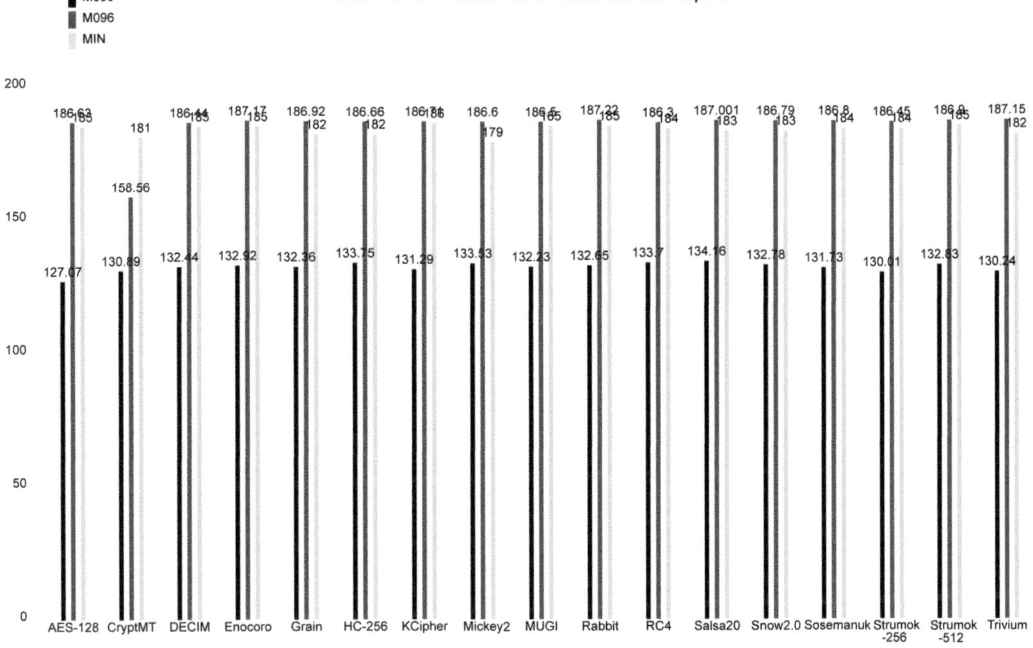

This chart visualizes the results of the NIST STS for various modern stream ciphers. It shows the mean number of passed statistical tests (M099, M096) and the minimum number of passed tests (MIN) for each algorithm.

FIGURE 15.2 Statistical Robustness of Modern Stream Ciphers: An Analysis via NIST STS.

For this analysis, each cipher, including AES, CryptMT, DECIM, Enocoro, Grain, HC, KCipher, Mickey2, MUGI, Rabbit, RC4, Salsa20, Snow 2, Sosemanuk, Strumok, and Trivium, was subjected to the DIEHARD suite. A total of 100 sequences were generated for each cipher using a random key and initialization vector. Each sequence, 10^6 bits in length, was treated as an independent test. The probabilities of passing the tests were calculated and categorized into three intervals:

- test failure ($P_j \leq 0.1$ or $P_j > 0.9$),
- dubious results ($0.1 < P_j \leq 0.25$ or $0.75 < P_j \leq 0.9$),
- and test passed ($0.25 < P_j \leq 0.75$).

The distribution of these probabilities provides an approximation of the sequence's randomness compared to a truly random sequence.

The results, summarized in Table 15.2, show the distribution of test outcomes across the defined intervals for each cipher. All ciphers demonstrated a sufficient level of statistical security. The majority of the test results fell within the interval $0.25 < P_j \leq 0.75$, indicating good statistical properties. However, it is noteworthy that Sosemanuk, despite showing minor deviations from uniform distribution, led in the number of tests passed. Conversely, KCipher, which ranked first in the NIST STS tests, showed the least favorable results in the DIEHARD tests.

Figure 15.3 presents a histogram illustrating the number of tests successfully passed by each cipher. This visual representation highlights the statistical strength of the ciphers, with Sosemanuk standing out for its high number of passed tests.

The DIEHARD test suite analysis provides a comprehensive evaluation of the statistical properties of modern stream ciphers. The results affirm the robustness and randomness of these ciphers,

TABLE 15.2
DIEHARD Test Suite Results for Modern Stream Ciphers

Encryption algorithm	$P_j \leq 0.1$	$0.1 < P_j \leq 0.25$	$0.25 < P_j \leq 0.75$	$0.75 < P_j \leq 0.9$	$P_j > 0.9$
AES-128	16	40	106	29	28
AES-256	15	33	119	31	21
CryptMT	21	33	99	36	25
DECIM	25	29	92	42	26
Enocoro	18	36	102	44	16
Grain	17	27	124	32	19
HC-128	17	39	103	36	24
HC-256	25	36	110	30	18
KCipher-2	25	43	90	43	25
Mickey2	23	36	104	33	23
MUGI	25	35	107	27	25
RC4	23	21	103	35	32
Salsa20	17	34	107	27	34
Snow 2	21	31	112	33	18
Sosemanuk	13	25	125	34	17
Strumok-256	20	29	113	35	18
Strumok-512	26	26	113	28	22
Trivium	24	27	107	35	22

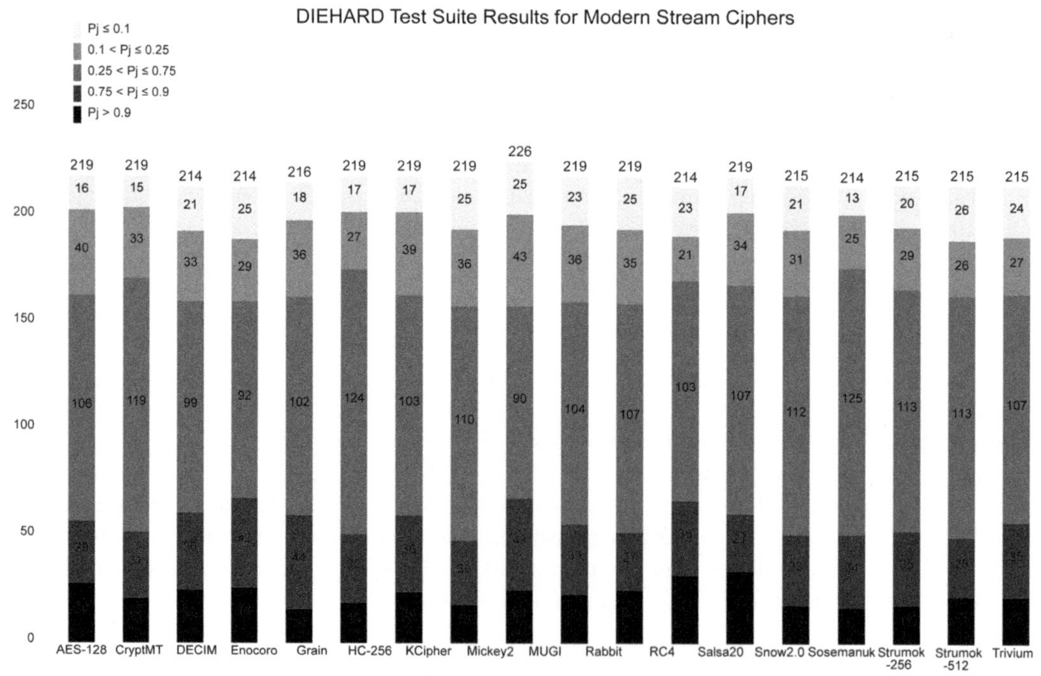

This chart visualizes the DIEHARD test suite results for modern stream ciphers, showcasing the distribution of probabilities across different ranges for each algorithm.

FIGURE 15.3   Evaluating the Randomness Efficacy of Stream Ciphers: Insights from DIEHARD Tests.

essential qualities for cryptographic algorithms used in wireless communication systems. This analysis not only validates the security of these ciphers but also offers insights into their relative strengths and weaknesses, guiding their selection and implementation in secure communication protocols.

15.6 DISCUSSION OF RESULTS

In the quest to fortify wireless communication systems, the cryptographic strength of stream ciphers plays a pivotal role. The results obtained from the NIST STS and DIEHARD test suites provide a comprehensive evaluation of various modern stream ciphers, offering insights into their statistical properties and randomness. This discussion aims to dissect these results, comparing the performance of different ciphers and drawing conclusions about their suitability for secure wireless communications.

The NIST STS results revealed a high degree of cryptographic robustness across the examined ciphers. Notably, KCipher-2 emerged as a standout, demonstrating a superior ability to pass statistical tests, which suggests a high level of randomness and unpredictability in its output. This characteristic is crucial in cryptographic applications, as it enhances the cipher's resistance to various forms of cryptanalysis. Similarly, ciphers like AES-128 and AES-256, known for their widespread use and robust security features, also showed strong performance, reinforcing their reliability in secure communications.

On the other hand, the DIEHARD test suite results painted a slightly different picture. While all ciphers displayed a sufficient level of statistical security, Sosemanuk distinguished itself with the highest number of tests passed. This result indicates that Sosemanuk, despite its minor deviations in uniform distribution, maintains excellent statistical properties, making it a strong candidate for applications requiring high randomness. In contrast, KCipher-2, which excelled in the NIST STS, showed a lower number of tests in the desired range in the DIEHARD suite. This discrepancy highlights the importance of using multiple test suites to thoroughly evaluate cryptographic algorithms, as different tests can reveal various aspects of a cipher's statistical behavior.

The comparative analysis of these ciphers underscores a crucial point: no single cipher is universally superior in all aspects. Each cipher exhibits unique strengths and weaknesses, making them suitable for different applications within wireless communication systems. For instance, AES variants, with their balance of speed and security, are well-suited for general-purpose wireless applications. In contrast, ciphers like Sosemanuk and Grain, with their high randomness, are more appropriate for scenarios where unpredictability is paramount.

In conclusion, the results from the NIST STS and DIEHARD test suites provide valuable insights into the statistical properties of modern stream ciphers. This analysis is instrumental in guiding the selection of appropriate cryptographic solutions for wireless communication systems, ensuring that data remains secure against evolving cyber threats. As the field of wireless communication continues to advance, the role of robust and statistically sound cryptographic algorithms becomes increasingly crucial in safeguarding the integrity and confidentiality of wireless communications.

15.7 CONCLUSIONS AND FUTURE SCOPE

15.7.1 Conclusions

The exploration of cryptographic security in advanced wireless communication systems, particularly through the lens of stream ciphers, has revealed several key insights. The rigorous statistical testing of these ciphers, using methodologies like NIST STS and DIEHARD, has affirmed their robustness and randomness, which are crucial for maintaining the integrity and confidentiality of wireless communications. Computational modeling has emerged as an indispensable tool, bridging the gap between theoretical cryptography and its practical applications. It has enabled the simulation of real-world scenarios, providing a deeper understanding of the performance and resilience of cryptographic algorithms under various conditions.

The analysis has also underscored the dynamic nature of cryptographic challenges in wireless networks. As these networks evolve, so too must the cryptographic solutions that protect them. The balance between computational efficiency and cryptographic strength remains a critical consideration, especially in environments with limited resources.

15.7.2 FUTURE SCOPE

Looking ahead, several areas warrant further exploration and development:

- Adaptation to Emerging Technologies: As wireless communication technologies continue to advance, future research should focus on developing cryptographic solutions that are adaptable to new protocols and standards, such as 5G and beyond.
- Quantum-Resistant Cryptography: With the advent of quantum computing, there is a growing need for cryptographic algorithms that are resistant to quantum attacks. Future research should explore the development of quantum-resistant stream ciphers.
- Machine Learning and AI in Cryptography: The integration of machine learning and artificial intelligence in cryptographic processes presents a promising area of research. These technologies could be used to enhance the security of wireless networks, detect vulnerabilities, and automate the testing and optimization of cryptographic algorithms.
- Energy-Efficient Cryptography: In wireless networks, especially in IoT and mobile devices, energy efficiency is crucial. Future research should focus on developing cryptographic algorithms that provide strong security while consuming minimal power.
- User Privacy and Data Protection: As data privacy concerns continue to grow, future research should also focus on enhancing user privacy in wireless communications, ensuring that cryptographic measures protect not only the data but also the privacy of the individuals involved.

In conclusion, the field of cryptographic security in wireless communication systems is ever-evolving, with new challenges and opportunities emerging continuously. The ongoing research and development in this field are vital for ensuring the security and reliability of wireless networks in our increasingly connected world.

ACKNOWLEDGEMENTS

- This project has received funding from the European Union's Horizon 2020 research and innovation programme under the Marie Skłodowska-Curie grant agreement No. 101007820 – TRUST. This publication reflects only the author's view and the REA is not responsible for any use that may be made of the information it contains.
- This research was funded by the European Union – NextGenerationEU under the Italian Ministry of University and Research (MIUR), National Innovation Ecosystem grant ECS00000041-VITALITY-CUP D83C22000710005.

REFERENCES

[1] E. Conrad, S. Misenar, and J. Feldman, "Chapter 4 – Domain 3: Security Architecture and Engineering," in *CISSP® Study Guide (Fourth Edition)*, E. Conrad, S. Misenar, and J. Feldman, Eds., Syngress, 2023, pp. 107–223. doi: 10.1016/B978-0-443-18734-6.00007-6.

[2] B. Sklar and F. J. Harris, *Digital Communications: Fundamentals and Applications*, 3 edition. Hoboken: Prentice Hall, 2020.

[3] S.-M. M. Yang, *Modern Digital Radio Communication Signals and Systems*. Springer International Publishing, 2019. doi: 10.1007/978-3-319-71568-1.

[4] Y. Qian, F. Ye, and H.-H. Chen, "Cryptographic Techniques," in *Security in Wireless Communication Networks*, IEEE, 2022, pp. 51–76. doi: 10.1002/9781119244400.ch4.

[5] A. A. Kuznetsov, O. V. Potii, N. A. Poluyanenko, Y. I. Gorbenko, and N. Kryvinska, Stream Ciphers in Modern Real-time IT Systems. in *Studies in Systems, Decision and Control*. Cham: Springer Nature, 2022. doi: 10.1007/978-3-030-79770-6.

[6] D. Fang, Y. Qian, and R. Q. Hu, "Introduction to 5G Wireless Systems," in *5G Wireless Network Security and Privacy*, IEEE, 2024, pp. 1–6. doi: 10.1002/9781119784340.ch1.

[7] E. Conrad, S. Misenar, and J. Feldman, "Chapter 5 – Domain 4: Communication and Network Security," in *CISSP® Study Guide (Fourth Edition)*, E. Conrad, S. Misenar, and J. Feldman, Eds., Syngress, 2023, pp. 225–293. doi: 10.1016/B978-0-443-18734-6.00003-9.

[8] H. J. Madi, "A review of new developments in Reliability and Connectivity of wireless sensor Technology over recent years," in *Reference Module in Materials Science and Materials Engineering*, Elsevier, 2023. doi: 10.1016/B978-0-323-96020-5.00070-4.

[9] H. Sun, R. Q. Hu, and Y. Qian, "5G Wireless Networks with Underlaid D2D Communications," in *5G and Beyond Wireless Communication Networks*, IEEE, 2024, pp. 10–20. doi: 10.1002/9781119089469. ch2.

[10] A. J. Menezes, P. C. van Oorschot, S. A. Vanstone, P. C. van Oorschot, and S. A. Vanstone, *Handbook of Applied Cryptography*. CRC Press, 2018. doi: 10.1201/9780429466335.

[11] "ISO/IEC 18033-4:2011 Information technology — Security techniques — Encryption algorithms — Part 4: Stream ciphers." Accessed: Jan. 18, 2024. [Online]. Available: www.iso.org/standard/54532.html

[12] "Home," E-Stream. Accessed: Jan. 18, 2024. [Online]. Available: https://e-stream.it/

[13] L. Deng, B. Wang, Y. Gao, Z. Chen, and S. Li, "Certificateless Anonymous Signcryption Scheme With Provable Security in the Standard Model Suitable for Healthcare Wireless Sensor Networks," *IEEE Internet of Things Journal*, vol. 10, no. 18, pp. 15953–15965, Sep. 2023, doi: 10.1109/ JIOT.2023.3266335.

[14] J. C. Gonzalez-Arango, D. C. Ocampo-Munera, L. F. Castano-Londono, G. D. Goez-Sanchez, and R. A. Velasquez-Velez, "Performance Evaluation of Symmetric Cryptographic Algorithms in Resource Constrained Hardware for Wireless Sensor Networks," *IEEE Latin America Transactions*, vol. 19, no. 10, pp. 1632–1639, Oct. 2021, doi: 10.1109/TLA.2021.9477225.

[15] B. Halak, Y. Yilmaz, and D. Shiu, "Comparative Analysis of Energy Costs of Asymmetric vs Symmetric Encryption-Based Security Applications," *IEEE Access*, vol. 10, pp. 76707–76719, 2022, doi: 10.1109/ ACCESS.2022.3192970.

[16] A. Halbouni, L.-Y. Ong, and M.-C. Leow, "Wireless Security Protocols WPA3: A Systematic Literature Review," *IEEE Access*, vol. 11, pp. 112438–112450, 2023, doi: 10.1109/ACCESS.2023.3322931.

[17] L. Harn, C.-F. Hsu, Z. Xia, and Z. He, "Lightweight Aggregated Data Encryption for Wireless Sensor Networks (WSNs)," *IEEE Sensors Letters*, vol. 5, no. 4, pp. 1–4, Apr. 2021, doi: 10.1109/ LSENS.2021.3063326.

[18] M. Helmy, E.-S. M. El-Rabaie, I. Eldokany, and F. E. Abd El-Samie, "Proposed Hybrid Encryption Framework for Robust 3D Image Communication Over Wireless Channels," *Optik*, vol. 273, p. 170205, Feb. 2023, doi: 10.1016/j.ijleo.2022.170205.

[19] N. Hrovatin, A. Tošić, M. Mrissa, and J. Vičič, "A General Purpose Data and Query Privacy Preserving Protocol for Wireless Sensor Networks," *IEEE Transactions on Information Forensics and Security*, vol. 18, pp. 4883–4898, 2023, doi: 10.1109/TIFS.2023.3300524.

[20] Y. Hua and A. Maksud, "Continuous Encryption Functions for Security Over Networks," *Signal Processing*, vol. 203, p. 108807, Feb. 2023, doi: 10.1016/j.sigpro.2022.108807.

[21] O. A. Khashan, R. Ahmad, and N. M. Khafajah, "An Automated Lightweight Encryption Scheme for Secure and Energy-Efficient Communication in Wireless Sensor Networks," *Ad Hoc Networks*, vol. 115, p. 102448, Apr. 2021, doi: 10.1016/j.adhoc.2021.102448.

[22] L. Li, S. Li, H. Peng, and J. Bi, "An Efficient Secure Data Transmission and Node Authentication Scheme for Wireless Sensing Networks," *Journal of Systems Architecture*, vol. 133, p. 102760, Dec. 2022, doi: 10.1016/j.sysarc.2022.102760.

[23] H. S. Park and S. K. Hong, "Encryption Device Based on Wave-Chaos for Enhanced Physical Security of Wireless Wave Transmission," *IEEE Access*, vol. 11, pp. 102917–102925, 2023, doi: 10.1109/ ACCESS.2023.3317052.

[24] P. Ramadevi, S. Ayyasamy, Y. Suryaprakash, C. Anilkumar, S. Vijayakumar, and R. Sudha, "Security for Wireless Sensor Networks Using Cryptography," *Measurement: Sensors*, vol. 29, p. 100874, Oct. 2023, doi: 10.1016/j.measen.2023.100874.

[25] S. Urooj, S. Lata, S. Ahmad, S. Mehfuz, and S. Kalathil, "Cryptographic Data Security for Reliable Wireless Sensor Network," *Alexandria Engineering Journal*, vol. 72, pp. 37–50, Jun. 2023, doi: 10.1016/j.aej.2023.03.061.

[26] F. Xu *et al.*, "Beyond Encryption: Exploring the Potential of Physical Layer Security in UAV Networks," *Journal of King Saud University – Computer and Information Sciences*, vol. 35, no. 8, p. 101717, Sep. 2023, doi: 10.1016/j.jksuci.2023.101717.

[27] V. Rudnytskyi, O. Korchenko, N. Lada, R. Ziubina, L. Wieclaw, and L. Hamera, "Cryptographic Encoding in Modern Symmetric and Asymmetric Encryption," *Procedia Computer Science*, vol. 207, pp. 54–63, Jan. 2022, doi: 10.1016/j.procs.2022.09.037.

[28] H. Du, J. Wang, D. Niyato, J. Kang, Z. Xiong, M. Guizani, D.I. Kim, "Rethinking Wireless Communication Security in Semantic Internet of Things," in *IEEE Wireless Communications*, vol. 30, no. 3, pp. 36–43, June 2023. doi: 10.1109/MWC.011.2200547.

[29] S. Rubinstein-Salzedo, Cryptography. in *Springer Undergraduate Mathematics Series*. Cham: Springer International Publishing, 2018. doi: 10.1007/978-3-319-94818-8.

[30] "eSTREAM Optimized Code HOWTO." Accessed: Jan. 18, 2024. [Online]. Available: https://www.ecrypt.eu.org/stream/perf/

[31] H. Kopetz, Ed., "Modeling Real-Time Systems," in *Real-Time Systems: Design Principles for Distributed Embedded Applications*, in The International Series in Engineering and Computer Science., Boston, MA: Springer US, 1997, pp. 71–96. doi: 10.1007/0-306-47055-1_4.

[32] I. Gorbenko, A. Kuznetsov, V. Tymchenko, Y. Gorbenko, and O. Kachko, "Experimental Studies of the Modern Symmetric Stream Ciphers," in *2018 International Scientific-Practical Conference Problems of Infocommunications. Science and Technology (PIC S T)*, Oct. 2018, pp. 125–128. doi: 10.1109/INFOCOMMST.2018.8632058.

[33] I. Gorbenko, A. Kuznetsov, M. Lutsenko, and D. Ivanenko, "The Research of Modern Stream Ciphers," in *2017 4th International Scientific-Practical Conference Problems of Infocommunications. Science and Technology (PIC S T)*, Oct. 2017, pp. 207–210. doi: 10.1109/INFOCOMMST.2017.8246381.

[34] D. Johnston, *Random Number Generators—Principles and Practices: A Guide for Engineers and Programmers*. De|G Press, 2018. doi: 10.1515/9781501506062.

[35] E. Barker and J. Kelsey, "Recommendation for Random Number Generation Using Deterministic Random Bit Generators," National Institute of Standards and Technology, NIST Special Publication (SP) 800-90A Rev. 1, Jun. 2015. doi: https://doi.org/10.6028/NIST.SP.800-90Ar1.

[36] I. T. L. Computer Security Division, "NIST SP 800-22: Documentation and Software – Random Bit Generation | CSRC | CSRC," CSRC | NIST. Accessed: Mar. 12, 2021. [Online]. Available: https://csrc.nist.gov/projects/random-bit-generation/documentation-and-software

[37] "The Marsaglia Random Number CDROM Including the Diehard Battery of Tests." Accessed: Jan. 20, 2024. [Online]. Available: https://web.archive.org/web/20160125103112/http://stat.fsu.edu/pub/diehard/

[38] "GINARTeam/Diehard-Statistical-Test." GINAR Team, Jun. 27, 2023. Accessed: Jan. 20, 2024. [Online]. Available: https://github.com/GINARTeam/Diehard-statistical-test

[39] J. Daemen and V. Rijmen, "Specification of Rijndael," in *The Design of Rijndael: The Advanced Encryption Standard (AES)*, J. Daemen and V. Rijmen, Eds., in Information Security and Cryptography., Berlin, Heidelberg: Springer, 2020, pp. 31–51. doi: 10.1007/978-3-662-60769-5_3.

[40] J. Daemen and V. Rijmen, "Rijndael/AES," in *Encyclopedia of Cryptography and Security*, H. C. A. van Tilborg, Ed., Boston, MA: Springer US, 2005, pp. 520–524. doi: 10.1007/0-387-23483-7_358.

[41] M. Ota, "magurosan/CryptMT." Oct. 02, 2021. Accessed: Jan. 20, 2024. [Online]. Available: https://github.com/magurosan/CryptMT

[42] M. Matsumoto, M. Saito, T. Nishimura, and M. Hagita, "CryptMT3 Stream Cipher," in *New Stream Cipher Designs: The eSTREAM Finalists*, M. Robshaw and O. Billet, Eds., in Lecture Notes in Computer Science., Berlin, Heidelberg: Springer, 2008, pp. 7–19. doi: 10.1007/978-3-540-68351-3_2.

[43] "CryptMT." Accessed: Jan. 20, 2024. [Online]. Available: www.math.sci.hiroshima-u.ac.jp/m-mat/MT/CRYPTMT/index.html

[44] C. Berbain *et al.*, "Decimv2," in *New Stream Cipher Designs: The eSTREAM Finalists*, M. Robshaw and O. Billet, Eds., in Lecture Notes in Computer Science., Berlin, Heidelberg: Springer, 2008, pp. 140–151. doi: 10.1007/978-3-540-68351-3_11.

[45] "Enocoro: Hitachi Crypto Technology: Hitachi." Accessed: Jan. 20, 2024. [Online]. Available: www. hitachi.com/rd/yrl/crypto/enocoro/index.html

[46] D. Watanabe, T. Owada, K. Okamoto, Y. Igarashi, and T. Kaneko, "Update on Enocoro Stream Cipher," in *2010 International Symposium On Information Theory & Its Applications*, Oct. 2010, pp. 778–783. doi: 10.1109/ISITA.2010.5649627.

[47] L. Ding, C. Jin, and J. Guan, "Slide Attack on Standard Stream Cipher Enocoro-80 in the Related-Key Chosen IV Setting," *Pervasive and Mobile Computing*, vol. 24, pp. 224–230, Dec. 2015, doi: 10.1016/ j.pmcj.2015.08.002.

[48] M. Hell, T. Johansson, A. Maximov, and W. Meier, "The Grain Family of Stream Ciphers," in *New Stream Cipher Designs: The eSTREAM Finalists*, M. Robshaw and O. Billet, Eds., in Lecture Notes in Computer Science., Berlin, Heidelberg: Springer, 2008, pp. 179–190. doi: 10.1007/978-3-540-68351-3_14.

[49] "Grain-128AEAD." Accessed: Jan. 20, 2024. [Online]. Available: https://grain-128aead.github.io/

[50] H. Wu, "A New Stream Cipher HC-256," in *Fast Software Encryption*, B. Roy and W. Meier, Eds., in Lecture Notes in Computer Science. Berlin, Heidelberg: Springer, 2004, pp. 226–244. doi: 10.1007/ 978-3-540-25937-4_15.

[51] H. Wu, "A New Stream Cipher HC-256." 2004. Accessed: Jan. 18, 2024. [Online]. Available: https:// eprint.iacr.org/2004/092

[52] S. Kiyomoto and W. Shin, "A Description of the KCipher-2 Encryption Algorithm," Internet Engineering Task Force, Request for Comments RFC 7008, Aug. 2013. doi: 10.17487/RFC7008.

[53] T. Hibiki *et al.*, "Chosen-IV Correlation Power Analysis on KCipher-2 and a Countermeasure," in *Constructive Side-Channel Analysis and Secure Design*, E. Prouff, Ed., in Lecture Notes in Computer Science. Berlin, Heidelberg: Springer, 2013, pp. 169–183. doi: 10.1007/978-3-642-40026-1_11.

[54] "Product Outline | KCipher-2 | KDDI Research." Accessed: Jan. 20, 2024. [Online]. Available: www. kddi-research.jp/english/products/kcipher2.html

[55] S. Babbage and M. Dodd, "The MICKEY Stream Ciphers," in *New Stream Cipher Designs: The eSTREAM Finalists*, M. Robshaw and O. Billet, Eds., in Lecture Notes in Computer Science., Berlin, Heidelberg: Springer, 2008, pp. 191–209. doi: 10.1007/978-3-540-68351-3_15.

[56] M. Henricksen and E. Dawson, "Rekeying Issues in the MUGI Stream Cipher," in *Selected Areas in Cryptography*, B. Preneel and S. Tavares, Eds., in Lecture Notes in Computer Science. Berlin, Heidelberg: Springer, 2006, pp. 175–188. doi: 10.1007/11693383_12.

[57] D. Watanabe, S. Furuya, H. Yoshida, K. Takaragi, and B. Preneel, "A New Keystream Generator MUGI," in *Fast Software Encryption*, J. Daemen and V. Rijmen, Eds., in Lecture Notes in Computer Science. Berlin, Heidelberg: Springer, 2002, pp. 179–194. doi: 10.1007/3-540-45661-9_14.

[58] H. Ltd, "MUGI: Multi Giga Cipher: Research & Development: Hitachi." Accessed: Jan. 20, 2024. [Online]. Available: www.hitachi.com/rd/glossary/m/mugi.html

[59] "RC4 Encryption Algorithm," GeeksforGeeks. Accessed: Jan. 20, 2024. [Online]. Available: www. geeksforgeeks.org/rc4-encryption-algorithm/

[60] T. D. B. Weerasinghe, "An Effective RC4 Stream Cipher." 2014. Accessed: Jan. 20, 2024. [Online]. Available: https://eprint.iacr.org/2014/171

[61] "RC4." Accessed: Jan. 20, 2024. [Online]. Available: www.thalesdocs.com/gphsm/ptk/5.7/docs/Cont ent/PTK-J/Ciphers/RC4.htm

[62] D. J. Bernstein, "The Salsa20 Family of Stream Ciphers," in *New Stream Cipher Designs: The eSTREAM Finalists*, M. Robshaw and O. Billet, Eds., in Lecture Notes in Computer Science., Berlin, Heidelberg: Springer, 2008, pp. 84–97. doi: 10.1007/978-3-540-68351-3_8.

[63] T. R. Campbell, "Daence: Salsa20 and ChaCha in Deterministic Authenticated Encryption with no noNCEnse." 2020. Accessed: Jan. 18, 2024. [Online]. Available: https://eprint.iacr.org/2020/067

[64] S. S. P. Goswami and G. Trivedi, "FPGA Implementation of Modified SNOW 3G Stream Ciphers Using Fast and Resource Efficient Substitution Box," *IEEE Embedded Systems Letters*, vol. 15, no. 4, pp. 238–241, Dec. 2023, doi: 10.1109/LES.2023.3298743.

[65] A. N. Alekseychuk, S. M. Koniushok, and M. V. Poremskyi, "Security Evaluation for Snow 2.0-like Stream Ciphers Against Correlation Attacks over Extension Fields." 2019. Accessed: Jan. 18, 2024. [Online]. Available: https://eprint.iacr.org/2019/287

[66] P. Ekdahl, T. Johansson, A. Maximov, and J. Yang, "A New SNOW Stream Cipher Called SNOW-V." 2018. Accessed: Jan. 18, 2024. [Online]. Available: https://eprint.iacr.org/2018/1143

[67] C. Berbain *et al.*, "Sosemanuk, a Fast Software-Oriented Stream Cipher," in *New Stream Cipher Designs: The eSTREAM Finalists*, M. Robshaw and O. Billet, Eds., in Lecture Notes in Computer Science., Berlin, Heidelberg: Springer, 2008, pp. 98–118. doi: 10.1007/978-3-540-68351-3_9.

[68] "The eSTREAM Portfolio Page." Accessed: Jan. 18, 2024. [Online]. Available: www.ecrypt.eu.org/stream/e2-sosemanuk.html

[69] A. A. Kuznetsov, O. V. Potii, N. A. Poluyanenko, Y. I. Gorbenko, and N. Kryvinska, "Stream Symmetric Cipher 'Strumok,'" in *Stream Ciphers in Modern Real-time IT Systems: Analysis, Design and Comparative Studies*, A. A. Kuznetsov, O. V. Potii, N. A. Poluyanenko, Y. I. Gorbenko, and N. Kryvinska, Eds., in Studies in Systems, Decision and Control., Cham: Springer International Publishing, 2022, pp. 467–516. doi: 10.1007/978-3-030-79770-6_16.

[70] I. Gorbenko, O. Kuznetsov, Y. Gorbenko, A. Alekseychuk, and V. Tymchenko, "Strumok Keystream Generator," in *2018 IEEE 9th International Conference on Dependable Systems, Services and Technologies (DESSERT)*, Kyiv, Ukraine: IEEE, May 2018, pp. 294–299. doi: 10.1109/DESSERT.2018.8409147.

[71] Y. Tian, G. Chen, and J. Li, "On the Design of Trivium." 2009. Accessed: Jan. 18, 2024. [Online]. Available: https://eprint.iacr.org/2009/431

[72] C. De Cannière, "Trivium: A Stream Cipher Construction Inspired by Block Cipher Design Principles," in *Information Security*, S. K. Katsikas, J. López, M. Backes, S. Gritzalis, and B. Preneel, Eds., in Lecture Notes in Computer Science. Berlin, Heidelberg: Springer, 2006, pp. 171–186. doi: 10.1007/11836810_13.

[73] "The eSTREAM Portfolio Page." Accessed: Jan. 18, 2024. [Online]. Available: www.ecrypt.eu.org/stream/e2-trivium.html

16 Computational Modeling of Enhanced Spread Spectrum Codes for Asynchronous Wireless Communication

Oleksandr Kuznetsov, Emanuele Frontoni, Natalia Kryvinska, Oleksii Smirnov and Agbotiname Lucky Imoize

16.1 INTRODUCTION

In the rapidly evolving landscape of wireless communication, the quest for efficient and robust network systems has never been more pressing [1]. The burgeoning number of mobile devices and the exponential growth in data traffic demand innovative solutions to optimize network performance and reliability. This chapter delves into the realm of Spread Spectrum Multiple-Access (SSMA) techniques, a cornerstone in the architecture of modern wireless communication systems [2]–[4]. Our focus is on the computational modeling and simulation of enhanced spread spectrum codes tailored for asynchronous wireless communication environments [5], [6].

Spread spectrum technology, renowned for its efficient bandwidth utilization and resilience to interference, forms the backbone of various wireless communication systems, including cellular networks, satellite communications, and wireless local area networks [6], [7]. The asynchronous nature of these systems, where multiple users transmit without coordination, poses unique challenges in maintaining signal integrity and network efficiency [7], [8]. Traditional spread spectrum codes, such as Gold codes and Kasami sequences, have been instrumental in addressing these challenges [9], [10]. However, the limited cardinality of these code sets restricts the capacity for multiple access, a limitation that becomes increasingly pronounced in the context of next-generation wireless networks [11].

In response to these challenges, this chapter introduces computationally derived improved codes for SSMA asynchronous techniques [12], [13]. These codes are designed to expand the capacity of multiple access while maintaining, or even enhancing, the inherent advantages of spread spectrum technology [14], [15]. The computational approach to modeling these codes involves a rigorous exploration of their cross-correlation properties, particularly for arbitrarily random starting points. This exploration is grounded in the fundamental theoretical framework provided by the Welch bound, which links the cardinality, length, and cross-correlation of codes.

Our computational models extend beyond traditional analytical methods, employing advanced techniques such as Laurent and Puiseux series approximations. These models provide a deeper understanding of the behavior of spread spectrum codes in finite-length scenarios, a critical aspect in practical wireless communication systems. The computational simulations demonstrate that with a marginal increase in the correlation modulus, a significant expansion in the cardinality of spreading codes is achievable. This property is pivotal for future radio control systems and smart grids, where the flexibility to adjust subscriber capacity with minimal impact on service quality is essential.

DOI: 10.1201/9781003457428-17

Moreover, this chapter presents a comparative analysis of traditional and improved spread spectrum codes, highlighting the computational advancements in our proposed models. The analysis underscores the potential of these enhanced codes in supporting the burgeoning demands of advanced wireless communication systems, particularly in terms of network coverage, throughput, and user mobility.

In summary, this chapter contributes to the field of wireless communication by presenting a comprehensive computational framework for the development and analysis of enhanced spread spectrum codes. These codes are poised to play a crucial role in the next generation of wireless networks, offering a scalable and efficient solution to the challenges posed by the ever-increasing demand for high-quality wireless communication services. Through a blend of theoretical insights and practical simulations, we aim to provide a valuable resource for researchers and practitioners in the field, paving the way for more resilient and efficient wireless communication systems.

16.1.1 Key Contributions of the Chapter

This chapter presents a significant advancement in the field of spread spectrum communication systems, introducing innovative spreading sequences with a five-level correlation function. It bridges the gap in existing literature by providing a comprehensive algebraic framework for sequence generation, coupled with a detailed computational analysis. The chapter's comparative analysis of correlation characteristics and cardinality against established codes like Gold and Kasami sets underscores its contribution to enhancing the capacity and security of communication systems.
Key Contributions:

i. Innovative Spreading Sequence Development: This chapter introduces new spreading sequences with a unique five-level correlation function, marking a significant leap in spread spectrum communication technology.
ii. Algebraic Framework and Sequence Generation: A comprehensive algebraic framework for generating spreading sequences is presented, offering a novel approach to sequence generation that leverages the algebraic structure of Galois Fields and the cyclical properties of group codes.
iii. Comparative Analysis with Established Codes: The chapter provides a detailed comparative analysis of the newly developed sequences against established codes like Gold and Kasami sequences. This comparison focuses on correlation characteristics and cardinality, highlighting the superiority of the new sequences in terms of capacity and security enhancements in communication systems.
iv. For the early prediction of certain diseases, a framework based on explainable deep learning has been suggested, and a real-world medical dataset has been utilised to assess the effectiveness of the proposed framework.

16.1.2 Chapter Organization

The chapter begins with a thorough review of related work and a gap analysis in Section 16.2, setting the stage for our contributions. Section 16.3 delves into materials and methods, discussing the concept of Direct-Sequence Spread Spectrum and formalizing tasks in spread spectrum systems, followed by an exploration of requirements for spreading sequences in synchronous and asynchronous communication systems, and theoretical limits including the Welch Bound. Theoretical foundations are laid out in Section 16.4, where we discuss the algebraic framework for sequence generation, the role of Galois Fields, cyclical properties of group codes, with practical application examples. Section 16.5 details the computational modeling process for extended spread spectrum codes. Section 16.6

offers a computational analysis of spreading sequences, while Section 16.7 compares the correlation characteristics and cardinality of different sequences. The chapter concludes with a discussion and future scope in Section 16.8, summarizing research findings and outlining potential future directions, before wrapping up with a conclusion in Section 16.9.

16.2 RELATED WORK AND LITERATURE GAP ANALYSIS

This section provides a comprehensive review of existing literature in the field of spread spectrum communication systems. It encompasses a diverse range of studies, from the application of chaotic sequences and deep learning in signal detection to the exploration of novel spreading sequences for enhanced communication performance. The review critically examines the contributions of each study, highlighting their significance and the context within which they operate.

16.2.1 ANALYSIS OF RELATED WORKS

Yang, Zhou, Lou, Qiao, Ahmed, and He (2021) [16]: The research presents a double-differential coded M-ary DSSS system for mobile underwater acoustic communication, addressing challenges like low SNR and severe Doppler effect. While this study offers advancements in underwater communication, it does not explore the potential of novel spreading sequences in other communication environments, particularly in terms of multi-user capacity, which our research addresses.

Zhang and Helleseth (2022) [17]: This paper develops sequences with good correlation properties based on circular Florentine arrays. The authors connect sequences to combinatorial objects, leading to the construction of sequences with optimal correlation. While this research provides significant insights into sequence construction, it does not explore their application in spread spectrum communication systems, particularly in enhancing multi-user capacity and efficiency, which is the focus of our research.

Murillo-Escobar et al. (2022) [18]: This research introduces a multibiosignal chaotic encryption scheme using spread spectrum and global diffusion processes for e-health applications. It emphasizes the importance of data privacy and security in e-health, utilizing chaos-based cryptography and spread spectrum technology. While this study advances the security of multibiosignal transmission in e-health, it does not explore the potential of novel spreading sequences in enhancing the efficiency of multiuser networks, which our research addresses.

Chen, Hu, and Feng (2022) [19]: This article investigates direct-sequence spread spectrum time division multiple access in passive optical networks, focusing on reducing latency and improving performance. The study emphasizes the interaction between multiple access control schemes and transport layers. However, it primarily concentrates on the physical layer aspects, leaving room for further exploration of the impact of spreading sequences on system performance, especially in terms of signal detection and interference management.

Chu and Barry (2022) [20]: The authors challenge the necessity of alphabet awareness in detecting direct-sequence spread spectrum signals. They propose a multi-antenna likelihood-ratio-test detector that outperforms alphabet-unaware detectors in certain scenarios. This study contributes to the understanding of signal detection in spread spectrum systems but does not extensively explore the design and optimization of spreading sequences themselves, particularly in relation to their correlation properties and impact on detection performance.

Ghafoor et al. (2022) [21]: This comprehensive survey on NOMA in the context of 5G and B5G wireless networks highlights the potential of NOMA to meet the growing demands of connectivity and user capacity. The paper thoroughly discusses various aspects of NOMA, including its network architecture, mobility management, and energy efficiency. However, it does not delve into the specific role of spreading sequences in optimizing NOMA performance, particularly in asynchronous communication environments, which is a gap our research aims to address.

Shao, Karsilayan, Rodenbeck, and Silva-Martinez (2022) [22]: The paper presents an interference-tolerant synchronization scheme for DSSS wireless communication systems. It focuses on overcoming in-band interferences using a three-level synchronization approach. However, the study does not address the potential of novel spreading sequences in further enhancing interference tolerance, which is a key aspect of our research.

Tian, Liu, and Li (2022) [23]: This study focuses on constructing binary Golay spreading sequences for uplink grant-free NOMA, aiming to achieve low coherence and low peak-to-average power ratio (PAPR). The authors utilize a framework of spreading matrices from Boolean functions, leading to new constructions that significantly enhance the capacity for accommodating overloaded devices. While this research advances the field of NOMA, it does not explore the potential of these sequences in asynchronous communication systems, which is a gap our research aims to fill.

Wei, Zheng, Zhou, Zhang, Lou, Zhao, and Yang (2022) [24]: This paper investigates the detection of DSSS signals using deep learning techniques. The authors propose a convolutional neural network-based scheme to automatically learn the features of DSSS signals, demonstrating significant improvements over traditional methods. However, the study primarily focuses on signal detection and does not delve into the optimization of spreading sequences themselves, an area our research contributes to.

Hu et al. (2023) [25]: This study introduces a deep learning-based DSSS underwater acoustic communication system, focusing on improving performance under Doppler effects and low SNR conditions. The use of a Temporal Convolutional Network model as the system's receiver marks a significant advancement. However, the research primarily concentrates on the receiver's architecture and does not explore the potential of novel spreading sequences in enhancing system performance, an area our research contributes to.

Huang, Shen, and Wang (2023) [26]: This paper reviews the application of Terahertz (THz) technology in imaging and wireless communications, identifying key developments and common challenges in both fields. The integration of THz imaging/sensing with wireless communications is considered, offering insights into future technological advancements. However, the role of spreading sequences in THz wireless communications, particularly in terms of spectrum efficiency and interference management, is not addressed, which is a focus of our research.

Lu and Jiao (2023) [27]: The study presents an efficiently all-digital code tracking method for band-limited DSSS systems, highlighting its lower computational complexity and constant tracking performance. While this research advances the code tracking techniques in DSSS systems, it does not explore the impact of different spreading sequences on the tracking performance, an aspect our research investigates.

Roque and Poulliat (2023) [28]: This research proposes SNR-optimal spreading sequences for faster-than-Nyquist (FTN) signaling, aiming to mitigate cyclostationary features in covert communications. While the study offers significant advancements in FTN transmission, it does not explore the application of these sequences in asynchronous communication systems, an area our research aims to enhance.

16.2.2 Literature Gap Analysis

The comprehensive review of existing literature in spread spectrum communication systems reveals several critical gaps that our research aims to address:

- Integration of Advanced Techniques: While there have been significant advancements in the use of chaotic sequences and deep learning for signal detection and performance enhancement, there is a noticeable gap in their integration into the development of new spreading sequences. This integration could potentially optimize system performance, particularly in complex and asynchronous communication environments.

- Broadening Application Scope: Current research has made notable strides in specific areas such as underwater and mobile communication systems. However, there is a need for extending these findings to more diverse and congested communication networks. Addressing the challenges of multi-user capacity and interference in such environments remains a largely unexplored area.
- Optimization of Spreading Sequences: While there have been efforts in constructing sequences with optimal correlation properties, the application of these sequences in practical spread spectrum systems, especially in enhancing multi-user capacity and reducing interference, is still underdeveloped.
- Holistic Solutions for Asynchronous Systems: A significant gap exists in research that provides comprehensive solutions for optimizing spread spectrum systems in asynchronous communication scenarios. Such solutions are crucial for the advancement of modern communication networks.

Our research contributes to filling these gaps by developing new spreading sequences with enhanced multi-level correlation functions. We aim to explore their application in improving the performance of spread spectrum communication systems across various environments, with a particular focus on asynchronous systems. By providing both theoretical insights and practical solutions, our work seeks to improve multi-user capacity, reduce interference, and enhance overall system efficiency in a range of communication scenarios. This approach positions our research as a pivotal contribution to the ongoing evolution of spread spectrum communication technologies.

16.3 MATERIALS AND METHODS

The realm of wireless communication is witnessing an unprecedented surge in the volume of data and the number of connected devices. This growth has brought to the fore a critical challenge: optimizing the use of the electromagnetic spectrum while ensuring robust and efficient communication. The significance of this challenge cannot be overstated, as it directly impacts the performance and scalability of wireless networks, which are integral to numerous applications ranging from mobile telephony to global positioning systems. The development of advanced spread spectrum techniques, particularly in asynchronous environments, is pivotal in addressing this challenge. Computational modeling and simulation play a crucial role in this endeavor, offering insights and solutions that are vital for the evolution of wireless communication systems.

16.3.1 THE CONCEPT OF DIRECT-SEQUENCE SPREAD SPECTRUM

Direct-Sequence Spread Spectrum (DSSS) is a pivotal technique in modern wireless communication, characterized by its method of spreading a signal over a wider frequency band than the minimum required for transmission [27], [29]. The core idea of DSSS involves modulating the original data signal with a pseudo-random code sequence, which is typically much wider in bandwidth than the data signal [1]. This process, known as 'spreading', transforms a narrowband signal into a wideband noise-like signal [30].

Mathematically, the spreading of the signal can be represented as:

$$S(t) = D(t) \times C(t),$$

where $S(t)$ is the spread signal, $D(t)$ is the original data signal, and $C(t)$ is the pseudo-random code sequence.

The wider bandwidth of the spread signal offers several advantages, including resistance to interference and eavesdropping, and the ability to share the same frequency band with multiple users without significant mutual interference.

Implications in Wireless Communication Systems:

1. Multiple Access Communication Systems. In systems like CDMA (Code-division multiple access) [31], [32], DSSS allows multiple users to share the same frequency band. Each user is assigned a unique pseudo-random code, enabling the receiver to selectively demodulate the desired signal using a correlator. This is computationally modeled as:

$$R(t) = S(t) \times C_{user}(t),$$

where $R(t)$ is the received signal, and $C_{user}(t)$ is the user-specific pseudo-random code.

2. Global Positioning Systems (GPS). DSSS is integral to GPS, where satellite signals are spread over a wide bandwidth, enhancing accuracy and reliability [33], [34]. The GPS receiver correlates the received signal with a locally generated pseudo-random code to compute precise timing information, crucial for location determination.

3. Wireless Networks. In Wi-Fi and other wireless networks [8], [26], DSSS aids in mitigating interference from other sources, ensuring robust and stable connectivity. This is particularly important in environments with multiple overlapping networks.

Thus, DSSS stands as a cornerstone in the architecture of modern wireless communication systems. Its ability to spread a signal across a wide frequency band offers significant advantages in terms of interference resistance, security, and multiple access capabilities.

16.3.2 Formalization of the Task in Spread Spectrum Systems

In the context of spread spectrum communication systems, the task revolves around the design and analysis of pseudo-random sequences that form the core of the system's performance. These sequences are defined within an ensemble $S = S_0, S_1, ..., S_{M-1}$, where each $S_i = (S_{i_0}, S_{i_1}, ..., S_{i_{N-1}})$ represents a pseudo-random sequence (code signal) [1], [30]. Each element of the sequence, S_{i_z}, takes a value of either +1 or -1, where $z = 0, ..., N-1$.

The effectiveness of these sequences in a spread spectrum system is largely determined by their correlation properties. The correlation functions, both auto-correlation and cross-correlation, are pivotal in assessing the performance of the system.

Auto-Correlation Function: This function measures the similarity of a sequence with a shifted version of itself. It is defined as:

$$R_{ii}(\tau) = \frac{1}{N} \sum_{z=0}^{N-1} S_{i_z} S_{i_{(z+\tau) \bmod N}},$$

where τ is the shift and $R_{ii}(\tau)$ is the auto-correlation of sequence S_i.

Cross-Correlation Function: This function measures the similarity between two different sequences. It is defined as:

$$R_{ij}(\tau) = \frac{1}{N} \sum_{z=0}^{N-1} S_{i_z} S_{j_{(z+\tau) \bmod N}},$$

where $R_{ij}(\tau)$ is the cross-correlation between sequences S_i and S_j.

The maximum absolute correlation, $\rho_{\max}$, is a critical parameter in the design of spread spectrum systems. It is defined as the highest absolute value of the cross-correlation function among all pairs of sequences in the ensemble for any shift τ:

$$\rho_{\max} = \max_{i \neq j, \tau} \mid R_{ij}(\tau) \mid .$$

In synchronous systems, where signals are aligned, $\rho_{\max}$ directly impacts the system's resistance to interference and noise [26]. A lower $\rho_{\max}$ indicates better performance as it minimizes the potential for signal overlap and interference.

In asynchronous systems, $\rho_{\max}$ is crucial for determining the system's capacity to handle multiple users transmitting simultaneously without coordination. A lower $\rho_{\max}$ allows for a higher number of users to coexist in the system without significant mutual interference.

Computational modeling and simulation play a vital role in analyzing and optimizing the correlation properties of these sequences. By simulating various scenarios and sequence combinations, researchers can computationally assess the performance of the system under different conditions, guiding the development of more efficient and robust spread spectrum systems. This computational approach is essential for pushing the boundaries of current technology and adapting to the ever-increasing demands of wireless communication.

16.3.3 Requirements for Spreading Sequences in Synchronous and Asynchronous Communication Systems

In the realm of radio communication systems, the design and selection of spreading sequences are of paramount importance, both for synchronous and asynchronous techniques. These sequences, which are central to the functionality of spread spectrum systems, must adhere to a set of stringent requirements to ensure optimal system performance:

- Low Cross-Correlation: To minimize interference between different users in a shared spectrum, spreading sequences must exhibit low cross-correlation. This is crucial in dense network environments where multiple users are communicating simultaneously.
- Good Auto-Correlation Properties: A spreading sequence should have a sharp auto-correlation peak. This property is essential for accurate timing synchronization and signal detection, especially in environments with multipath propagation.
- Balance and Uniformity: The sequences should be balanced, with an equal number of positive and negative chips, and uniformly distributed. This balance helps in maintaining the power efficiency of the transmitted signal.
- Randomness: The sequences should resemble random noise as closely as possible. This randomness aids in securing the communication and making it resistant to interception and jamming.
- Ease of Generation and Implementation: The sequences should be easily generated and implemented in hardware or software. This practical consideration is vital for the widespread adoption of the technology.

In synchronous systems, where signals from different users are aligned in time, the emphasis is on the orthogonality of the spreading sequences. Orthogonal sequences ensure that cross-correlation is zero at zero delay, which is ideal for minimizing interference in such systems. However, maintaining orthogonality becomes challenging as the number of users increases, especially in systems with limited bandwidth.

Asynchronous systems, such as CDMA [35], [36], do not require alignment of signals from different users. In these systems, the spreading sequences need to have good cross-correlation properties not only at zero delay but also at various non-zero delays. This requirement is critical because the signals from different users can arrive at the receiver with different time delays. The sequences should be designed to ensure that these time shifts do not lead to a significant increase in cross-correlation, which could otherwise lead to increased interference and degradation of system performance.

The design of spreading sequences is also guided by theoretical limits and boundaries, such as the Welch Bound [30], [37], [38].

16.3.4 THEORETICAL LIMITS AND WELCH BOUND

The Welch Bound, first introduced in [37], sets a fundamental limit on the correlation properties of spreading sequences in spread spectrum systems. This bound provides a fundamental limit on the correlation properties of the sequences and is a crucial factor in the design of both synchronous and asynchronous systems. It helps in understanding the trade-offs between the length of the sequences, their number, and their cross-correlation properties.

In synchronous systems, where alignment of signals is precise, the Welch Bound provides a limit on the square of the maximum correlation $(\rho_{max}{}^2)$ between different sequences of a given length N and cardinality M. Let $x_1, \ldots, x_M$ be binary vectors of length N (each element of vector x_i is either 1 or -1). The maximum correlation is defined as [30]:

$$\rho_{max} = \frac{\max_{i \neq j} \langle x_i, x_j \rangle}{N},$$

where $\langle x_i, x_j \rangle$ is the scalar product of vectors x_i and x_j. The Welch Bound for $k = 1, 2, \ldots$ is then given by [30]:

$$\rho_{max}{}^{2k} \geq \frac{1}{M-1}\left[\frac{M}{\binom{N+k-1}{k}} - 1\right].$$

For $k = 1$, this simplifies to [30]:

$$\rho_{max}{}^2 \geq \frac{M-N}{(M-1)N}. \tag{16.1}$$

This bound sets a theoretical limit on the mutual correlation of sequences, guiding the design of spreading codes in synchronous systems.

In asynchronous systems, like CDMA, all sequences x_i must be statistically uncorrelated for arbitrarily random starting points. This means, in addition to each sequence x_i, all cyclically shifted copies of x_i must also be considered. The Welch Bound in this context modifies equation (16.1) by replacing M with MN, yielding [30]:

$$\rho_{max}{}^2 \geq \frac{MN-N}{(MN-1)N} = \frac{M-1}{MN-1}. \tag{16.2}$$

This bound defines the fundamental limit below which the square of the mutual correlation between any cyclically shifted copies of different sequences cannot fall. For $M \gg 1$, the bound simplifies to [30]:

$$\rho_{max}{}^2 \geq \frac{1}{N}. \tag{16.3}$$

Typically, the mutual correlation magnitude of generated signals is compared with

$$\frac{1}{\sqrt{N}}. \tag{16.4}$$

In mobile communication, ensuring precise coordination of communication lines between mobile devices is challenging. This implies that correlation may need to be computed for arbitrarily random starting points, necessitating the use of asynchronous techniques. Traditional sets of spreading sequences, such as Gold codes and Kasami sequences, are statistically uncorrelated for random starting points but often have limited cardinality.

In this chapter, we present new sets of spreading sequences for asynchronous techniques, including CDMA. Our sets boast a significantly larger cardinality $M \gg N$, with sequences that are statistically uncorrelated. The correlation function is multivalued, and the maximum absolute value ρ_{max} asymptotically approaches the Welch Bound as $N \to \infty$.

This advancement represents a significant leap in the computational modeling and simulation of spread spectrum systems. By pushing the limits of traditional Welch Bound constraints, our sequences offer enhanced capacity and robustness for asynchronous communication systems, paving the way for more efficient and scalable wireless communication technologies.

16.4 THEORETICAL FOUNDATIONS OF SPREADING SEQUENCE GENERATION

In the pursuit of advancing wireless communication technologies, the generation of spreading sequences stands as a critical component. The proposed approach in this chapter is rooted in the exploitation of algebraic and structural properties inherent in group codes over finite fields. This exploration is not merely an academic exercise but a practical endeavor to enhance the efficiency and robustness of radio communication systems, particularly in the context of spread spectrum techniques.

16.4.1 ALGEBRAIC FRAMEWORK FOR SPREADING SEQUENCE GENERATION

The cornerstone of our proposed technique is the utilization of finite fields, denoted as $GF(q)$, and the corresponding vector spaces $GF^n(q)$. These spaces represent sets of n-sequences of elements from $GF(q)$, with component-wise addition and scalar multiplication. Within this framework, a linear (n, k, d) code V is conceptualized as a subspace of $GF^k(q)$ in the space $GF^n(q)$. This subspace comprises non-empty sets of n-sequences (codewords) over $GF(q)$, where k denotes the dimension of the linear subspace, and d represents the minimum code distance (the minimum weight of a non-zero codeword).

A cyclic code, a specific case of subspace, possesses an additional property of cyclicity. Each vector in $GF^n(q)$ can be represented by a polynomial of a formal variable x of degree no higher than $n-1$. The components of the vector correspond to the coefficients of the polynomial. The set of these polynomials forms a vector space structure identical to that of $GF^n(q)$, as well as the structure of the ring of polynomials $GF(q)[x] / (x^n - 1)$. In this ring, polynomial multiplication is defined as

$$p_1(x) \cdot p_2(x) = R_{x^n - 1}[p_1(x) \cdot p_2(x)],$$

where $R_b[a]$ denotes the remainder of polynomial a divided by polynomial b. A cyclic shift by $\tau \in 0,...,n-1$ in algebraic terms is expressed as

$$x^\tau \cdot p(x) = R_{x^n - 1}[x^\tau \cdot p(x)].$$

If codewords of a (n,k,d) code over $GF(q)$ are defined as polynomials, then the code V is a subset of the ring $GF(q)[x] / (x^n - 1)$. The code V is cyclic if, along with the codeword $p(x)$, it also contains the polynomial $x \cdot p(x)$. A unique, reduced, non-zero polynomial $g(x)$ of the lowest degree $r = n - k$ uniquely defines a (n,k,d) cyclic code over $GF(q)$ and is termed the generator polynomial. This polynomial can be expressed as

$$g(x) = \prod_i (x - \beta^i),$$

where $\beta^i \in GF(q^m)$. It is related to the check polynomial $h(x)$ by the relationship

$$g(x) \cdot h(x) = x^n - 1,$$

or equivalently,

$$R_{x^n - 1}[g(x) \cdot h(x)] = 0.$$

We further delve into the structure of the finite field $GF(q^m)$, conceptualized as a set of polynomials of degree $\leq m$ with coefficients from $GF(q)$, that is, the structure of the ring of polynomials $GF(q)[x] / (x^m - 1)$. Following the general principles of Galois field theory, the ring of polynomials $GF(q)[x] / (x^m - 1)$ with operations modulo an irreducible polynomial constitutes the extended Galois field $GF(q^m)$. This field comprises a collection of cyclotomic classes $\alpha^{i(q^s)}$, where $s = 0,1,...,m_i - 1$, and m_i is the smallest positive integer such that $iq^{m_i} = (i) \mod (q^m - 1)$.

16.4.2   Algebraic Structure of Galois Fields in Sequence Generation

The algebraic structure of the Galois field $GF(q^m)$ plays a pivotal role in the generation of spreading sequences. This field is represented in a structured format, as illustrated in Table 16.1, where each conjugate class of elements defines a minimal polynomial $f_i(x)$. The product of all minimal polynomials $f_j(x)$ in the finite field yields a polynomial, leading to the equation:

$$(x^{q^m - 1} - 1) = \prod_{\forall i \in 0,...,q^m - 1} f_i(x) = \prod_{\forall i \in 0,...,q^m - 1} (x - \alpha^i),$$

where α is a primitive element of the field $GF(q^m)$. From this, we deduce:

$$g(x) = \mathrm{LCM}\left(\prod_j f_j(x) \right) = \mathrm{LCM}\left(\prod_j \prod_{s=0}^{m_j} (x - \alpha^{j(q^s)}) \right),$$

TABLE 16.1

Conjugate Classes and Corresponding Minimal Polynomials

Conjugate Elements	Minimal Polynomials
α^0	$f_0(x) = (x - \alpha^0)$
$\alpha^1, \alpha^q, \alpha^{q^2}, \ldots, \alpha^{q^m}$	$f_1(x) = f_q(x) = f_{q^2}(x) = \ldots$
	$= f_{q^m}(x) = (x - \alpha^1)(x - \alpha^q)(x - \alpha^{q^2})\ldots(x - \alpha^{q^m})$
$\ldots$	$\ldots$
$\alpha^i, \alpha^{iq}, \alpha^{iq^2}, \ldots, \alpha^{iq^m}$	$f_i(x) = f_{iq}(x) = f_{iq^2}(x) = \ldots$
	$= f_{iq^m}(x) = (x - \alpha^i)(x - \alpha^{iq})(x - \alpha^{iq^2})\ldots(x - \alpha^{iq^m})$
$\ldots$	$\ldots$

and

$$h(x) = \frac{x^n - 1}{g(x)} = \mathrm{LCM}\left(\prod_{i \neq j} f_i(x)\right) = \mathrm{LCM}\left(\prod_{i \neq j}\prod_{s=0}^{m_i}(x - \alpha^{i(q^s)})\right),$$

where LCM denotes the Least Common Multiple.

This table encapsulates the essence of the algebraic structure of $GF(q^m)$, where each conjugate class through its roots defines a unique minimal polynomial. These polynomials, when combined, construct the generator polynomial $g(x)$ and the check polynomial $h(x)$, which are fundamental in defining the cyclic codes over $GF(q^m)$. The Least Common Multiple approach in constructing these polynomials ensures the generation of efficient and robust spreading sequences, suitable for advanced wireless communication systems.

16.4.3 Cyclical Properties of Group Codes in Sequence Generation

In exploring the structure of group (n, k, d) codes V over $GF(q)$, a key concept is the cyclical orbit V_ξ – a set of sequences with elements from $GF(q)$ that are equivalent under the operation of cyclic shift. This concept is vital in understanding the cyclic nature of the sequences that constitute the code.

Consider sequences $C_i = (c_0^i, c_1^i, \ldots, c_{n-1}^i)$ and $C_j = (c_0^j, c_1^j, \ldots, c_{n-1}^j)$, with $c_v^i, c_v^j \in GF(q)$. These sequences are part of the same cyclical orbit if they satisfy the condition:

$$(c_0^i, c_1^i, \ldots, c_{n-1}^i) = (c_{(\tau \bmod n)}^j, c_{(\tau+1) \bmod n}^j, \ldots, c_{(\tau+n-1) \bmod n}^j),$$

for some $\tau \in 0, \ldots, n-1$. This relationship, expressed in polynomial algebra, becomes:

$$p_i(x) = x^\tau \cdot p_j(x) = R_{x^n-1}[x^\tau \cdot p_j(x)],$$

where:

$$p_i(x) = c_0^i + c_1^i x + \ldots + c_{n-1}^i x^{n-1},$$
$$p_j(x) = c_0^j + c_1^j x + \ldots + c_{n-1}^j x^{n-1}.$$

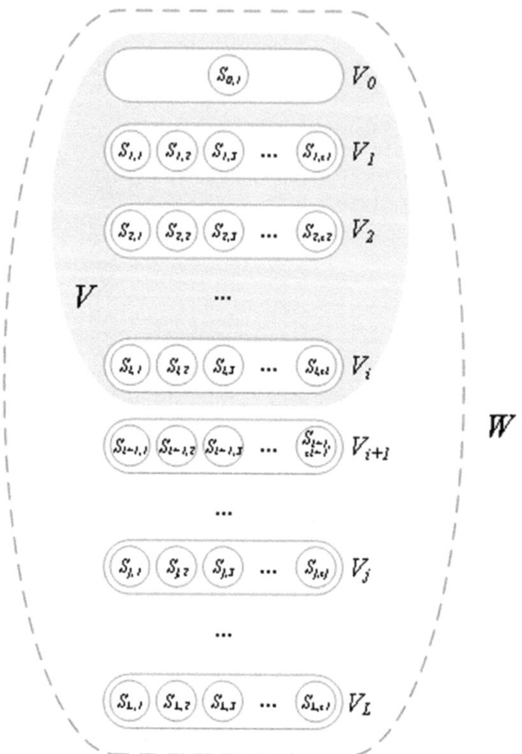

FIGURE 16.1 Decomposition of the Vector Space $GF^n(q)$ into Non-Intersecting Cyclical Orbits, V_i, $i = 0,...,i, L$

The set W of all n-sequences with elements from $GF(q)$ forms the so-called "full code," structurally equivalent to the vector space $GF^n(q)$ with component-wise addition and scalar multiplication. This set W is partitioned into subsets of orbits $V_0, V_1,...,V_L$, each containing sequences equivalent under cyclic shift. This partitioning results in a decomposition of the vector space into non-intersecting cyclical orbits, as depicted in Figure 16.1.

In Figure 16.1, sequences $S_{i,j} \in GF^n(q)$, where $i = 0,...,L$ and $j = 1,...,Z_i$, are schematically represented as elements of the set W. These sequences are grouped based on their equivalence under cyclic shift, forming the set V_i, with each set constituting an orbit in W. Each sequence $S_{i,j}$ can be represented as a sequence $_L$ in the aforementioned notation. The cardinality of the set V_i (the number of elements in the i-th orbit of W) is Z_i. For the set W formed by n-sequences from $GF^n(q)$, the cardinality of V_i for any does not exceed n, that is, $Z_i \leq n$, and the number of non-zero orbits is bounded below by [14], [39]:

$$L \geq \frac{q^n - 1}{n}. \tag{16.5}$$

Cyclical codes, by definition, encompass all cyclic shifts of their constituent sequences (codewords). In terms of orbits, this implies that a cyclical code is a union of a finite number of orbits, that is, a union of finite subsets of sequences equivalent under cyclic shift operations. This union is schematically represented in Figure 16.1, illustrating that a cyclical code can be uniquely determined by elements (leaders) of the orbits constituting the code.

Analyzing the cyclical properties, we infer that each orbit contains n-sequences with a fixed weight, as the operation of cyclic shift does not alter the sequence's weight. Concurrently, the code V does not include sequences of weight less than d. Precisely, the weights of sequences forming code V are determined exclusively by the code's weight spectrum.

Suppose the considered code V has a weight spectrum denoted as:

$$A(W), \quad W = 0,\ldots,n,$$

where $A(W)$ represents the number of codewords in code V with weight W. For an (n,k,d) code V, the weight spectrum is characterized as:

$$\begin{cases} A(w) = 1, & w = 0; \\ A(w) = 0, & w = 1,\ldots,d-1; \\ A(w), & w = d,\ldots,n; \end{cases} \tag{16.6}$$

implying that non-zero components of the weight spectrum $A(w) \neq 0$ (excluding the single zero sequence $A(0) = 1$) are concentrated in the weight range $W = d,\ldots,n$.

Considering the above discussions and the decomposition of the vector space into non-intersecting cyclical orbits $V_i, i = 0,\ldots,L$, we can represent this decomposition schematically as shown in Figure 16.2. It is evident that the "full code" is represented as a union of a finite number of non-intersecting orbits of sequences with fixed weight, equivalent under cyclic shift operations [14], [39]. Code V, as a subset of the space, is a union of a finite number of orbits, where the weights of the sequences in these orbits are determined exclusively by the weight spectrum, corresponding to $w \neq 0$.

The number of orbits with a fixed weight is also determined by the code's weight spectrum, that is, the number of sequences within these orbits. Thus, the number of orbits of weight $w \neq 0$ in code V with weight spectrum (16.6) is bounded below by the expression (analogous to (16.5)):

$$L(w) \geq \frac{A(w)}{n}.$$

16.4.4   Example: Application to BCH Codes in Finite Field $GF(2)$

Let us consider a practical example by fixing the finite field $GF(2)$ and examining a BCH code of length $n = 2^m - 1$. For instance, with $m = 3$, we define the generator polynomial $g(x) = f_j(x) = \prod_{j=1,2,4} (x - a^j)$, where a is a primitive element of the field $GF(2^3)$, and $f_j(x)$ represents the corresponding primitive polynomial.

Consequently, we obtain a code V with parameters $n = 7$, $d \geq 3$, and $k \geq 4$. This code is the well-known binary $(7, 4, 3)$ Hamming code. Its weight spectrum is characterized as follows:

$$\begin{cases} A(w) = 1, & w = 0; \\ A(w) = 0, & w = 1,2; \\ A(w) = 7, & w = 3; \\ A(w) = 7, & w = 4; \\ A(w) = 0, & w = 5,6; \\ A(w) = 1, & w = 7. \end{cases}$$

The corresponding check polynomial:

$$h(x) = \prod_{j=0,3,5,6} f_j(x)$$

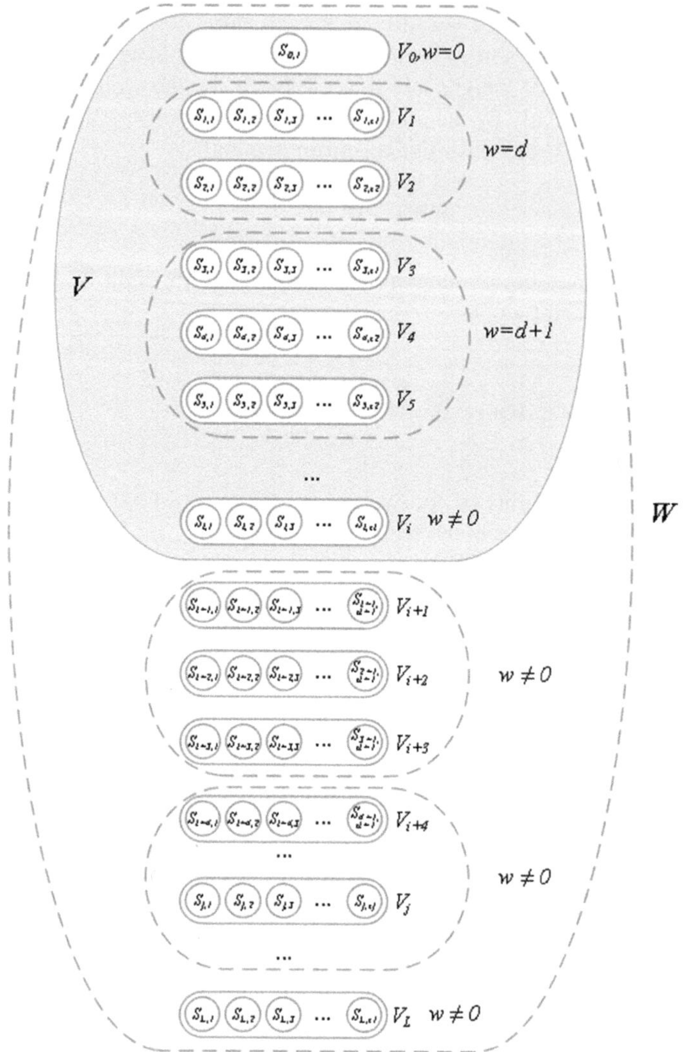

FIGURE 16.2   Decomposition of the Vector Space into Non-Intersecting Cyclical Orbits, V_i, $i = 0, ..., L$.

defines the dual code $V^\perp$ with parameters $n = 7$, $d \geq 4$, $k \geq 3$. The weight spectrum of this dual code is given by:

$$\begin{cases} A(w) = 1, & w = 0; \\ A(w) = 0, & w = 1, 2, 3; \\ A(w) = 7, & w = 4; \\ A(w) = 0, & w = 5, 6, 7. \end{cases}$$

The decomposition of the binary seven-dimensional vector space $GF^7(2)$ into non-intersecting orbits and the corresponding representation of the group (7, 4, 3) and (7, 3, 4) binary codes are illustrated in Figure 16.3. This representation highlights the intricate structure of the vector space and how it is partitioned into distinct orbits, each representing a unique set of sequences within the

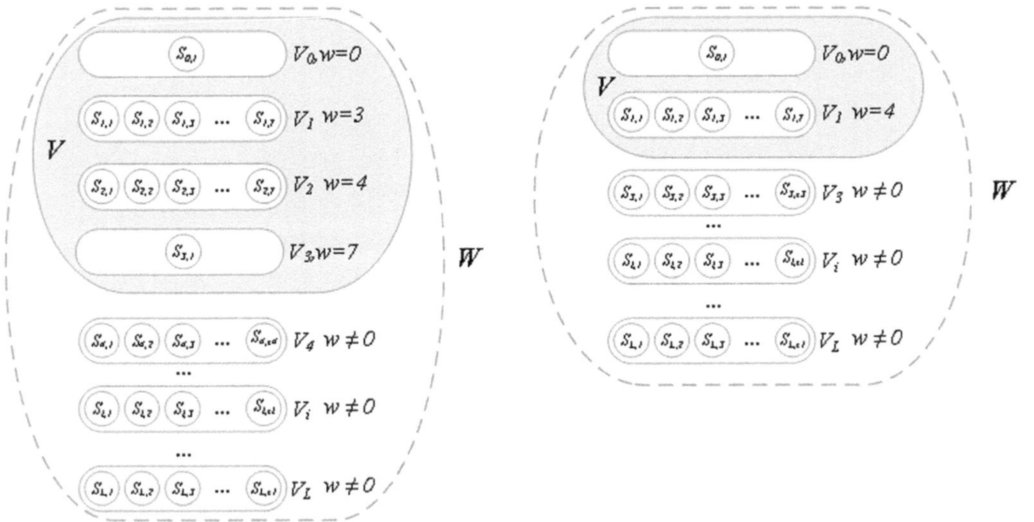

FIGURE 16.3 Decomposition of the vector space $GF^7(2)$ into non-intersecting cyclical orbits V_i, $i = 0, ..., L$, and the corresponding representations of the group (7, 4, 3) (on the left) and (7, 3, 4) (on the right) binary codes.

code. The orbits are grouped based on their equivalence under cyclic shift operations, providing a clear visualization of the code's structure and its cyclical properties.

From the diagrams in Figure 16.3, it is evident that the group code (7, 4, 3) comprises four orbits: one zero orbit (a sequence of all zeros), one orbit of weight three, and one orbit of weight four (each containing seven sequences equivalent under cyclic shift operations), as well as one orbit consisting of a single sequence of all ones. The group code (7, 3, 4) consists of two orbits: one zero orbit and one orbit with sequences of weight four.

This example vividly demonstrates the cyclical structure of the examined group codes: to define them, it is sufficient to specify one leader for each orbit. For instance, the code (7, 4, 3) is uniquely determined by four sequences: a zero sequence, a sequence of all ones, and one sequence each of weight three and four, representing their respective orbits. The code (7, 3, 4) is uniquely determined by two sequences: a zero sequence and a sequence of weight four – the leaders of their respective orbits.

The representation of a group code through the union of a finite number of cyclical orbits, as shown in Figure 16.3, forms the basis of the proposed algebraic method for generating discrete spreading signals. This approach not only simplifies the understanding and construction of group codes but also provides a clear framework for developing efficient spreading sequences. By leveraging the cyclical properties and the concept of orbit leaders, we can efficiently generate codes that are both robust and adaptable for various applications in wireless communication.

16.4.5 THE PROPOSED METHOD FOR GENERATING DISCRETE SPREADING SEQUENCES BASED ON A SECTION OF CYCLIC ORBITS OF A GROUP CODE

The distance (correlation) properties of the orbit leaders are determined by the distance properties of the group codes, with no equivalence under cyclic shift operations by definition. This property forms the basis for forming an ensemble of discrete signals: by selecting a leader (representative) from each non-zero cyclical orbit, we create a set of discrete sequences with the distance properties

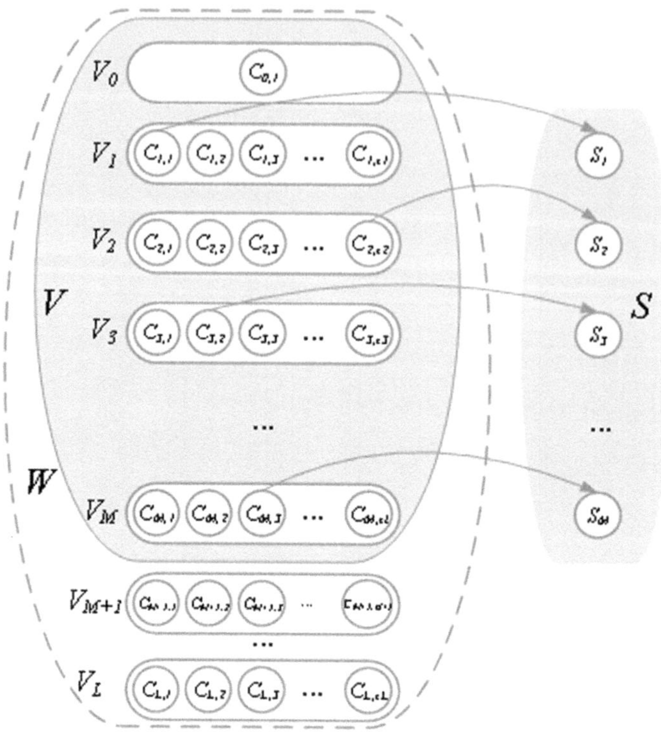

FIGURE 16.4   Scheme for Intersecting Non-Zero Cyclical Orbits of a Group Code to Form an Ensemble of Discrete Signals.

of the original group code and not equivalent to each other under cyclic shift operations. Figure 16.4 illustrates the scheme for intersecting non-zero cyclical orbits of a group code.

Figure 16.4 shows the decomposition of the vector space $GF^n(q)$ into sets of non-intersecting orbits $V_\xi, \xi = 0,...,L$, the representation of the group code V through the union of a finite number of orbits, and the scheme for selecting orbit leaders – one arbitrary representative from each cyclical subset $V_\xi, \xi = 0,...,M$. For convenience, the codewords $C_{v,u} = (c_0^{v,u}, c_1^{v,u}, ... c_{n-1}^{v,u})$ are denoted with two indices: $v-$ the orbit number V_v of code V, $v = 1,...,M$; $u-$ the codeword number in the orbit, $u = 1,...,z_v$, where z_v is the number of codewords in orbit V_v, $z_v \le n-1$.

From the selected orbit representatives, we form the set $S = (S_1, S_2,...,S_M)$, where $S_v = C_{v,u}$, $v = 1,...,M$, and the choice of index u for the corresponding $C_{v,u}$ is determined by the rule for intersecting the v-th cyclical orbit of the group code.

In the binary case, that is, limiting our study to the properties of the set $S = (S_1, S_2,...,S_M)$ formed by intersecting the cyclical orbits of a binary group code, we define the elements of the formed discrete sequences (discrete signals) $S_v = (s_0^v, s_1^v, ... s_{n-1}^v)$ based on the elements of the selected codewords (orbit leaders) $C_{v,u} = (c_0^{v,u}, c_1^{v,u}, ... c_{n-1}^{v,u})$ as follows:

$$s_i^v = \begin{cases} 1, & c_i^{v,u} = 1; \\ -1, & c_i^{v,u} = 0. \end{cases}$$

Assuming that the code V under consideration has a weight spectrum as follows:

$$\begin{cases} A(0) = 1; \\ A(1) = 0; \\ A(2) = 0; \\ \ldots \\ A(d-1) = 0; \\ A(d); \\ A(d+1); \\ \ldots \\ A(n) \end{cases}$$

for $w = 0, \ldots, n$, where $A(w)$ is the number of codewords in code V with weight w.

The set of binary discrete signals $S = (S_1, S_2, \ldots, S_M)$ formed by intersecting the cyclical orbits of code V possesses correlation and ensemble properties defined as follows:

- The side lobes of the periodic auto-correlation function (PACF) and cross-correlation function (PCCF) of the signal ensemble $S = (S_1, S_2, \ldots, S_M)$ take the following values:

$$\rho_{max} = \frac{n - 2w}{n}, \quad \text{for such } w = d, d+1, \ldots, n, \text{ that } A(w) \neq 0. \tag{16.7}$$

- For all such $w = d, d+1, \ldots, n$, where $A(w) = 0$, the side lobes of PACF and PCCF never take the values $\frac{n - 2w}{n}$.
- The power M of the ensemble $S = (S_1, S_2, \ldots, S_M)$ is determined by the number of non-zero orbits of code V and is bounded below by the expression:

$$M \geq \frac{2^k - 1}{n}.$$

Equality is achieved in the case of the maximum period of sequences of all orbits forming the code, that is, if code V consists of a set of orbits formed by sequences of maximum length (m-sequences).

16.4.5 EXAMPLE: BINARY (7, 3, 4) GROUP CODE AND ITS DISCRETE SIGNAL FORMATION

Consider the binary (7, 3, 4) group code from the previous example. It contains one non-zero cyclical orbit, the intersection of which yields a sequence with a periodic auto-correlation function taking two values:

$$\begin{cases} n = 7; \\ n - 2d = -1, \end{cases}$$

aligning perfectly with expression (16.7).

Furthermore, as evident from the example, the obtained sequence belongs to the group of sub-orthogonal sequences, which includes the well-known m-sequences. Indeed, any sequence from the non-zero cyclical orbit of the binary (7, 3, 4) group code is an m-sequence with a two-level periodic auto-correlation function. Thus, we have derived a known class of discrete signals (m-sequences) in a new way – through the specification of discrete sequences as leaders of cyclical orbits of a group code.

The algebraic intersection of cyclical orbits of a group code allows for the formation of a set of sequences $S = (S_1, S_2, ..., S_M)$ that are not cyclically shifted copies of each other and possess pre-defined distance properties of the group code. This enables the algebraic construction of large ensembles of discrete signals with multi-level auto- and cross-correlation functions.

The key step in forming discrete signals is the synthesis of a linear group code with required distance and structural properties. A group (n, k, d) code over $GF(q)$ is defined by its check $h(x)$ and/or generating $g(x)$ polynomials through the product of a finite number of minimal polynomials $f_i(x)$ of the finite field $GF(q^m)$ (hereafter, we consider group codes of primitive length $n = q^m - 1$). Each minimal polynomial is defined through its roots, grouped in corresponding classes of conjugate elements, that is, we write:

$$h(x) = \text{LCM}\left(\prod_{i \neq j} f_i(x)\right) = \text{LCM}\left(\prod_{i \neq j} \prod_{s=0}^{m_i} (x - \alpha^{i(q^s)})\right).$$

Adding minimal polynomials as factors in the decomposition of the check polynomial leads to an increase in the power of the code (increase in the number of codewords) and a corresponding decrease in distance properties. Conversely, reducing the number of minimal polynomials as factors in the decomposition of $h(x)$ leads to a decrease in the power of the code (reduction in the number of codewords) and a corresponding improvement in distance properties.

16.4.6 Summary and Implications of Theoretical Findings

The theoretical section of this study has provided a comprehensive exploration of the algebraic structure and properties of group codes, particularly focusing on their application in the generation of discrete signals. By examining various aspects such as the weight spectrum of codes, the decomposition of vector spaces into cyclical orbits, and the selection of orbit leaders, we have gained a deeper understanding of the intricate relationships within these codes and their practical implications.

Key insights include the realization that the distance properties of orbit leaders are intrinsically linked to the distance properties of the group codes themselves. This understanding has led to the development of a novel method for forming discrete signal ensembles by intersecting cyclical orbits of group codes. Such an approach allows for the creation of signal sets that are not merely cyclically shifted versions of one another but possess distinct correlation properties defined by the group code.

In summary, the theoretical framework established in this section lays a solid foundation for further research and application in the field of digital signal processing and communications. The algebraic approach to signal generation, rooted in the properties of group codes, offers a robust and versatile tool for designing advanced communication systems.

16.5 COMPUTATIONAL MODELING PROCESS FOR ENHANCED SPREAD SPECTRUM CODES

The computational modeling of enhanced spread spectrum codes involves a systematic and detailed process, crucial for the effective simulation and analysis of asynchronous wireless communication systems. This section outlines the structured approach employed in the simulation, highlighting each step in the process and its significance in the overall modeling framework.

The Figure 16.5 presents a comprehensive structural framework for the computational modeling of spread spectrum codes, essential in asynchronous wireless communication systems. It outlines a

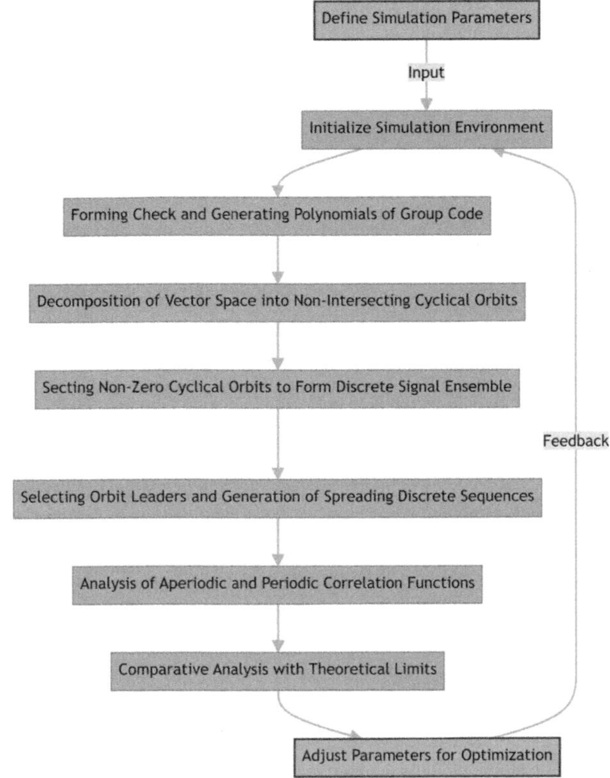

FIGURE 16.5   Structural Framework for Computational Modeling of Spread Spectrum Codes.

sequential, multi-stage process, starting from the initial definition of simulation parameters to the final optimization of the codes. Each step is intricately linked, ensuring a cohesive and systematic approach to the simulation.

The process begins with the crucial task of defining the simulation parameters, setting the foundation for the entire modeling exercise. Following this, the simulation environment is initialized, preparing the necessary computational tools and software. The subsequent steps involve the intricate formation of check and generating polynomials, and the decomposition of the vector space into non-intersecting cyclical orbits, which are pivotal in understanding the cyclic nature and behavior of the codes under various conditions. The selection of non-zero cyclical orbits for forming discrete signal ensembles and the generation of spreading discrete sequences are key steps that lead to the creation of the actual codes for simulation. These codes are then rigorously analyzed for their aperiodic and periodic correlation functions, providing valuable insights into their performance. A comparative analysis with theoretical limits, such as the Welch bound, is conducted to evaluate the effectiveness of the codes. Based on this analysis, parameters are adjusted in an iterative process to optimize the codes' performance, ensuring they are fine-tuned for the best possible outcomes in the simulation environment.

16.5.1 Step-by-Step Simulation Process

- Define Simulation Parameters: The first step involves defining the key parameters that will govern the simulation. These parameters include the length of the sequences, the characteristics of the finite field, and the specific requirements of the spread spectrum codes. This foundational step sets the stage for the entire simulation process.

- Initialize Simulation Environment: With the parameters set, the next step is to initialize the simulation environment. This involves setting up the computational tools and software necessary for the simulation, ensuring that all components are aligned with the defined parameters.
- Forming Check and Generating Polynomials of Group Code: This critical step involves the creation of check and generating polynomials for the group code. These polynomials are essential in defining the structure and properties of the codes that will be used in the simulation.
- Decomposition of Vector Space into Non-Intersecting Cyclical Orbits: The vector space is decomposed into non-intersecting cyclical orbits. This decomposition is key to understanding the cyclic nature of the codes and their behavior under various conditions.
- Selecting Non-Zero Cyclical Orbits to Form Discrete Signal Ensemble: From the decomposed vector space, specific non-zero cyclical orbits are selected to form an ensemble of discrete signals. This selection is based on the requirements of the spread spectrum system being modeled.
- Selecting Orbit Leaders and Generation of Spreading Discrete Sequences: Orbit leaders are selected from each cyclical subset, and spreading discrete sequences are generated. This step is crucial in creating the actual codes that will be used in the simulation.
- Analysis of Aperiodic and Periodic Correlation Functions: The generated codes are then analyzed for their aperiodic and periodic correlation functions. This analysis provides insights into the performance and suitability of the codes for specific applications in wireless communication.
- Comparative Analysis with Theoretical Limits: The codes are compared against theoretical limits, such as the Welch bound, to evaluate their performance. This comparative analysis is vital for assessing the effectiveness of the codes in practical scenarios.
- Adjust Parameters for Optimization: Based on the analysis, parameters are adjusted to optimize the performance of the codes. This iterative process ensures that the codes are fine-tuned for the best possible performance in the simulation environment.

This structured approach to the computational modeling of enhanced spread spectrum codes ensures a thorough and detailed analysis, crucial for the development of efficient and robust codes for asynchronous wireless communication systems. The process not only aids in the understanding of the behavior of these codes but also provides a framework for their optimization and application in real-world scenarios.

16.5.2 Simulation Parameters for Computational Modeling of Spread Spectrum Codes

The computational modeling of spread spectrum codes for asynchronous wireless communication systems involves a detailed analysis of various parameters. Table 16.2 summarizes the key parameters used in the simulation of discrete spreading sequences:

- Type of Sequence: This column categorizes the sequences into three types based on their correlation functions: suborthogonal binary signals, three-level binary signals, and five-level binary signals. Each type represents a distinct class of spreading sequences with unique correlation properties.
- Parameter m: This parameter signifies the order of the Galois field $GF(2^m)$, which plays a pivotal role in the construction of the group codes. The value of m directly influences the length of the spreading sequences, with $n = 2^m - 1$.

TABLE 16.2
Parameters for Modeling Discrete Spreading Sequences

Type of Sequence	Parameter m	Check Polynomial $h(x)$	Group Code Parameters (n, k, d)
Suborthogonal Binary Signals	3	$\prod_{s=0}^{2}\left(x-\alpha^{(2^s)}\right)$	(7, 3, 4)
	5	$\prod_{s=0}^{4}\left(x-\alpha^{(2^s)}\right)$	(31, 5, 16)
	7	$\prod_{s=0}^{6}\left(x-\alpha^{(2^s)}\right)$	(127, 7, 64)
	11	$\prod_{s=0}^{10}\left(x-\alpha^{(2^s)}\right)$	(2047, 11, 1024)
Three-Level Binary Signals	5	$\prod_{s=0}^{4}\left(x-\alpha^{(2^s)}\right)\left(x-\alpha^{3(2^s)}\right)$	(31, 10, 12)
	7	$\prod_{s=0}^{6}\left(x-\alpha^{(2^s)}\right)\left(x-\alpha^{3(2^s)}\right)$	(127, 14, 56)
	11	$\prod_{s=0}^{10}\left(x-\alpha^{(2^s)}\right)\left(x-\alpha^{3(2^s)}\right)$	(2047, 22, 992)
Five-Level Binary Signals	5	$\prod_{s=0}^{4}\prod_{j=1,3,5}\left(x-\alpha^{j(2^s)}\right)$	(31, 15, 8)
	7	$\prod_{s=0}^{6}\prod_{j=1,3,5}\left(x-\alpha^{j(2^s)}\right)$	(127, 21, 48)
	11	$\prod_{s=0}^{10}\prod_{j=1,3,5}\left(x-\alpha^{j(2^s)}\right)$	(2047, 33, 960)

- Check Polynomial $h(x)$: The check polynomial is a fundamental component in the formation of group codes. It is derived from the product of factors based on the elements of the finite field. The structure of $h(x)$ varies across different types of sequences, impacting the correlation characteristics of the resulting codes.
- Group Code Parameters (n, k, d): These parameters define the specific group code used in the generation of spreading sequences. n represents the length of the code, k is the dimension, and d is the minimum distance. These parameters are crucial in determining the performance and robustness of the spreading sequences in communication systems.

These parameters are critical in defining the characteristics and performance of the generated codes. The simulation parameters outlined in this table are integral to the computational modeling process. They provide a structured approach to generating and analyzing spread spectrum codes, ensuring that the sequences meet the desired correlation and performance criteria for efficient asynchronous wireless communication.

16.6   RESULTS OF COMPUTATIONAL MODELING FOR SPREAD SPECTRUM CODE GENERATION

In this section, we delve into the outcomes of our computational modeling efforts focused on the generation of Spread Spectrum Codes. The results presented here are the culmination of extensive simulations based on the parameters and methodologies outlined in the previous sections. This analysis aims to demonstrate the efficacy and robustness of the generated codes in terms of their correlation properties and suitability for asynchronous wireless communication systems. The findings are crucial in validating the theoretical constructs and in providing insights into the practical applicability of these codes in real-world scenarios.

We will evaluate the properties of discrete signals for various rules of forming the group code, that is, for different numbers of factors in the expressions of generating and check polynomials:

$$g(x) = \mathrm{LCM}\left(\prod_{j} f_j(x) \right) = \mathrm{LCM}\left(\prod_{j} \prod_{s=0}^{m_j} (x - \alpha^{j(q^s)}) \right),$$

and

$$h(x) = \frac{x^n - 1}{g(x)} = \mathrm{LCM}\left(\prod_{i \neq j} f_i(x) \right) = \mathrm{LCM}\left(\prod_{i \neq j} \prod_{s=0}^{m_i} (x - \alpha^{i(q^s)}) \right).$$

Let us consider the first, most straightforward case, where a binary group (n,k,d) code over $GF(2)$ is defined through a check polynomial of the form:

$$h(x) = f_i(x) = \prod_{s=0}^{m-1} (x - \alpha^{i(2^s)}),$$

where $f_i(x)$ is an arbitrary minimal polynomial of the element α^i in $GF(2^m)$, and the order of the element α^i equals the order of the multiplicative group of the finite field $GF(2^m)$, $n = 2^m - 1$, with α being the primitive element of the field $GF(2^m)$. Assuming $i = 1$, the check polynomial is defined through its roots – all powers of the primitive element of the field as follows:

$$h(x) = \prod_{s=0}^{m-1} (x - \alpha^{(2^s)}). \tag{16.8}$$

The corresponding generating polynomial is also defined through its roots – all other elements of the field $GF(2^m)$:

$$g(x) = \frac{x^n - 1}{h(x)} = \prod_{j \neq 1} \prod_{s=0}^{m_j} (x - \alpha^{j(2^s)}).$$

The process of forming check and generating polynomials is schematically represented in Figure 16.6. The symbol v denotes the number of classes of conjugate elements constituting the multiplicative group of the finite field $GF(2^m)$. The first class (elements $\alpha^1, \alpha^2, \ldots, \alpha^{2^{m-1}}$) contains m conjugate elements, defining the primitivity of the element α. Subsequent classes (elements $\alpha^j, \alpha^{2j}, \ldots, \alpha^{j2^{m_j}}$) contain m_j conjugate elements, where m_j divides m evenly, and j ranges from 1 to v.

For each j in the range $[1..v]$, the corresponding m_j is determined as the smallest positive integer for which the following equation holds true:

$$j = (j2^{m_j}) \bmod (2^m - 1).$$

For instance, for $m = 4$ and $j = 5$, it corresponds to $m_5 = 2$. If the order of the multiplicative group is a prime number, that is, when $2^m - 1$ is a prime number, then:

$$\forall j : m_j = m.$$

The unit element of the field $\alpha^0 = 1$ forms an additional conjugate class of one element. Figure 16.7 illustrates the corresponding distribution of elements of the finite field $GF(2^m)$ across the polynomials $h(x)$ and $g(x)$. The entire set of elements of the field $GF(2^m)$ constitutes the set of roots of the binomial $x^n - 1 = h(x)g(x)$.

As illustrated in Figures 16.6 and 16.7, the check and generating polynomials are determined by the product of corresponding factors $(x - X)$, defined by elements of the finite field $X \in GF(2^m)$.

Our interest lies in the maximum number of consecutive elements in the representation of these polynomials, as shown in Figure 16.7. Specifically, the number of consecutive elements serving as roots of the generating polynomial (exactly $2t$) defines the distance properties of the corresponding group code $d = 2t + 1$ and its weight spectrum.

Examining the structure of the finite field $GF(2^m)$ and the distribution of its elements in the check and generating polynomials (Figure 16.7), it is evident that the roots of the check polynomial $h(x)$ include elements of the finite field ranging from α to $\alpha^{2^{m-1}}$, excluding elements from $\alpha^{2^{m-1}+1}$ to $\alpha^0 = 1$. This observation utilizes the cyclic nature of the field's multiplicative group, expressed in the equality $\alpha^z = \alpha^{(z)\bmod(2^m - 1)}$.

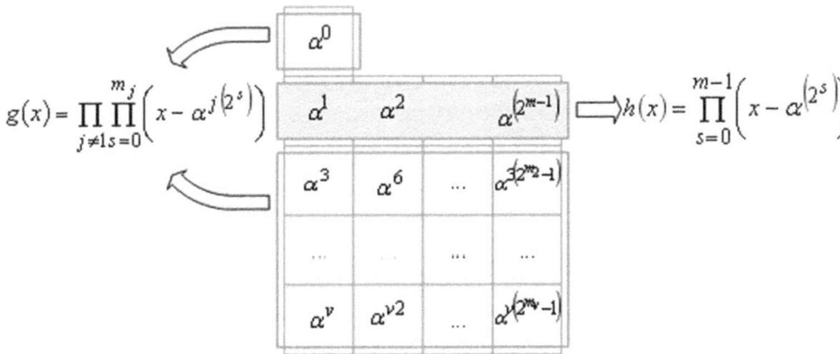

FIGURE 16.6 Scheme for Forming Check and Generating Polynomials of Group Code with Parameters $(n = 2^m - 1, k = m, d = 2^{m-1})$.

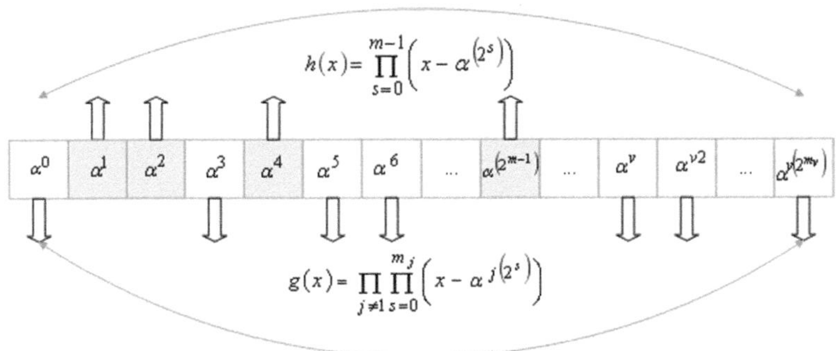

FIGURE 16.7 Distribution of Finite Field Elements in Check and Generating Polynomials of Group Code with Parameters $(n = 2^m - 1, k = m, d = 2^{m-1})$.

Thus, for m being a prime number, the number of consecutive elements α^i of the finite field $GF(2^m)$ serving as roots of the check polynomial is at least:

$$\underbrace{\alpha^i,\ldots,\alpha^{i(2^{m-1})}}_{2^{m-1}},$$

leading to the following estimation when $z = 2^{m-1}$:

$$2t = (2^m - 1) - 2^{m-1} = 2^{m-1} - 1,$$

from which we derive the estimate:

$$d = 2t + 1 = 2^{m-1}.$$

The number of check elements r of the group code is determined by the degree of the generating polynomial, which in turn is defined by the number of its roots. In this case, we have the estimate:

$$r = 2^m - m - 1,$$

leading directly to:

$$k = m$$

and the corresponding code parameters of the group code are:

$$(n = 2^m - 1, \quad k = m, \quad d = 2^{m-1}). \tag{16.9}$$

The group code defined by the check polynomial (16.8) with code parameters (16.9) contains a single non-zero cyclical orbit. The weight spectrum of the code is as follows:

$$A(w) = \begin{cases} 1, w = 0; \\ 0, w = 1,\ldots,2^{m-1} - 1; \\ \neq 0, w = 2^{m-1}; \\ 0, w = 2^{m-1} + 1,\ldots,2^m - 1. \end{cases}$$

To evaluate the correlation properties of discrete signals, we use expression (16.7), yielding:

$$\rho_{\max} = \frac{-1}{2^m - 1}.$$

The power M of the ensemble of formed discrete signals is estimated considering the total of $2^k - 1 = 2^m - 1$ non-zero codewords of the group code. As each binary sequence, being a codeword of the group code, has the maximum period (m being a prime number), it follows that each cyclical orbit contains at least $2^m - 1$ sequences. The intersection of orbits yields the desired ensemble of discrete signals with a power of:

$$M = \frac{2^m - 1}{2^m - 1} = 1.$$

These estimates correspond to sub-orthogonal discrete sequences, known in literature as m-sequences. They possess the maximum period of $2^m - 1$ and are constructed using linear feedback shift registers defined by a primitive polynomial of degree m. In Figure 16.8 and 16.9 show graphs of non-periodic and periodic correlation functions for the case $m = 7$. These figures correspond to the considered rule for generating spreading discrete signals (suborthogonal sequences) of length $n = 127$ through a section of cyclic orbits of a register code of maximum length. Figure 16.8 shows the Aperiodic Autocorrelation Function, and Figure 16.9 – Periodic Autocorrelation Function.

To construct more complex classes of signals, let us consider a case where a binary group (n, k, d) code over $GF(2)$ is defined through a check polynomial of the form:

$$h(x) = f_{i1}(x) f_{i_2}(x) = \prod_{s=1}^{m-1} \left(x - \alpha^{i_1(2^s)} \right) \left(x - \alpha^{i_2(2^s)} \right), \qquad (16.10)$$

where $f_{i_1}(x)$ and $f_{i_2}(x)$ are two consecutive minimal polynomials of elements α^{i_1} and α^{i_2} in $GF(2^m)$, respectively. The order of elements α^{i_1} and α^{i_2} equals the order of the multiplicative group of the finite field $GF(2^m)$, with α being the primitive element of the field. Assuming $i_1 = 1$, the check and generating polynomials are defined through their roots as follows:

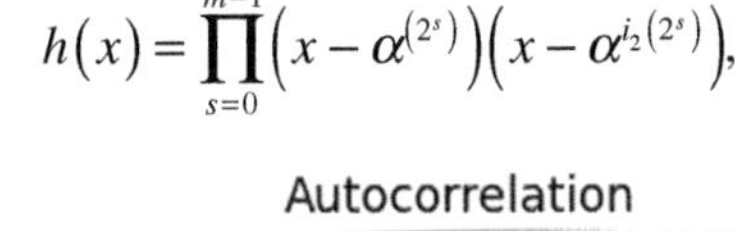

$$h(x) = \prod_{s=0}^{m-1} \left(x - \alpha^{(2^s)} \right) \left(x - \alpha^{i_2(2^s)} \right),$$

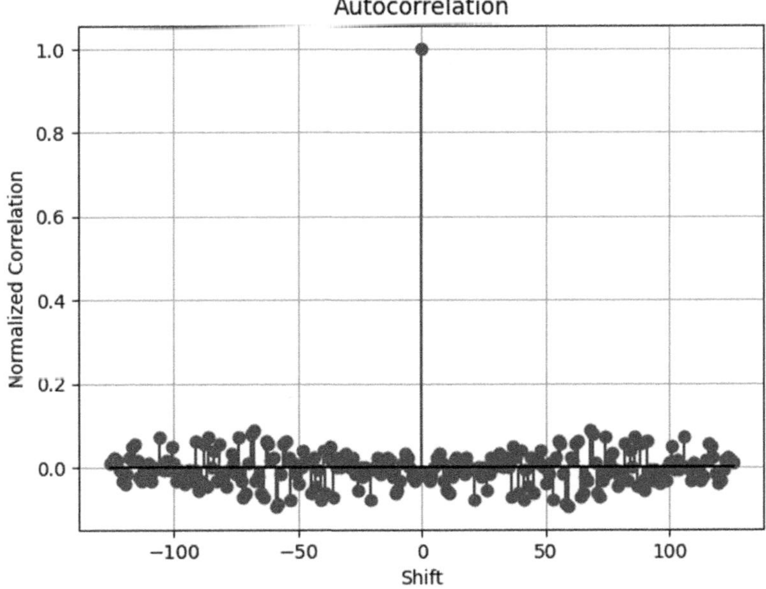

FIGURE 16.8　Aperiodic Autocorrelation Function of Sub-Orthogonal Discrete Signals of Length $n = 127$.

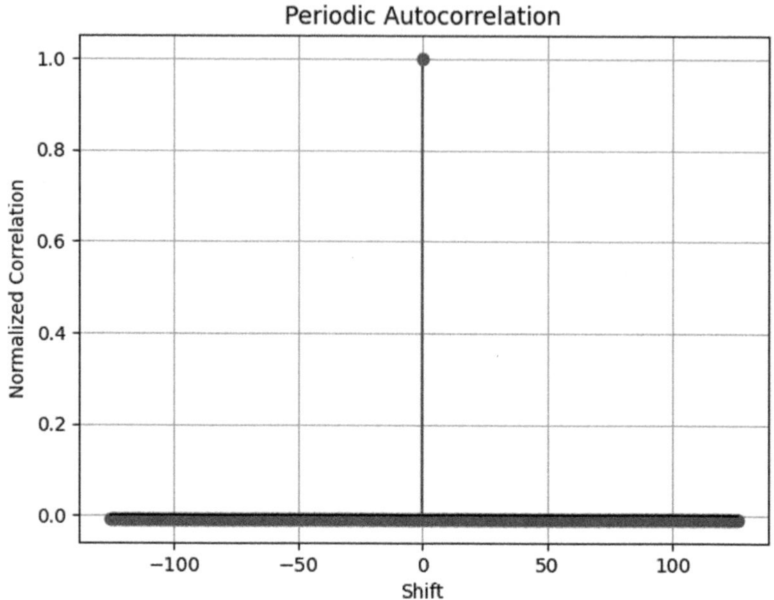

FIGURE 16.9   Periodic Autocorrelation Function of Sub-Orthogonal Discrete Signals of Length $n = 127$.

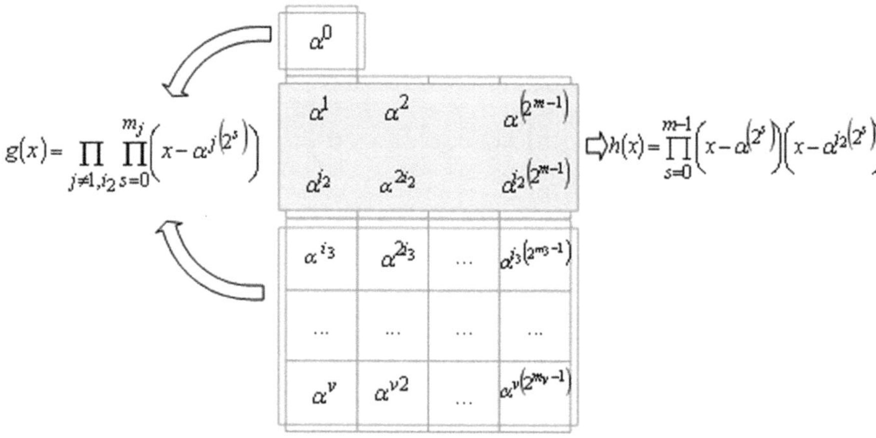

FIGURE 16.10 Scheme for Forming Check and Generating Polynomials of Group Code with Parameters $(n = 2^m - 1, k = 2m, d = 2^{m-1} - 2^{\frac{m+1}{2}-1})$.

$$g(x) = \frac{x^n - 1}{h(x)} = \prod_{j \neq 1, i_2} \prod_{s=0}^{m_j} \left(x - \alpha^{j(2^s)}\right).$$

The process of forming check and generating polynomials is schematically represented in Figure 16.10, while Figure 16.11 shows the corresponding distribution of elements of the finite field $GF(2^m)$ across the polynomials $h(x)$ and $g(x)$. All notations are analogous to the previously considered case. This case corresponds to the formation of binary spreading discrete signals with a three-level correlation function.

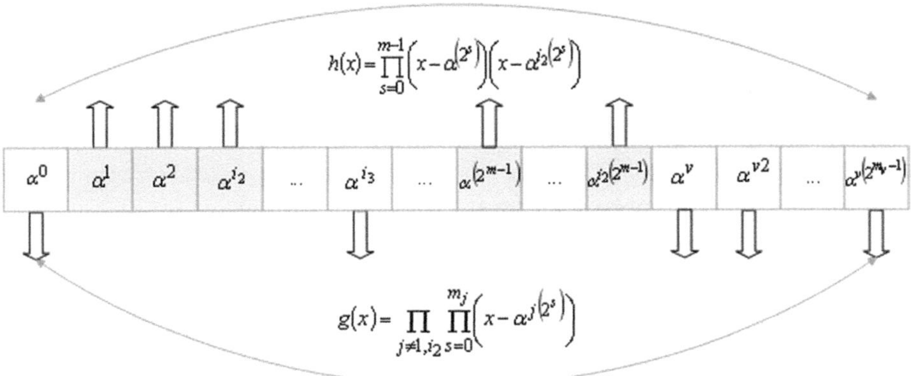

FIGURE 16.11 Distribution of Finite Field Elements in Check and Generating Polynomials of Group Code with Parameters $(n = 2^m - 1, k = 2m, d = 2^{m-1} - 2^{\frac{m+1}{2}-1})$.

The roots of the check polynomial $h(x)$ include elements of the finite field from the first two classes of conjugate elements. Consequently, the range of elements from α to α^z in the finite field $GF(2^m)$, where the roots of the polynomial $h(x)$ lie, is determined by the largest value of z for which the condition $\alpha^z = \alpha^{(z)\,\mathrm{mod}\,(2^m-1)}$ holds, that is,

$$z = \max_{s=0,\dots,m-1}\left\{(2^s)\,\mathrm{mod}\,(2^m-1),(i_2 2^s)\,\mathrm{mod}\,(2^m-1)\right\}.$$

In cases where m is a prime number, and the roots of the polynomials $f_{i_1}(x)$ and $f_{i_2}(x)$ are chosen such that they lie in the range:

$$\underbrace{\alpha^{i_1},\dots,\alpha^{i_1\,(2^{m-1})},\dots,\alpha^{i_2},\dots,\alpha^{i_2\,(2^{m-1})}}_{2^{m-1}+2^{\frac{m+1}{2}-1}},$$

that is, when $z = 2^{m-1} + 2^{\frac{m+1}{2}-1}$, we have:

$$2t = (2^m - 1) - 2^{m-1} - 2^{\frac{m+1}{2}-1} = 2^{m-1} - 2^{\frac{m+1}{2}-1} - 1,$$

leading to the following estimation:

$$d = 2t + 1 = 2^{m-1} - 2^{\frac{m+1}{2}-1}.$$

The number of check elements r of the group code is determined by the degree of the generating polynomial, that is, $r = 2^m - 2m - 1$, leading to: $k = 2m$. The corresponding code parameters of the group code are:

$$\left(n = 2^m - 1, k = 2m, d = 2^{m-1} - 2^{\frac{m+1}{2}-1}\right). \tag{16.11}$$

The weight spectrum of the group code, defined by the check polynomial (16.10) with code parameters (16.11), includes the check polynomial of the previously discussed code with parameters (16.9). This implies that all codewords of the maximal length register code with parameters (16.9) are also codewords of the group code with parameters (16.11). Thus, the first non-zero component $A(w) \neq 0, w = 2^{m-1}$ in the weight spectrum $A(w)$ coincides.

Adding the minimal polynomial $f_{i_2}(x)$ as a factor in the check polynomial (see expression (16.8)) results in two additional non-zero components in the weight spectrum:

$$A(w) \neq 0, w = 2^{m-1} - 2^{\frac{m+1}{2}-1},$$

and

$$A(w) \neq 0, w = 2^{m-1} + 2^{\frac{m+1}{2}-1}.$$

Hence, the weight spectrum of the code is estimated as:

$$A(w) = \begin{cases} 1, w = 0; \\ 0, w = 1, ..., 2^{m-1} - 2^{\frac{m+1}{2}-1} - 1; \\ \neq 0, w = 2^{m-1} - 2^{\frac{m+1}{2}-1}; \\ 0, w = 2^{m-1} - 2^{\frac{m+1}{2}-1} + 1, ..., 2^{m-1} - 1; \\ \neq 0, w = 2^{m-1}; \\ 0, w = 2^{m-1} + 1, ..., 2^{m-1} + 2^{\frac{m+1}{2}-1} - 1; \\ \neq 0, w = 2^{m-1} + 2^{\frac{m+1}{2}-1}; \\ 0, w = 2^{m-1} + 2^{\frac{m+1}{2}-1} + 1, ..., 2^m - 1. \end{cases}$$

Using reasoning from the previous section, we derive expressions for the normalized correlation functions of the formed discrete signals relative to length n:

- For $w = 2^{m-1} - 2^{\frac{m+1}{2}-1}$: $\dfrac{n - 2i}{n} = \dfrac{2^m - 1 - 2(2^{m-1} - 2^{\frac{m+1}{2}-1})}{2^m - 1} = \dfrac{-1 + 2^{\frac{m+1}{2}}}{2^m - 1}$;

- For $w = 2^{m-1}$: $\dfrac{n - 2i}{n} = \dfrac{2^m - 1 - 2(2^{m-1})}{2^m - 1} = \dfrac{-1}{2^m - 1}$;

- For $w = 2^{m-1} + 2^{\frac{m+1}{2}-1}$: $\dfrac{n - 2i}{n} = \dfrac{2^m - 1 - 2(2^{m-1} + 2^{\frac{m+1}{2}-1})}{2^m - 1} = \dfrac{-1 - 2^{\frac{m+1}{2}}}{2^m - 1}$.

From the analysis, we derive expressions for estimating possible levels of sidelobes in the periodic correlation function:

$$\begin{cases} \dfrac{-1 + 2^{\frac{m+1}{2}}}{2^m - 1}, w = 2^{m-1} - 2^{\frac{m+1}{2}-1}; \\ \dfrac{-1}{2^m - 1}, w = 2^{m-1}; \\ \dfrac{-1 - 2^{\frac{m+1}{2}}}{2^m - 1}, w = 2^{m-1} + 2^{\frac{m+1}{2}-1}; \end{cases}$$

and

$$\rho_{max} = \left| \frac{-1 - 2^{\frac{m+1}{2}}}{2^m - 1} \right|.$$

We estimate the power M of the ensemble of formed discrete signals. For the used group code, the number of non-zero codewords is estimated as:

$$2^k - 1 = 2^{2m} - 1.$$

Each codeword has a maximum period, thus each cyclic orbit of the group code contains exactly $2^m - 1$ codewords. The intersection of cyclic orbits of the considered code yields the desired ensemble of discrete signals. The power of the ensemble formed in this way is determined by:

$$M = \frac{2^{2m} - 1}{2^m - 1} = 2^m + 1.$$

This estimate corresponds to Gold sequences, known in literature for their construction procedures through preferred pairs of polynomials defining the rules of linear feedback shift registers [1], [30]. Thus, the proposed approach allows synthesizing known classes of Gold discrete signals in a new way – through the intersection of cyclic orbits of the group code.

Figures 16.12 and 16.13 show the graphs of the corresponding correlation functions for signals of length $n = 127$. These spreading sequences are generated from Group Code with Parameters $(n = 2^m - 1, k = 2m, d = 2^{m-1} - 2^{\frac{m+1}{2}-1})$. As can be seen from Figure 16.13, the periodic cross-correlation function has three levels of side lobes, which corresponds to the well-known Gold codes.

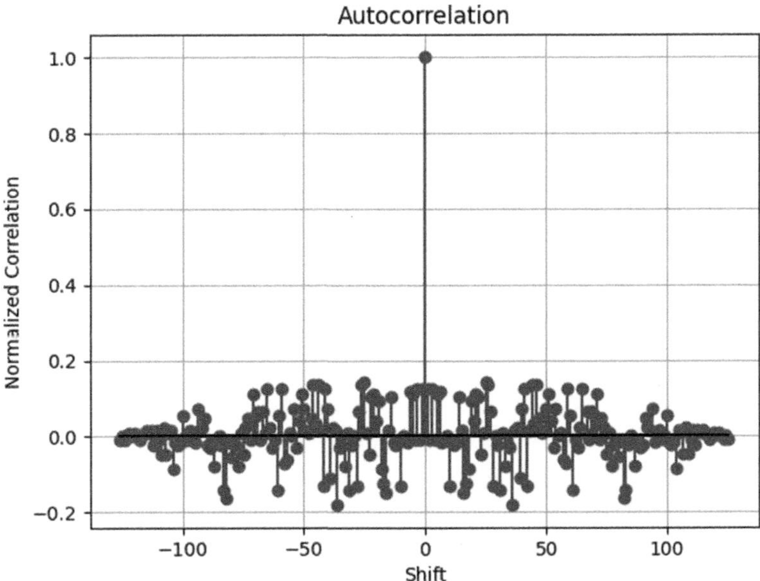

FIGURE 16.12 Aperiodic Autocorrelation Function of Generated Discrete Signals of Length $n = 127$.

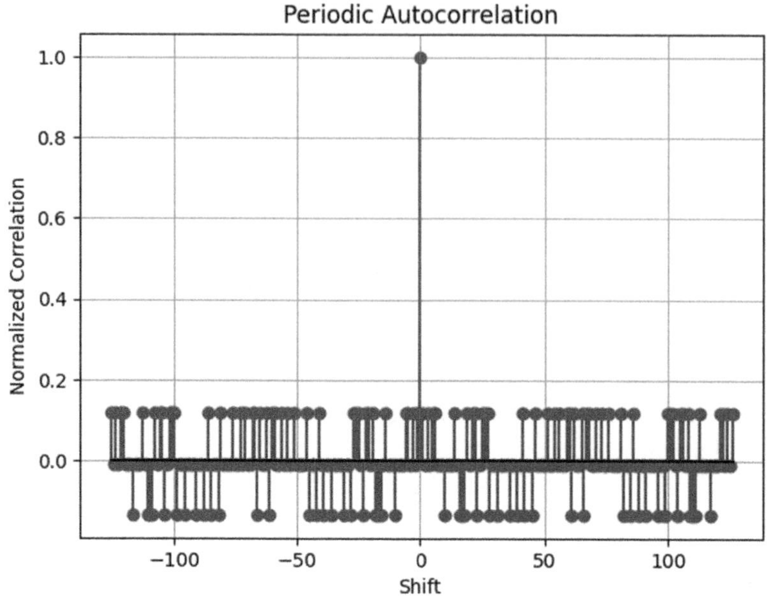

FIGURE 16.13   Periodic Autocorrelation Function of Three-Level Discrete Signals of Length $n = 127$.

To construct new classes of discrete signals, let us consider a binary group code (n,k,d) over $GF(2)$ defined by a check polynomial of the form:

$$h(x) = f_{i_1}(x) f_{i_2}(x) f_{i_3}(x) = \prod_{s=1}^{m-1} \left(x - \alpha^{i_1 (2^s)}\right)\left(x - \alpha^{i_2 (2^s)}\right)\left(x - \alpha^{i_3 (2^s)}\right), \tag{16.12}$$

where $f_{i_1}(x)$, $f_{i_2}(x)$, and $f_{i_3}(x)$ are three consecutive minimal polynomials of elements α^{i_1}, α^{i_2}, and α^{i_3} in $GF(2^m)$, respectively. The order of elements α^{i_1}, α^{i_2}, and α^{i_3} equals the order of the multiplicative group of the finite field $GF(2^m)$, with $n = 2^m - 1$ and α as the primitive element of $GF(2^m)$.

Assuming without loss of generality that $i_1 = 1$, we define the check and generating polynomials as follows:

$$h(x) = \prod_{s=0}^{m-1} \left(x - \alpha^{(2^s)}\right)\left(x - \alpha^{i_2 (2^s)}\right)\left(x - \alpha^{i_3 (2^s)}\right),$$

$$g(x) = \frac{x^n - 1}{h(x)} = \prod_{j \neq 1, i_2, i_3} \prod_{s=0}^{m_j} \left(x - \alpha^{j(2^s)}\right).$$

The schematic process of forming the check and generating polynomials is illustrated in Figure 16.14. Figure 16.15 shows the corresponding distribution of elements of the finite field $GF(2^m)$ across the polynomials $h(x)$ and $g(x)$. All notations are similar to the previously considered case.

The roots of the check polynomial $h(x)$ consist of elements from the first three conjugate classes of the finite field $GF(2^m)$. Consequently, the range of elements from α to α^z in the finite field $GF(2^m)$, which are the roots of the polynomial $h(x)$, is determined by the largest value of z that satisfies the condition $\alpha^z = \alpha^{(z) \bmod (2^m - 1)}$. This can be expressed as:

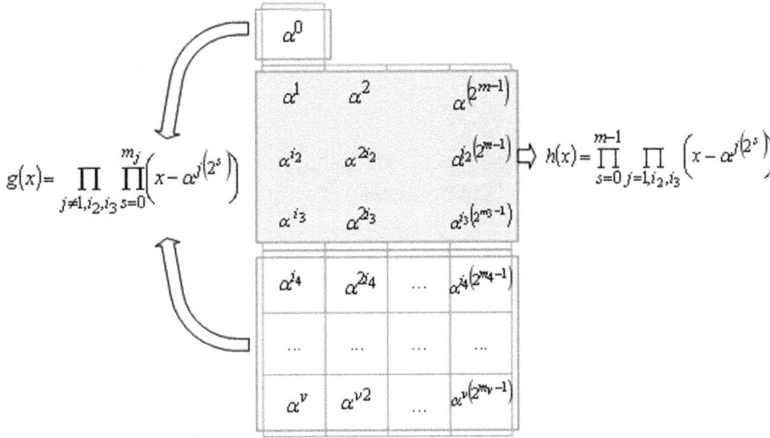

FIGURE 16.14 Scheme for Forming Check and Generating Polynomials of Group Code. with Parameters ($n = 2^m - 1, k = 3m, d = 2^{m-1} - 2^{\frac{m+1}{2}}$).

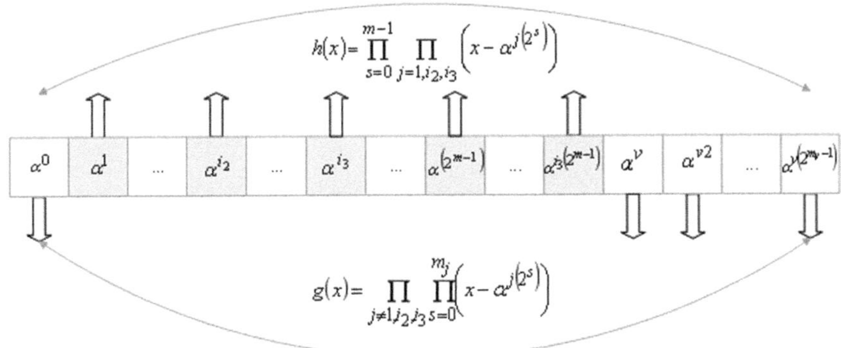

FIGURE 16.15 Distribution of Finite Field Elements in Check and Generating Polynomials of Group Code with Parameters ($n = 2^m - 1, k = 3m, d = 2^{m-1} - 2^{\frac{m+1}{2}}$).

$$z = \max_{s=0,\dots,m-1} \left\{ \left(2^s\right) \bmod \left(2^m - 1\right), \left(i_2 2^s\right) \bmod \left(2^m - 1\right), \left(i_3 2^s\right) \bmod \left(2^m - 1\right) \right\}$$

When m is a prime number and the roots of the polynomials $f_{i_1}(x)$, $f_{i_2}(x)$, and $f_{i_3}(x)$ are chosen such that they lie in the range:

$$\underbrace{\alpha^{i_1},\dots,\alpha^{i_1(2^{m-1})},\dots,\alpha^{i_2},\dots,\alpha^{i_2(2^{m-1})},\dots,\alpha^{i_3},\dots,\alpha^{i_3(2^{m-1})}}_{2^{m-1}+2^{\frac{m+1}{2}}}$$

and when $z = 2^{m-1} + 2^{\frac{m+1}{2}}$, we have:

$$2t = (2^m - 1) - 2^{m-1} - 2^{\frac{m+1}{2}} = 2^{m-1} - 2^{\frac{m+1}{2}} - 1,$$

leading to the estimate:

$$d = 2t + 1 = 2^{m-1} - 2^{\frac{m+1}{2}}.$$

The number of check elements r of the group code is determined by the degree of the generating polynomial, which in turn is defined by the number of its roots. In this case, we have the estimate:

$$r = 2^m - 3m - 1,$$

yielding:

$$k = 3m.$$

The corresponding code parameters of the group code are:

$$(n = 2^m - 1, k = 3m, d = 2^{m-1} - 2^{\frac{m+1}{2}}). \tag{16.13}$$

Estimating the weight spectrum of the code, the check polynomial of the code with parameters (16.13) includes as a factor the check polynomial of the previously discussed codes with parameters (16.11) and (16.9). Consequently, all codewords of the maximum length register code with parameters (16.9) and the codewords of the code with parameters (16.11) are codewords of the group code with parameters (16.13). Therefore, the first non-zero components:

$$A(w) \neq 0, w = 2^{m-1},$$

$$A(w) \neq 0, w = 2^{m-1} - 2^{\frac{m+1}{2}-1},$$

$$A(w) \neq 0, w = 2^{m-1} + 2^{\frac{m+1}{2}-1}$$

of the weight spectrum $A(w)$ coincide. Adding the minimal polynomial $f_{i_3}(x)$ as a factor to the check polynomial leads to two additional non-zero components in the weight spectrum.

When the equality $h(x) = f_{i_1}(x)f_{i_2}(x)f_{i_3}(x)$ is satisfied (see expression (16.12)), we obtain new non-zero components in the weight spectrum of the code:

$$A(w) \neq 0, \quad w = 2^{m-1} - 2^{\frac{m+1}{2}}$$

and

$$A(w) \neq 0, \quad w = 2^{m-1} + 2^{\frac{m+1}{2}}$$

Thus, we have the following estimation of the code's weight spectrum:

$$A(w) = \begin{cases} 1, w = 0; \\ 0, w = 1,\ldots,2^{m-1} - 2^{\frac{m+1}{2}} - 1; \\ \neq 0, w = 2^{m-1} - 2^{\frac{m+1}{2}}; \\ 0, w = 2^{m-1} - 2^{\frac{m+1}{2}} + 1,\ldots,2^{m-1} - 2^{\frac{m+1}{2}-1} - 1; \\ \neq 0, w = 2^{m-1} - 2^{\frac{m+1}{2}-1}; \\ 0, w = 2^{m-1} - 2^{\frac{m+1}{2}-1} + 1,\ldots,2^{m-1} - 1; \\ \neq 0, w = 2^{m-1}; \\ 0, w = 2^{m-1} + 1,\ldots,2^{m-1} + 2^{\frac{m+1}{2}-1} - 1; \\ \neq 0, w = 2^{m-1} + 2^{\frac{m+1}{2}-1}; \\ 0, w = 2^{m-1} + 2^{\frac{m+1}{2}-1} + 1,\ldots,2^{m-1} + 2^{\frac{m+1}{2}} - 1; \\ \neq 0, w = 2^{m-1} + 2^{\frac{m+1}{2}}; \\ 0, w = 2^{m-1} + 2^{\frac{m+1}{2}} + 1,\ldots,2^{m} - 1. \end{cases}$$

The expressions for estimating the correlation properties are as follows:

For $w = 2^{m-1} - 2^{fracm+12}$, the normalized correlation is given by:

$$\frac{n - 2w}{n} = \frac{2^m - 1 - 2\left(2^{m-1} - 2^{fracm+12}\right)}{2^m - 1} = \frac{-1 - 2^{fracm+12+1}}{2^m - 1}$$

For $w = 2^{m-1} - 2^{fracm+12-1}$, the normalized correlation is:

$$\frac{n - 2w}{n} = \frac{2^m - 1 - 2\left(2^{m-1} - 2^{fracm+12-1}\right)}{2^m - 1} = \frac{-1 - 2^{fracm+12}}{2^m - 1}$$

For $w = 2^{m-1}$, it is calculated as:

$$\frac{n - 2w}{n} = \frac{2^m - 1 - 2\left(2^{m-1}\right)}{2^m - 1} = \frac{-1}{2^m - 1}$$

For $w = 2^{m-1} + 2^{fracm+12-1}$, the expression is:

$$\frac{n - 2w}{n} = \frac{2^m - 1 - 2\left(2^{m-1} + 2^{fracm+12-1}\right)}{2^m - 1} = \frac{-1 + 2^{fracm+12}}{2^m - 1}$$

Lastly, for $w = 2^{m-1} + 2^{fracm+12}$:

$$\frac{n - 2w}{n} = \frac{2^m - 1 - 2\left(2^{m-1} + 2^{fracm+12}\right)}{2^m - 1} = \frac{-1 + 2^{fracm+12+1}}{2^m - 1}$$

This gives us an estimate of the levels of the side lobes of the periodic cross-correlation function:

$$\left\{ \begin{array}{l} \dfrac{-1-2^{\frac{m+1}{2}+1}}{2^m-1}, \; w = 2^{m-1} - 2^{\frac{m+1}{2}}; \\[2ex] \dfrac{-1-2^{\frac{m+1}{2}}}{2^m-1}, \; w = 2^{m-1} - 2^{\frac{m+1}{2}-1}; \\[2ex] \dfrac{-1}{2^m-1}, \; w = 2^{m-1}; \\[2ex] \dfrac{-1+2^{\frac{m+1}{2}}}{2^m-1}, \; w = 2^{m-1} + 2^{\frac{m+1}{2}-1}; \\[2ex] \dfrac{-1+2^{\frac{m+1}{2}+1}}{2^m-1}, \; w = 2^{m-1} + 2^{\frac{m+1}{2}}. \end{array} \right. \tag{16.14}$$

The analysis of expression (16.14) reveals that employing group codes defined through the check polynomial of the form (16.12) enables the formation of a new class of discrete signals with a five-level correlation function. Notably, three levels of the side lobes correspond to the previously discussed three-level discrete signals (Gold signals), while the two additional levels are aligned with the non-zero components in the weight spectrum of the group code with code parameters (16.14).

An estimation of the power of the ensemble of generated discrete signals is derived from the used code, which possesses $2^{3m}-1$ non-zero code words. Each code word exhibits a maximum period, and each cyclic orbit of the group code contains exactly 2^m-1 code words. Consequently, the power of the ensemble of generated signals is estimated as

$$M = \frac{2^{3m}-1}{2^m-1} = 2^{2m} + 2^m + 1.$$

Figures 16.16 and 16.17 present the graphs of the corresponding correlation functions. These spreading sequences are generated from Group Code with Parameters ($n = 2^m - 1, k = 3m, d = 2^{m-1} - 2^{\frac{m+1}{2}}$). As can be seen from Figure 16.17, the periodic cross-correlation function has five levels of side lobes. This is a new class of spreading sequences that was synthesized through computational modeling in this work.

In conclusion, the comprehensive computational analysis conducted in this section has successfully demonstrated the versatility and robustness of group codes in generating discrete signals with distinct correlation properties. By methodically varying the check polynomials and examining their impact on the weight spectrum and correlation functions, we have unlocked new possibilities in signal design. The exploration of codes with different parameters has revealed a rich landscape of signal classes, each with unique attributes suitable for specific applications in communication systems. This journey through the computational modeling of spreading sequences in communication systems has not only provided valuable insights but also set the stage for further explorations in this dynamic field.

16.7 COMPARATIVE ANALYSIS OF SPREADING SEQUENCE CORRELATION CHARACTERISTICS AND CARDINALITY

This section is dedicated to a comparative analysis of various spreading sequences, focusing on their correlation characteristics and cardinality. We aim to juxtapose well-known Gold codes and Kasami

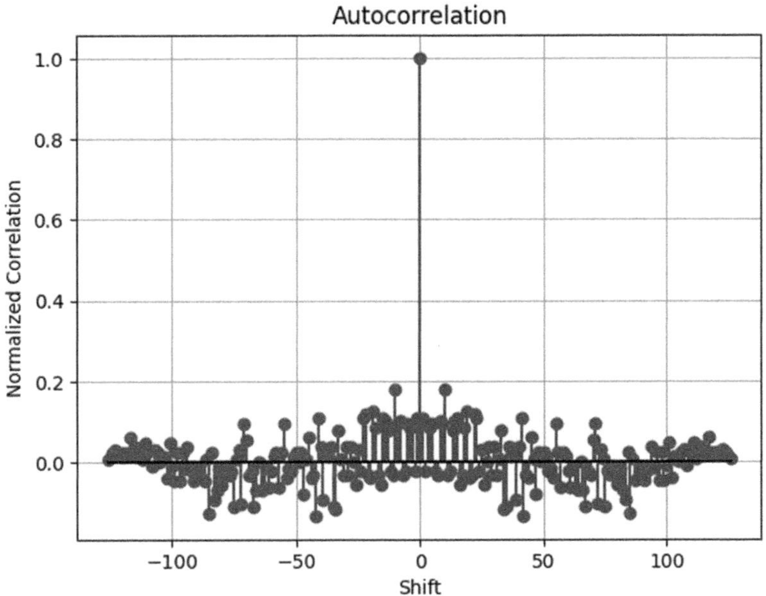

FIGURE 16.16 Aperiodic autocorrelation function of generated discrete signals of length $n = 127$.

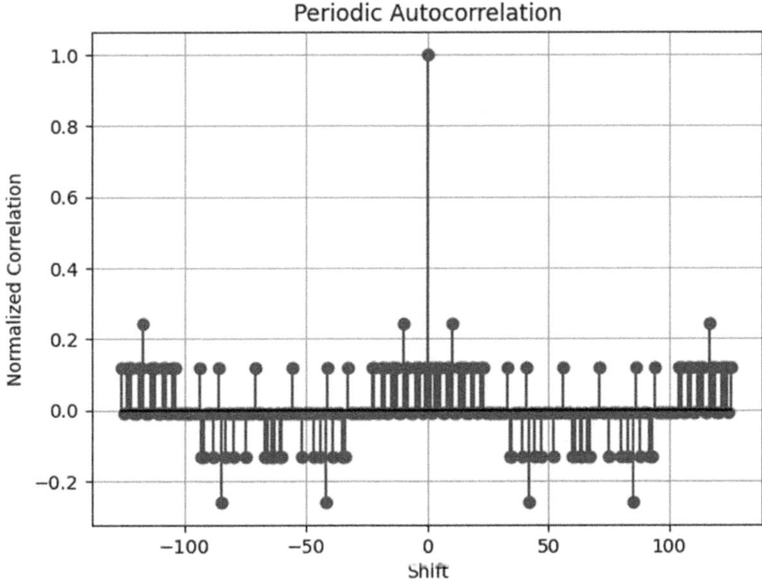

FIGURE 16.17 Periodic autocorrelation function of five-level discrete signals of length $n = 127$.

sequences with the newly generated spreading sequences featuring a five-level correlation function. The comparison is anchored on the theoretical Welch bound, providing a comprehensive view of the performance of these sequences in terms of correlation and capacity for multiple access systems.

In our comparative study, we first consider the renowned Gold codes. The Welch bound for Gold codes parameters is given by:

$$N = 2^n - 1,$$
$$M = 2^n + 1 = N + 2.$$

Substituting into the Welch bound formula, we derive:

$$\left(\rho_{\max}\right)^2 \geq \frac{N+1}{(N+2)N-1},$$

with the limit as N approaches infinity:

$$\lim_{N \to \infty} \left(\rho_{\max}\right)^2 = \lim_{N \to \infty} \left(\frac{N+1}{(N+2)N-1}\right) = 0.$$

The Laurent series expansion is [12], [13]:

$$\left(\rho_{\max}\right)^2 \approx \frac{1}{N} - \left(\frac{1}{N}\right)^2 + \left(\frac{3}{N}\right)^3 - \left(\frac{7}{N}\right)^4 + \left(\frac{17}{N}\right)^5 - \left(\frac{41}{N}\right)^6 +$$
$$+ \left(\frac{99}{N}\right)^7 - \left(\frac{239}{N}\right)^8 + \left(\frac{577}{N}\right)^9 - \left(\frac{1393}{N}\right)^{10} + O\left(\left(\frac{1}{N}\right)^{11}\right)$$

which approximates the bound (16.4) effectively.
The actual mutual correlation for Gold codes (for odd n) is:

$$\left(\rho_{Gold\max}\right)^2 = \left(\frac{1+2^{\frac{n+1}{2}}}{2^n-1}\right)^2 = \left(\frac{\sqrt{2(N+1)}+1}{N}\right)^2.$$

The Puiseux series approximation yields [12], [13]: P

$$\left(\rho_{Gold\max}\right)^2 \approx \frac{2}{N} + 2\sqrt{2}\left(\frac{1}{N}\right)^{(3/2)} + \frac{3}{N^2} + \sqrt{2}\left(\frac{1}{N}\right)^{(5/2)} - \frac{\left(\frac{1}{N}\right)^{(7/2)}}{2\sqrt{2}} + \frac{\left(\frac{1}{N}\right)^{(9/2)}}{4\sqrt{2}} + O\left(\left(\frac{1}{N}\right)^{(11/2)}\right).$$

where using only the first term, we get

$$\left(\rho_{Gold\max}\right)^2 \approx \frac{2}{N}.$$

This implies that for large N, the mutual correlation of Gold codes is approximately $\dfrac{1.4}{\sqrt{N}}$, closely approaching the Welch bound of (16.3).

In our comparative analysis, we next consider the well-known Kasami codes in relation to the Welch bound. For these codes, the parameters are defined as follows:

$$N = 2^n - 1$$

and

$$M = 2^{\frac{n}{2}} = \sqrt{N+1}.$$

Substituting these values into the Welch bound formula, we derive:

$$\left(\rho_{\max}\right)^2 \geq \frac{\sqrt{N+1}-1}{\sqrt{N+1}\,N-1},$$

where the limit as N approaches infinity for $\left(\rho_{\max}\right)^2$ is zero. This suggests that the correlation properties improve significantly for larger values of N. The Puiseux series expansion for this bound is given by [12], [13]:

$$\left(\rho_{\max}\right)^2 \approx \frac{1}{N} - \left(\frac{1}{N}\right)^{3/2} + \frac{3}{2}\left(\frac{1}{N}\right)^{5/2} - \left(\frac{1}{N}\right)^3 - \frac{7}{8}\left(\frac{1}{N}\right)^{7/2} + \frac{2}{N^4} - \frac{5}{16}\left(\frac{1}{N}\right)^{9/2} - \frac{2}{N^5} + O\left(\left(\frac{1}{N}\right)^{11/2}\right),$$

which approximates the bound (16.4) quite closely.

For the actual mutual correlation values of small set Kasami codes, the expression is:

$$\left(\rho_{SKasami_{\max}}\right)^2 = \left(\frac{\sqrt{N+1}+1}{N}\right)^2,$$

which, when approximated using the Puiseux series, yields [12], [13]:

$$\left(\rho_{SKasami\,\max}\right)^2 \approx \frac{1}{N} + 2\left(\frac{1}{N}\right)^{3/2} + \frac{2}{N^2} + \left(\frac{1}{N}\right)^{5/2} - \frac{1}{4}\left(\frac{1}{N}\right)^{7/2} + \frac{1}{8}\left(\frac{1}{N}\right)^{9/2} + O\left(\left(\frac{1}{N}\right)^{11/2}\right).$$

By considering only the first term of the series, we approximate $\left(\rho_{SKasami_{\max}}\right)^2 \approx \dfrac{1}{N}$, indicating that for large values of N, the mutual correlation of Kasami codes is approximately (16.3), aligning closely with the theoretical Welch bound.

For the large set of Kasami codes, the cardinality is defined as

$$M = 2^n + 2^{n/2} = N + 1 + \sqrt{N+1},$$

where

$$N = 2^n - 1.$$

This leads to the following Welch bound expression [12], [13]:

$$\left(\rho_{\max}\right)^2 \geq \frac{N + \sqrt{N+1}}{(N+1+\sqrt{N+1})N - 1} \approx$$

$$\approx \frac{1}{N} - \left(\frac{1}{N}\right)^2 + \left(\frac{1}{N}\right)^{5/2} + \left(\frac{1}{N}\right)^3 - \frac{3}{2}\left(\frac{1}{N}\right)^{7/2} - \frac{1}{N^4} + \frac{23}{8}\left(\frac{1}{N}\right)^{9/2} + O\left(\left(\frac{1}{N}\right)^{11/2}\right) \approx \frac{1}{N}.$$

This approximation closely aligns with the theoretical Welch bound of (16.4).

For the actual mutual correlation values of the large set of Kasami codes, as per s [40], [41], we have:

$$\rho_{GKasami\,\max} = \frac{1 + 2^{\frac{n+2}{2}}}{N}, \text{ where } n = 2p.$$

This leads to the following approximation [12], [13]:

$$\left(\rho_{GKasami\,\max}\right)^2 = \left(\frac{1 + 2\sqrt{N+1}}{N}\right)^2 \approx$$

$$\approx \frac{4}{N} + 4\left(\frac{1}{N}\right)^{3/2} + \frac{5}{N^2} + 2\left(\frac{1}{N}\right)^{5/2} - \frac{1}{2}\left(\frac{1}{N}\right)^{7/2} + \frac{1}{4}\left(\frac{1}{N}\right)^{9/2} + O\left(\left(\frac{1}{N}\right)^{11/2}\right) \approx \frac{4}{N}.$$

This approximation suggests that for large values of N, the mutual correlation of the large set of Kasami codes is approximately $\dfrac{4}{\sqrt{N}}$, which is a significant deviation from the ideal Welch bound (16.3).

In our exploration of new spreading sequences characterized by a five-level correlation function, we encounter a distinct correlation dynamic. The maximum absolute value of the correlation function for these sequences is given by:

$$\rho_{New\,\max} = \frac{1 + 2^{\frac{n+1}{2}+1}}{2^n - 1}, n = 2p + 1. \tag{16.15}$$

For these sequences, the parameters are defined as follows:

- Total number of sequences, $N = 2^n - 1$;
- Cardinality of the set, $M = 2^{2n} + 2^n + 1 = (N+1)^2 + N + 2$.

Substituting these parameters into the Welch bound equation (16.2), we derive:

$$\left(\rho_{max}\right)^2 \geq \frac{(N+1)^2 + N + 1}{((N+1)^2 + N + 2)N - 1}.$$

This expression, when approximated using the Laurent series, yields [12], [13]:

$$\left(\rho_{max}\right)^2 \approx \frac{1}{N} - \left(\frac{1}{N}\right)^3 + \left(\frac{4}{N}\right)^4 - \left(\frac{9}{N}\right)^5 + \left(\frac{14}{N}\right)^6 - \\ -\left(\frac{11}{N}\right)^7 - \left(\frac{18}{N}\right)^8 + \left(\frac{101}{N}\right)^9 - \left(\frac{260}{N}\right)^{10} + O\left(\left(\frac{1}{N}\right)^{11}\right).$$

In the limit as N approaches infinity, the correlation bound approaches zero:

$$\lim_{N \to \infty} \left(\rho_{max}\right)^2 = \lim_{N \to \infty} \left(\frac{(N+1)^2 + N + 1}{((N+1)^2 + N + 2)N - 1}\right) = 0.$$

Utilizing the real value as specified in equation (16.15), we can express the maximum correlation of the new spreading sequences as follows:

$$\left(\rho_{Newmax}\right)^2 = \left(\frac{1 + 2^{\frac{n+1}{2}+1}}{2^n - 1}\right)^2 = \left(\frac{\sqrt{8(N+1)} + 1}{N}\right)^2.$$

When approximated using the Puiseux series, this yields [12], [13]:

$$\left(\rho_{Newmax}\right)^2 \approx \frac{8}{N} + 4\sqrt{2}\left(\frac{1}{N}\right)^{(3/2)} + \frac{9}{N^2} + 2\sqrt{2}\left(\frac{1}{N}\right)^{(5/2)} - \\ -\frac{\left(\frac{1}{N}\right)^{(7/2)}}{\sqrt{2}} + \frac{\left(\frac{1}{N}\right)^{(9/2)}}{2\sqrt{2}} + O\left(\left(\frac{1}{N}\right)^{(11/2)}\right).$$

This implies that for large values of N, the mutual correlation of these codes approximates to about $\approx \frac{2.8}{\sqrt{N}}$ when only the first term of the series is considered. In the limit as N approaches infinity, the correlation bound approaches zero:

$$\lim_{N \to \infty} \left(\rho_{max}\right)^2 = \lim_{N \to \infty} \left(\frac{\sqrt{8(N+1)} + 1}{N}\right)^2 = 0,$$

This asymptotic behavior (as $N \to \infty$) indicates that the new sets of sequences conform to the Welch bound, demonstrating their efficacy in systems where large N values are prevalent and robust correlation properties are essential.

The final results of the cardinality and correlation modulus estimates for the considered sequences are summarized in Table 16.3 [12], [13].

It is evident that even a slight increase in the cardinality of spreading sequence sets leads to a significant increase in the correlation modulus. For instance, only the small Kasami set with $M = \sqrt{N} + 1$ has a correlation estimate comparable to the Welch bound. For the large set, the correlation modulus doubles. While asymptotically (as $N \to \infty$), all sets in Table 16.3 satisfy the Welch bound $\lim_{N \to \infty} \left(\rho_{max} \right)^2 = 0$.

We propose new sets of spreading sequences for asynchronous Spread Spectrum Multiple-Access systems. Our codes exhibit enhanced properties, particularly in terms of significantly increased cardinality. This advantage facilitates subscriber capacity expansion and the implementation of Soft Capacity, as well as in specialized defense systems. For example, compared to Gold codes, the cardinality of the new sets increases more than N times, while the correlation modulus approximately doubles.

In conclusion, our research demonstrates that asymptotically, all considered code sets conform to the Welch bound, that is, as $N \to \infty$, the square of the correlation approaches zero. By approximating with the first term of the series, we have shown differences in the correlation characteristics of the spreading sets. Specifically, all sets are close to the theoretical limit of $\dfrac{1}{\sqrt{N}}$ in correlation modulus. The new expanded set of sequences with a five-level correlation function also asymptotically satisfies the Welch bound. With a first-term approximation, the correlation modulus is limited to $\dfrac{2.8}{\sqrt{N}}$, which is twice the value for Gold codes. However, the cardinality of the new set is significantly (N times) higher. This substantial advantage can significantly enhance the capacity of asynchronous CDMA and reduce communication service costs. For instance, at $N = 1000$, the expansion of the set by more than three orders of magnitude could be significant. The new sets of spreading signals can be used for Soft Capacity, allowing base stations to increase subscriber capacity with minimal service quality reduction.

These findings can also be beneficial for the advancement of future wireless communication systems and smart grids. The high cardinality of the new code sets enables their use in signal imitation protection, enhancing communication security and reliability.

16.8   DISCUSSION AND FUTURE SCOPE

16.8.1   Overview of Research Findings

The research presented in this paper provides a comprehensive analysis of spreading sequences, with a particular focus on the development of new sequences characterized by a five-level correlation

---

TABLE 16.3

Comparison of Correlation and Cardinality Across Various Classes of Spreading Sequences

Sequence Type	Set Cardinality, M	Correlation Modulus Estimate (First Term of Series)
Gold Codes	$N + 2$	$\dfrac{1.4}{\sqrt{N}}$
Small Kasami Set	$\sqrt{N} + 1$	$\dfrac{1}{\sqrt{N}}$
Large Kasami Set	$N + 1 + \sqrt{N+1}$	$\dfrac{2}{\sqrt{N}}$
New Spreading Sequences with Five-Level Correlation Function	$(N+1)^2 + N + 2$	$\dfrac{2.8}{\sqrt{N}}$

function. This exploration encompasses theoretical foundations, computational modeling, experimental analysis, and comparative studies.

- Theoretical Foundations: The study began with a detailed exploration of the construction of cyclic codes and their transformation into spreading sequences. The algebraic methods employed here were crucial in generating sequences with unique correlation properties, distinct from mere cyclic shifts.
- Computational Modeling: Utilizing Python programming, the study effectively visualized auto-correlation and cross-correlation functions. These models were instrumental in demonstrating the practical implications of the theoretical constructs, especially in terms of correlation peaks and their variations.
- Experimental Analysis: This phase brought theoretical and computational models into a real-world context. The generation and analysis of new spreading sequences provided empirical evidence that reinforced theoretical predictions and offered insights into potential applications and limitations.
- Comparative Analysis: Comparing the new sequences against established ones like Gold codes and Kasami sets revealed their relative strengths and weaknesses. The new sequences showed a significant increase in cardinality, a crucial factor in spread spectrum communication systems, albeit with a higher correlation.

16.8.2 FUTURE SCOPE

Looking ahead, the research opens several avenues for further exploration and application:

- Optimization for Specific Applications: Future studies could focus on tailoring these sequences for particular applications. This involves balancing the trade-off between cardinality and correlation to optimize performance in specific communication system scenarios.
- Enhancing System Capacity and Security: The increased cardinality of the new sequences presents opportunities for enhancing the capacity and security of communication systems. Further research could explore how these sequences can be integrated into existing systems to improve performance.
- Exploring Variants and Modifications: Investigating variants of these sequences or modifications to the existing models could lead to the discovery of sequences with even better properties, suitable for more demanding applications.
- Application in Emerging Technologies: The potential of these sequences in emerging technologies like 5G networks, IoT, and smart grids is vast. Future research could focus on customizing these sequences for such cutting-edge applications.
- Addressing Practical Implementation Challenges: While the theoretical and computational models show promise, practical implementation in real-world systems may pose challenges. Future research should aim to bridge this gap, ensuring that the theoretical advantages translate effectively into practical benefits.

In summary, the research conducted provides a solid foundation for further exploration in the field of spreading sequences. The potential applications in communication systems, particularly in enhancing capacity and security, are significant. Future research in this area promises to yield even more innovative solutions, contributing to the advancement of communication technology.

16.9   CONCLUSION

This study embarked on a comprehensive journey through the realm of spreading sequences, culminating in the development of new sequences with a five-level correlation function. The research traversed various domains, from theoretical underpinnings to practical applications, offering a holistic view of the potential and challenges in the field of spread spectrum communication systems. Key highlights of the study include:

- Theoretical Insights: The research provided a deep dive into the algebraic construction of cyclic codes and their transformation into spreading sequences. This theoretical exploration was pivotal in understanding the core principles that govern the behavior and properties of these sequences.
- Computational Modeling and Experimental Validation: The use of Python for computational modeling brought theoretical concepts to life, offering a visual and practical perspective on the correlation properties of the sequences. The experimental analysis further validated the theoretical models, demonstrating their applicability in real-world scenarios.
- Comparative Analysis: The study's comparative approach, juxtaposing the newly developed sequences against established ones like Gold codes and Kasami sets, highlighted the trade-offs between cardinality and correlation. This comparison was instrumental in positioning the new sequences within the broader context of existing technologies.
- Practical Implications and Future Directions: The research underscored the potential of the new sequences in enhancing the capacity and security of communication systems. Looking forward, the study opens avenues for further research in optimizing these sequences for specific applications, exploring their variants, and integrating them into emerging technologies.

In conclusion, this research marks a significant step forward in the field of spreading sequences. The new sequences, with their unique correlation properties and increased cardinality, present promising opportunities for advancing communication technologies. The study lays a foundation for future research aimed at harnessing these properties for practical applications, potentially revolutionizing the way we approach communication systems in various domains.

ACKNOWLEDGEMENTS

- This project has received funding from the European Union's Horizon 2020 research and innovation programme under the Marie Skłodowska-Curie grant agreement No. 101007820 – TRUST. This publication reflects only the author's view and the REA is not responsible for any use that may be made of the information it contains.
- This research was funded by the European Union – NextGenerationEU under the Italian Ministry of University and Research (MIUR), National Innovation Ecosystem grant ECS00000041-VITALITY-CUP D83C22000710005

REFERENCES

[1] B. Sklar and F. J. Harris, *Digital Communications: Fundamentals and Applications*, 3 edition. Hoboken: Prentice Hall, 2020.

[2] G. L. Stüber, "Spread Spectrum Techniques," in *Principles of Mobile Communication*, G. L. Stüber, Ed., Cham: Springer International Publishing, 2017, pp. 449–499. doi: 10.1007/978-3-319-55615-4_9.

[3] M. R. Rahman, S. Bojja-Venkatakrishnan, E. A. Alwan, and J. L. Volakis, "Spread Spectrum Techniques with Channel Coding for Wideband Secured Communication Links," in *2020 IEEE International*

Symposium on Antennas and Propagation and North American Radio Science Meeting, Jul. 2020, pp. 1783–1784. doi: 10.1109/IEEECONF35879.2020.9329603.

[4] M. B. Pursley, "Spread-spectrum Multiple-access Communications," in *Multi-User Communication Systems*, pp. 139–199. Vienna: Springer Vienna, 1981.

[5] D. A. Visan, I. Lita, M. Jurian, and M. Gherghe, "Pseudorandom Sequence Generator for Spread Spectrum Communications," in *2018 10th International Conference on Electronics, Computers and Artificial Intelligence (ECAI)*, Jun. 2018, pp. 1–4. doi: 10.1109/ECAI.2018.8678968.

[6] D. Torrieri, *Principles of Spread Spectrum Communication Systems*. Cham: Springer International Publishing, 2018. doi: 10.1007/978-3-319-70569-9.

[7] S.-M. M. Yang, *Modern Digital Radio Communication Signals and Systems*. Cham: Springer International Publishing, 2019. doi: 10.1007/978-3-319-71568-1.

[8] C. Apostolas, G. Sfikas, and R. Tafazolli, "24 – Wireless Asynchronous Transfer Mode (ATM) in Data Networks for Mobile Systems," in *Database and Data Communication Network Systems*, C. T. Leondes, Ed., San Diego: Academic Press, 2002, pp. 859–900. doi: 10.1016/B978-012443895-8/50026-X.

[9] L. Novosel, G. Šišul, Ž. Ilić, and J. Božek, "Performance Enhancement of LR WPAN Spread Spectrum System Using Chaotic Spreading Sequences," *AEU – International Journal of Electronics and Communications*, vol. 118, p. 153131, May 2020, doi: 10.1016/j.aeue.2020.153131.

[10] H. Zhu and M. Shi, "The b-Symbol Weight Hierarchy of the Kasami codes," *Discrete Mathematics*, vol. 347, no. 2, p. 113774, Feb. 2024, doi: 10.1016/j.disc.2023.113774.

[11] C. Sandeepa, B. Siniarski, N. Kourtellis, S. Wang, and M. Liyanage, "A Survey on Privacy for B5G/6G: New Privacy Challenges, and Research Directions," *Journal of Industrial Information Integration*, vol. 30, p. 100405, Nov. 2022, doi: 10.1016/j.jii.2022.100405.

[12] A. Kuznetsov, S. A. Rajba, O. Veselska, M. Bagmut, R. Ziubina, and O. Peshkova, "Codes for Multiple-Access Asynchronous Techniques," *Procedia Computer Science*, vol. 207, pp. 2192–2201, Jan. 2022, doi: 10.1016/j.procs.2022.09.279.

[13] A. Kuznetsov, V. Krasnobaev, M. Bagmut, and Y. Kuznetsova, "Investigation of the Correlation Properties of Spreading Sequences for Asynchronous CDMA," in *2021 11th IEEE International Conference on Intelligent Data Acquisition and Advanced Computing Systems: Technology and Applications (IDAACS)*, Sep. 2021, pp. 510–515. doi: 10.1109/IDAACS53288.2021.9660947.

[14] A. Kuznetsov, O. Smirnov, D. Kovalchuk, M. Pastukhov, K. Kuznetsova, and D. Prokopovych-Tkachenko, "Discrete Signals with Special Correlation Properties," in *Proceedings of the Second International Workshop on Computer Modeling and Intelligent Systems (CMIS-2019), Zaporizhzhia, Ukraine, April 15-19, 2019*, D. Luengo, S. Subbotin, P. Arras, Y. Bodyanskiy, K. Henke, I. Izonin, V. G. Levashenko, V. Lytvynenko, A. Parkhomenko, A. Pester, N. Shakhovska, A. Sharpanskykh, G. Tabunshchyk, C. Wolff, H.-D. Wuttke, and E. Zaitseva, Eds., in CEUR Workshop Proceedings, vol. 2353. CEUR-WS.org, 2019, pp. 618–629. Accessed: Jul. 02, 2020. [Online]. Available: http://ceur-ws.org/Vol-2353/paper49.pdf

[15] A. Kuznetsov, O. Smirnov, Y. Zaicenko, O. Oleshko, M. Pastukhov, and K. Kuznetsova, "Formation of Discrete Signals with Special Correlation Properties," in *2019 International Conference on Information and Telecommunication Technologies and Radio Electronics (UkrMiCo)*, Odessa, Ukraine: IEEE, Sep. 2019, pp. 1–6. doi: 10.1109/UkrMiCo47782.2019.9165331.

[16] G. Yang, F. Zhou, Y. Lou, G. Qiao, N. Ahmed, and Y. He, "Double-Differential Coded M-ary Direct Sequence Spread Spectrum for Mobile Underwater Acoustic Communication System," *Applied Acoustics*, vol. 183, p. 108303, Dec. 2021, doi: 10.1016/j.apacoust.2021.108303.

[17] D. Zhang and T. Helleseth, "Sequences With Good Correlations Based on Circular Florentine Arrays," *IEEE Transactions on Information Theory*, vol. 68, no. 5, pp. 3381–3388, May 2022, doi: 10.1109/TIT.2022.3147073.

[18] M. A. Murillo-Escobar, C. Cruz-Hernández, L. Cardoza-Avendaño, D. Murillo-Escobar, and R. M. López-Gutiérrez, "Multibiosignal Chaotic Encryption Scheme Based on Spread Spectrum and Global Diffusion Process for e-Health," *Biomedical Signal Processing and Control*, vol. 78, p. 104001, Sep. 2022, doi: 10.1016/j.bspc.2022.104001.

[19] X. Chen, W. Hu, and D. Feng, "Direct-Sequence Spread Spectrum Time Division Multiple Access with Direct Detection for Latency Optimized Passive Optical Network," *Optics Communications*, vol. 510, p. 127955, May 2022, doi: 10.1016/j.optcom.2022.127955.

[20] D. A. Chu and J. R. Barry, "The Case Against Alphabet Awareness for Detection of Direct-Sequence Spread-Spectrum Signals," *Signal Processing*, vol. 194, p. 108449, May 2022, doi: 10.1016/j.sigpro.2021.108449.

[21] U. Ghafoor, M. Ali, H. Z. Khan, A. M. Siddiqui, and M. Naeem, "NOMA and Future 5G & B5G Wireless Networks: A Paradigm," *Journal of Network and Computer Applications*, vol. 204, p. 103413, Aug. 2022, doi: 10.1016/j.jnca.2022.103413.

[22] J. Shao, A. I. Karsilayan, C. T. Rodenbeck, and J. Silva-Martinez, "An Interference-Tolerant Synchronization Scheme for Wireless Communication Systems Based on Direct Sequence Spread Spectrum," *IEEE Transactions on Circuits and Systems I: Regular Papers*, vol. 69, no. 1, pp. 415–427, Jan. 2022, doi: 10.1109/TCSI.2021.3113862.

[23] L. Tian, T. Liu, and Y. Li, "New Constructions of Binary Golay Spreading Sequences for Uplink Grant-Free NOMA," *IEEE Communications Letters*, vol. 26, no. 10, pp. 2480–2484, Oct. 2022, doi: 10.1109/LCOMM.2022.3194158.

[24] F. Wei *et al.*, "Detection of Direct Sequence Spread Spectrum Signals Based on Deep Learning," *IEEE Transactions on Cognitive Communications and Networking*, vol. 8, no. 3, pp. 1399–1410, Sep. 2022, doi: 10.1109/TCCN.2022.3174609.

[25] Y. Hu, S. Han, H. Li, H. Zhao, G. Yang, and J. Xu, "TCN-Based M-ary Mobile Spread Spectrum Underwater Acoustic Communication," *Applied Acoustics*, vol. 211, p. 109457, Aug. 2023, doi: 10.1016/j.apacoust.2023.109457.

[26] Y. Huang, Y. Shen, and J. Wang, "From Terahertz Imaging to Terahertz Wireless Communications," *Engineering*, vol. 22, pp. 106–124, Mar. 2023, doi: 10.1016/j.eng.2022.06.023.

[27] Z. Lu and Y. Jiao, "Efficiently All-Digital Code Tracking for Band-Limited DSSS Systems," *IEEE Communications Letters*, vol. 27, no. 2, pp. 686–690, Feb. 2023, doi: 10.1109/LCOMM.2022.3222296.

[28] D. Roque and C. Poulliat, "SNR-Optimal Spreading Sequences for Chip-Wise Faster-Than-Nyquist Signaling," *IEEE Communications Letters*, vol. 27, no. 6, pp. 1594–1598, Jun. 2023, doi: 10.1109/LCOMM.2023.3265070.

[29] C. Ma and L. Zhang, "A Viterbi Algorithm for Blind Estimation of Periodic Long-Code DSSS Signals," in *2018 2nd IEEE Advanced Information Management, Communicates, Electronic and Automation Control Conference (IMCEC)*, May 2018, pp. 128–132. doi: 10.1109/IMCEC.2018.8469687.

[30] V. P. Ipatov, *Spread Spectrum and CDMA: Principles and Applications*. Chichester, UK: John Wiley & Sons, Ltd, 2005. doi: 10.1002/0470091800. www.semanticscholar.org/paper/Spread-Spectrum-and-CDMA%3A-Principles-and-Ipatov/5c775d5d240c19d4a6a640d874d6378530b959b5

[31] R. M. Buehrer, *Code Division Multiple Access (CDMA)*. in Synthesis Lectures on Communications. Cham: Springer International Publishing, 2006. doi: 10.1007/978-3-031-01673-8.

[32] R. Rao and S. Dianat, *Basics of Code Division Multiple Access (CDMA)*. Bellingham: SPIE, 2005. doi: 10.1117/3.626142.

[33] J. Khalife and Z. M. Kassas, "Navigation With Cellular CDMA Signals—Part II: Performance Analysis and Experimental Results," *IEEE Transactions on Signal Processing*, vol. 66, no. 8, pp. 2204–2218, Apr. 2018, doi: 10.1109/TSP.2018.2799166.

[34] A. O. El-Rayis, T. Arslan, and A. T. Erdogan, "A Processing Engine for GPS Correlation," in *2010 IEEE 8th Symposium on Application Specific Processors (SASP)*, Jun. 2010, pp. 44–49. doi: 10.1109/SASP.2010.5521149.

[35] T. Song, K. Zhou, and T. Li, "CDMA System Design and Capacity Analysis Under Disguised Jamming," *IEEE Transactions on Information Forensics and Security*, vol. 11, no. 11, pp. 2487–2498, Nov. 2016, doi: 10.1109/TIFS.2016.2585089.

[36] X. Wang, X. Liu, H.-H. Chen, and W. Meng, "Complementary Coded CDMA Systems With CP-Free OFDM," *IEEE Transactions on Vehicular Technology*, vol. 69, no. 10, pp. 11515–11528, Oct. 2020, doi: 10.1109/TVT.2020.3013155.

[37] L. Welch, "Lower Bounds on the Maximum Cross Correlation of Signals (Corresp.)," *IEEE Transactions on Information Theory*, vol. 20, no. 3, pp. 397–399, May 1974, doi: 10.1109/TIT.1974.1055219.

[38] J. L. Massey and T. Mittelholzer, "Welch's Bound and Sequence Sets for Code-Division Multiple-Access Systems," in *Sequences II*, R. Capocelli, A. De Santis, and U. Vaccaro, Eds., New York, NY: Springer, 1993, pp. 63–78. doi: 10.1007/978-1-4613-9323-8_7.

[39] A. Kuznetsov, A. Kiian, K. Kuznetsova, M. Zub, Y. Zaburmekha, and E. Lyshchenko, "Pseudorandom Sequences with Multi-Level Correlation Function for Direct Spectrum Spreading," in *2019 IEEE International Conference on Advanced Trends in Information Theory (ATIT)*, Dec. 2019, pp. 232–237. doi: 10.1109/ATIT49449.2019.9030436. https://ieeexplore.ieee.org/document/9030436f

[40] T. Kasami, "Weight Distribution Formula for Some Class of Cyclic Codes," Apr. 1966, Accessed: Aug. 01, 2020. [Online]. Available: www.ideals.illinois.edu/handle/2142/74439

[41] S. W. Golomb, *Signal Design for Good Correlation: For Wireless Communication, Cryptography, and Radar*, 1st edition. Cambridge; New York: Cambridge University Press, 2005.

Index